Active Flow Control: Recent Advances in Fundamentals and Applications

Active Flow Control: Recent Advances in Fundamentals and Applications

Editors

Hui Tang
Xin Wen

MDPI

Basel • Beijing • Wuhan • Barcelona • Belgrade • Novi Sad • Cluj • Manchester

Editors

Hui Tang
Department of Mechanical Engineering
The Hong Kong Polytechnic University
Hong Kong, China

Xin Wen
School of Mechanical Engineering
Shanghai Jiao Tong University
Shanghai, China

Editorial Office
MDPI
St. Alban-Anlage 66
4052 Basel, Switzerland

This is a reprint of articles from the Special Issue published online in the open access journal *Actuators* (ISSN 2076-0825) (available at: https://www.mdpi.com/journal/actuators/special_issues/active_flow_controls).

For citation purposes, cite each article independently as indicated on the article page online and as indicated below:

Lastname, A.A.; Lastname, B.B. Article Title. *Journal Name* **Year**, *Volume Number*, Page Range.

ISBN 978-3-0365-8900-8 (Hbk)
ISBN 978-3-0365-8901-5 (PDF)
doi.org/10.3390/books978-3-0365-8901-5

Contents

actuators

MDPI

Article

Deep Reinforcement Learning for Flow Control Exploits Different Physics for Increasing Reynolds Number Regimes

Pau Varela [1], Pol Suárez [2,3], Francisco Alcántara-Ávila [3], Arnau Miró [2], Jean Rabault [4], Bernat Font [2], Luis Miguel García-Cuevas [1], Oriol Lehmkuhl [2] and Ricardo Vinuesa [3,*]

1 CMT—Motores Térmicos, Universitat Politècnica de València, 46022 Valencia, Spain
2 Barcelona Super Computing Center—Centro Nacional de Supercomputación (BSC-CNS), 08034 Barcelona, Spain
3 FLOW, Engineering Mechanics, KTH Royal Institute of Technology, 114 28 Stockholm, Sweden
4 Norwegian Meteorological Institute, 0313 Oslo, Norway
* Correspondence: rvinuesa@mech.kth.se

Abstract: The increase in emissions associated with aviation requires deeper research into novel sensing and flow-control strategies to obtain improved aerodynamic performances. In this context, data-driven methods are suitable for exploring new approaches to control the flow and develop more efficient strategies. Deep artificial neural networks (ANNs) used together with reinforcement learning, i.e., deep reinforcement learning (DRL), are receiving more attention due to their capabilities of controlling complex problems in multiple areas. In particular, these techniques have been recently used to solve problems related to flow control. In this work, an ANN trained through a DRL agent, coupled with the numerical solver Alya, is used to perform active flow control. The Tensorforce library was used to apply DRL to the simulated flow. Two-dimensional simulations of the flow around a cylinder were conducted and an active control based on two jets located on the walls of the cylinder was considered. By gathering information from the flow surrounding the cylinder, the ANN agent is able to learn through proximal policy optimization (PPO) effective control strategies for the jets, leading to a significant drag reduction. Furthermore, the agent needs to account for the coupled effects of the friction- and pressure-drag components, as well as the interaction between the two boundary layers on both sides of the cylinder and the wake. In the present work, a Reynolds number range beyond those previously considered was studied and compared with results obtained using classical flow-control methods. Significantly different forms of nature in the control strategies were identified by the DRL as the Reynolds number Re increased. On the one hand, for $Re \leq 1000$, the classical control strategy based on an opposition control relative to the wake oscillation was obtained. On the other hand, for $Re = 2000$, the new strategy consisted of energization of the boundary layers and the separation area, which modulated the flow separation and reduced the drag in a fashion similar to that of the drag crisis, through a high-frequency actuation. A cross-application of agents was performed for a flow at $Re = 2000$, obtaining similar results in terms of the drag reduction with the agents trained at $Re = 1000$ and 2000. The fact that two different strategies yielded the same performance made us question whether this Reynolds number regime ($Re = 2000$) belongs to a transition towards a nature-different flow, which would only admits a high-frequency actuation strategy to obtain the drag reduction. At the same time, this finding allows for the application of ANNs trained at lower Reynolds numbers, but are comparable in nature, saving computational resources.

Keywords: numerical simulation; wake dynamics; flow control; machine learning; deep reinforcement learning

Citation: Varela, P.; Suárez, P.; Alcántara-Ávila, F.; Miró, A.; Rabault, J.; Font, B.; García-Cuevas, L.M.; Lehmkhul, O.; Vinuesa, R. Deep Reinforcement Learning for Flow Control Exploits Different Physics for Increasing Reynolds Number Regimes. *Actuators* **2022**, *11*, 359. https://doi.org/10.3390/act11120359

Academic Editor: Luigi de Luca

Received: 31 October 2022
Accepted: 29 November 2022
Published: 2 December 2022

1. Introduction

In transport applications (especially, in aeronautics), drag reduction is directly related to a decrease in fuel use, which translates into reducing pollution and greenhouse-gas emissions [1]. In the past decades, several techniques have been developed and used to reduce drag, both passively (Bechert and Bartenwerfer [2]) and actively (Gad-el Hak [3]). Passive methods typically rely on fixed geometric changes without using actuators. An example of this technology would be the widespread use of winglets in aircraft. Inspired by bird wings, winglets consist of small wing extensions at the wing tip with an angle relative to the wing-span direction. Using winglets, lift-induced drag is decreased by reducing the size and formation of vortices at the wing tip [4], incurring (hopefully) small increases in the structural weight and parasitic drag. Regarding active methods, diverse techniques are used to reduce aircraft drag and associated emissions. In the work of Tiseira et al. [5] and Serrano et al. [6,7], the use of distributed electric propulsion is combined with boundary-layer ingestion, setting small propellers along the wing near the trailing edge. Thanks to both technologies, it is possible to increase the aerodynamic efficiency of small aircraft or unmanned air vehicles (UAVs). In the same way, the location and use of both jet pumps in wings and synthetic jets are studied to reduce the drag of different aerodynamic bodies. An example of blowing and suction used to control turbulent boundary layers is provided by Kametani and Fukagata [8], and a number of studies have shown the feasibility of this approach in turbulent wings [9–12].

Additional examples can be found in the work of Voevodin et al. [13] and Yousefi and Saleh [14], where the placement of the suction and ejection in wing airfoils is studied to reduce aerodynamic drag or achieve efficient control of the flow around the wing in specific flight-operating conditions. In the same way, the use of synthetic jets has been studied to achieve this same objective, as displayed in the work of Cui et al. [15] on an Ahmed body, or the investigation of Park et al. [16] using an array of synthetic jets applied to a car. A comprehensive review of active flow control in a turbulent flow is provided by Choi et al. [17].

Active-control methods rely on complicated control strategies due to their variable behaviors and dependencies. As shown by Muddada and Patnaik [18], even the control of simple actuators to reduce the drag of a cylinder immersed in a low Reynolds number flow can be complicated due to the interaction between boundary layers, separating shear layers and wake. In this case, by correctly developing a control algorithm, the drag is reduced by about 53 %, leaving ample room for improvement in the case of achieving better control. However, thanks to the application of artificial neural networks (ANNs) and deep reinforcement learning (DRL), it is possible to develop functional control strategies at an affordable computational cost in similar problems, as shown by the work by Rabault et al. [19]. An ANN is a non-parametric tool formed by layers of connected processing nodes or neurons and it can be trained to solve complex problems [20,21]. The ANN training can take place through different types of learning, where DRL is one of the fastest-growing in solving a wide range of cases [22,23], including flow control in complex geometries [24,25].

DRL is based on maximizing a reward by means of an agent interacting with the environment through actions, based on partial observations of the environment. Note that the combination of DRL and ANN was successfully carried out to solve active flow control in the work of Rabault et al. [19]. A proximal policy optimization (PPO) is used to obtain the control policy of two synthetic jets on a two-dimensional cylinder in a low Reynolds number flow. PPO parameterizes a policy function using an ANN with a set of neuron weights given an observation state. The ANN then produces a set of moments that describe a distribution function from which actions are sampled. It is possible to obtain an expression for the estimation of the gradient of the reward as a function of the neuron weights. Additionally, the PPO uses a critic network that estimates the new reward function, which is helpful with stochastic data. Moreover, there is a limit to the maximum update at every training step, preventing rare events from producing large rewards [26]. It was shown that DRL was feasible and enabled finding a control strategy, such that the cylinder drag was reduced by 8 %. Rabault and Kuhnle [27] developed a framework to parallelize environments, enabling speeding up computation and learning, thus reaching a better solution faster. Additionally, Tang et al. [28] validated the DRL approach in a more complex problem using four jets on the cylinder and extending the Reynolds number range. The results of Rabault et al. in [19,27] have been utilized in a multitude of studies in recent years. For example, the works of Tokarev et al. [29] and Xu et al. [30], where oscillatory rotary controls were applied to the cylinder; the study of linear stability and low sensitivity that allows a better understanding of DRL by Li and Zhang [31], or the effort of Ren et al. [32] where the same DRL approach is used with a solver that allows calculations at higher Reynolds numbers, with the additional problem of controlling the increase in turbulence around the cylinder. Applications to square cylinders, which are relevant to civil engineering, have also been presented recently [?], as well as using traditional modal analysis methods for defining effective reward functions [34]. Regarding these works, there is room for improvement through the application of high-performance computing (HPC), enabling conducting faster calculations at higher Reynolds numbers not previously calculated.

The main objective of this work is, indeed, to demonstrate the capabilities of the HPC resources available today to perform aerodynamic optimization with DRL-based control at much higher Reynolds numbers than the ones already studied in the literature. This project proposes the use of the numerical code Alya, developed in the Barcelona Supercomputer Center (BSC-CNS) [35], employing the HPC MareNostrum IV cluster. Alya is a tier-0 massively parallel code and is designed to solve discretized partial differential equations using finite elements. This code has been successfully used in different problems of fluid mechanics, including simulations of turbulent flows [36,37].

In the present work, the methodology is first detailed in Section 2, dedicated to the numerical setup and the application of the DRL and the ANN. The results are then summarized and discussed in Section 3, which starts with the validation of the Alya solver in the combination of an HPC cluster to solve the DRL problem in the cylinder. Then, the results derived from the increase in the Reynolds number, which exceed the conditions reported in the literature, are shown. Finally, the main conclusions are collected in Section 4.

2. Methods

This section is divided into two parts: (1) the problem description and the domain setup along the methodology (regarding the numerical simulations) and (2) the framework, including the DRL algorithm.

2.1. Problem Configuration and Numerical Setup

The domain simulation is two-dimensional (2D) and consists of a cylinder immersed in a flat rectangular channel without any curvature, in the same configuration as that described in the work by Rabault et al. [19]. The domain is normalized using a cylinder with diameter D as the reference length scale. The channel has a dimension of $L = 22$ in length, aligned with the main flow direction, and a height equal to $H = 4.1$. The origin of the coordinate system is set on the left-bottom corner of the channel, and the center of the cylinder displaced towards the bottom by $0.05D$ units, to develop the vortex-shedding behind it. A schematic representation of the domain is depicted in Figure 1. Aligned with the vertical axis and symmetrically on the cylinder wall, one at the top and one at the bottom, there are two synthetic jets with total openings of $\omega = 10°$ each. These positions of the jets were chosen so their actuations were normal to the flow and drag reduction came from effective actuations rather than the injection of momentum.

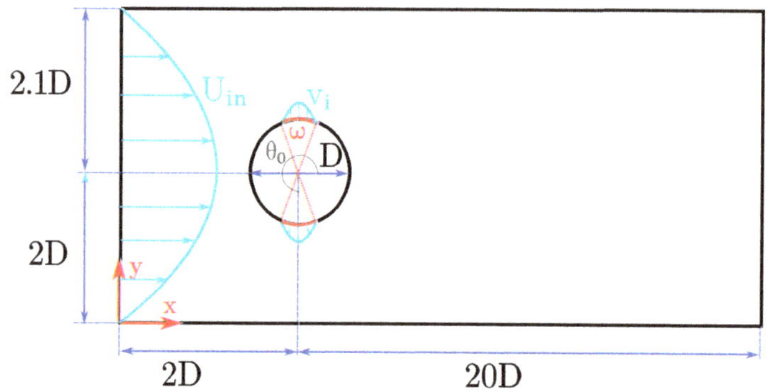

Figure 1. Main domain dimensions in terms of the cylinder diameter D, where ω represents the jet width and θ_0 is the angle of the center of the jet. In light blue, the parabolic velocity profile of the inlet and the jet are represented. The domain representation is not to scale.

Regarding the boundary conditions of the problem, an inlet parabolic velocity condition was imposed and expressed as in Equation (1):

$$U_{\text{in}}(y) = U_{\text{max}}\left[1 - \left(2\frac{y - H/2}{H}\right)^2\right], \tag{1}$$

where $[U_{\text{in}}(y), V_{\text{in}}(y) = 0]$ is the velocity vector and U_{max} is the maximum velocity reached at the middle of the channel, which is equal to 1.5 times the mean velocity $\overline{U}$, defined as shown in Equation (2):

$$\overline{U} = \frac{1}{H}\int_0^H U_{\text{in}}(y)\mathrm{d}y = \frac{2}{3}U_{\text{max}}. \tag{2}$$

A value of $U_{\text{max}} = 1.5$ is used so that the scaling velocity of the problem is $\overline{U} = 1$. A no-slip condition is imposed on the cylinder solid wall, while a smooth-wall condition is imposed on the channel walls. The right boundary of the channel is set as a free outlet with zero velocity gradient and constant pressure. The jet velocity (v_i) is a function of both the set jet angle (θ) and the mass flow rate (Q) determined by the ANN, as described in Equation (3):

$$v_i = Q\frac{\pi}{2\omega R^2}\cos\left[\frac{\pi}{\omega}(\theta - \theta_0)\right], \tag{3}$$

where θ_0 corresponds to the angle where the jet is centered and R is the cylinder radius. The scaling factor $\pi/(2\omega R^2)$ is used so that the integration of the jet velocity over the jet width gives the desired mass flow rate Q. More details about the intensity parameters are provided in Section 2.2.

The Reynolds number $Re = \overline{U}D/\nu$ of the simulation is varied between 100, 1000 and 2000, where ν is the fluid kinematic viscosity. For the mesh, an unstructured mesh of triangular elements with refinement near the cylinder wall and the jets was used, as shown in Figure 2. The number of elements changes with the Reynolds number as expressed in Table 1.

Table 1. Parameters of the simulations for each Reynolds number considered in this work.

Re	100	1000	2000
Mesh cells (approximately)	11,000	19,000	52,000
Number of witness points	151	151	151
$\lvert Q \rvert_{\mathrm{max}}$	0.088	0.04	0.04
s_{norm}	1.7	2	2
C_{offset}	3.17	3.29	3.29
r_{norm}	5	1.25	1.25
w	0.2	1	1
T_k	3.37	3.04	4.39
T_a	0.25	0.2	0.2
Actions per episode	80	100	100
Number of episodes	350	1000	1400
CPUs per environment	46	46	46
Environments	1	1 or 20	20
Total CPUs	46	46 or 920	920
Baseline duration	100	250	100

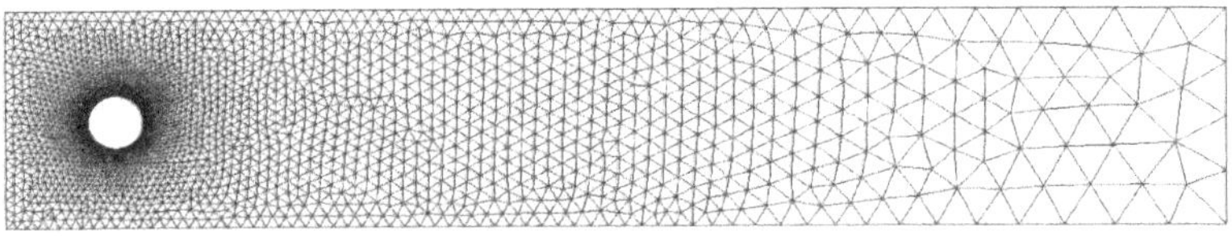

(**a**) Unstructured triangular mesh employed in all of the domains.

(**b**) Detailed mesh around the cylinder.

Figure 2. Computational grid used for the $Re = 100$ calculations.

Alya is used to simulate the flow. This solver assumes that the flow is viscous and incompressible, where the governing Navier–Stokes equations can be written for a domain Ω as in Equation (5):

$$\partial_t \boldsymbol{u} + (\boldsymbol{u} \cdot \nabla)\boldsymbol{u} - \nabla \cdot (2\nu\boldsymbol{\epsilon}) + \nabla p = \boldsymbol{f} \qquad \text{in } \Omega \in (t_0, t_f), \tag{4}$$

$$\nabla \cdot \boldsymbol{u} = 0 \qquad \text{in } \Omega \in (t_0, t_f), \tag{5}$$

where $\boldsymbol{\epsilon}$ is a function of the velocity $\boldsymbol{u}$, which defines the velocity strain-rate tensor ($\boldsymbol{\epsilon} = 1/2(\nabla \boldsymbol{u} + \nabla^T \boldsymbol{u})$) and $\boldsymbol{f}$ is the external body force. In Equation (4), the convective form of the nonlinear term, $C_{\text{nonc}}(\boldsymbol{u}) = (\boldsymbol{u} \cdot \nabla)\boldsymbol{u}$, is expressed as a term conserving energy, momentum and angular momentum [38,39]. This form is known as EMAC (energy-, momentum-, and angular-momentum-conserving equation), and its expression appears in Equation (6). The EMAC form (Equation (6)) adds to the equation the conservation of energy as well as linear and angular momentum at the discrete level:

$$C_{\text{emac}} = 2\boldsymbol{u} \cdot \boldsymbol{\epsilon} + (\nabla \cdot \boldsymbol{u})\boldsymbol{u} - \frac{1}{2}\nabla|\boldsymbol{u}|^2. \tag{6}$$

The spatial discretization of the Navier-Stokes equations is performed by means of the finite-element method (FEM). Meanwhile, the time discretization uses a semi-implicit Runge–Kutta scheme of second order for the convective term, and a Crank–Nicolson scheme for the diffusive term [40]. In the time integration, Alya uses an eigenvalue time-step estimation as described by Trias and Lehmkuhl [41]. The complete formulation of the flow solver is described in the work by Lehmkuhl et al. [37].

At each time step, the numerical solution is obtained, and the drag F_D and lift F_L forces are integrated over the cylinder surface S as follows (Equation (7)):

$$\boldsymbol{F} = \int (\boldsymbol{\varsigma} \cdot \boldsymbol{n}) \cdot \boldsymbol{e}_j \, \mathrm{d}\boldsymbol{S}, \tag{7}$$

where ς is the Cauchy stress tensor, $\boldsymbol{n}$ is the unit vector normal to the cylinder and $\boldsymbol{e}_j$ is a unit vector in the direction of the main flow velocity when calculating the drag and a vector perpendicular to the velocity of flow for the calculation of the lift force. The drag C_D and lift C_L coefficients are computed as described in Equations (8) and (9):

$$C_D = \frac{2F_D}{\rho \overline{U}^2 D}, \tag{8}$$

$$C_L = \frac{2F_L}{\rho \overline{U}^2 D}. \tag{9}$$

2.2. DRL Setup

As discussed in the introduction, the DRL interacts with the domain through three channels. The first channel is the observation state (s), based on the extraction of pressure values at a series of predefined points along the domain. These points, known as witness points or probes, are located in the same positions as in Rabault et al. [19]. Moreover, 151 witness points were distributed around the cylinder and along the wake, as shown in Figure 3. The values of the pressure obtained at the witness points are normalized by a factor s_{norm} so that the state values given to the agent are between -1 and 1, approximately. The values of s_{norm} for each Reynolds number case are given in Table 1.

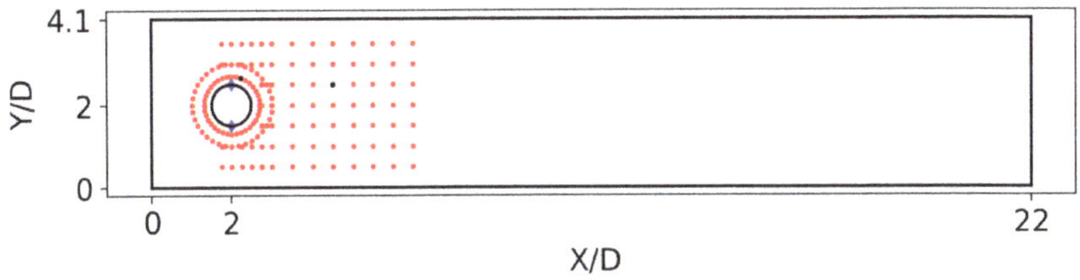

Figure 3. Schematic representation of the computational domain, where the red dots correspond to the location of the probes. The position of two probes is remarked as black dots for further analysis in Section 3.3.

The second channel of interaction between the DRL and the numerical simulation is the action that is given by the control of the jets on the cylinder (a). The value of action a is directly related to the control value of the upper jet intensity Q_1, while the bottom jet will do the opposite control to ensure a global zero mass flow rate between both jets, i.e., $Q_2 = -Q_1$. This is a more realistic control and helps to make the numerical scheme more stable as reflected by Rabault et al. [19]. During the training, the maximum value of $|Q_1|$ is limited to $|Q_1| < 0.06\ Q_{\mathrm{ref}} \approx 0.88$ for the $Re = 100$ case, as in the work by Rabault et al. [19], to avoid unrealistically large actuations. Note that Q_{ref} is the mass flow rate intercepting the cylinder, and it is calculated as in Equation (10):

$$Q_{\mathrm{ref}} = \int_{-D/2}^{D/2} \rho U_{\mathrm{in}}(y)\, \mathrm{d}y\,. \tag{10}$$

For higher Reynolds numbers, this clipping value of $|Q_1|$ is reduced to 0.04. An observation should be made here about the energy consumption of the jets. If one reduces the drag, it would lead to a performance improvement. Nevertheless, the positive energy balance should be obtained when comparing the energy saved by reducing the drag compared with the energy used to actuate with the jets. However, one cannot determine the actual energy consumption of the jets in a numerical simulation, since this highly depends on the systems used on a real setup and, therefore, it is related to experiments and not numerical simulations. An estimation of the energy consumption would consist of a combination of the kinetic energy of the flow injected and the opposing to the local pressure when injecting. Nevertheless, by restricting the mass flow rate of the jets to only 6% of the mass flow intercepting the cylinder, the energy consumption of the jets is guaranteed to be negligible compared with the drag reduction obtained. In addition, even though the value of Q selected by the ANN is unique in each action, Q is not imposed as a constant value during the entire action duration. In particular, Q is assessed as a curve to prevent significant changes in the boundary condition between actions that could lead to numerical discrepancies, similar to how the smooth control was presented by Tang et al. [28]. This way, the imposed mass flow starts from the previous value Q_0 and increases or decreases linearly until the new value Q_1 is achieved during the entire action time $T_a = t_1 - t_0$. Consequently, Q in Equation (3) is presented as a function of time in Equation (11) and is illustrated in Figure 4.

$$Q(t) = \frac{Q_1 - Q_0}{T_a}(t - t_0) + Q_0. \tag{11}$$

It should be noted that the first action of each episode (an episode is understood as a sequence of interactions between the neural network and the simulation, which generates the input data for the agent algorithm) always starts from a baseline case, i.e., a periodically stable flow without jet actuation. This baseline case uses the same domain, and the flow is fully developed without applying jet control.

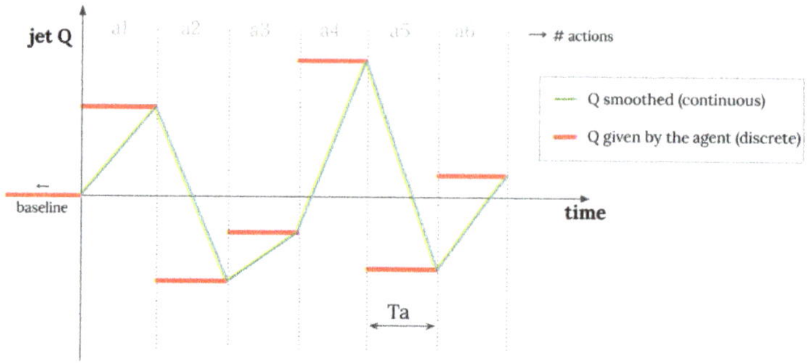

Figure 4. Flow rate of the jet Q smoothed and applied in the numerical simulation (green) versus the discrete Q decided by the DRL agent (red).

The third interaction channel between the DRL and the numerical simulation is the reward or goal (r), which in this case is aimed at minimizing the cylinder drag; it is defined in Equation (12):

$$r = r_{\text{norm}}(-\langle\overline{C_D}\rangle - w\langle\overline{C_L}\rangle + C_{\text{offset}}), \tag{12}$$

where $\langle\cdot\rangle$ indicates averaging over a baseline vortex-shedding period, w is a lift penalization factor, C_{offset} is a coefficient to center the initial reward around 0, obtained from the value of r at the end of the baseline simulation, and r_{norm} is used to normalize this reward between 0 and 1 approximately. In such a way, the agent receives the reward in an optimal range. Note that the value of r_{norm} is set from an a priori guess of the expected maximum reward. The lift penalization factor is set in such a way that the drag is minimized, and at the same time, the effect of the possible growth of induced lift is mitigated. If this lift penalization is not introduced into the reward function, the agent can find a strategy where both jets blow in the same direction at their maximum strength, as discussed in Rabault and Kuhnle [27]. The values of w, C_{offset}, and r_{norm} are given in Table 1 for the different Reynolds numbers.

To summarize, the agent will choose an action (a) given a specific state (s) in order to maximize a reward (r). The function that determines what will be the reward is a normal distribution. During the training process, the agent will choose an action around the average of this normal distribution. This is known as the exploration noise and it helps the method to converge toward a better solution. Once the training has been performed, the agent can be tested in a deterministic mode. Here, the most probable action in the normal distribution is chosen in order to maximize the reward with the learning obtained during the training.

Following the work by Rabault et al. [19], the ANN is designed with two dense layers of 512 neurons and a PPO agent is selected to carry out the control. The PPO agent follows a policy-gradient method in order to obtain the weights of the ANN. The implementation of the DRL is done through the open-source Tensorforce library [42], which is built from the TensorFlow open-source library [43], and it includes defining and creating both the ANN and the control agent. The selection of the initial parameters of the DRL depends intrinsically on the problem itself. In this case, the total number of actions is related to the vortex-shedding period $T_k = 1/f_k$ through the Strouhal number $St = f_k \cdot D/\overline{U}$.

Figure 5 shows the lift coefficient of the baseline case. This figure shows that the vortex-shedding period is $T_k = 3.37$ time units for the $Re = 100$ case and $St = 0.29$; this is in agreement with both experiments [44] and simulations [19]. Taking the vortex-shedding period as reference, the action time T_a is defined as 7.5% of T_k, as in Rabault et al. [19]. This period of actuation was found to be large enough so the consequences of the actuation can be perceived by the flow, and small enough so that the actuations can anticipate and adapt to the needs of the control. In the higher-Re cases, this actuation period is reduced to $T_a = 0.2$ due to their more chaotic flow behavior and, therefore, the necessity to adapt the actuations faster. Note that because this is a 2D flow, the word chaotic and not turbulent is employed, which restricts more to 3D flows. Based on this simulation, it can be concluded that before starting the DRL, the baseline case needs to be run for over 50 time units in order to obtain a periodic stabilized flow. As suggested in the work of Rabault et al. [19], at $Re = 100$ the typical time of an episode should be between 6 and 8 vortex sheddings so that the agent has enough time to learn the suitable control policy. A total of 80 actuations are carried out on each episode, resulting in an episode duration of 20 time units. In the case of a higher Reynolds number, a total of 100 actions are conducted in each episode, since the duration of each actuation is reduced.

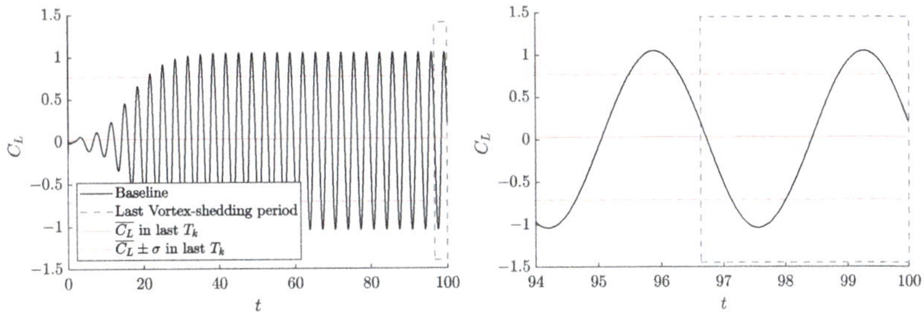

(a) (b)

Figure 5. Temporal evolution of the lift coefficient in the baseline case (without jet actuation). (**a**) Baseline lift-coefficient results. The vortex-shedding period is 3.37 time units. (**b**) Zoom-in view of the vortex-shedding period in the baseline case.

For the high Reynolds numbers of 1000 and 2000, a parallel environment framework was adopted to speed up the learning process using a total of 20 environments. Moreover, in the case at $Re = 1000$, a comparison of the results was done when a single environment was employed. A total of 46 MareNostrum IV central-processing units (CPUs) were used in each environment. Therefore, a total of 46 or 920 CPUs were used depending on whether a single environment or 20 environments were considered in the case, respectively. All of these parameters are presented in Table 1. The general DRL framework is summarized in Figure 6.

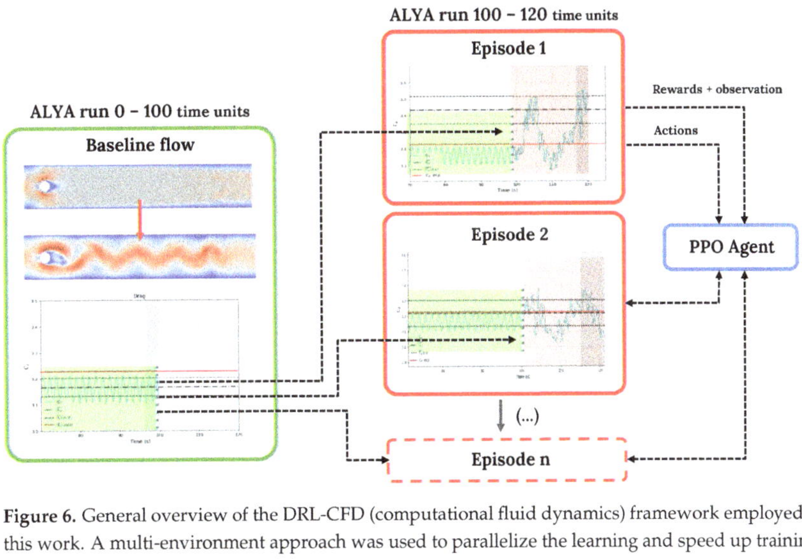

Figure 6. General overview of the DRL-CFD (computational fluid dynamics) framework employed in this work. A multi-environment approach was used to parallelize the learning and speed up training.

3. Results and Discussion

This section is divided into three parts. First, the results obtained using the Alya solver and the DRL applied are validated using literature data. Once validated, the higher Reynolds number results are discussed, for which no precedent has been found in the literature. Finally, a cross-application of agents is analyzed to save computational resources in resolving cases with high Reynolds numbers.

3.1. CFD and DRL Code Validation

As can be seen in the work by Rabault et al. [19], for a $Re = 100$, through the correct actuation of the DRL, it was possible to achieve a decrease in the cylinder drag of 8 %. These results were validated in other simulations, such as Li and Zhang [31], and applied in similar research [29,45,46].

Figure 7 shows eleven different "trainings" launched and the average C_D obtained for the last vortex-shedding period of the different episodes. It can be seen that all cases follow the same learning trend, where most of the said learning is observed in the first hundred episodes. The main trend is obtained by adjusting a fourth-degree polynomial fit using the average coefficient data, and its purpose is to help to visualize the data. From this point on, the slope of the C_D decrease is not so pronounced, reaching a stable solution for 350 episodes.

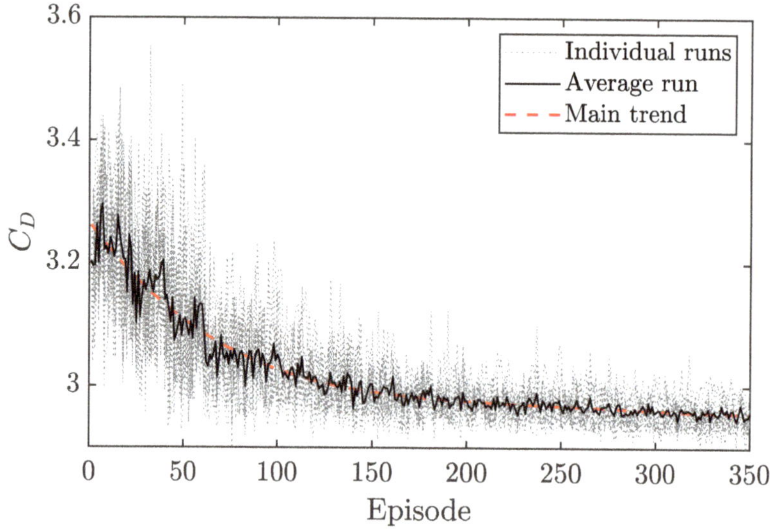

Figure 7. Evolution of C_D in the last vortex-shedding period as a function of the episode.

The obtained learning is then applied to the previously-shown baseline. To this end, one of the 11 trained agents is randomly selected and is run in a deterministic mode. This deterministic simulation is initiated from the conditions of the baseline case at 100 time units from the start. The C_D and C_L trends are shown in Figure 8, which indicates that a value of $C_D = 2.95$ is obtained after applying the control, i.e., a decrease of 8.9 % compared with the baseline case. The C_D improvement is calculated as a function of the baseline C_D:

$$\text{improvement} = \left[1 - \frac{{C_D}^{\text{controlled}}}{{C_D}^{\text{baseline}}}\right] \times 100. \quad (13)$$

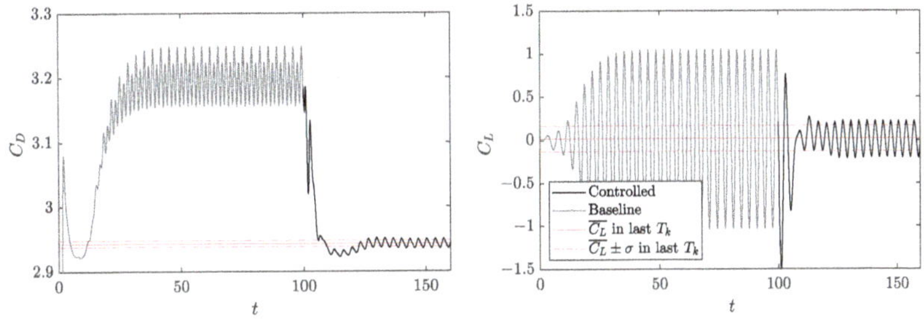

Figure 8. Temporal evolution of C_D (**left**) and C_L (**right**) obtained through the application of the DRL control (at $t = 100$), run in a deterministic mode, for $Re = 100$.

This result is rapidly achieved from the moment the control is applied. In the case of C_L, the mean remains approximately at 0. However, the amplitude of the variation of this coefficient has been reduced, as shown by the lower standard deviation obtained. This may indeed have great beneficial consequences from the structural and stability points of view.

The control imposed on each jet is represented in Figure 9. As discussed above, the control starts at 100 time units of simulation, so the values of injected or suctioned mass prior to this time are null. Having imposed the synthetic-jet condition, everything injected by one jet (positive values) will be equivalent to what is suctioned by the other (negative values). In the first 10 time units approximately, the control of greater amplitude leads to a significant reduction in the drag, maximizing the reward. Next, the transitional control seeks to obtain a more stable actuation. After 25 time units, the solution is practically periodic.

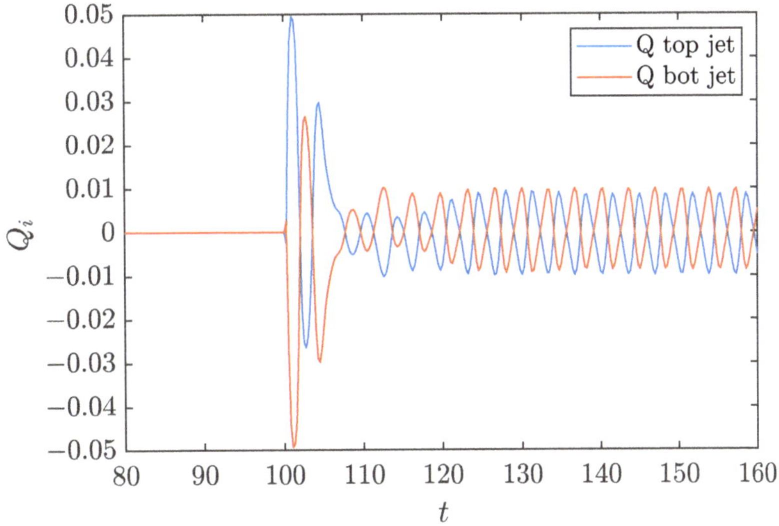

Figure 9. Flow rate through each jet as a function of time after applying the DRL control.

More information about the control is obtained by observing the contours of instantaneous velocity and pressure for the controlled and uncontrolled cases, as shown in Figure 10. The DRL agent reduces C_D by manipulating the wake vortex: it increases the size of the recirculation region and reduces both the frequency and the amplitude of the von Kármán street downstream of the cylinder. A similar conclusion is obtained from the instantaneous pressure field, which exhibits a decrease in the pressure maxima after applying the control. Comparing our results with those by Rabault et al. [19], it can be seen that the response of the control is practically the same except for very small differences that can come from the different numerical schemes to resolve the flow.

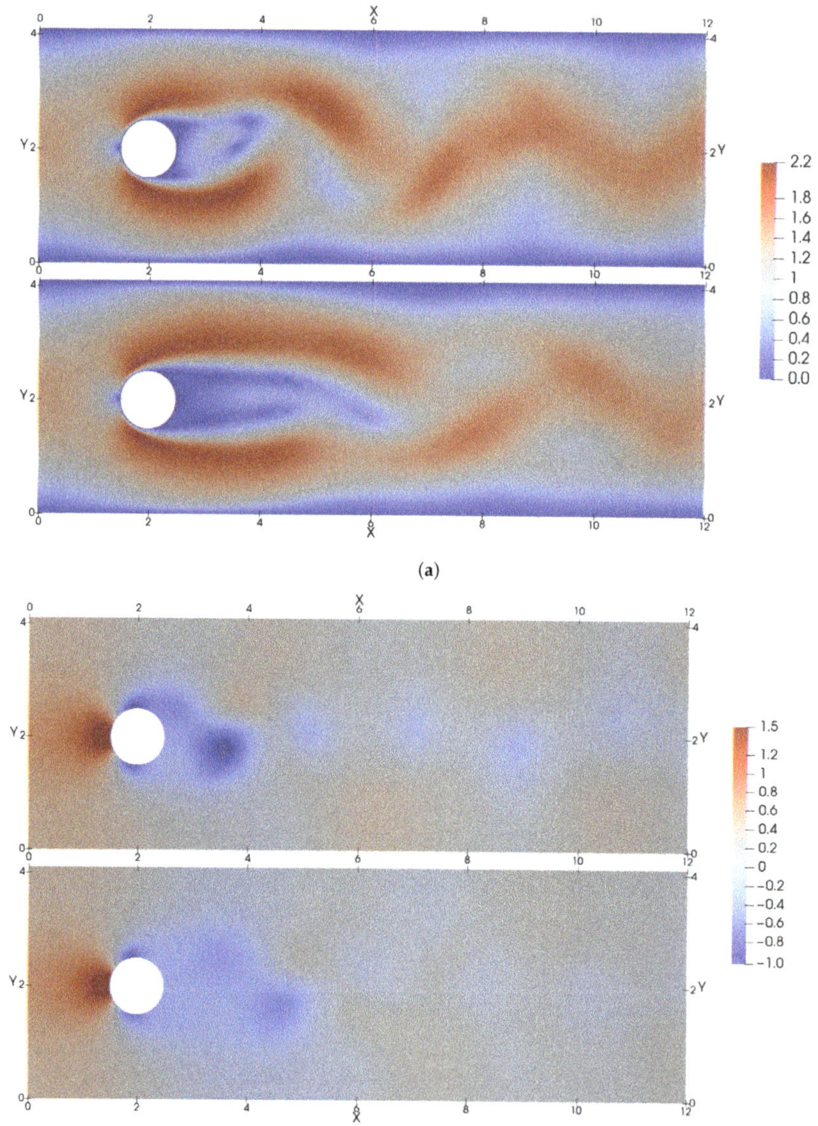

(**a**)

(**b**)

Figure 10. Instantaneous flow fields at $Re = 100$, where for each pair of images, the baseline case without control is depicted on the top and the controlled case is depicted on the bottom. (**a**) Instantaneous magnitude of the velocity fields. (**b**) Instantaneous pressure fields.

Increasing the Reynolds number also requires reducing the period of actuation, as mentioned in Section 2.2. Additionally, increasing the Reynolds number leads to a more complex flow, thus a higher number of episodes are expected to be necessary to complete the training. Therefore, parallelization using 20 environments has been adopted when simulating higher-Re cases. The parallelization using a multi-environment approach is fully detailed in the work of Rabault and Kuhnle [27].

In Figure 11, a comparison between employing a single environment and a 20 multi-environment approach is conducted for a case of $Re = 1000$. It can be observed that if the Reynolds increases an order of magnitude, in this case, the number of episodes for the DRL to act to the same degree is three times greater using only one environment, as compared with Figure 7. The same solution can be obtained by parallelizing through 20 environments, using 50 episodes in each environment, which helps to reduce the real calculation time significantly.

As mentioned before, the flow is more complex, and its structures are less periodic and more chaotic. To analyze the flow, the instantaneous and average velocity fields are represented in Figure 12.

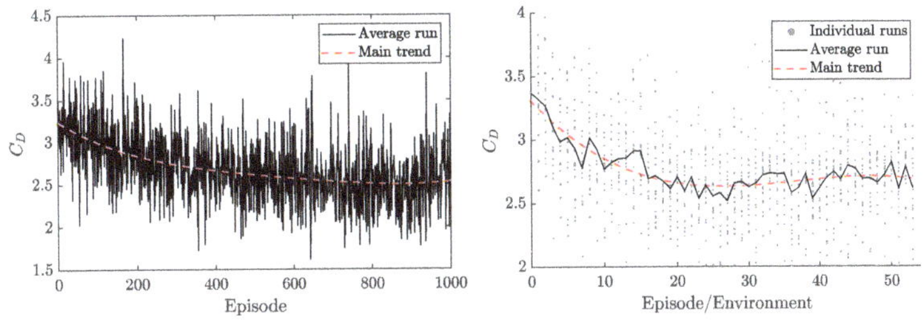

Figure 11. Evolution of C_D in the last vortex-shedding for each episode employing a single environment (**left**) or a 20-multi-environment approach (**right**) showing the learning curve indexed by the episode number for one of the 20 environments.

The obtained results are consistent with those of Ren et al. [32], which report an extensive flow analysis of the same Reynolds number. The control produces an elongation of the recirculation bubble behind the cylinder, reducing at the same time the hydrodynamic drag. For this Reynolds number, the control strategy found by the PPO agent is based on synchronous blowing. This strategy is similar to that found in lower Reynolds numbers: the jets produce an ejection or suction, generating a flow opposite to that caused by the wake.

3.2. DRL Application at Reynolds Number 2000

Through the use of parallelization with a multi-environment strategy, even higher Reynolds numbers can be achieved. The results obtained for training with $Re = 2000$ are discussed below and, to the authors' knowledge, there is no record of applying DRL control to such a high Reynolds number in the literature.

The C_D and C_L obtained through the application of learning in a deterministic mode are represented in Figure 13. This figure contrasts sharply with the one observed with $Re = 100$ (Figure 8). In this case, the starting baseline flow is much more chaotic, which can be seen in the magnitude of the peaks of both variables. The average drag coefficient of the non-controlled flow is $C_D = 3.39$. After the control, the average in this period is $C_D = 2.79$, which means more than 17% of C_D improvement. In this case, it has been chosen to represent the mean in the entire controlled period and not during the last vortex-shedding since, as the solution behaves more erratically, it could lead to an average that is far from reality. Regarding the lift coefficient, the applied control manages to set the average C_L with a close value to 0, maximizing the reward function as desired and avoiding a systematic biased strategy (see Appendix B in Rabault and Kuhnle [27]). Moreover, the absolute value of the peaks during the oscillating periods is reduced but not as much as in the case of Reynolds number 100, due to the chaotic nature of the flow at this higher Reynolds number.

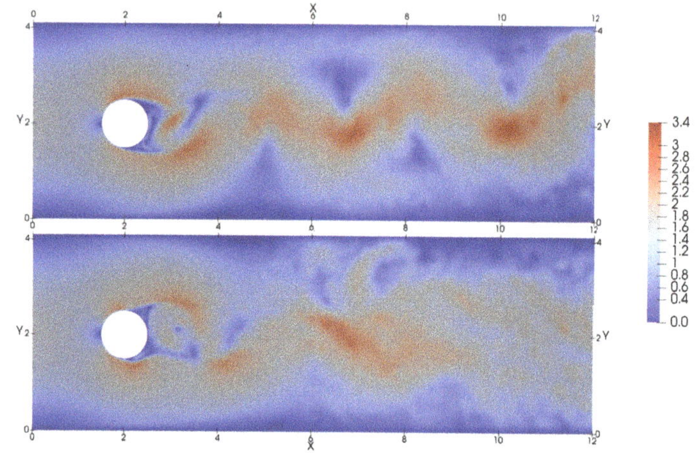

(**a**)

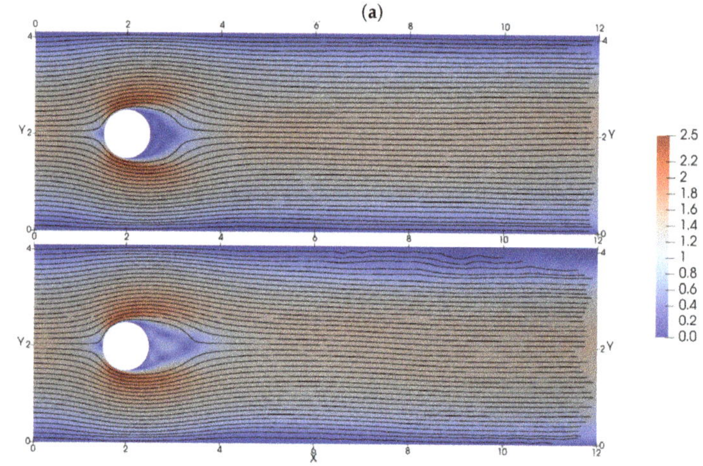

(**b**)

Figure 12. Instantaneous and average flow fields at $Re = 1000$, where for each pair of images, the baseline case without control is depicted on the top and the controlled case is depicted on the bottom. (**a**) Instantaneous magnitude of velocity fields. (**b**) Average magnitude of velocity fields, where the mean streamlines are indicated in black.

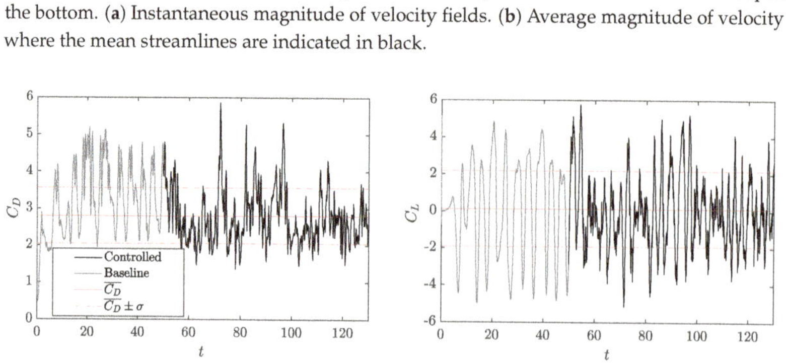

Figure 13. Temporal evolution of C_D (**left**) and C_L (**right**) obtained through the application of the DRL control (at $t = 50$), run in a deterministic mode, for $Re = 2000$.

In Figure 14 the average velocity and pressure fields are shown both for the baseline and the controlled deterministic case. As can be seen in the average velocity field comparison, the separation point in the controlled case is further downstream of the cylinder. This way, the average wake is narrowed faster in the controlled case, lowering the drag produced. This phenomenon is easier to observe if the difference between the baseline and the controlled case is considered as shown in Figure 15. This figure shows that the more significant difference between the controlled and uncontrolled cases occurs in the separation zone. At the same time, as observed for lower-Reynolds number control, the pressure drop behind the cylinder is reduced when controlled.

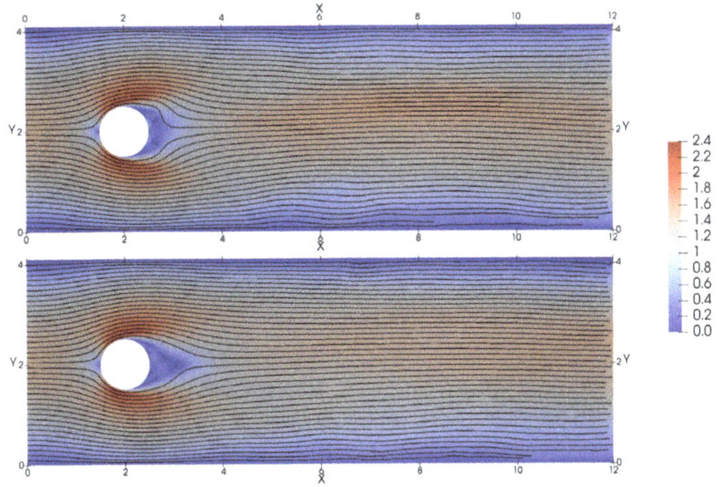

(**a**)

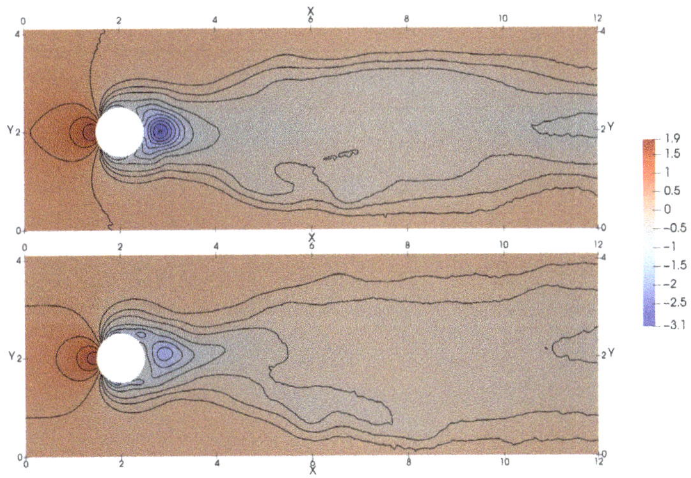

(**b**)

Figure 14. Average flow fields at $Re = 2000$, where each pair of images, the baseline case without control is depicted on the top and the controlled case is depicted on the bottom. (**a**) Average magnitude of velocity fields. (**b**) Average pressure fields.

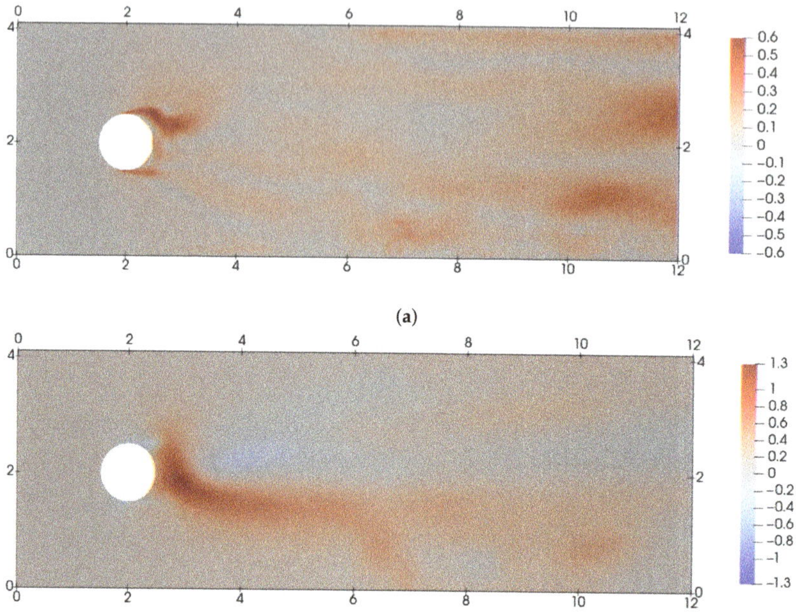

(**a**)

(**b**)

Figure 15. Difference in the mean flow fields at $Re = 2000$, between the controlled and baseline case. (**a**) Difference in the mean magnitude of velocity fields. (**b**) Difference in mean pressure fields.

This control strategy is completely different from that used low at lower Reynolds numbers of values $Re = 100$ and 1000. In order to better understand this control strategy, a chronological velocity and pressure field snapshot is plotted in Figure 16. In this case, the control strategy does not involve the elongation of the recirculation bubble behind the cylinder. The flow separation is controlled and energized by a high-frequency actuation of the jets. The separation point is moved behind the cylinder lowering the drag of the cylinder, similar to the Eiffel paradox phenomenon, also known as the drag crisis, as defined by Stabnikov and Garbaruk [47]. The agent attempts to minimize the drag with this high frequency, breaking the vortices produced after the cylinder into smaller and less-energetic ones.

Additionally, one video (of the velocity and pressure fields) is shared to aid the visualization. The corresponding link can be found at the end of the document in the "Data Availability Statement".

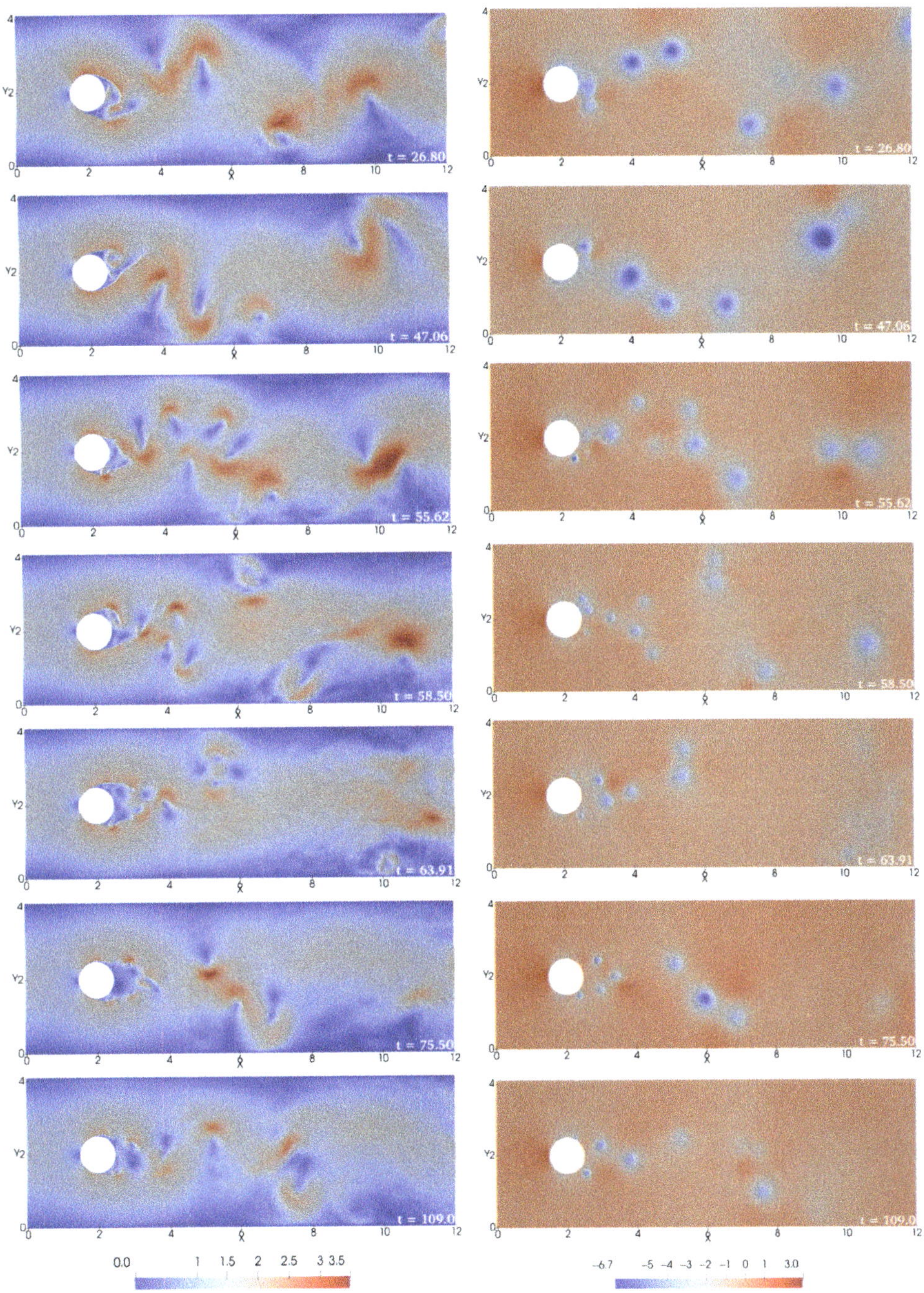

Figure 16. Magnitude of the velocity (**left**) and pressure (**right**) field progression at $Re = 2000$. The control starts at $t = 50$, therefore the first two panels of each column correspond to the uncontrolled case and the rest (involving smaller vortices) correspond to the controlled flow.

3.3. Cross-Application of Agents

Once the C_D improvement results have been obtained for all the calculated Reynolds numbers thanks to the use of DRL, the cross-application of agents for a flow at $Re = 2000$ is studied.

Cross-application of agents involves applying a previously trained ANN with a different Reynolds number flow to solve a similar problem, using the achieved learning in other conditions. The final objective is to reduce the computational cost of training the DRL agent, since training an ANN with a lower Reynolds number is less expensive. This approach can be very beneficial as long as the results produced by the agent are comparable, i.e., the physics are similar enough between the two different cases.

Here it is investigated the drag reduction obtained by the agents previously trained for Reynolds numbers 100 and 1000 when they are applied in a deterministic mode to the flow of $Re = 2000$, as can be seen in Figure 17.

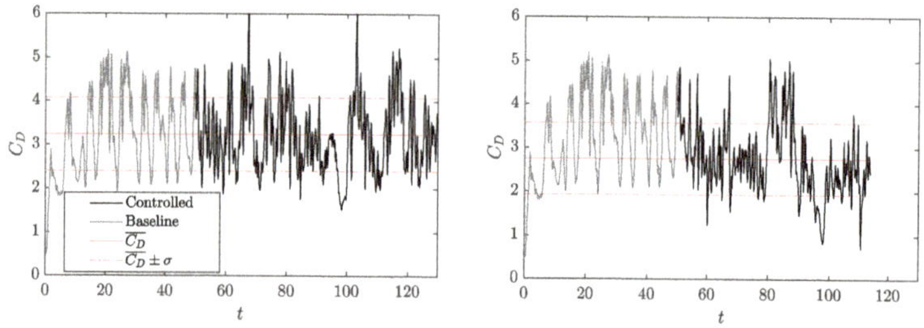

Figure 17. Temporal evolution of C_D obtained through cross-application of agents trained at $Re = 100$ (**left**) and $Re = 1000$ (**right**) in a deterministic mode to a case with a Reynolds number of 2000.

The cross-application of the agent trained at $Re = 100$ does not yield a result that reduces drag in a flow at $Re = 2000$. No drag improvement is obtained because the nature of the flow where the agent is applied is too different from the one where it was trained. However, when the trained agent at $Re = 1000$ is applied for the case of $Re = 2000$, a significant drag reduction is produced. An average C_D value of 2.74 is obtained, translating into a 19 % drag reduction improvement compared with the baseline flow. This is a slightly larger drag reduction than that obtained by the agent trained at $Re = 2000$. This small difference can be attributed to the chaotic nature of the flow, from evaluation run to evaluation run, and sub-optimal trade-off with respect to the actual reward function due to less good control of the lift and drag fluctuations, as will be discussed later. The fact that two similar results are obtained with different natures of control strategies, as explained in Section 3.2, may indicate that the flow at $Re = 2000$ belongs to a transition regime toward a higher Reynolds number flow, which only admits a high-frequency control in order to obtain the drag reduction. Therefore, being in this transition regime would admit both controls yielding comparable results. In order to go deep into this topic, higher Reynolds numbers must be simulated, but that goes beyond the scope of this article. At the same time, this is a doubly positive result since it confirms the fact that it is possible to apply deep-learning models previously trained at lower Reynolds number but within the same Re regime (similar dynamics) while saving time and computational resources [48].

The average C_D comparisons for each agent used in the cross-application at $Re = 2000$ are shown in Figure 18 (left). As detailed in the previous section, the control strategy varies significantly as the Reynolds number increases. At $Re = 100$ and 1000, the control attempts to keep the recirculation bubble behind the cylinder, at $Re = 2000$, high-frequency suction and blowing of the jets is applied to quickly force the flow transition, breaking the vortices into smaller ones. Despite these differences in the control strategies between the agents trained at $Re = 1000$ and 2000, both are capable to obtain a comparable drag reduction in a flow at $Re = 2000$. As shown in Figure 18 (right), the C_L bias obtained using the policy from $Re = 1000$ is higher than that obtained using the policy of $Re = 2000$. Moreover, C_L

fluctuations are larger when employing the agent trained at $Re = 1000$ (not shown here for the sake of brevity). Note that the $Re = 2000$ strategy reduces the drag almost as much as using cross-application of agents and, at the same time, exhibits a slightly better performance in the lift; since the lift is part of the reward function, the agent trained at $Re = 2000$ achieves a higher reward at $Re = 2000$ than the agent trained at $Re = 1000$. Nevertheless, the small differences between both "trainings" may be reversed for different baseline flows (changing the initial condition). This is similar to the results observed by Ren et al. [32].

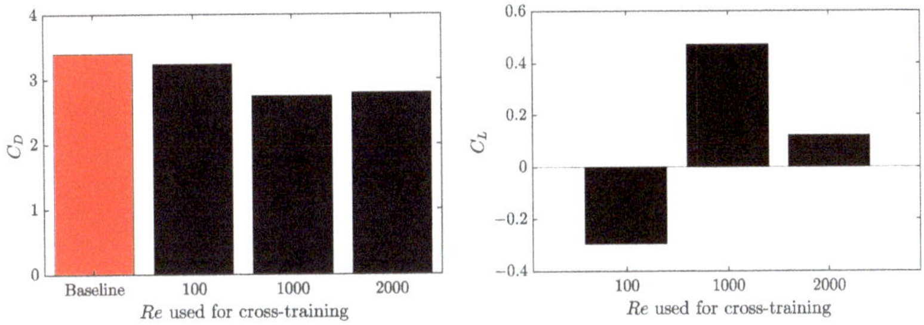

Figure 18. Average C_D (**left**) and average C_L (**right**) comparison between three different cross-application of agents.

In Figure 19, the average magnitude of velocity field at $Re = 2000$ is shown for the cross-application of the agent trained at $Re = 1000$. At first glance, the existent bias in the average velocity field, which is directly linked with the asymmetric lift behavior, is noticeable.

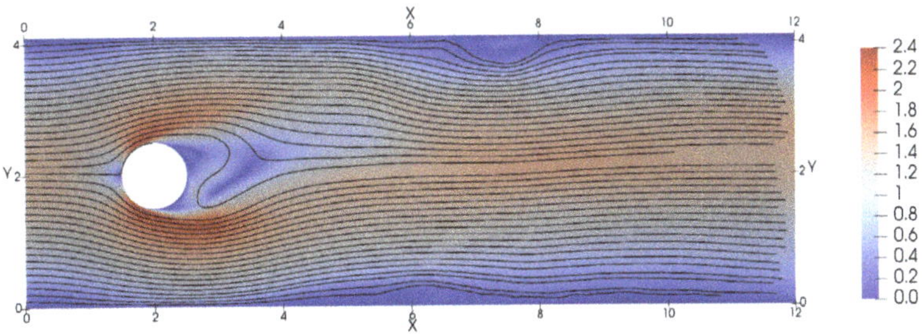

Figure 19. Average flow fields at $Re = 2000$, using the training obtained for $Re = 1000$.

In order to quantitatively observe the different control strategies, the power-spectral density (PSD) of the actions is plotted for the policy of $Re = 2000$, and the cross-application of agents, $Re = 1000$ and 100, in Figure 20 (left). The latter two agents share the same strategy, which consists of a low-frequency action with no significant content at medium and high frequencies. In contrast, with the policy of $Re = 2000$, the frequency spectrum is distributed, with a peak at a frequency of 1.1. In Figure 20 (right), the frequency spectrum of the pressure at the probe in the detachment region indicated in Figure 3 is shown. A low frequency of 0.24 governs the baseline flow, corresponding to the vortex-shedding frequency. This frequency is mitigated by the policy of $Re = 2000$ in higher frequencies. Specifically, the frequency peak at 1.1 is a consequence of the corresponding actuation at this frequency, which is responsible for breaking the flow into smaller vortices as seen in Figure 16. On the other hand, it can be seen how the agent trained at $Re = 1000$ tends towards a completely different strategy in terms of frequency, actuating with a low-frequency strategy. The agent trained at $Re = 100$, also uses this low-frequency actuation, but it does not have the capabilities to influence the pressure field. As mentioned before, in those cases, the agent attempts an opposition control strategy increasing the recirculation bubble to minimize the drag.

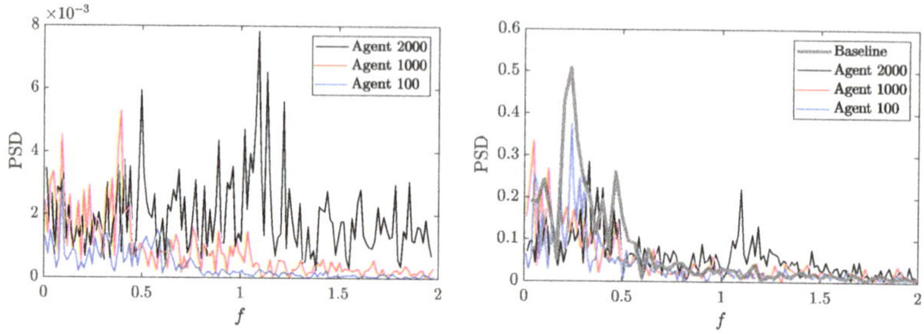

Figure 20. Power-spectral density of the action output, Q, for the $Re = 2000$ case, with its own policy and two cross-application policies (**left**) and for pressure in the probe near the cylinder in the separation zone (**right**).

4. Conclusions

In this work, the high-performance CFD solver Alya was coupled with an ANN DRL agent to simulate and control the flow around a 2D cylinder with two jets attached to the cylinder surface. The main control objective was set on the reduction of the average drag. For the first time, the Reynolds number of the canonical 2D cylinder control problem from Rabault et al. [19] was extended to $Re = 2000$, providing new insights into the DRL control strategies for highly complex and dynamic flows. Additionally, the $Re = 100$ and $Re = 1000$ cases were studied for validation and comparison purposes.

The most striking outcome of the $Re = 2000$ case is that the DRL agent uses a radically different strategy from those obtained at lower Reynolds numbers while still being able to provide a 17 % drag reduction. In the new strategy, the agent attempts to delay the detachment point in the cylinder surface using a high-frequency signal in the actuation of the jets, similar to what can be observed in drag crisis phenomena. It is shown that the cylinder wake is narrowed by the breakdown of the detaching vortices into smaller structures. This strategy contrasts with the one obtained at a lower Reynolds number, where the agent acts at a lower frequency to perform opposition control and to elongate the recirculation bubble behind the cylinder. These results are further verified with the spectral analysis of the jet actuating signals and a pressure witness point in the wake. Table 2 summarizes the results from this paper and provides an overview of the results in previous works. Information about drag reduction, the control strategy, and the configuration of the jet location is also provided.

The applications of agents trained at $Re = 100$ and $Re = 1000$ on the $Re = 2000$ case were investigated (namely cross-applications). It was shown that the $Re = 100$ agent is not able to reduce the drag in the high Reynolds number regime due to the different dynamics of the system. On the other hand, the $Re = 1000$ provides satisfactory results, which even beat the $Re = 2000$ agent itself (19 % drag reduction), even though the mean wake displays an asymmetric pattern in this case, which corresponds to an overall reward (including lift bias penalization) that is lower than with the agent trained at $Re = 2000$. Still, this opens up the door to accelerating the training of an agent by first exposing it to a lower Reynolds number flow, which demands a lower computational effort, and then finalize the training at the target Reynolds number condition.

As a final observation, the actual time that the agent takes to calculate the output action once it receives the input state is negligible for an ANN consisting of 2 layers of 512 neurons each, and it could be considered instantaneous. Therefore, it will be completely possible to implement these schemes in real time. In the experiments, there can be some non-negligible delay times depending on the hardware systems used in the experiment. However, if the training is performed in the same conditions, i.e., the delay is also present during the training, the agent will learn a strategy that works under such circumstances. In addition, if the agent is trained in a numerical environment and applies it on the experimental side, then the state sent as an input must consider the known delay in the experiment.

Table 2. Summary of the results and strategies. In the strategy, column E states the elongation of the bubble of recirculation and D states the delay of the detachment point through a high-frequency actuation. Note that for the four cases of Tang et al. [28], a unique agent is trained through transfer learning for all four Reynolds numbers and jets have small x-velocity components.

Re	Work	CD Reduction	Strategy	Configuration
100	Present work	8.9%	E	2 jets (1 top & 1 bottom)
100	Rabault et al. [19]	8%	E	2 jets (1 top & 1 bottom)
100	Tang et al. [28]	5.7%	E	4 jets (2 top & 2 bottom)
200	Tang et al. [28]	21.6%	E	4 jets (2 top & 2 bottom)
300	Tang et al. [28]	32.7%	E	4 jets (2 top & 2 bottom)
400	Tang et al. [28]	38.7%	E	4 jets (2 top & 2 bottom)
1000	Present work	20.0%	E	2 jets (1 top & 1 bottom)
1000	Ren et al. [32]	30%	E	2 jets (1 top & 1 bottom)
2000	Present work	17.7%	D	2 jets (1 top & 1 bottom)

Author Contributions: Conceptualisation: R.V., J.R. and O.L.; investigation: all authors; data curation: P.V., P.S., F.A.-Á., A.M., J.R. and B.F.; software: J.R., A.M., B.F. and O.L.; writing—original draft: P.V., P.S. and F.A.-Á.; editing: all authors; funding: L.M.G.-C., O.L. and R.V. All authors have read and agreed to the published version of the manuscript.

Funding: Pau Varela is partially supported through a grant for the mobility of doctoral students provided by Universitat Politècnica de València and the program Erasmus Prácticas E+ 2020-1. R.V. acknowledges funding from the ERC through grant no. "2021-CoG-101043998, DEEPCONTROL".

Institutional Review Board Statement: Not applicable.

Informed Consent Statement: Not applicable.

Data Availability Statement: The data presented in this study are available upon request from the corresponding author. The YouTube link to "Deep reinforcement learning for flow control on a cylinder at Re = 2000" is https://youtu.be/8R3adCQmeEA, accessed on 28 November 2022.

Acknowledgments: The authors acknowledge the contribution of Maxence Deferrez to this work. R.V. acknowledges funding from the ERC through grant no. "2021-CoG-101043998, DEEPCONTROL".

Conflicts of Interest: The authors declare no conflict of interest. The funders had no role in the design of the study; in the collection, analyses, or interpretation of data; in the writing of the manuscript, or in the decision to publish the results.

Abbreviations

The following abbreviations are used in this manuscript:

Abbreviations

AFC	active flow control
ANN	artificial neural network
BSC-CNS	Barcelona Supercomputing Center—Centro Nacional de Supercomputación
CFD	computational fluid dynamics
CFL	Courant–Friedrichs—Lewy
CPU	central processing unit
DRL	deep reinforcement learning
EMAC	energy-, momentum-, and angular-momentum-conserving equation
FEM	finite-element method
HPC	high-performance computing
PPO	proximal policy optimization
PSD	power-spectral density
UAV	unmanned aerial vehicle

Roman letters

a	action
C_L	lift coefficient
C_D	drag coefficient
C_{offset}	offset coefficient of the reward
D	cylinder diameter
e_j	vector used in force calculation
f	external forces, frequency
f_k	vortex shedding frequency
F	Force
F_D	drag force
F_L	lift force
H	channel height
L	channel length
n	unit vector normal to the cylinder
Q	mass flow rate
Q^*	normalized mass flow rate
Q_{ref}	reference mass flow rate
p	pressure
r	reward
r_{norm}	reference value of the reward after control
R	cylinder radius
Re	Reynolds number
S	surface
s	observation state
s_{norm}	reference pressure in the observation state
St	Strouhal number
t	time
t_0	initial time
t_f	final time
T_a	action period
T_k	vortex-shedding period
$\mathbf{u}$	flow speed
$\bar{U}$	mean velocity
U_{in}	inlet boundary velocity in x direction
U_{max}	inlet boundary velocity in the middle of the channel
V_{in}	inlet boundary velocity in y direction
v_i	jet velocity
w	lift penalization
x	horizontal coordinate
y	vertical coordinate

Greek letters

ϵ	velocity strain-rate tensor
ν	kinematic viscosity
Ω	domain
ω	jet angular opening
ρ	density
σ	standard deviation
ς	Cauchy stress tensor
θ	jet angle
θ_0	center jet angle

References

1. Howell, J.P. *Aerodynamic Drag Reduction for Low Carbon Vehicles*; Woodhead Publishing Limited: Sawston, UK, 2012; pp. 145–154. [CrossRef]
2. Bechert, D.W.; Bartenwerfer, M. The viscous flow on surfaces with longitudinal ribs. *J. Fluid Mech.* **1989**, *206*, 105–129. [CrossRef]
3. Gad-el Hak, M. *Active, and Reactive Flow Management*; Cambridge University Press: Cambridge, UK, 2000. [CrossRef]

4. Guerrero, J.; Sanguineti, M.; Wittkowski, K. CFD Study of the Impact of Variable Cant Angle Winglets on Total Drag Reduction. *Aerospace* **2018**, *5*, 126. [CrossRef]
5. Tiseira, A.O.; García-Cuevas, L.M.; Quintero, P.; Varela, P. Series-hybridisation, distributed electric propulsion and boundary layer ingestion in long-endurance, small remotely piloted aircraft: Fuel consumption improvements. *Aerosp. Sci. Technol.* **2022**, *120*, 107227. [CrossRef]
6. Serrano, J.R.; García-Cuevas, L.M.; Bares Moreno, P.; Varela Martínez, P. Propeller Position Effects over the Pressure and Friction Coefficients over the Wing of an UAV with Distributed Electric Propulsion: A Proper Orthogonal Decomposition Analysis. *Drones* **2022**, *6*, 38. [CrossRef]
7. Serrano, J.R.; Tiseira, A.O.; García-Cuevas, L.M.; Varela, P. Computational Study of the Propeller Position Effects in Wing-Mounted , Distributed Electric Propulsion with Boundary Layer Ingestion in a 25 kg Remotely Piloted Aircraft. *Drones* **2021**, *5*, 56. [CrossRef]
8. Kametani, Y.; Fukagata, K. Direct numerical simulation of spatially developing turbulent boundary layers with uniform blowing or suction. *J. Fluid Mech.* **2011**, *681*, 154–172. [CrossRef]
9. Fan, Y.; Atzori, M.; Vinuesa, R.; Gatti, D.; Schlatter, P.; Li, W. Decomposition of the mean friction drag on an NACA4412 airfoil under uniform blowing/suction. *J. Fluid Mech.* **2022**, *932*, A31. [CrossRef]
10. Atzori, M.; Vinuesa, R.; Schlatter, P. Control effects on coherent structures in a non-uniform adverse-pressure-gradient boundary layer. *Int. J. Heat Fluid Flow* **2022**, *97*, 109036. [CrossRef]
11. Atzori, M.; Vinuesa, R.; Stroh, A.; Gatti, D.; Frohnapfel, B.; Schlatter, P. Uniform blowing and suction applied to nonuniform adverse-pressure-gradient wing boundary layers. *Phys. Rev. Fluids* **2021**, *6*, 113904. [CrossRef]
12. Fahland, G.; Stroh, A.; Frohnapfel, B.; Atzori, M.; Vinuesa, R.; Schlatter, P.; Gatti, D. Investigation of Blowing and Suction for Turbulent Flow Control on Airfoils. *AIAA J.* **2021**, 4422–4436. [CrossRef]
13. Voevodin, A.V.; Kornyakov, A.A.; Petrov, A.S.; Petrov, D.A.; Sudakov, G.G. Improvement of the take-off and landing characteristics of wing using an ejector pump. *Thermophys. Aeromech.* **2019**, *26*, 9–18. [CrossRef]
14. Yousefi, K.; Saleh, R. Three-dimensional suction flow control and suction jet length optimization of NACA 0012 wing. *Meccanica* **2015**, *50*, 1481–1494. [CrossRef]
15. Cui, W.; Zhu, H.; Xia, C.; Yang, Z. *Comparison of Steady Blowing and Synthetic Jets for Aerodynamic Drag Reduction of a Simplified Vehicle*; Elsevier B.V.: Amsterdam, The Netherlands, 2015; Volume 126, pp. 388–392. [CrossRef]
16. Park, H.; Cho, J.H.; Lee, J.; Lee, D.H.; Kim, K.H. Experimental study on synthetic jet array for aerodynaic drag reduction of a simplified car. *J. Mech. Sci. Technol.* **2013**, *27*, 3721–3731 [CrossRef]
17. Choi, H.; Moin, P.; Kim, J. Active turbulence control for drag reduction in wall-bounded flows. *J. Fluid Mech.* **1994**, *262*, 75–110. [CrossRef]
18. Muddada, S.; Patnaik, B.S. An active flow control strategy for the suppression of vortex structures behind a circular cylinder. *Eur. J. Mech. B/Fluids* **2010**, *29*, 93–104. [CrossRef]
19. Rabault, J.; Kuchta, M.; Jensen, A.; Réglade, U.; Cerardi, N. Artificial neural networks trained through deep reinforcement learning discover control strategies for active flow control. *J. Fluid Mech.* **2019**, *865*, 281–302. [CrossRef]
20. Ghraieb, H.; Viquerat, J.; Larcher, A.; Meliga, P.; Hachem, E. Optimization and passive flow control using single-step deep reinforcement learning. *Phys. Rev. Fluids* **2021**, *6*. [CrossRef]
21. Pino, F.; Schena, L.; Rabault, J.; Mendez, M. Comparative analysis of machine learning methods for active flow control. *arXiv* **2022**, arXiv:2202.11664.
22. Garnier, P.; Viquerat, J.; Rabault, J.; Larcher, A.; Kuhnle, A.; Hachem, E. A review on deep reinforcement learning for fluid mechanics. *Comput. Fluids* **2021**, *225*, 104973. [CrossRef]
23. Rabault, J.; Ren, F.; Zhang, W.; Tang, H.; Xu, H. Deep reinforcement learning in fluid mechanics: A promising method for both active flow control and shape optimization. *J. Hydrodyn.* **2020**, *32*, 234–246. [CrossRef]
24. Vinuesa, R.; Brunton, S.L. Enhancing computational fluid dynamics with machine learning. *Nat. Comput. Sci.* **2022**, 2, 358–366. [CrossRef]
25. Vinuesa, R.; Lehmkuhl, O.; Lozano-Durán, A.; Rabault, J. Flow Control in Wings and Discovery of Novel Approaches via Deep Reinforcement Learning. *Fluids* **2022**, *7*, 62. [CrossRef]
26. Belus, V.; Rabault, J.; Viquerat, J.; Che, Z.; Hachem, E.; Reglade, U. Exploiting locality and translational invariance to design effective deep reinforcement learning control of the 1-dimensional unstable falling liquid film. *AIP Adv.* **2019**, *9*, 125014. [CrossRef]
27. Rabault, J.; Kuhnle, A. Accelerating deep reinforcement learning strategies of flow control through a multi-environment approach. *Phys. Fluids* **2019**, *31*, 094105. [CrossRef]
28. Tang, H.; Rabault, J.; Kuhnle, A.; Wang, Y.; Wang, T. Robust active flow control over a range of Reynolds numbers using an artificial neural network trained through deep reinforcement learning. *Phys. Fluids* **2020**, *32*, 053605. [CrossRef]
29. Tokarev, M.; Palkin, E.; Mullyadzhanov, R. Deep reinforcement learning control of cylinder flow using rotary oscillations at low reynolds number. *Energies* **2020**, *13*, 5920. [CrossRef]
30. Xu, H.; Zhang, W.; Deng, J.; Rabault, J. Active flow control with rotating cylinders by an artificial neural network trained by deep reinforcement learning. *J. Hydrodyn.* **2020**, *32*, 254–258. [CrossRef]
31. Li, J.; Zhang, M. Reinforcement-learning-based control of confined cylinder wakes with stability analyses. *J. Fluid Mech.* **2022**, *932*, A44. [CrossRef]

32. Ren, F.; Rabault, J.; Tang, H. Applying deep reinforcement learning to active flow control in weakly turbulent conditions. *Phys. Fluids* **2021**, *33*, 037121. [CrossRef]
33. Wang, Q.; Yan, L.; Hu, G.; Li, C.; Xiao, Y.; Xiong, H.; Rabault, J.; Noack, B.R. DRLinFluids: An open-source Python platform of coupling deep reinforcement learning and OpenFOAM. *Phys. Fluids* **2022**, *34*, 081801. [CrossRef]
34. Qin, S.; Wang, S.; Rabault, J.; Sun, G. An application of data driven reward of deep reinforcement learning by dynamic mode decomposition in active flow control. *arXiv* **2021**, arXiv:2106.06176.
35. Vazquez, M.; Houzeaux, G.; Koric, S.; Artigues, A.; Aguado-Sierra, J.; Aris, R.; Mira, D.; Calmet, H.; Cucchietti, F.; Owen, H.; et al. Alya: Towards Exascale for Engineering Simulation Codes. *arXiv* **2014**, arXiv:1404.4881.
36. Owen, H.; Houzeaux, G.; Samaniego, C.; Lesage, A.C.; Vázquez, M. Recent ship hydrodynamics developments in the parallel two-fluid flow solver Alya. *Comput. Fluids* **2013**, *80*, 168–177. [CrossRef]
37. Lehmkuhl, O.; Houzeaux, G.; Owen, H.; Chrysokentis, G.; Rodriguez, I. A low-dissipation finite element scheme for scale resolving simulations of turbulent flows. *J. Comput. Phys.* **2019**, *390*, 51–65. [CrossRef]
38. Charnyi, S.; Heister, T.; Olshanskii, M.A.; Rebholz, L.G. On conservation laws of Navier–Stokes Galerkin discretizations. *J. Comput. Phys.* **2017**, *337*, 289–308. [CrossRef]
39. Charnyi, S.; Heister, T.; Olshanskii, M.A.; Rebholz, L.G. Efficient discretizations for the EMAC formulation of the incompressible Navier–Stokes equations. *Appl. Numer. Math.* **2019**, *141*, 220–233. [CrossRef]
40. Crank, J.; Nicolson, P. A practical method for numerical evaluation of solutions of partial differential equations of the heat-conduction type. *Adv. Comput. Math.* **1996**, *6*, 207–226. [CrossRef]
41. Trias, F.X.; Lehmkuhl, O. A self-adaptive strategy for the time integration of navier-stokes equations. *Numer. Heat Transf. Part B Fundam.* **2011**, *60*, 116–134. [CrossRef]
42. Kuhnle, A.; Schaarschmidt, M.; Fricke, K. Tensorforce: A TensorFlow Library for Applied Reinforcement Learning. 2017. Available online: https://tensorforce.readthedocs.io (accessed on 28 November 2022).
43. Abadi, M.; Agarwal, A.; Barham, P.; Brevdo, E.; Chen, Z.; Citro, C.; Corrado, G.S.; Davis, A.; Dean, J.; Devin, M.; et al. TensorFlow: Large-Scale Machine Learning on Heterogeneous Systems. 2015. Available online: tensorflow.org (accessed on 28 November 2022).
44. Schäfer, M.; Turek, S.; Durst, F.; Krause, E.; Rannacher, R. *Benchmark Computations of Laminar Flow Around a Cylinder*; Vieweg+Teubner Verlag: Wiesbaden, Germany, 1996; pp. 547–566. [CrossRef]
45. Elhawary, M. Deep reinforcement learning for active flow control around a circular cylinder using unsteady-mode plasma actuators. *arXiv* **2020**, arXiv:2012.10165.
46. Han, B.Z.; Huang, W.X.; Xu, C.X. Deep reinforcement learning for active control of flow over a circular cylinder with rotational oscillations. *Int. J. Heat Fluid Flow* **2022**, *96*, 109008. [CrossRef]
47. Stabnikov, A.; Garbaruk, A. Prediction of the drag crisis on a circular cylinder using a new algebraic transition model coupled with SST DDES. *J. Phys. Conf. Ser.* **2020**, *1697*, 012224. [CrossRef]
48. Guastoni, L.; Güemes, A.; Ianiro, A.; Discetti, S.; Schlatter, P.; Azizpour, H.; Vinuesa, R. Convolutional-network models to predict wall-bounded turbulence from wall quantities. *J. Fluid Mech.* **2021**, *928*, A27. [CrossRef]

actuators

Article

Experimental Study of Dynamical Airfoil and Aerodynamic Prediction

Zheyu Shi [1,2], Kaiwen Zhou [3,4,*], Chen Qin [2] and Xin Wen [3,4,*]

[1] College of Aerospace Engineering, Nanjing University of Aeronautics and Astronautics, Nanjing 211100, China; shizheyu@cardc.cn
[2] Low Speed Aerodynamics Institute of China Aerodynamics Research and Development Center, Mianyang 621000, China; qinchen@cardc.cn
[3] Key Lab of Education Ministry for Power Machinery and Engineering School of Mechanical Engineering, Shanghai Jiao Tong University, 800 Dongchuan Road, Shanghai 200240, China
[4] Gas Turbine Research Institute, Shanghai Jiao Tong University, 800 Dongchuan Road, Shanghai 200240, China
* Correspondence: zhoukaiwen@sjtu.edu.cn (K.Z.); wenxin84@sjtu.edu.cn (X.W.)

Abstract: Dynamic stall is a critical limiting factor for airfoil aerodynamics and a challenging problem for active flow control. In this experimental study, dynamic stall was measured by high-frequency surface pressure tapes and pressure-sensitive paint (PSP). The influence of the oscillation frequency was examined. Dynamic mode decomposition (DMD) with time-delay embedding was proposed to predict the pressure field on the oscillating airfoil based on scattered pressure measurements. DMD with time-delay embedding was able to reconstruct and predict the dynamic stall based on scattered measurements with much higher accuracy than standard DMD. The reconstruction accuracy of this method increased with the number of delay steps, but this also prolonged the computation time. In summary, using the Koopman operator obtained by DMD with time-delay embedding, the future dynamic pressure on an oscillating airfoil can be accurately predicted. This method provides powerful support for active flow control of dynamic stall.

Keywords: dynamic mode decomposition; prediction; dynamical airfoil; time-delay embedding

Citation: Shi, Z.; Zhou, K.; Qin, C.; Wen, X. Experimental Study of Dynamical Airfoil and Aerodynamic Prediction. *Actuators* **2022**, *11*, 46. https://doi.org/10.3390/act11020046

Academic Editors: André Preumont and Luigi de Luca

Received: 6 December 2021
Accepted: 31 January 2022
Published: 2 February 2022

1. Introduction

Stall is one of the most critical limiting factors for airfoil aerodynamics. Static stall occurs when the airfoil reaches a certain angle of attack. However, in real-world applications, dynamic stall caused by an unsteady motion is more common. An example of such motion is the rapid pitch-up of the airfoil in rotor blades and wind turbines. Compared with static stall, dynamic stall is a more complicated unsteady phenomenon due to the flow separation caused by the rapid change in the angle of attack and the variable stall frequency. Thus, it leads to an abrupt decay of aerodynamic forces and a fluctuation of the structural load on the airfoil. McCroskey et al. [1] comprehensively studied dynamic stall in an oscillating airfoil and dynamic vortex shedding from the airfoil. Further research [2–5] has since provided a detailed understanding of the characteristics of dynamic stall with respect to the lift and pitch moment coefficient. Carr et al. [6] reviewed the significant progress in dynamic stall and concluded that the key influencing factors of this process were oscillation amplitude and frequency, Mach number, and the three-dimensional effects of the airfoil.

Dynamic stall is common in practical engineering and detrimental to the performance of air vehicles. Many researchers employed vortex generators to realize effective control of the dynamic stall [7,8]. Vortex generator control is a passive control method, which is easy-implemented and does not require extra energy injection. However, passive control cannot adapt to the change in working conditions. Recently, many researchers focused on active flow control, which is adjustable based on a feedback control loop [9–12]. Therefore, real-time sensing of the complicated and unsteady flow, e.g., the dynamic stall, is critical in the control loop.

The use of pressure tapes for local measurements is considered a promising method to investigate the flow conditions on an airfoil due to their reliability and ease of implementation. To reduce the complexity of the algorithm or system, some researchers applied some conventional and easy-implemented methods to identify the aerodynamic parameters of the airfoils. Therefore, it is important to predict the flow dynamic based on scattered pressure data. Patel et al. [13] used a pressure sensor array on the upper surface of an airfoil to detect the onset of flow separation. Saini et al. [14] predicted the aerodynamic parameters (i.e., inflow velocity and angle of attack) using discrete surface pressures measured at ports in the vicinity of the leading edge of an airfoil. Provost et al. [15] used a time-shift linear model to predict the roll moment using a sparse set of surface pressure measurements. Juliano et al. [16] found that the pressure field on an oscillating airfoil can have distinct patterns at different locations that are difficult to capture using pressure sensors. Similarly, An et al. [17] develop a method for estimating the instantaneous lift coefficient on a rapidly pitching airfoil that uses a small number of pressure sensors and a measurement of the angle of attack. Gao et al. [18] detected unsteady boundary layer transition on a pitching airfoil using a statistical criterion calculated from thirty surface-pressure transducer measurements. With the development of artificial intelligence, the machine learning method has been successfully applied to sense or predict the aerodynamic parameters [19,20]. However, machine learning methods, such as a blacked box, cannot provide an interpretable physical explanation. Therefore, some researchers employed data-driven techniques to realize the system or parameters identification [21–23]. Zhou et al. [24] explained the physical relationship of pressure data and then developed an offline-online method to sense the real-time aerodynamic parameters. Saini et al. [25] developed the leading-edge flow sensing technique, which uses a few pressures in the airfoil leading-edge to identify and sense various flow events associated with vortex shedding in unsteady airfoils.

Most of the researches mentioned above focused on the flow field around static objects. The associated and available methods typically combine proper orthogonal decomposition (POD) with other algorithms, such as linear stochastic estimation [26] and particle filter [27]. However, it is difficult to predict the evolution of the aerodynamics of dynamic airfoil over time. In this regard, the dynamical system concept should be taken into consideration, and dynamic mode decomposition (DMD) can provide some inspiration, which is an equation-free and data-driven method to model dynamical systems. DMD was initially proposed by Schmid [28] to decompose the flow field into both spatially and temporally coherent structures and has the capability of predicting dynamic systems in the temporal domain. Berger et al. [29] successfully applied DMD to control robotic movement parameters by predicting human–robot interactions based on initial-state input data. However, it is difficult for standard DMD to make predictions based on scattered measurements because the method relies on an infinite-dimensional linear dynamic approximation for a finite-dimensional nonlinear dynamic system. It, therefore, may not provide an accurate result or may even fail when applied to highly nonlinear systems with low-dimensional measurement data [30]. To solve this problem, some data-driven methods have been proposed. The theory of DMD with delay embedding was proposed by Brunton et al. [31] to decompose a three-dimensional chaotic Lorenz system into a linear model in the delay coordinates using a single variable. Mohammad et al. [32] utilized DMD in time delay coordinates to capture the dynamics of a compressible signal containing oscillation perfectly. In addition, Susuki et al. [33] conducted this method to realize the prediction of the voltage dynamics of the rudimentary model directly from the single-bus measurement. The time delay embedding [34] has been investigated in detail on the structure and conditioning of linear models of nonlinear dynamics. Yuan et al. [35] developed DMD with a delay embedding method to predict high temporal and spatial resolved flow fields based on local particle image velocimetry (PIV) measurement. Arbabi et al. [36] applied the delay embedding method to sparse measurements and realized predictive control of nonlinear flow based on a Koopman model framework.

In the current work, the scattered data obtained from pressure tapes on the surface of an airfoil were used to predict the surface pressure of the airfoil during the oscillation process. Pressure-sensitive paint (PSP) surface measurement provided a qualitative visualization to verify the dynamic stall on the oscillating airfoil. Based on the scattered pressure tape measurements, DMD with time-delay embedding was applied to reconstruct the spatiotemporal pressure distribution on the oscillating airfoil and predict the future data in real-time.

2. Experimental Setup and Methodology

2.1. Measurement Setup

The experiment was conducted in an open-circuit low-speed wind tunnel at the China Aerodynamics Research and Development Center. The test section of the wind tunnel has a cross-section of 1.8 m × 1.4 m. In the experiment, the freestream velocity was $U_\infty = 34$ m/s. The two-dimensional airfoil model OA309, made of high-strength carbon fiber composite materials, with a chord length of $c = 0.4$ m and span length of $s = 1.8$ m, was mounted on a pitching testing platform. The freestream Reynolds number based on the chord length c was $\text{Re} = 0.9 \times 10^6$. The airfoil model was subjected to an unsteady periodic motion. As shown in Figure 1, the pitching motion mechanism was composed of a servo motor, planetary reducer, and encoder. The mechanism allowed control over motion parameters, such as mean angle of attack α_0, pitching amplitude α_1, and oscillation frequency f. In the experiment, α_0 was set as 10°, α_1 was set as 10°, and f was varied as 1, 2, 3, and 4 Hz. The reduced frequency was defined as $k = \pi f c / U_\infty$, with values of 0.037, 0.074, 0.111, and 0.148. The formula for the angle of attack α can be expressed as follows:

$$\alpha = \alpha_0 + \alpha_1 sin(2\pi f t) \tag{1}$$

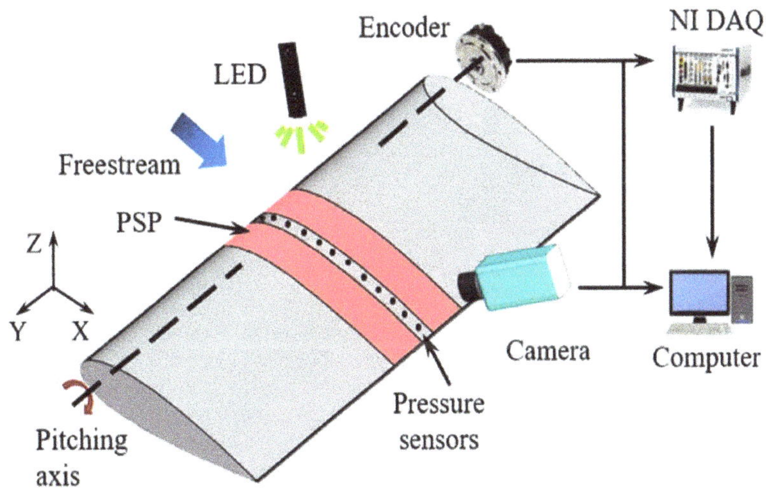

Figure 1. Schematic of a measurement system using pressure and PSP.

The PSP measurement system and pressure tapes are shown in Figure 1. Part of the upper surface of the airfoil was painted with PSP to evaluate the changes in pressure distribution when the airfoil was oscillating. The PSP was excited by a UV LED (UHP-T-385-EP from Prizmatix Ltd., Givat Shmuel, Israel), and images were captured by a camera (PCO 1600) with a filter. Using a synchronizer (BNC-575) and potentiometer, PSP images were captured during certain phases of the oscillating airfoil. During the PSP measurement process, exposure signals from the camera were acquired simultaneously with measurements from the potentiometer, which contained the angle of attack of the airfoil. Using these two signals, every PSP frame was assigned to a certain angle of attack. Then, the phase-locked pressure distribution of the upper surface of the airfoil could be obtained by PSP measurement.

The pressure tape measurements and PSP measurements were taken simultaneously. Forty dynamic differential pressure sensors (8510B, ENDVECO) were arranged chordwise at the midspan of the airfoil model. The measurement range of these pressure sensors is 1 psi, and the measurement inaccuracy is ±0.05% of full scale. Locations of these 40 pressure sensors are detailed in Table 1. These pressure sensors adopted a dynamic sampling frequency to sample 256 data points per oscillation cycle of the dynamic airfoil. The acquisition time was approximately 30 s. The aerodynamic parameters, such as lift and pitching moments, were obtained by the integration of the pressure data. Aerodynamic loads can be calculated based on the integration of the surface pressure coefficient. The pressure coefficients can be calculated from the following expression:

$$C_{p_i} = \frac{P_i - P_\infty}{\frac{1}{2}\rho U_\infty} \tag{2}$$

where ρ is the air density, P_i is the static pressure obtained from a single pressure tap, P_∞ is the static pressure of the incoming flow. The detailed process of the integration can refer to previous material [37].

Table 1. Locations of the pressure transducers on the OA209 airfoil.

Upper Surface				Lower Surface			
No.	***x/c***	**No.**	***x/c***	**No.**	***x/c***	**No.**	***x/c***
1	0.01	11	0.24	21	0	31	0.25
2	0.02	12	0.3	22	0.0125	32	0.3
3	0.03	13	0.36	23	0.025	33	0.35
4	0.04	14	0.42	24	0.0375	34	0.4
5	0.05	15	0.48	25	0.05	35	0.5
6	0.06	16	0.57	26	0.075	36	0.6
7	0.09	17	0.66	27	0.1	37	0.7
8	0.12	18	0.75	28	0.125	38	0.8
9	0.15	19	0.84	29	0.15	39	0.9
10	0.18	20	0.93	30	0.2	40	1

2.2. *Dynamic Mode Decomposition with Time-Delay Embedding*

DMD is an efficient method of decomposing a dynamical system into specific spatial modes that evolve at a certain frequency and growth–decay rate in time. In DMD, the measurement data obtained by sensors at time j are arranged into a column vector $\boldsymbol{x}_j = (x_{1,j}, x_{2,j}, \ldots, x_{n,j})$, where $\boldsymbol{x}_j$ is a data snapshot and n denotes the number of measurement points (in this case, the number of sensors). All m snapshots were sampled with a constant time interval (sampling time step) Δt. DMD describes a discrete-time system as:

$$\boldsymbol{x}_{j+1} = A\boldsymbol{x}_j \tag{3}$$

where matrix A reflects the dynamical characteristics of the system. The first step of DMD is to arrange all m snapshots into two matrices, $\boldsymbol{X}_1 = (\boldsymbol{x}_1, \boldsymbol{x}_2, \ldots, \boldsymbol{x}_{m-1})$ and $\boldsymbol{X}_2 = (\boldsymbol{x}_2, \boldsymbol{x}_3, \ldots, \boldsymbol{x}_m)$. Then, the linear approximation A may be written in terms of matrices based on Equation (4) as:

$$\boldsymbol{X}_2 \approx A\boldsymbol{X}_1 \tag{4}$$

Using matrix A, known as the Koopman operator, future data can be predicted from current data. Thus, the core of DMD is to search for the optimal linear approximation A of the nonlinear dynamical system. The crucial step of the DMD algorithm is obtaining the rank-reduced representation of A in terms of a POD-projected matrix $\widetilde{A}$, which can be written as follows:

$$\widetilde{A} = U^T A U = U^T X_2 V \Sigma^{-1} \tag{5}$$

where $U \in \mathbf{R}^{n \times r}$, $\Sigma \in \mathbf{R}^{r \times r}$, and $V \in \mathbf{R}^{m \times r}$ are obtained by singular value decomposition (SVD). Here, r is defined as the truncated number and represents the rank of approximation to $\mathbf{X_1}$ using SVD reduction. The next step is to compute the eigen-decomposition of $\widetilde{A}$, which is similar to A:

$$\widetilde{A} W = W \Lambda \tag{6}$$

where the columns of W are eigenvectors of $\widetilde{A}$ and Λ is a diagonal matrix containing the eigenvalues λ_k, which capture the time dynamics of the corresponding DMD modes. These eigenvalues and eigenvectors of $\widetilde{A}$ can be related back to the similarity-transformed eigenvalues and eigenvectors of A to reconstruct the DMD modes given by:

$$\Phi = X_2 V \Sigma^{-1} W \tag{7}$$

where Φ is the matrix whose columns each represent a DMD mode. Then, a snapshot at any time can be reconstructed from the DMD modes and eigenvalues:

$$\boldsymbol{x}_j = \Phi \Sigma^{j-1} \boldsymbol{b} \tag{8}$$

where $\boldsymbol{b} = \Phi^{-1} \times {}_1$ contains the initial values of DMD modes. As shown above, the DMD method relies on the linear approximation of a nonlinear dynamic system. When the subdomains are of limited spatial dimensionality, it is difficult for DMD to provide accurate predictions. In this study, the direct use of discrete sensor measurements as the observables creates a system that is far removed from the infinite-dimensional linear approximation of highly nonlinear systems. To solve this problem, extended DMD [38] and kernel DMD [39] have previously been proposed. These methods use new observables to provide a set of coordinates in which the dynamical systems appear linear. However, a universal and systematic approach needs to be found. The details of time delay embedding methods can refer to the works of Brunton et al. [31]. Time-delay embedding augments the limited spatial observables by embedding future measurements into the current measurements. Therefore, more information can be captured to construct the linear Koopman operator.

A requirement of time-delay embedding is to have access to at least $n_d + 1$ consecutive snapshots, where n_d is the number of delay steps. The time-delay-embedded matrices $\mathbf{X_1}$ and $\mathbf{X_2}$ can be transformed into $\boldsymbol{H_1}$ and $\boldsymbol{H_2}$ as follows:

$$\boldsymbol{H_1} = \begin{bmatrix} x_1 & x_2 & \cdots & x_{m-n_d} \\ x_2 & x_3 & \cdots & x_{m-n_d+1} \\ \vdots & \vdots & \ddots & \vdots \\ x_{n_d} & x_{n_d+1} & \cdots & x_{m-1} \end{bmatrix} = [y_1, y_2 \cdots y_{m-n_d}], \tag{9}$$

$$\boldsymbol{H_2} = \begin{bmatrix} x_2 & x_3 & \cdots & x_{m-n_d+1} \\ x_3 & x_4 & \cdots & x_{m-n_d+2} \\ \vdots & \vdots & \ddots & \vdots \\ x_{n_d+1} & x_{n_d+2} & \cdots & x_m \end{bmatrix} = [y_2, y_3 \cdots y_{m-n_d+1}], \tag{10}$$

where y_j is a new snapshot after delay embedding, and $\boldsymbol{x_m}$ occupies the last n rows y_{m-n_d+1}.

The dimensions of the matrices after delay embedding $\boldsymbol{H_1}$ and $\boldsymbol{H_2}$, are $(n \times n_{d)} \times (m - n_d)$. When $n_d = 1$, the matrix after delay embedding is the same as the original matrix. Analogously to Equation (4), the linear mapping of $\boldsymbol{H_1}$ and $\boldsymbol{H_2}$ can be written as follows:

$$\boldsymbol{H_2} \approx B\boldsymbol{H_1} \tag{11}$$

Then, the time-delay-embedded matrices $\boldsymbol{H_1}$ and $\boldsymbol{H_2}$ are used to build the DMD model. The matrix B is the Koopman operator and reflects the dynamical characteristics. Therefore, standard DMD can be applied to find the corresponding eigenvalues and eigenvectors of B. Using the matrix B as follows:

$$y_{j+1} \approx By_j \tag{12}$$

The current new snapshot y_j at time $n_d + j - 1$ can be used to predict the future new snapshot y_{j+1}, which contains the data at time $n_d + j$.

3. Results and Discussion

3.1. Dynamic Stall

3.1.1. Flow Visualization

To examine the two-dimensional pressure field on the oscillating airfoil, the images from PSP measurement were phase-averaged over 15 periods. The evolution of the pressure field during the oscillation of the clean airfoil is shown in Figure 2 at two typical AoAs under an oscillation frequency of f = 1 Hz. It should be noted that these PSP measurements only provided a qualitative visualization because the PSP data was influenced by the temperature of the incoming flow.

The upper panels represent the downstroke, and the lower panels represent the upstroke. The low-pressure region is indicated in blue, and high pressures are shown in yellow. At α = 9°, the pressure distribution in the downstroke is similar to that in the upstroke. However, when α increases to 13°, an obvious low-pressure region forms at the leading edge of the airfoil in the upstroke, as can be seen in Figure 2d. The color intensity shows that the pressure at the leading edge in the downstroke is higher than that in the upstroke at the same angle of attack, which suggests the onset of flow separation and the decrease in lift. The phenomenon that flow separation occurs at the leading edge in the downstroke but does not occur in the upstroke is consistent with the dynamic stall was also investigated by Heine et al. using PIV measurement [8].

3.1.2. Aerodynamic Loads

The angle of attack of the airfoil was varied to examine the influence of the VGs mounted at the leading edge of the airfoil on the static aerodynamic loads. In our study, the aerodynamic loads of static and dynamical airfoil were measured to verify the phenomenon of dynamic stall. The measurement results of static airfoil are shown in Figure 3. Note that there is no oscillation frequency for this static case. As shown in Figure 3, the curves of the lift and pitch moment show a relatively constant trend until $\alpha \approx 13.5°$. With a further increase in the angle of attack, the lift and pitch moment drop abruptly. An increase of 0.5° in the angle of attack leads to a drop of approximately 0.39 in the lift coefficient and 0.08 in the pitch moment coefficient. Therefore, α = 13.5° can be considered as the stall angle with respect to flow separation. These results were in line with our common perception and are consistent with the findings of previous studies [8,40].

The influence of the oscillation frequency f of the dynamical airfoil on the aerodynamic loads was examined in this study. The lift and moment coefficients under different AoA for the oscillating airfoil at different f are shown in Figure 4. It should be reminded that the relevant aerodynamic load data were phase-averaged from 10 measurement periods, and the resolution was improved by interpolation. In the present work, instantaneous pressure data at different periods are similar, so phase-averaged data based on 10 periods is credible.

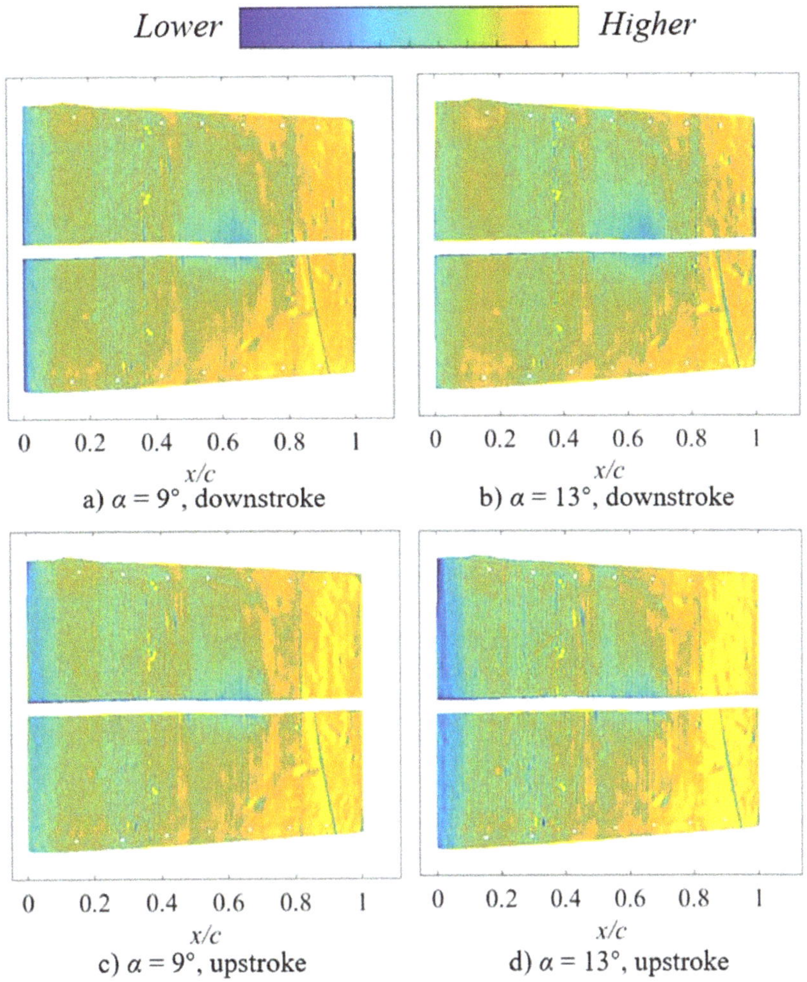

Figure 2. PSP visualization of the dynamic pressure on the oscillating airfoil with f = 1 Hz. (The horizontal direction is the full chord length, and the vertical direction is the spanwise direction).

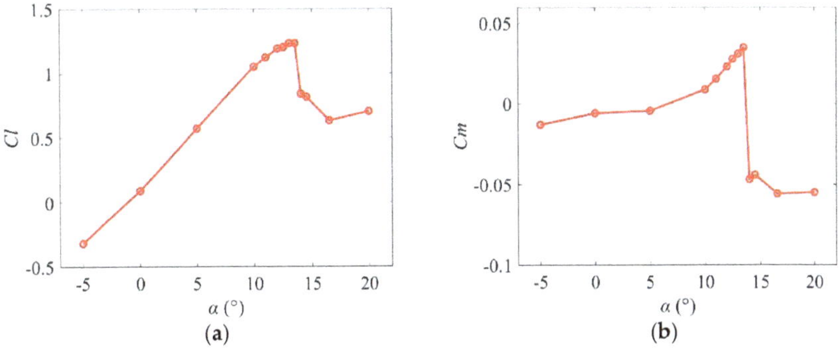

Figure 3. Static stall characteristics are shown in coefficients of (**a**) lift (Cl) and (**b**) pitch moment (Cm) for the static airfoil.

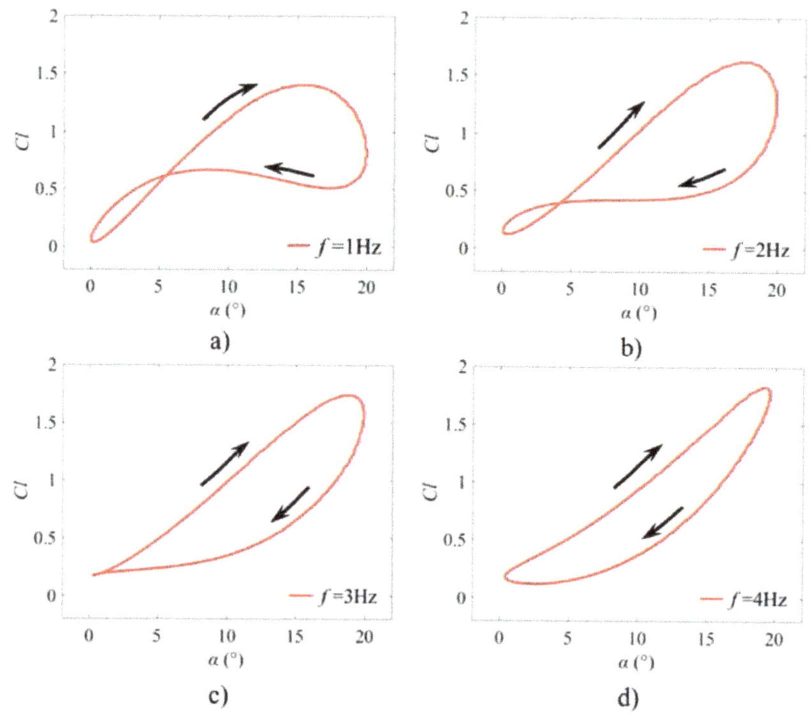

Figure 4. Phase-averaged lift coefficient (*Cl*) at frequencies of (**a**) $f = 1$ Hz, (**b**) $f = 2$ Hz, (**c**) $f = 3$ Hz, and (**d**) $f = 4$ Hz for oscillating airfoils.

When $f = 1$ Hz, the lift coefficient of the upstroke is higher than that of the downstroke by up to 0.8 under the same angle of attack. There was a crucial characteristic feature in the curve of loop hysteresis, which was related to a certain $\alpha = 6°$ when $f = 1$ Hz. When $\alpha < 6°$, the lift coefficients of the upstroke is lower than that of the downstroke under the same AoA. Once $\alpha > 6°$, the lift coefficients of the upstroke is higher than that of the downstroke. Therefore, we can say that there are two hysteresis loops during the oscillating process. It can be assumed that the pressure of the lower surface of airfoil can be higher under small AoA due to the downstroke of the airfoil. When increasing the oscillation frequency to 2 Hz (Figure 4b), the certain AoA which decided the two-loop hysteresis became smaller. Continuing to improve the oscillation frequency f to 3 Hz (Figure 4c), only one loop hysteresis is left. However, when $\alpha = 0, 1, 2°$, the lift moments under upstroke and downstroke were close. When the oscillation frequency f was increased to 4 Hz (Figure 4d), the lift coefficients under even a small angle of attack during the downstroke were smaller than that during the upstroke. With increasing oscillation frequency f, the loop hysteresis becames more pronounced, and the shapes of the hysteretic loops changed. In detail, the increase of oscillation frequency f induced the slight increase of upstroke and downstroke in the lift coefficient under larger AoAs, but the lift coefficient under smaller AoAs in the downstroke decreased slightly.

In addition to the lift coefficients, the characteristics of pitch moment were also investigated, as shown in Figure 5. For $f = 1$ Hz, there is no abrupt drop in the moment coefficient such as the static case during the upstroke. The moment coefficient of both upstroke and downstroke was very close, i.e., the difference of moment coefficient between upstroke and downstroke was small. It is worth noting that compared to the static case, the moment coefficient under large AoAs decreased dramatically, which suggested that dynamic stall influenced the moment coefficient. For example, when $\alpha = 20°$, the moment coefficient in the static case was about −0.055 but it decreased to about −0.085 in the dynamical case. With increasing oscillation frequency *f*, the curve of moment coefficient formed two-loop hysteresis. The moment coefficient of upstroke and downstroke showed a huge difference, suggesting that high oscillation frequency leads to a large fluctuation in pitch moment. Combined with the change in lift coefficient, it can be inferred that not only did the dynamic stall exert a negative influence on the aerodynamic load, the oscillation frequency *f* also did. More importantly, high oscillation frequency *f* makes the aerodynamic performance more intricate (i.e., two-loop hysteresis in a coefficient curve). This problem is detrimental to the performance of real-time active flow control. Therefore, real-time perception, even the prediction of aerodynamic parameters, is challenging and meaningful. In Section 3.2, we will use a novel data-driven method to reconstruct the time-resolved pressure data and then realize the prediction of future pressure.

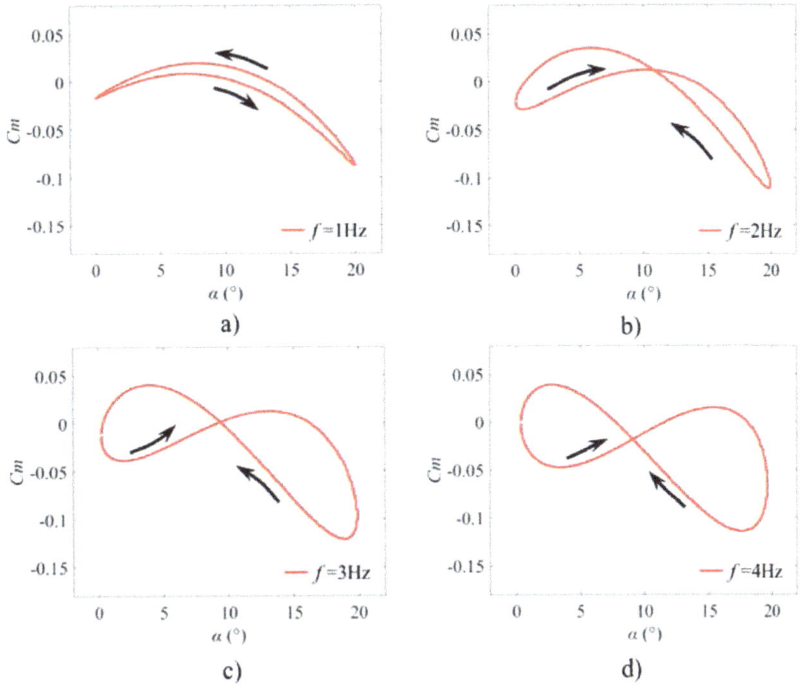

Figure 5. Phase-averaged moment coefficient (*Cm*) for (**a**) *f* = 1 Hz, (**b**) *f* = 2 Hz, (**c**) *f* = 3 Hz, and (**d**) *f* = 4 Hz for the oscillating airfoil.

3.2. Prediction of Future Data

In this section, DMD with time delay embedding was used to reconstruct the history of pressure data of the oscillating airfoil. The high time-resolved pressure data collected by 20 sensors on the upper surface (see Table 1) were used in this method because the leading edge is the region with the most obvious pressure fluctuations. To examine the performance of this method, the error rate *e* between reconstructed data and true data is defined as:

$$e = \frac{1}{m} \sum_{t=1}^{m} \frac{||P_{reconstructed,t} - P_{true,t}||_2}{||P_{true,t}||_2} \quad (13)$$

where m is the number of used snapshots, $P_{reconstructed,t}$ is the reconstructed pressure field at time t, and $P_{true,t}$ is the true velocity field on the subdomain at time t.

POD analysis was applied to the time delay embedded dataset, and the mode energy spectrum is shown in Figure 6. The core of time delay embedding is the improvement of data dimensionality to capture more phase information. Therefore, the first three POD modes contained more than 98% energy. With the increase of the number of delay steps n_d, more POD modes were needed to reconstruct the data accurately. It should be noted that when the number of delay steps was increased to 200, the POD energy spectrum tended to be similar. In this experiment, one period contains 256 snapshots. Therefore, when $n_d > 200$, most of the phase information was included in the time delay embedded matrix.

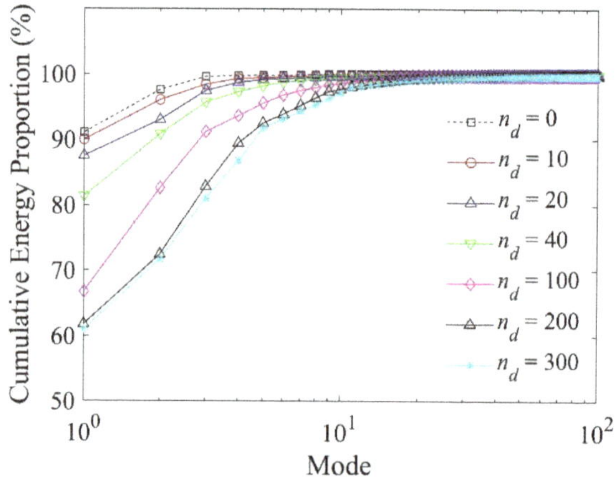

Figure 6. POD mode energy spectrum of the time-delay embedded data with different n_d.

The raw and reconstructed pressure data of sensor #1 were used to compare the reconstruction performance of the standard DMD and DMD with delay embedding. Sensor #1 is located at the leading edge, where the most obvious pressure fluctuations happen. The standard DMD (i.e., $n_d = 0$) and DMD with time-delay embedding were then applied to reconstruct the whole-time sequence of the pressure data. As shown in Figure 7a, standard DMD cannot accurately predict the dynamic pressure on sensor #1 even at the initial stage of the oscillation cycle. With time-delay embedding, the prediction performance is clearly improved. With $n_d = 50$ delay steps, DMD can capture the periodic trend, but with an underprediction of the oscillation amplitude and e $\approx$ 43%. With n_d increased to 200, the predicted values agree closely with the ground truth both in periodic trend and amplitude for more than six cycles and e $\approx$ 7%. Thus, DMD with time-delay embedding demonstrates the ability to reconstruct the periodic single-point pressure data over a long period. With an increasing number of delay steps n_d, the accuracy of the reconstruction is improved. However, an excessive number of delay steps will lead to too large a dimensionality of the matrix, which will not only consume computing and storage resources but negatively affect the real-time performance of the prediction algorithm. Therefore, the duration of the prediction cycle should be balanced with the requirement for computation time. In fact, the accurate prediction of just one cycle, that is, the next cycle, is usually helpful for effective decision-making in real-time control systems.

Based on the predicted pressures at single measurement points, DMD with time-delay embedding was then used to reconstruct the time-resolved pressure field on the upper surface of the airfoil. The ground truth of the pressure field obtained by 20 pressure sensors is shown in Figure 8a. The values reconstructed using DMD with time-delay embedding are shown in Figure 8b and match well with the measured data and accurately show the shape and peak of the three-dimensional surface.

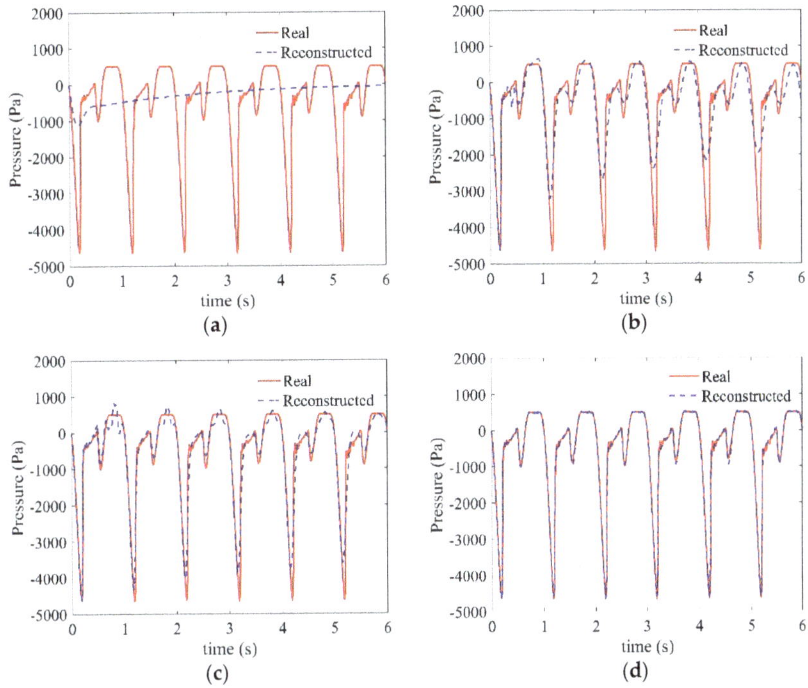

Figure 7. Reconstruction of pressure data measured by sensor #1 (see Table 1) using (**a**) standard DMD and time-delay embedding DMD with number of delay steps (**b**) n_d = 50, (**c**) n_d = 100, and (**d**) n_d = 200.

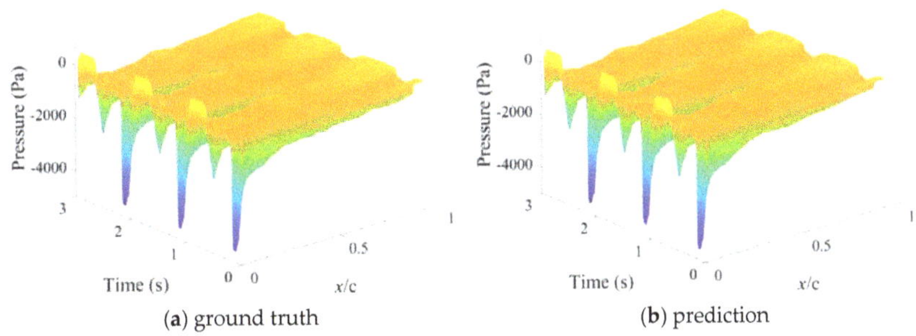

Figure 8. Pressure field on the upper surface of the oscillating airfoil: (**a**) ground-truth data from pressure sensors, (**b**) reconstruction using DMD with time-delay embedding.

The core of DMD is to find the Koopman operator, which allows the prediction of future data based on past data. The above section has demonstrated the accurate reconstruction of time-resolved pressure data using DMD with time-delay embedding, from which an optimal Koopman operator was obtained. Using the Koopman operator, real-time prediction can be realized. In Figure 9, the black line represents the ground-truth pressure data measured by sensor #1 over one oscillation period. The blue markers represent the past sampled pressure data that were used to build a new snapshot y_j. The future new snapshot y_{j+1} was obtained based on the Koopman operator B provided by DMD with time-delay embedding, which had been verified in the above cases. Then, the future pressure data at the position of sensor #1 could be extracted from y_{j+1}. Furthermore, future snapshot y_{j+2}

can also be calculated with the help of a predicted snapshot y_{j+1} in the same way. When this operation is repeated, the data for a long period of time in the future can be predicted. In Figure 9, 256 snapshots (a whole oscillation period) were predicted. The predicted next-step value accurately matches the ground-truth value, indicating the feasibility and potential of DMD with time delay embedding. The computation was implemented in MATLAB, and all the cases are run on an Intel Xeon E5-1620 at 3.50 GHz. Every prediction took 0.09 s. Admittedly, it is difficult to meet real-time requirements, but high-performance industrial computers can realize real-time prediction in practical applications. Work is ongoing to further improve the rapidity and adaptability of this method.

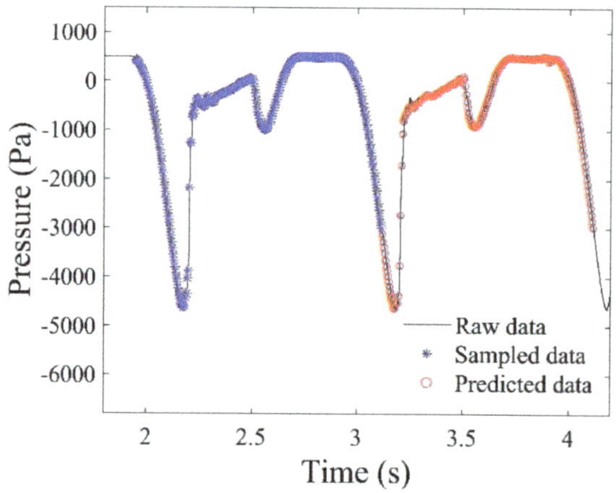

Figure 9. Prediction of future pressure data based on Koopman operator obtained by DMD with time-delay embedding.

4. Conclusions

In this study, dynamic stall of oscillating airfoils was measured by dynamic pressure tapes and PSP visualization. The influence of the oscillation frequency was examined. Dynamic mode decomposition (DMD) with time-delay embedding was proposed to predict the pressure field on the oscillating airfoil based on scattered measurements. The major findings are as follows:

(1) Under the dynamic conditions of the oscillating airfoil, with increasing oscillation frequency, although lift coefficient can be improved at large AoAs, the aerodynamic characteristics showed a huge difference between upstroke and downstroke. The specific performance is that the loop hysteresis of both the lift and pitch moment coefficients were enlarged, which is detrimental to aerodynamic performance.

(2) DMD with time-delay embedding was able to reconstruct and predict dynamic stall based on scattered measurements with much higher accuracy than standard DMD. The reconstruction accuracy of this method increased with the number of delay steps, but this also prolonged the computation time. With the help of the Koopman operator obtained by DMD with time-delay embedding, the future dynamic pressure for a long time of an oscillating airfoil can be accurately predicted.

Author Contributions: Writing—original draft, Z.S.; Methodology and Investigation, K.Z.; Resources, C.Q.; Writing—review & editing, X.W. All authors have read and agreed to the published version of the manuscript.

Funding: This research was funded by the National Natural Science Foundation of China (grant number: 12072196), the Science and Technology Commission of Shanghai Municipality, China (grant number: 19JC1412900), the Oceanic Interdisciplinary Program of Shanghai Jiao Tong University (grant number: SL2020MS006), Advanced Jet Propulsion Innovation Center, China (grant number: AEAC

HKCX2020-02-028), Aeronautical Science Foundation of China (grant number: 2020Z006057001) and State Key Laboratory of Aerodynamics, China (grant number: SKLA-20200303).

Acknowledgments: All individuals included in this section have consented to the acknowledgement. The authors gratefully thank for the support of Lingrui Jiao and Di Peng from Shanghai Jiao Tong University, who provided some help and suggestion for the pressure-sensitive paint measurement.

Conflicts of Interest: The authors declare no conflict of interest.

Nomenclature

α	angle of attack, °
α_0	mean angle of attack, °
α_1	pitching amplitude, °
c	chord length, m
C_l	lift coefficient
C_m	pitching moment coefficient
e	the error rate between reconstructed data and true data
f	pitching frequency, Hz
n_d	the number of delay steps
U_∞	freestream velocity, m/s
s	span length, m
k	reduced frequency, Hz
Re	Reynolds number
ρ	air density, $kg/m^3 P_i$ = the static pressure obtained from a pressure tap, Pa
P_∞	the static pressure of the incoming flow, Pa
H_1, H_2	time-delay-embedded matrices
X_1, X_2	raw data matrices
A, B	Koopman operator
Acronyms	
DMD	dynamic mode decomposition
PIV	particle image velocimetry
POD	proper orthogonal decomposition
PSP	pressure sensitive paint

References

1. McCroskey, W.J. The Phenomenon of Dynamic Stall. *NASA Tech. Memo.* **1981**, *81264*, 1–28.
2. Mulleners, K.; Raffel, M. The onset of dynamic stall revisited. *Exp. Fluids* **2012**, *52*, 779–793. [CrossRef]
3. Chandrasekhara, M.S.; Carr, L.W.; Wilder, M.C. Interferometric investigations of compressible dynamic stall over a transiently pitching airfoil. *AIAA J.* **1994**, *32*, 586–593. [CrossRef]
4. Visbal, M.R. Dynamic stall of a constant-rate pitching airfoil. *J. Aircraft* **1990**, *27*, 400–407. [CrossRef]
5. Rival, D.; Tropea, C. Characteristics of pitching and plunging airfoils under dynamic-stall conditions. *J. Aircraft* **2010**, *47*, 80–86. [CrossRef]
6. Carr, L.W. Progress in analysis and prediction of dynamic stall. *J. Aircraft* **1988**, *25*, 6–17. [CrossRef]
7. Geissler, W.; Dietz, G.; Mai, H.; Bosbach, J.; Richard, H. Dynamic Stall and its passive control investigations on the OA209 airfoil section. In Proceedings of the 31st European Rotorcraft Forum 2005 (ERF31), Florence, Italy, 13–15 September 2005.
8. Heine, B.; Mulleners, K.; Joubert, G.; Raffel, M. Dynamic stall control by passive disturbance generators. *AIAA J.* **2013**, *51*, 2086–2097. [CrossRef]
9. Konstantinidis, E.; Balabani, S. Symmetric vortex shedding in the near wake of a circular cylinder due to streamwise perturbations. *J. Fluids Struct.* **2007**, *23*, 1047–1063. [CrossRef]
10. Melton, L.P.; Koklu, M.; Andino, M.; Lin, J.C. Active flow control via discrete sweeping and steady jets on a simple-hinged flap. *AIAA J.* **2018**, *56*, 2961–2973. [CrossRef]
11. Schmidt, H.J.; Woszidlo, R.; Nayeri, C.N.; Paschereit, C.O. Separation control with fluidic oscillators in water. *Exp. Fluids* **2017**, *58*, 1–17. [CrossRef]
12. Gao, D.; Chen, G.; Chen, W.; Huang, Y.; Li, H. Active control of circular cylinder flow with windward suction and leeward blowing. *Exp. Fluids* **2019**, *60*, 26. [CrossRef]
13. Patel, M.P.; Carver, R.; Lisy, F.J.; Prince, T.S.; Ng, T. Detection and control of flow separation using pressure sensors and micro-vortex generators. In Proceedings of the 40th AIAA Aerospace Sciences Meeting & Exhibit, Reno, NV, USA, 14–17 January 2002. [CrossRef]

14. Saini, A.; Gopalarathnam, A. Leading-edge flow sensing for aerodynamic parameter estimation. *AIAA J.* **2018**, *56*, 4706–4718. [CrossRef]
15. Le Provost, M.; He, X.; Williams, D.R. Real-time roll moment identification with distributed surface pressure sensors on a rolling UCAS wing. In Proceedings of the AIAA Aerospace Sciences Meeting, Kissimmee, FL, USA, 8–12 January 2018.
16. Juliano, T.J.; Peng, D.; Jensen, C.; Gregory, J.W.; Liu, T.; Montefort, J.; Palluconi, S.; Crafton, J.; Fonov, S. PSP measurements on an oscillating NACA 0012 airfoil in compressible flow. In Proceedings of the 41st AIAA Fluid Dynamics Conference and Exhibit, Honolulu, HI, USA, 27–30 June 2011. [CrossRef]
17. An, X.; Williams, D.R.; Eldredge, J.D.; Colonius, T. Lift coefficient estimation for a rapidly pitching airfoil. *Exp. Fluids* **2021**, *62*, 1–12. [CrossRef]
18. Gao, Y.; Zhu, Q.; Wang, L. Measurement of unsteady transition on a pitching airfoil using dynamic pressure sensors. *J. Mech. Sci. Technol.* **2016**, *30*, 4571–4578. [CrossRef]
19. Sabater, C.; Stürmer, P.; Bekemeyer, P. Fast Predictions of Aircraft Aerodynamics using Deep Learning Techniques. In Proceedings of the AIAA AVIATION 2021 FORUM, Online, 2–6 August 2021. [CrossRef]
20. Singh, A.P.; Medida, S.; Duraisamy, K. Machine-learning-augmented predictive modeling of turbulent separated flows over airfoils. *AIAA J.* **2017**, *55*, 2215–2227. [CrossRef]
21. Meyer, K.E.; Pedersen, J.M.; Özcan, O. A turbulent jet in crossflow analysed with proper orthogonal decomposition. *J. Fluid Mech.* **2007**, *583*, 199–227. [CrossRef]
22. Dawson, S.T.M.; Schiavone, N.K.; Rowley, C.W.; Williams, D.R. A data-driven modeling framework for predicting forces and pressures on a rapidly pitching airfoil. In Proceedings of the 45th AIAA Fluid Dynamics Conference, Dallas, TX, USA, 22–26 June 2015. [CrossRef]
23. Bright, I.; Lin, G.; Kutz, J.N. Compressive sensing based machine learning strategy for characterizing the flow around a cylinder with limited pressure measurements. *Phys. Fluids* **2013**, *25*, 127102. [CrossRef]
24. Zhou, K.; Zhou, L.; Zhao, S.; Qiang, X.; Liu, Y.; Wen, X. Data-Driven Method for Flow Sensing of Aerodynamic Parameters Using Distributed Pressure Measurements. *AIAA J.* **2021**, *59*, 3504–3516. [CrossRef]
25. Saini, A.; Narsipur, S.; Gopalarathnam, A. Leading-edge flow sensing for detection of vortex shedding from airfoils in unsteady flows. *Phys. Fluids* **2021**, *33*, 21–28. [CrossRef]
26. Pinier, J.T.; Ausseur, J.M.; Glauser, M.N.; Higuchi, H. Proportional closed-loop feedback control of flow separation. *AIAA J.* **2007**, *45*, 181–190. [CrossRef]
27. Zhao, X.; Kania, R.; Kebbie-Anthony, A.; Azarm, S.; Balachandran, B. Dynamic data-driven aeroelastic response prediction with discrete sensor observations. In Proceedings of the 2018 AIAA Non-Deterministic Approaches Conference, Kissimmee, FL, USA, 8–12 January 2018; pp. 1–12. [CrossRef]
28. Schmid, P.J. Dynamic mode decomposition of numerical and experimental data. *J. Fluid Mech.* **2010**, *656*, 5–28. [CrossRef]
29. Berger, E.; Sastuba, M.; Vogt, D.; Jung, B.; Ben Amor, H. Dynamic Mode Decomposition for perturbation estimation in human robot interaction. In Proceedings of the The 23rd IEEE International Symposium on Robot and Human Interactive Communication, Edinburgh, UK, 25–29 August 2014; pp. 593–600. [CrossRef]
30. Le Clainche, S.; Vega, J.M. Higher order dynamic mode decomposition. *SIAM J. Appl. Dyn. Syst.* **2017**, *16*, 882–925. [CrossRef]
31. Brunton, S.L.; Brunton, B.W.; Proctor, J.L.; Kaiser, E.; Nathan Kutz, J. Chaos as an intermittently forced linear system. *Nat. Commun.* **2017**, *8*, 1–8. [CrossRef]
32. Murshed, M.N.; Monir Uddin, M. Time delay coordinate based dynamic mode decomposition of a compressible signal. In Proceedings of the 2019 22nd International Conference on Computer and Information Technology, ICCIT, Dhaka, Bangladesh, 18–20 December 2019; pp. 18–20.
33. Susuki, Y.; Sako, K. Data-Based Voltage Analysis of Power Systems via Delay Embedding and Extended Dynamic Mode Decomposition. *IFAC-PapersOnLine* **2018**, *51*, 221–226. [CrossRef]
34. Pan, S.; Duraisamy, K. On the structure of time-delay embedding in linear models of non-linear dynamical systems. *Chaos* **2020**, *30*, 073135. [CrossRef]
35. Yuan, Y.; Zhou, K.; Zhou, W.; Wen, X.; Liu, Y. Flow prediction using dynamic mode decomposition with time-delay embedding based on local measurement. *Phys. Fluids* **2021**, *33*, 095109. [CrossRef]
36. Arbabi, H.; Korda, M.; Mezić, I. A data-driven Koopman model predictive control framework for nonlinear ows. In Proceedings of the The 57th IEEE Conference on Decision and Control, Miami Beach, FL, USA, 17–19 December 2018; pp. 6409–6414.
37. Miller, S.D. Lift, Drag and Moment of a NACA 0015 Airfoil. Available online: https://smiller.sbyers.com/temp/AE510_03%20NACA%200015%20Airfoil.pdf (accessed on 6 December 2021).
38. Williams, M.O.; Kevrekidis, I.G.; Rowley, C.W. A Data–Driven Approximation of the Koopman Operator: Extending Dynamic Mode Decomposition. *J. Nonlinear Sci.* **2015**, *25*, 1307–1346. [CrossRef]
39. Dellnitz, M.; Von Molo, M.H.; Ziessler, A. A kernel-based method for data-driven Koopman spectral analysis. *J. Comput. Dyn.* **2016**, *2*, 247–265. [CrossRef]
40. Mai, H.; Dietz, G.; Geißler, W.; Richter, K.; Bosbach, J.; Richard, H.; Groo, K. De Dynamic stall control by leading edge vortex generators. *Annu. Forum Proc. AHS Int.* **2006**, *III*, 1728–1739.

actuators

Article

Experimental and Numerical Study on Incident Shock Wave/Boundary Layer Interaction Control

Chuanbiao Zhang, Yanhao Luo, Hua Liang *, Shanguang Guo and Hesen Yang

Science and Technology on Plasma Dynamics Laboratory, Air Force Engineering University, Xi'an 710038, China; zhangcb2020@126.com (C.Z.); lyh_1415@163.com (Y.L.); shanguang_guo@163.com (S.G.); yanghesen96@126.com (H.Y.)
* Correspondence: lianghua82702@163.com

Abstract: This study was designed to explore the control effect of pulsed arc discharge plasma actuation on the incident shock wave/boundary layer interaction (ISWBLI). Research was conducted on an ISWBLI flow field with 10 kHz single-channel pulsed arc discharge plasma actuation and pulsed arc discharge plasma actuation array applied at Mach 2.0 experimentally and numerically. In the investigation, high-speed schlieren flow field visualization technology was adopted, focusing on the change in shock wave intensity caused by plasma actuation. Combined with the detached eddy simulation (DES) method, the numerical simulation focused on the regulating effect of plasma actuation on the separation zone. The key research results showed that, in terms of the spatial flow field, the simulation results were consistent with the experimental results. The single-channel actuation could only just achieve the control effect on the ISWBLI, while array actuation effectively weakened the shock wave intensity. Furthermore, the ISWBLI separation zone in the base flow field was crescent shaped. Its length at the middle of the flat plate was longer than that at the two sides. It was identified that, after applying single-channel actuation, the start of separation slightly moved forward. Similarly, after the application of array actuation, the start point of separation at the middle section in a spanwise direction moved forward by about 19 mm. The length of the separation zone increased by 30 mm but reduced at the two sides. Its influence, spanwise, was also significantly diminished.

Keywords: shock wave/boundary layer interaction; pulsed arc discharge plasma actuation; flow control; high-speed schlieren flow visualization; DES

Citation: Zhang, C.; Luo, Y.; Liang, H.; Guo, S.; Yang, H. Experimental and Numerical Study on Incident Shock Wave/Boundary Layer Interaction Control. *Actuators* **2022**, *11*, 148. https://doi.org/10.3390/act11060148

Academic Editor: Luigi de Luca

Received: 22 April 2022
Accepted: 26 May 2022
Published: 2 June 2022

1. Introduction

The supersonic inlet is an important part of an aircraft's air intake system. When an aircraft is flying at supersonic speed, the inlet acts to slow down the airflow and provide subsonic airflow for the engine, ensuring the airflow is even and stable [1]. This process is of great importance as the quality of airflow directly affects the flight performance [2]. During the flight, the multiple compression configuration to slow down the airflow can significantly reduce the total pressure loss [3], but the ISWBLI triggered by compression is inevitable at the inlet lip. Moreover, the unsteadiness of the interaction region brings extra structural load to the aircraft [4,5], which threatens flight safety. In addition, with the adjustment of the flight attitude, the operating conditions of the inlet may change extremely. If that is the case, the ISWBLI can thicken the boundary layer and cause flow separation, resulting in total pressure loss and inlet unstart [6–8].

The ultimate approach to solve the ISWBLI issue is to address the separation of the boundary layer caused by the reverse pressure gradient. Therefore, flow control technologies based on boundary layer control are commonly used in the inlet [9]. In recent years, significant achievements have been made in solving the ISWBLI issue, regardless of their passive or active forms [10–12]. Various types of vortex generator (VG) were studied at different incoming Mach numbers, configurations, and layouts, revealing groundbreaking outcomes [13–15]. Huang et al. [15] discovered that a VG can achieve better results in suppressing flow separation. Similarly, other studies [16–19] concluded that a micro-vortex generator (MVG), with a size of only a fraction of the thickness of the boundary layer, can achieve similar control effects to a VG. They found that an MVG can also reduce the drag of the aircraft during high-speed flights [20]. Furthermore, modern scholars focused on the control of the ISWBLI with MVGs through experiments and numerical simulation. Yan et al. [16] compared the ISWBLI with and without MVG control at Mach 2.5. In their studies, the MVG reduced the separation zone and improved the shape factor. Their experimental results also showed that vortex rings play an important role in the reduction of the separation zone. Mohd et al. [17] found that the presence of micro-ramps reduces the upstream interaction length by delaying the pressure rise, hence, suppressing the SBLI's effect. Their research proved that MVGs can also be efficiently applied in hypersonic ISWBLI control. However, with the continuous optimization of the aerodynamic shape of aircrafts, it is likely that the size of the inlet will be constantly reduced to reduce the drag and overall weight of the flight [21]. Besides, a flow control device with a convex shape could easily detach and damage the engine blades. For all of these reasons, higher requirements are imposed on flow control devices.

The plasma actuator is a kind of small-sized active flow control device without moving parts. It is easy to maintain and, therefore, meets the application needs in the inlet [22]. Plasma actuators have become a research hotspot in the field of flow control due to their advantageous features such as fast reactions [23,24], wide bandwidth, and tunable actuation intensity. The common forms of plasma actuator are dielectric barrier discharge (DBD), plasma synthetic jet (PSJ), and pulsed arc discharge (PAD). PSJs exhibit the unique ability to produce high-velocity (>300 m/s) pulsed jets at high frequency (>5 kHz); thus, they are widely used in high-speed flow control [25]. Venkat et al. [26] studied the control effect of a PSJ on the ISWBLI at Mach 3.0 and found that a PSJ can shift the dominant frequencies of the separated flow. Venkateswaran et al. [27] found that the position of the pulsed plasma jet can influence the control effect on the ISWBLI, suggesting that a more satisfactory effect is achieved when the actuator is located upstream of the separation zone rather than inside the separation zone. Kursat et al. [28] numerically tested the performance of a sweeping jet actuator on flow separation control. The simulation results suggested that the performance of the sweeping jet actuator, in terms of decreasing the size of the separation bubble, increases with the mass flow rate. Through numerical simulation, Kiyoshi et al. [29] studied two forms of DBD plasma actuators. They found that the electrode, in parallel with the incoming direction, increases the thermal effect of the plasma and exacerbates flow separation, but the form of the canted electrode generates a vortex and mitigates flow separation. Through experiments and simulations, Yan et al. [30] found that the thermal effect of arc plasma actuation can reduce the Mach number upstream of the shock wave. The shock wave intensity is also weakened. Upon the study of the streamwise array of pulsed spark discharge at different frequencies, Tang et al. [31] found that increased actuation frequency helps weaken shock intensity. Although studies on SWBLI control based on plasma actuators are widely available, many scholarly conclusions have focused on the configuration of the compression ramp. There is a research gap when it comes to the control of the ISWBLI based on plasma.

This paper focuses on the control effect of plasma actuation on shock intensity and the separation region. First, the control effect of the pulsed arc discharge plasma actuation on ISWBLI is verified by wind tunnel experiments. The difference in the control effect on the ISWBLI with single-channel and array pulsed arc discharge plasma actuation is studied by numerical simulation. Additionally, the spanwise control effect of plasma actuation on the separation region is also analyzed. The results will provide a reference for effective ISWBLI control in supersonic inlets.

2. Experimental Setup

2.1. Wind Tunnel and Experimental Model

The experiment was carried out in the supersonic wind tunnel in the Science and Technology on Plasma Dynamics Laboratory, China. The wind tunnel is a blowdown with a total temperature of 300 K, total pressure of 95.6 kPa, and unit Reynolds number of 1.19×10^7/m. It can provide Mach 2.0 supersonic flow for 2~3 s. The experimental section of the wind tunnel is equipped with a fixed bracket for model installation, allowing the position and angle of the model to be adjusted. There are two optical observation windows with a diameter of 250 mm on the side of the tunnel, providing light channels for observing the flow field. The installation of the experimental model in the wind tunnel is shown in Figure 1. The model consisted of a bottom flat plate and a top shock wave generator. The shock generator had a wedge angle of 15° at the front edge and was installed on the fixed bracket. It protruded downward through a strut and hung at the nozzle of the wind tunnel. Its lower surface formed an angle of 12.9° with the horizontal plane. The flat plate was made of acrylic plastic, and the upper surface was horizontal and remained at a spacing of 109.8 mm from the shock wave generator.

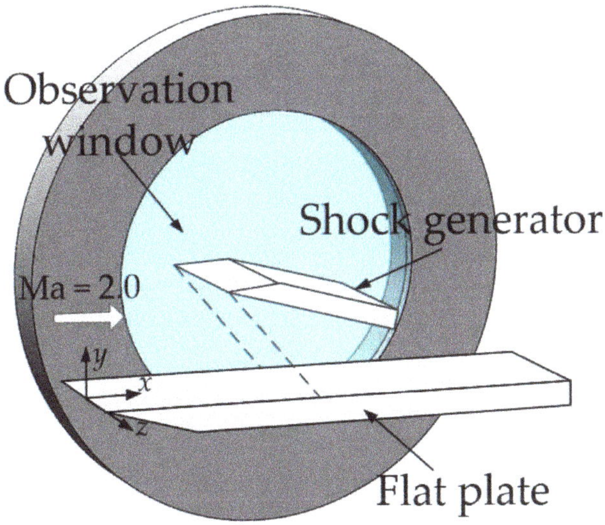

Figure 1. Schematic diagram of installation of experimental model in the wind tunnel.

As shown in Figure 2a, the plate model was 400 mm in length and 110 mm in width. The model was divided into three sections according to its functions, i.e., leading-edge section, actuation section, and interaction section. The first section was the zone from the leading edge of the plate to the most upstream plasma actuator. It was 70 mm in length and mainly used for adjusting the development of the boundary layer to control the downstream boundary layer state. The second section was 270 mm in length and was located in the downstream zone from the plasma actuator. The ISWBLI occurred in this zone by adjusting the angle of the shock generator. The coupling of plasma actuation with the interaction flow field also occurred in this area. The actuation section was the zone between the aforementioned two sections, and the pulsed arc discharge plasma actuator was equipped at this section. With reference to the relatively mature control mechanism developed previously through compression corner SWBLI control using pulsed arc discharge plasma, two kinds of actuator were set in the actuation section, i.e., a single-channel pulsed arc discharge plasma actuator and a pulsed arc discharge plasma actuator array. The single-channel actuator was placed 70 mm away from the leading edge of the plate, and the distance between the positive and negative electrodes was 5 mm. Five single-channel actuators formed an actuator array, and each kept a spacing of 15 mm in the flow direction. The connection of the pulsed arc discharge plasma actuator array is illustrated in Figure 2b. The positive electrode of the first actuator was connected to the output end of the plasma power supply while the negative electrode was connected to the positive electrode of the next actuator. The five actuators were connected in series to form a pulsed arc discharge plasma actuator array.

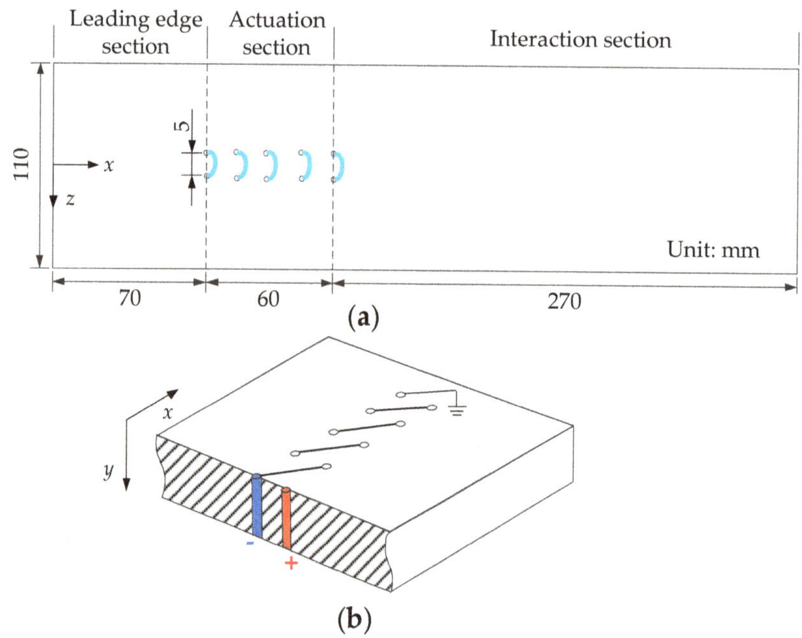

Figure 2. Schematic diagram of experimental model. (**a**) plate model (**b**) The connection of the pulsed arc discharge plasma actuator array.

2.2. Plasma Actuation System and High-Speed Schlieren System

As shown in Figure 3, the triggering of the pulsed arc discharge plasma actuation and the collection of electrical parameters were dependent on the plasma actuation system. Additionally, the high-speed schlieren system was used for the capture of the ISWBLI flow field. As mentioned above, the positive electrode of the array actuator was connected to the output end of the plasma power supply. The plasma power supply selected for this paper was a nanosecond pulse power supply that could provide a pulsed, high voltage ranging from 1 Hz to 100 kHz. The parameters of voltage and current were transmitted to the oscilloscope through the probes and then collected by the oscilloscope. The high-speed schlieren system was a typical, Z-type, optical path schlieren. Reflected by two concave mirrors, the beam was partially blocked at the knife edge and then received by a CCD camera. This process allowed the flow field information carried by the beam to be properly captured. The nanosecond pulse power supply was set with the output voltage as 20 kV, frequency as 10 kHz, and pulse width as 1 μs. The rising and falling edges were both 50 ns. The time interval between two consecutive arc plasma actuations was 100 μs. Additionally, the frame rate of the CCD camera was 25,000 fps; hence, the frame interval was 40 μs.

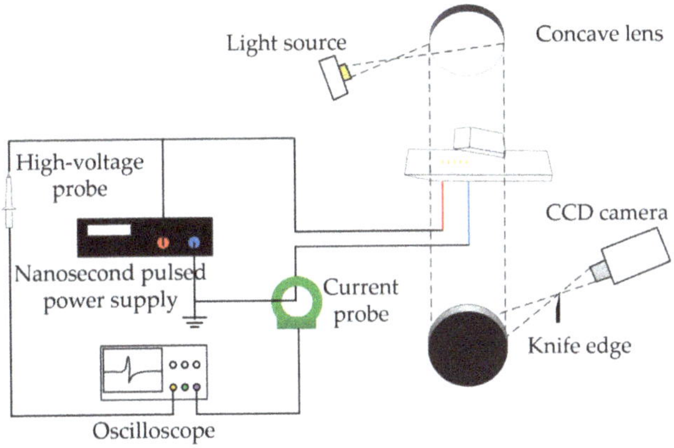

Figure 3. Schematic diagram of plasma actuation system and high-speed schlieren system.

3. Experimental Results and Discussion

As the schlieren image reflected the spanwise integral effect of the spatial flow field, the control effect of the single-channel pulsed arc discharge plasma actuation on the ISWBLI flow field could not be effectively refined in the schlieren image; it will not be listed herein. More flow field information is analyzed and discussed in the numerical simulation section. The experimental results of controlling the ISWBLI with array pulsed arc discharge plasma array actuation are shown in Figure 4. Figure 4a represents the instantaneous schlieren image of the base flow field. As illustrated, a strong incident shock wave was generated at the leading edge of the shock generator, and the incoming flow boundary layer was relatively stable. The interaction between the incident shock wave and the boundary layer induced a typical ISWBLI flow field structure; affected by the reverse pressure gradient after the shock wave, the boundary layer near the interaction point was thickened and lifted up, resulting in local flow field compression and a separation shock induced at the leading edge of this region. Then, the boundary layer was reattached to form local flow field compression at the wall surface, after which another shock wave appeared, namely, reattached shock.

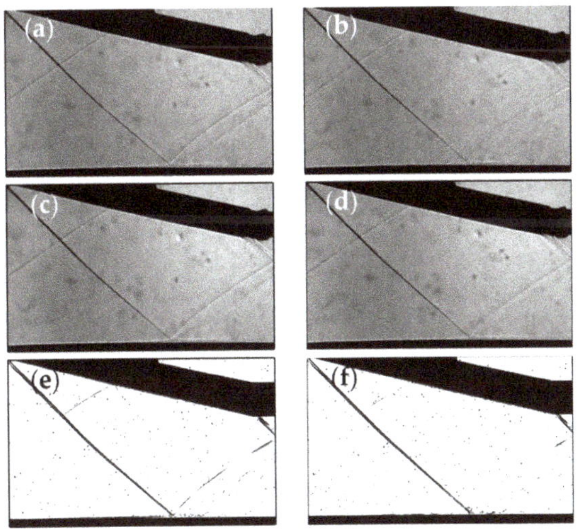

Figure 4. Experimental schlieren results: (**a**,**c**,**e**) the base flow field; (**b**,**d**,**f**) the actuation flow field.

Figure 4b is the instantaneous schlieren image with 10 kHz array actuation. As shown in the figure, the high-frequency actuation induced a series of precursor blast waves in the flow field, and the entire region of the shock wave was covered with the precursor blast waves. Being constantly impacted by the continuous precursor blast waves, the structure of the separation shock wave and the reattached shock wave were furcated. Compared with the base flow field, the reattached shock wave almost disappeared, and the shock wave intensity was considerably weakened. On the other hand, the actuation created a thermal gas bulb. When passing through the interaction region, the thermal gas bulb was coupled with the shock wave and the boundary layer to make the structure divergent, showing that the thermal gas bulb plays a positive role in promoting flow mixing and increasing the kinetic energy of the boundary layer. Although the zone covered by the precursor blast waves included the foot of the incident shock wave, no significant change was observed in this part of the incident shock wave.

The structure of an ISWBLI flow field can be better displayed through time-averaged results; hence, 1000 schlieren images of the base flow field and the actuation flow field were averaged, respectively, as shown in Figure 4c,d. In the time-averaged results, after the actuation was applied, the reattached shock wave almost completely disappeared, and the gray value of the separation shock also decreased. Furthermore, the intensity of the two shock waves was weakened, and the size of the interaction region decreased, indicating that the thermal gas bulb successfully inhibited flow separation. As represented in Figure 4e,f, a flow field with or without actuation can be compared more intuitively according to the threshold value of the spatial gradient of schlieren intensity. In the base flow field, the separation shock wave and the reattached shock wave were shown as two discontinuous strips. The strips disappeared after actuation, indicating that the shock wave intensity was weakened.

4. Numerical Results and Discussion

In the previous section, the role of single-channel actuation and array actuation for ISWBLI was experimentally studied, and the results show that the high frequency pulsed arc plasma actuation array can effectively weaken the intensity of the separation shock wave. In this section, the action of pulsed arc plasma actuation in controlling the ISWBLI is further analyzed through numerical simulation. The analysis focuses on the influence of actuation on the separation zone in the spanwise direction. The ultimate goal is to obtain a comprehensive understanding of pulsed arc plasma actuation in controlling the ISWBLI.

4.1. Calculation Method

4.1.1. Physical Model

The ISWBLI model studied in this section consisted of a shock wave generator and a flat plate, as shown in Figure 5. The length of the plate model was $L_s = 400$ mm. The shock generator was located above the plate, and its height from the surface of the plate was $H_s = 110$ mm. The width of the shock wave generator was $W_s = 110$ mm, corresponding with that of the plate. The angle was also consistent with the shock wave generator model studied in the experiment. It is worth specifying that the origin of the coordinates was located at the center of the plate's leading edge. Furthermore, the positive direction of the x-axis was the incoming flow direction; the positive direction of the y-axis was the normal direction of the wall surface; and the z-axis was the spanwise direction of the plate.

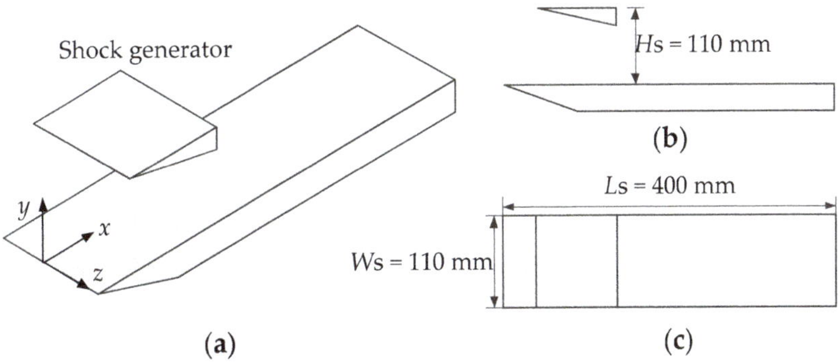

Figure 5. Computational model of ISWBLI: (**a**) 3D view; (**b**) front view; (**c**) top view.

4.1.2. Computational Grid and DES

ICEM was used to divide the computational grid into structured grids. At the same time, dense grids were located in the ISWBLI region, separation shock wave region, and plasma actuation region to obtain a more refined flow field structure. Three sets of grids were obtained. The number of the basic grids (Mesh_1) was 4.21 million. On this basis, two sets of denser meshes (Mesh_2, Mesh_3) were obtained, and the number of grids was 6.27 million and 7.63 million, respectively. Comparing the skin friction coefficients calculated by three sets of grids, as shown in Figure 3, Mesh_2 and Mesh_3 were in good agreement; hence, Mesh_2 (6.27 million) was chosen for further simulations, as shown in Figure 6. The height of the first layer of the skin was set to 0.007 mm to ensure $y^+ < 1$.

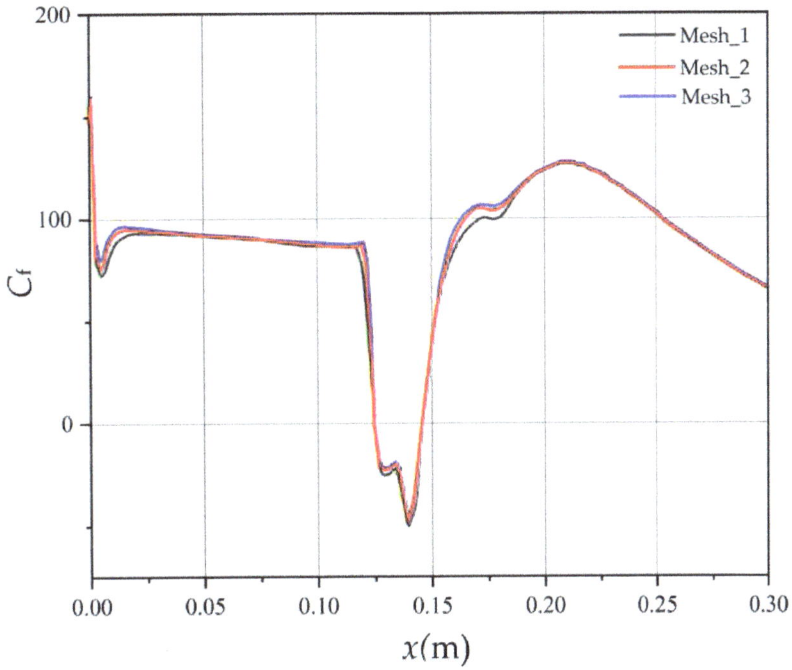

Figure 6. Verification of mesh independence.

The spanwise width of the flow field entrance was 5.275 L_S; the normal height was 5 L_S; and the distance from the leading edge of the plate was 3.75 L_S. The far-field boundary conditions of pressure were set based on the experimental conditions. As shown in Figure 7, the total pressure was 1.01×10^5 Pa; the total temperature was 300 K; and the incoming Mach number was Ma = 2.0. Adiabatic non-slip skin boundary conditions were adopted for the plate and shock wave generator.

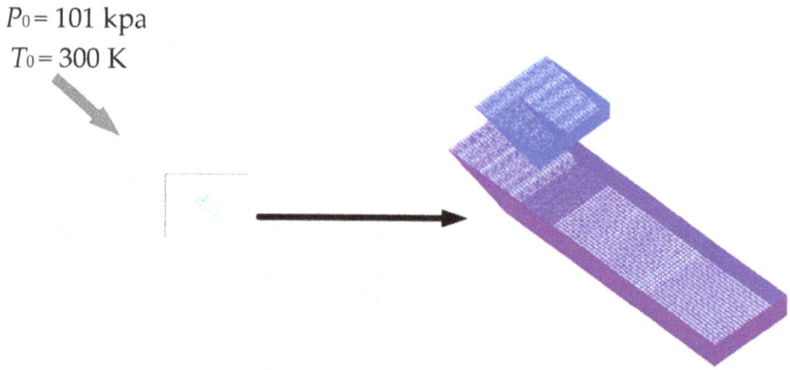

Figure 7. Model grid and boundary conditions.

The DES method was used to calculate the three-dimensional ISWBLI flow field and the characteristics of the flow field under the control of plasma actuation. Figure 8 shows the comparison of the numerical schlieren and the experimental schlieren. Figure 8a shows the flow field structure of the experimental mean schlieren. As mentioned in the previous section, near the foot of the incident shock wave, the boundary layer thickened, and a separation shock wave appeared. As the boundary layer reattached to the wall, a weak reattachment shock wave appeared. It can be seen that the angle of the incident shock wave was about 41.2°, that of the separation shock wave was about 38.4°, and that of the reattachment shock wave was about 33.6°. Upstream of the interaction region, the boundary layer was very thin and was a stable laminar boundary layer. The thickness of the boundary layer increased after passing through the interaction region. Figure 8b shows the simulation results calculated with the DES method. The figure clarifies that the angle of the incident shock wave was about 40.8°. The difference between the simulation and the experiment was about 0.1% of the experimental results. The separation shock wave was slightly smaller, about 37°, which is 3.6% less than the experimental results, and the reattachment shock wave was about 34.7°, which is 3.3% higher than that in Figure 8a. Furthermore, the structure of the interaction region was well simulated in the numerical schlieren, and the boundary layer thickened, passing through the interaction region, which is consistent with the experimental results. Based on the analysis above, it can be concluded that the simulation result was credible.

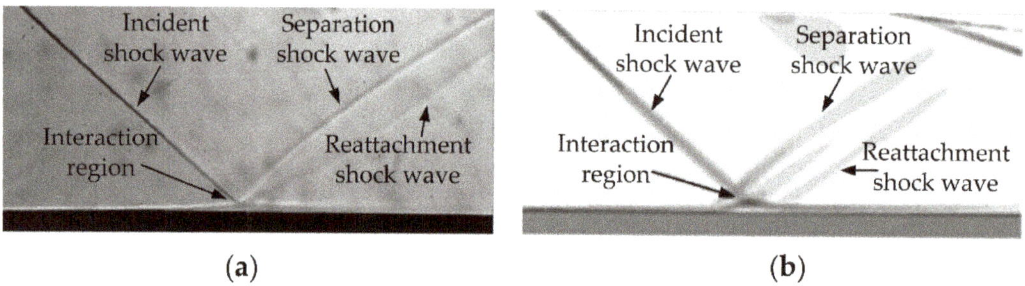

Figure 8. Comparison of experimental schlieren and DES simulation calculation: (**a**) experimental schlieren; (**b**) numerical schlieren.

4.1.3. Phenomenological Model of Plasma Actuation

In-house and foreign studies showed that the control effect of arc plasma actuation over the flow field is mainly a thermal effect. The phenomenological model of plasma actuation is a standard structure used to study plasma simulation. Differing from other physical models, this one simplified plasma flow control into the injection of kinetic and thermal energy into the flow field due to plasma discharge, which considerably reduced the amount of calculation, and can be successfully coupled with the NS equation. Therefore, in this study, the arc plasma actuation was simplified as a heat source model, and it was added to the flow field in the form of a heat source term. The energy of a single actuation was set to 10 mJ, and the density of the discharge heating power was 1×10^{11} W/m^3. This value was a rough estimate based on several wind tunnel experiments. The efficiency of the pulsed arc actuation in this paper was approximately 67%. It was the total thermal efficiency of actuators, including instantaneous rapid heating of gas, material heating of models, heat dissipation, and other thermal energies. The shape of the plasma heat source model was set as a square with a side length of 5 mm due to the fact that the spacing of the arc plasma actuation in the experiment was 5 mm.

In order to verify the accuracy of the plasma heat source model, the simulation results were compared with the experimental results obtained in the supersonic wind tunnel, as shown in Figure 9. To make the comparison more rigorous, the experimental five-channel array actuation was selected, as shown in Figure 9a. The precursor blast wave and thermal gas bulb produced by each actuation channel had similar structures, which showed that the experimental results were reliable. Figure 9b reveals that the calculated precursor blast wave and thermal gas bulb were consistent with the experiment. The simulation of the arc plasma actuation based on the proposed heat source model could achieve accurate results.

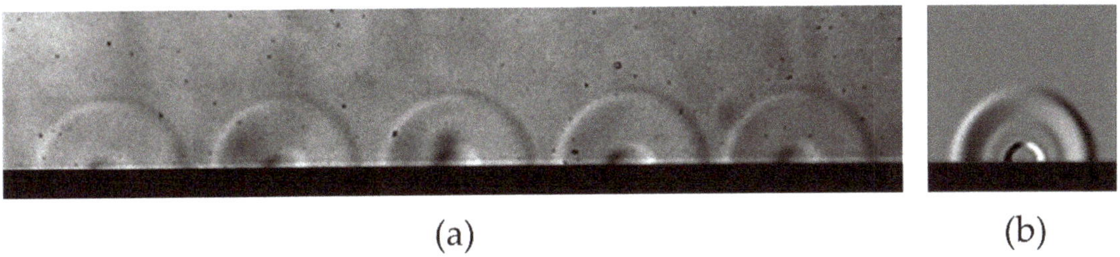

Figure 9. Plasma actuation evolution: (**a**) experimental schlieren; (**b**) numerical schlieren.

4.2. Simulation Result and Disscussion

4.2.1. Base Flow Field

Figure 10 shows the density gradient contours of different spanwise cross sections. At the spanwise middle cross section (z = 0), the structure of the separation shock wave, reattached shock wave, and the separation zone could be clearly identified in the density gradient contour. Furthermore, the separation zone still existed at z = 0.025 m. The strength of the shock wave was lower than that at the spanwise middle cross section. At the cross section at the edge of the plate (z = 0.054 m), the reattached shock wave and the separation zone disappeared, and the ISWBLI was a regular reflection because of the strong turbulence in the boundary layer at the edge of the plate. Based on the flow field structures at the three cross sections, it can be concluded that the ISWBLI demonstrates significant, spanwise, three-dimensional characteristics.

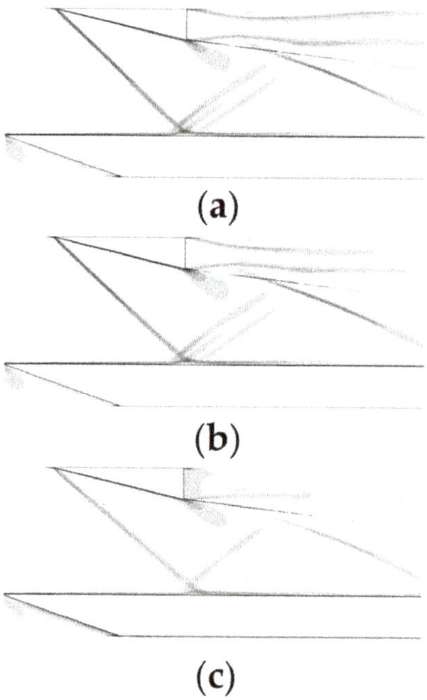

Figure 10. Density gradient contour of different cross sections: (**a**) z = 0; (**b**) z = 0.025 m; (**c**) z = 0.054 m.

The velocity contour in the ISWBLI zone was further analyzed. Figure 11 shows the velocity contour at the y = 0 cross section. Velocity u represents the velocity in the x-axis direction of the flow field, and the region where Velocity u is less than 0 is defined as the separation zone, which is showed in blue in Figure 11. The calculation results show that the separation zone induced by the incident oblique shock wave was crescent shaped, and the length of the separation zone reached its peak at the central section of the plate and decreased spanwise. The separation zone had a width smaller than that of the shock wave generator and the plate and basically disappeared at the edge of the plate. These outcomes demonstrate that the ISWBLI flow field has significant, spanwise, three-dimensional characteristics, which is consistent with the conclusion extracted from Figure 10.

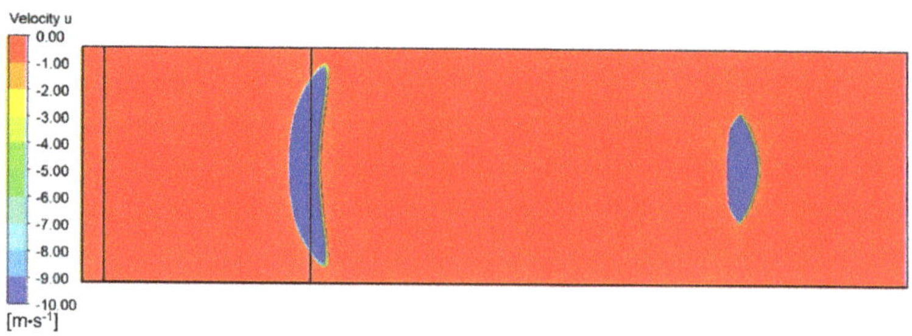

Figure 11. Velocity contour of ISWBLI at $y = 0$ cross section.

The specific location of the separation zone could be determined based on the skin friction coefficient C_f. Figure 12 represents the change curve of C_f at the $z = 0$ cross section and, as can be seen, at the central cross section of the plate, C_f had a negative value from x = 0.126 m to x = 0.146 m in the flow direction. This indicates that the length of the separation zone at the central cross section of the plate was 20 mm.

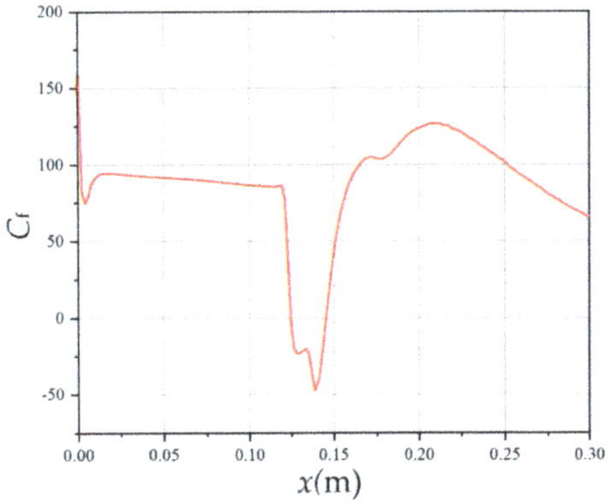

Figure 12. Skin friction coefficient curve of base flow field at $z = 0$ cross section.

Figure 13a,b shows the skin friction coefficient curves at the $z = 0.025$ m cross section and $z = 0.054$ m cross section, respectively. It can be interpreted from the figure that, at $z = 0.025$ m, the separation zone was about 16 mm in length from x = 0.0136 m to x = 0.0152 m. At $z = 0.054$ m cross section, the separation zone was located at x = 0.146 m~x = 0.153 m with a length of 7 mm. The separation moved backward with a smaller size. This result shows that the size of the separation zone at the edge of the plate was smaller than that at the middle of the plate.

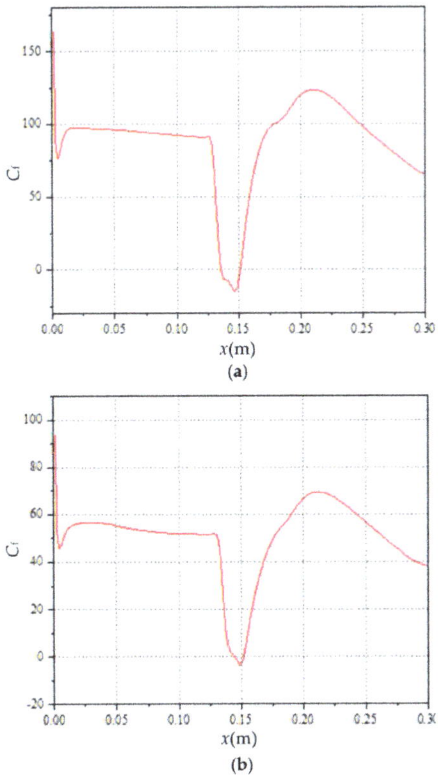

Figure 13. Skin friction coefficient curve of base flow field at different spanwise cross sections: (**a**) $z = 0.025$ m; (**b**) $z = 0.054$ m.

4.2.2. Single-Channel Actuation Flow Field

In this section, the coupling of single-channel arc plasma actuation and the ISWBLI flow field was studied. The single-channel actuation was set at the central cross section of the plate, 70 mm from its leading edge. The actuation frequency was set to 10 kHz, and the actuation energy was 10 mJ. Figure 14 shows the pressure contour of the ISWBLI flow field controlled by plasma actuation at different moments. The start time of discharge was recorded as $t = 0$. When $t = 0.5$ μs, a region with increased pressure appeared at the actuation position, because a local, high-pressure region was induced by high temperature due to the discharge. The high-pressure region was a square, similar to the shape of the plasma source term. When $t = 11$ μs, the precursor blast wave structure induced by actuation appeared in the flow field, which was essentially generated due to the outward transfer of local high pressure. When $t = 21$ μs~$t = 41$ μs, the precursor blast wave continued to spread. It can be interpreted from the figure that the precursor blast wave weakened as it spread and almost entirely disappeared when $t = 51$ μs.

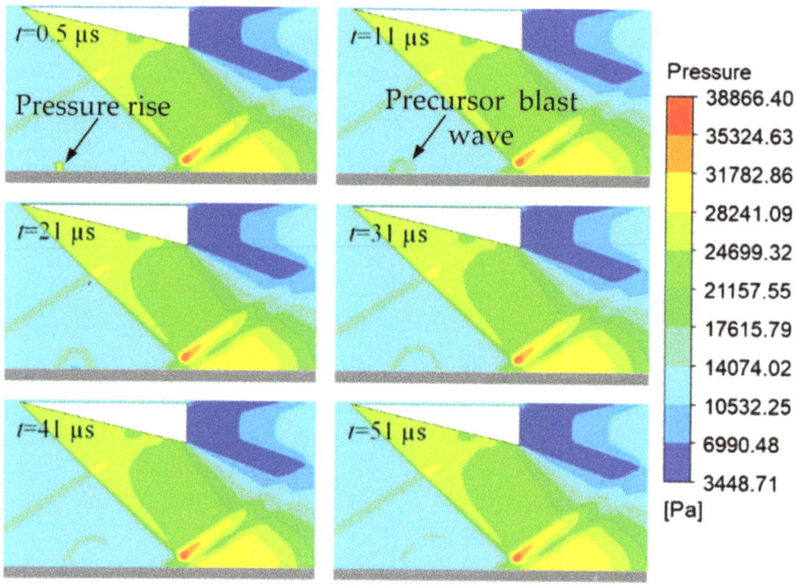

Figure 14. Pressure contour at z = 0 cross section at different moments.

Subsequently, the evolution of the thermal gas bulb generated by the arc plasma actuation was analyzed through the temperature contour of the flow field. As shown in Figure 15a, when $t = 0.5$ μs, a local, high-temperature zone appeared at the actuation position with a maximum temperature of 715 K. A thermal gas bulb was generated near the surface due to discharge heating. Then, the thermal gas bulb propagated downstream, accompanied by energy loss and structural diffusion. As shown in Figure 14c,d, since the structure of the thermal gas bulb was slightly higher than the boundary layer, the moving speed of the part far from the skin was higher while that of the part close to the skin was lower. This difference happened due to the mainstream shear force. When $t = 100.5$ μs, the second actuation was already triggered. At this time, the thermal gas bulb generated by the previous discharge propagated to the ISWBLI zone. As shown in Figure 15e, affected by the thermal gas bulb, the boundary layer was thickened. It can be concluded that the disturbance caused by the actuation was not only due to the shock, but also the thermal gas bulb.

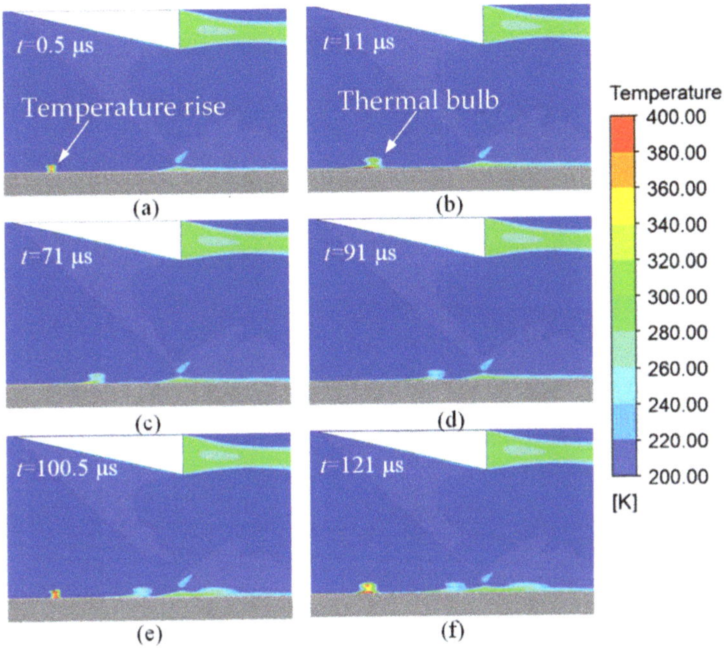

Figure 15. Temperature contour at $z = 0$ cross section at different moments.

Figure 16 shows the numerical schlieren time-averaged flow field with and without the single-channel actuation flow field at different spanwise cross sections. At the three selected cross sections ($z = 0$, $z = 0.025$ m, $z = 0.054$ m), no obvious changes in the shock wave structure and the separation zone were identified after the actuation was applied.

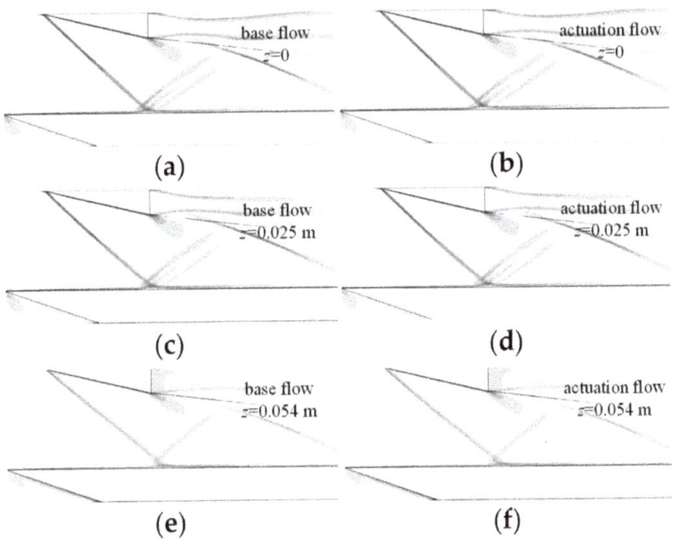

Figure 16. Numerical schlieren of three cross sections: (**a**) $z = 0$ base flow field; (**b**) $z = 0$ actuation flow field; (**c**) $z = 0.025$ m base flow field; (**d**) $z = 0.025$ m actuation flow field; (**e**) $z = 0.054$ m base flow field; (**f**) $z = 0.054$ m actuation field.

Figure 17 shows the curves of the skin friction coefficient at different spanwise cross sections. At three selected cross sections (z = 0, z = 0.025 m, z = 0.054 m), the curves of the skin friction coefficient of the base flow field and the actuation flow field were basically the same. At the z = 0 cross section, the length of the separation zone slightly increased in the actuation flow field, and the start of separation moves forward moderately. At the cross section away from the center of plate (z = 0.025 m, z = 0.054 m), the skin friction coefficient curved of the base flow field, and the actuation flow field completely overlapped.

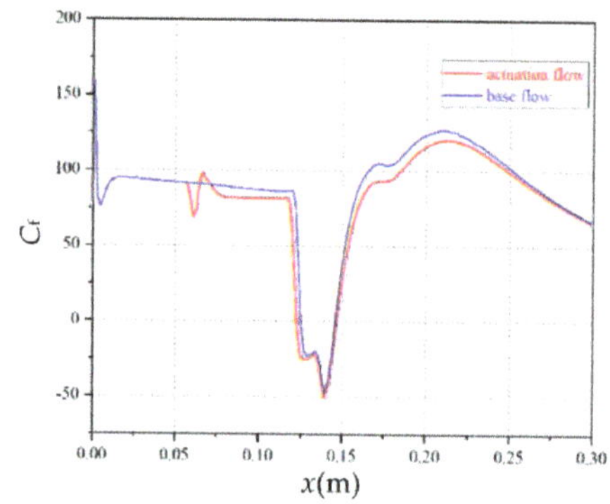

(a)

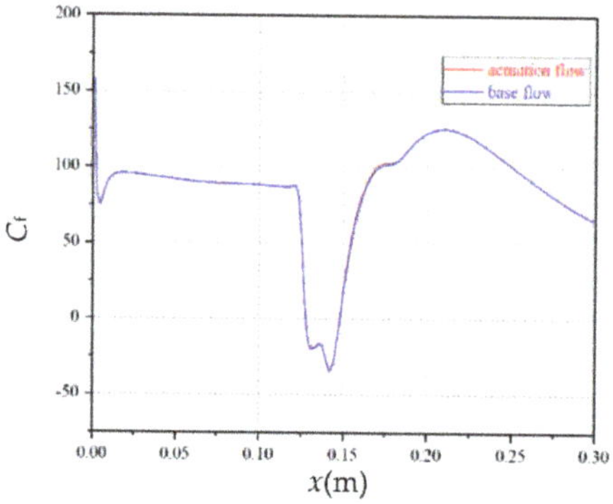

(b)

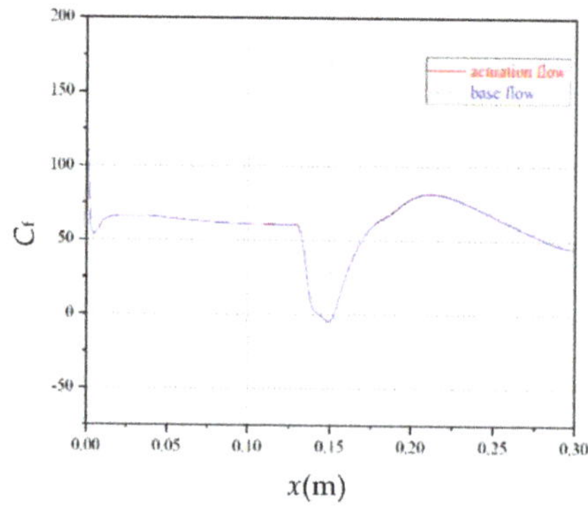

(c)

Figure 17. Skin friction coefficient curve at different spanwise cross sections: (**a**) z = 0; (**b**) z = 0.025 m; (**c**) z = 0.054 m.

The form of the separation zone with and without actuation was also taken into account. Figure 18 illustrates the velocity contour of the $y = 0$ cross section. Compared with the crescent-shaped separation zone in the base flow field, the length of the separation zone at the spanwise central cross section increased slightly with single-channel actuation, and the start of separation moved forward. However, no obvious changes were identified in the separation zone on both sides of the plate.

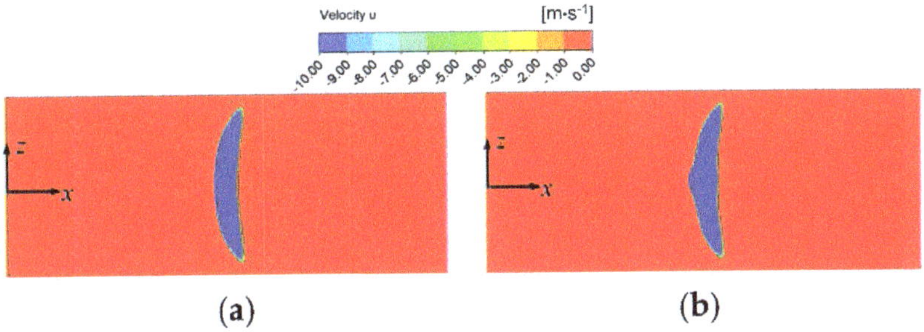

Figure 18. The form of separation zone at $y = 0$ cross section: (**a**) base flow field; (**b**) actuation flow field.

Based on the abovementioned results, when a 10 kHz single-channel actuation was applied, the intensity of the separation shock wave and the reattached shock wave in the ISWBLI flow field remained stable (barely weakened). Similarly, the size and shape of the separation zone did not change significantly either. Therefore, it was concluded that single-channel actuation can only just achieve the control effect over the ISWBLI flow field.

4.2.3. Array Actuation Flow Field

The effect of array actuation on the ISWBLI was further simulated. As shown in Figure 19, array actuation was applied at the central cross section of the plate; the single-channel actuation energy was set to 10 mJ; the actuation frequency was 10 kHz; the actuation point was 70 mm away from the plate's leading edge; and the actuation spacing was 15 mm. The array actuator was located upstream of the ISWBLI region, about 15~20 mm away from the interaction region.

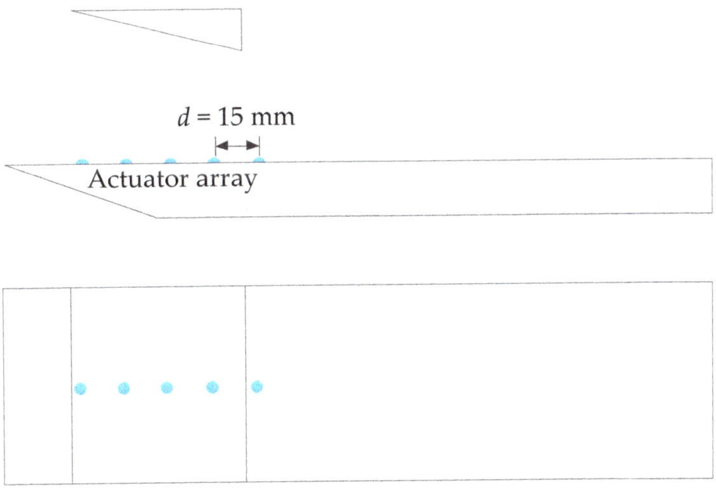

Figure 19. The schematic diagram of array actuation model.

Figure 20 shows the numerical schlieren image of the time-averaged actuation flow field and the time-averaged base flow field at the $z = 0$ cross section. As shown in Figure 20a, at the $z = 0$ cross section, after the array actuation was applied, the separation shock wave basically disappeared, and the angle of the reattached shock wave was reduced from 34.6° in the base flow field to 30.5°. These changes indicate that the array actuation can effectively weaken the intensity of the separation shock and reattached shock waves. Further, it was concluded that, after the application of actuation, the height of the separation zone at the $z = 0$ cross section was significantly increased. The height of the separation zone in the base flow field was about 5 mm, but, in the separation zone in the actuation flow field, it rose to 10 mm as the start of separation moved forward by about 19 mm. It can be concluded that the application of plasma actuation increases the size of the separation zone at the spanwise middle cross section and causes the start of separation to move forward. Upon further comparison of the time-averaged numerical schlieren, it was found that the intensity and angle of the incident shock wave remained unchanged, indicating that plasma actuation has no significant effect on the incident shock wave. This outcome is consistent with the experimental results presented in the previous section.

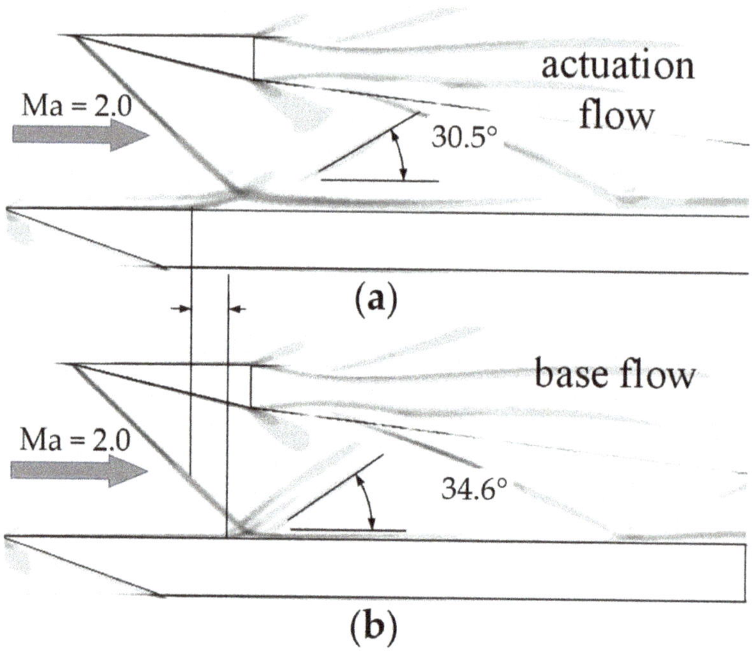

Figure 20. Numerical schlieren at $z = 0$ cross section: (**a**) 10 kHz actuation flow field; (**b**) base flow field.

The flow field at different cross sections was also analyzed. Figure 21a,b represents the numerical schlieren at the $z = 0.025$ m cross section and shows that the array actuation effectively weakened the intensity of the separation shock wave. Figure 21c,d show the numerical schlieren at the $z = 0.054$ m cross sections at the edge of the plate. The results show that the structure and intensity of the separation shock wave were not significantly affected after the actuation was applied. Hence, it can be concluded that actuation in the flow direction can hardly exert a full, spanwise control effect. In subsequent studies, the coupled layout of actuators in spanwise and flow directions should be considered to achieve a better control effect.

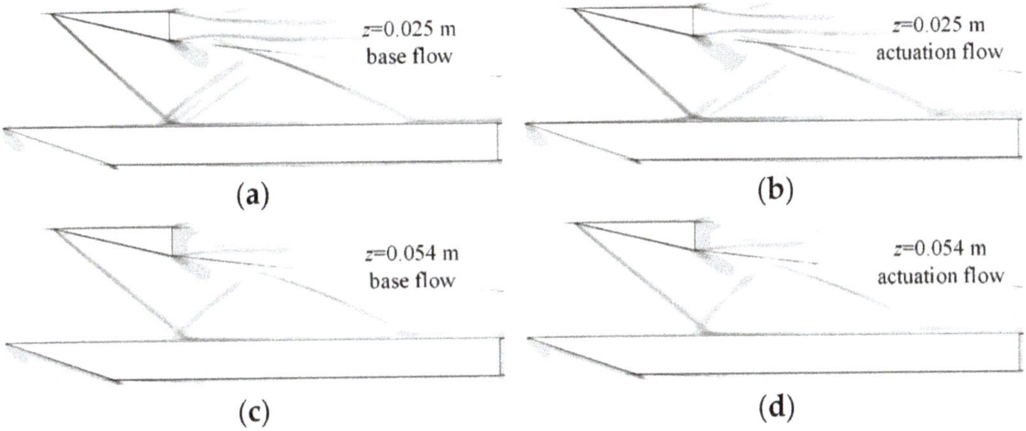

Figure 21. Numerical schlieren of two cross sections: (**a**) z = 0.025 m base flow field; (**b**) z = 0.025 m actuation flow field; (**c**) z = 0.054 m base flow field; (**d**) z = 0.054 m actuation field.

Complementarily, the interaction region of the ISWBLI at the y = 0 cross section was also examined. Figure 22a,b shows the velocity contours at the ISWBLI zone with and without actuation. The images show that, after the array actuation was applied, the start of separation at the spanwise middle position moved forward, and the length of the separation zone was increased. At the same time, the length of the separation zone near the edge of the plate was reduced, and the spanwise, affected area of the separation zone was also diminished.

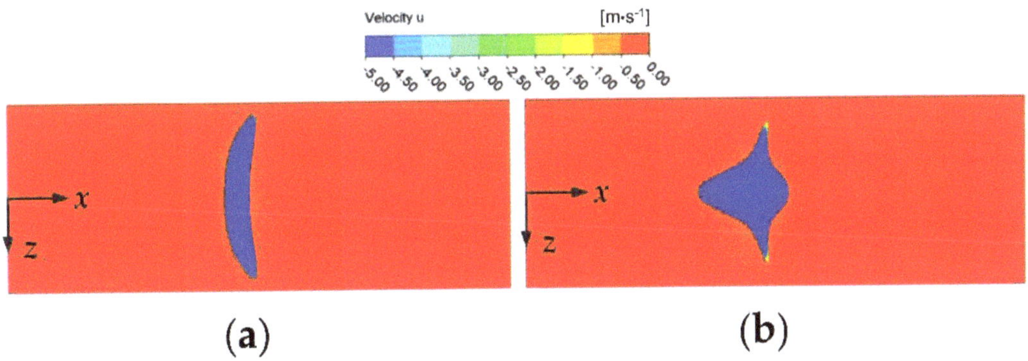

Figure 22. Velocity contour of ISWBLI y = 0 cross section: (**a**) base flow field; (**b**) actuation flow field.

Figure 23 shows the curve of the skin friction coefficient at the z = 0 cross section in the x-direction. The red and blue lines represent the base flow field and the actuation flow field, respectively. After the array actuation was applied, the length of the separation zone at the z = 0 cross section increased from 20 mm to 50 mm, which was an increase of 125%. The start of separation in the base flow field was about 120 mm from the plate's leading edge, and the start of separation in the actuation flow field moved forward, about 100 mm from the plate's leading edge. These changes show that array actuation is likely to increase the length of the separation zone at the z = 0 cross section, making the start point of separation move forward.

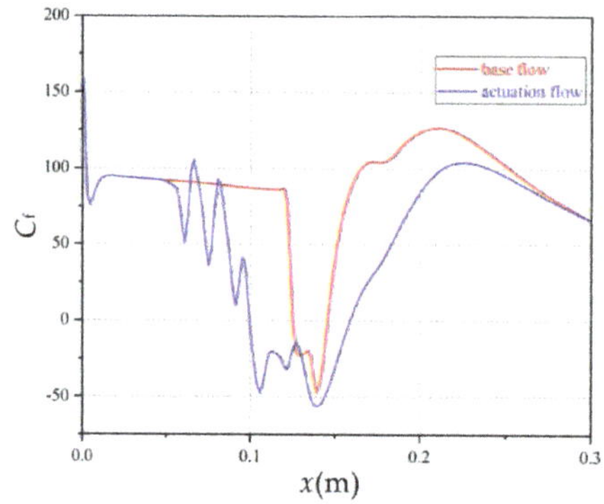

Figure 23. Skin friction coefficient curve at z = 0 cross section.

The other spanwise cross sections of the flow field were the last to be examined. Figure 24a,b shows the curves of the skin friction coefficient of the flow field with and without actuation at the z = 0.025 m cross section and z = 0.054 m cross section, respectively. At the z = 0.025 m cross section, array actuation resulted in the decreased length of the separation zone, making the separation point move backward. At the z = 0.054 m cross section, the length of the separation zone was close to 0, and the C_f curves with and without actuation tended to be consistent. This result indicates that plasma actuation has a poor control effect over the separation zone at the edge of the plate.

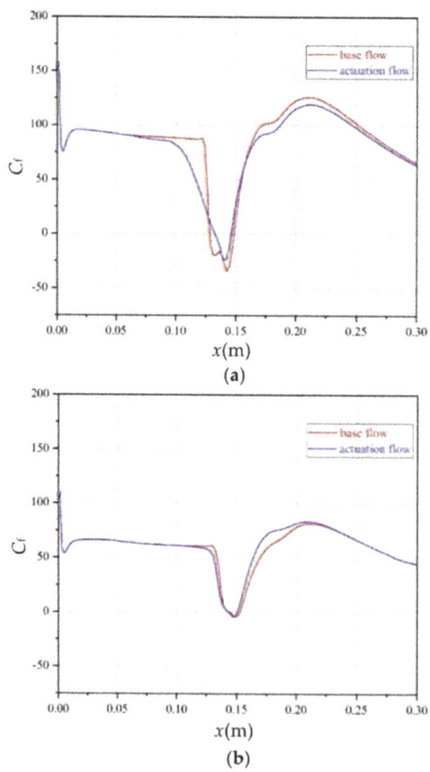

Figure 24. Skin friction coefficient curve at different spanwise cross sections: (**a**) z = 0.025 m; (**b**) z = 0.054 m.

5. Conclusions

In this research, experiments and simulations were carried out to closely examine the control of the ISWBLI through pulsed arc discharge plasma actuation. The high-speed schlieren technology was used to study the control effect of actuation on the shock intensity. The spanwise, three-dimensional characteristics of the flow field with and without actuation were analyzed using the DES method. The major conclusions are presented below:

(1) In the experimental study, the intensity of single-channel actuation was relatively weak. Comparatively, the five-channel array actuation effectively weakened the intensity of the separation shock wave and the reattached shock wave. The incident shock wave did not change significantly under these two forms of actuation;

(2) The numerical simulation results of the base flow field showed that the separation zone of the ISWBLI was crescent shaped, and the length of the separation zone at the middle cross section was longer than that at both sides. The length of the separation zone at the central cross section was about 20 mm; however, the separation zone almost entirely disappeared at the edge of the plate;

(3) The transient simulation results of the single-channel actuation showed that the pulsed arc discharge plasma actuation produced two disturbance forms in the supersonic flow field: the precursor blast wave and the thermal gas bulb. The pressure and temperature contours showed that the disturbance of the thermal gas bulb had a significant effect on the ISWBLI;

(4) Single-channel actuation can only just achieve a control effect over ISWBLI, as there was no significant change in the shock wave structure of the ISWBLI flow field with or without actuation. The length of the separation zone at the central cross section of the plate increased by 3 mm, and the start of separation moved forward by about 2 mm. There was no significant change in the separation zone on both sides of the plate;

(5) The five-channel pulsed arc discharge plasma array actuation could effectively weaken the intensity of the separation shock wave. After the application of actuation, the separation shock wave disappeared in the time-average flow field, and the angle of the reattached shock wave decreased from 34.6° to 30.5°. The application of the array actuation could change the shape of the separation zone; the start of separation moved forward by about 19 mm; and the length increased by 30 mm at the central cross section. It was also observed that the length of the separation zone on both sides was reduced. Furthermore, the array actuation could effectively mitigate the spanwise influence zone in the separation zone.

Author Contributions: C.Z. performed the experimental investigation. Y.L. performed the numerical simulations and analyzed the results; H.L. and S.G. supervised the work and reviewed and edited the manuscript; C.Z. and H.Y. wrote the main manuscript. All authors have read and agreed to the published version of the manuscript.

Funding: This research was founded by the National Science and Technology Major Project (Grant No. J2019-II-0014-0035).

Institutional Review Board Statement: Not applicable.

Informed Consent Statement: Not applicable.

Data Availability Statement: The data and materials are available upon request.

Conflicts of Interest: The authors declare no conflict of interest.

References

1. Yamamoto, J.; Kojima, Y.; Kameda, M.; Watanabe, Y.; Hashimoto, A.; Aoyama, T. Prediction of the onset of supersonic inlet buzz. *Aerosp. Sci. Technol.* **2020**, *96*, 105523. [CrossRef]
2. Ferrero, A. Control of a Supersonic Inlet in Off-Design Conditions with Plasma Actuators and Bleed. *Aerospace* **2020**, *7*, 32. [CrossRef]

3. Titchener, N.; Babinsky, H. Shock wave/boundary-layer interaction control using a combination of vortex generators and bleed. *AIAA J.* **2013**, *51*, 1221–1233. [CrossRef]
4. Trapier, S.; Duveau, P.; Deck, S. Experimental study of supersonic inlet buzz. *AIAA J.* **2006**, *44*, 2354–2365. [CrossRef]
5. Chen, H.; Tan, H.J.; Zhang, Q.F.; Zhang, Y. Throttling Process and Buzz Mechanism of a Supersonic Inlet at Overspeed Mode. *AIAA J.* **2018**, *56*, 1–12. [CrossRef]
6. Herrmann, C.; Koschel, W. Experimental investigation of the internal compression inside a hypersonic intake. In Proceedings of the 38th AIAA/ASME/SAE/ASEE Joint Propulsion Conference & Exhibit, Indianapolis, IN, USA, 7–10 July 2002; p. 4130.
7. Zhang, Y.; Tan, H.; Du, M.; Wang, D.P. Control of shock/boundary-layer interaction for hypersonic inlets by highly swept microramps. *J. Propuls. Power* **2015**, *31*, 133–143. [CrossRef]
8. Xie, W.Z.; Wu, Z.M.; Yu, A.Y.; Guo, S. Control of severe shock-wave/boundary-layer interactions in hypersonic inlets. *J. Propuls. Power* **2018**, *34*, 614–623. [CrossRef]
9. Babinsky, H.; Ogawa, H. SBLI control for wings and inlets. *Shock Waves* **2008**, *18*, 89–96. [CrossRef]
10. Gruhn, P.; Gülhan, A. Experimental investigation of a hypersonic inlet with and without sidewall compression. *J. Propuls. Power* **2011**, *27*, 718–729. [CrossRef]
11. Häberle, J.; Gülhan, A. Experimental investigation of a two-dimensional and a three-dimensional scramjet inlet at Mach 7. *J. Propuls. Power* **2008**, *24*, 1023–1034. [CrossRef]
12. Zhang, Y.; Tan, H.; Sun, S.; Rao, C.Y. Control of cowl shock/boundary-layer interaction in hypersonic inlets by bump. *AIAA J.* **2015**, *53*, 3492–3496. [CrossRef]
13. Vignesh Ram, P.S.; Das, S.; Kim, H.D. Influence of vortex generator on cylindrical protrusion aerodynamics at various Mach numbers. *Aerosp. Sci. Technol.* **2016**, *58*, 267–274.
14. Titchener, N.; Babinsky, H.; Loth, E. The effects of various vortex generator configurations on a normal shock wave/boundary layer interaction. In Proceedings of the 51st AIAA Aerospace Sciences Meeting including the New Horizons Forum and Aerospace Exposition, Grapevine, TX, USA, 7–10 January 2013; p. 18.
15. Huang, J.B.; Xiao, Z.X.; Fu, S.; Zhang, M. Study of control effects of vortex generators on a supercritical wing. *Sci. China Technol. Sci.* **2010**, *53*, 2038–2048. [CrossRef]
16. Yan, Y.; Chen, L.; Li, Q.; Liu, C. Numerical study of micro-ramp vortex generator for supersonic ramp flow control at Mach 2.5. *Shock. Waves* **2017**, *27*, 79–96. [CrossRef]
17. Saad, M.R.; Zare-Behtash, H.; Che-Idris, A.; Kontis, K. Micro-ramps for hypersonic flow control. *Micromachines* **2012**, *3*, 364–378. [CrossRef]
18. Grébert, A.; Bodart, J.; Jamme, S.; Joly, L. Simulations of shock wave/turbulent boundary layer interaction with upstream micro vortex generators. *Int. J. Heat Fluid Flow* **2018**, *72*, 73–85. [CrossRef]
19. Bagheri, H.; Mirjalily, S.A.A.; Oloomi, S.A.A.; Salimpour, M.R. Effects of micro-vortex generators on shock wave structure in a low aspect ratio duct, numerical investigation. *Acta. Astronaut.* **2021**, *178*, 616–624. [CrossRef]
20. Panaras, A.G.; Lu, F.K. Micro-vortex generators for shock wave/boundary layer interactions. *Prog. Aerosp. Sci.* **2015**, *74*, 16–47. [CrossRef]
21. Oorebeek, J.; Nolan, W.; Babinsky, H. Comparison of bleed and micro-vortex generator effects on supersonic boundary-layers. In Proceedings of the 50th AIAA Aerospace Sciences Meeting Including the New Horizons Forum and Aerospace Exposition, Nashville, TN, USA, 9–12 January 2012; p. 45.
22. Wang, J.J.; Choi, K.S.; Feng, L.H.; Jukes, T.N.; Whalley, R.D. Recent developments in DBD plasma flow control. *Prog. Aerosp. Sci.* **2013**, *62*, 52–78. [CrossRef]
23. Moreau, E. Airflow control by non-thermal plasma actuators. *J. Phys. D Appl. Phys.* **2007**, *40*, 605. [CrossRef]
24. Choi, K.S.; Jukes, T.; Whalley, R. Turbulent boundary-layer control with plasma actuators. *Philos. Trans. Royal Soc. A* **2011**, *369*, 1443–1458. [CrossRef] [PubMed]
25. Zong, H.; Chiatto, M.; Kotsonis, M.; De Luca, L. Plasma synthetic jet actuators for active flow control. *Actuators* **2018**, *7*, 77. [CrossRef]
26. Narayanaswamy, V.; Raja, L.L.; Clemens, N.T. Control of a shock/boundary-layer interaction by using a pulsed-plasma jet actuator. *AIAA J.* **2012**, *50*, 246–249. [CrossRef]
27. Narayanaswamy, V.; Raja, L.L.; Clemens, N.T. Control of unsteadiness of a shock wave/turbulent boundary layer interaction by using a pulsed-plasma-jet actuator. *Phys. Fluids* **2012**, *24*, 076101. [CrossRef]
28. Kara, K.; Kim, D.; Morris, P.J. Flow-separation control using sweeping jet actuator. *AIAA J.* **2018**, *56*, 4604–4613. [CrossRef]
29. Kinefuchi, K.; Starikovskiy, A.Y.; Miles, R.B. Numerical investigation of nanosecond pulsed plasma actuators for control of shock-wave/boundary-layer separation. *Phys. Fluids* **2018**, *30*, 106105. [CrossRef]
30. Yan, H.; Liu, F.; Xu, J.; Xue, Y. Study of oblique shock wave control by surface arc discharge plasma. *AIAA J.* **2018**, *56*, 532–541. [CrossRef]
31. Tang, M.; Wu, Y.; Guo, S.; Sun, Z.; Luo, Z. Effect of the streamwise pulsed arc discharge array on shock wave/boundary layer interaction control. *Phys. Fluids* **2020**, *32*, 076104. [CrossRef]

actuators

MDPI

Article

Transonic Buffet Active Control with Local Smart Skin

Kai Ren [1], Chuanqiang Gao [1,*], Fangqi Zhou [2] and Weiwei Zhang [1]

[1] Department of Fluid Mechanics, School of Aeronautics, Northwestern Polytechnical University, Xi'an 710072, China; renkai@mail.nwpu.edu.cn (K.R.); aeroelastic@nwpu.edu.cn (W.Z.)
[2] High Speed Aerodynamics Research Institute, China Aerodynamics Research and Development Center, Mianyang 621000, China; fqzhou@cardc.cn
* Correspondence: gaocq@nwpu.edu.cn

Abstract: Transonic flight has high economic benefits, but the appearance of transonic buffet limits the flight envelope. The shock control bump currently used for transonic buffet suppression tends to degrade the aerodynamic performance of the non-buffeting state. In this study, a smart skin system is used to eliminate the fluctuating load of transonic buffet by measuring the airfoil lift coefficient as the feedback signal and adjusting the local skin height using data-driven, model-free adaptive control. Since the actuator height is dynamically adjusted only after the occurrence of transonic buffet, the smart skin can completely suppress the fluctuating load and does not affect the aerodynamic performance in the non-buffeting state. The suppression effect of the proposed smart skin on transonic buffet is verified by numerical simulation of the flow. The simulation results show that due to the introduction of closed-loop control, the fluctuating load of transonic buffet can be effectively suppressed for different positions and maximum heights of the actuator. Even when the flow state changes, the robust smart skin system can also achieve the control goal. Therefore, smart skins combining flexible materials and control technologies have the potential to effectively improve the aerodynamic performance of aircraft.

Keywords: smart skin; transonic buffet; closed-loop flow control; fluctuating load suppression

Citation: Ren, K.; Gao, C.; Zhou, F.; Zhang, W. Transonic Buffet Active Control with Local Smart Skin. *Actuators* **2022**, *11*, 155. https://doi.org/10.3390/act11060155

Academic Editors: Ioan Ursu, Hui Tang and Xin Wen

Received: 29 April 2022
Accepted: 8 June 2022
Published: 10 June 2022

1. Introduction

Transonic flight has high economic benefits, but when the flow state enters above the transonic buffet boundary [1], which consists of Mach number (M) and angle of attack (α), transonic buffet flow [2] occurs on the wing surface. This is a kind of flow instability phenomenon caused by shock wave/boundary layer interaction, which is manifested as a large oscillation of shock wave on the wing surface, causing fluctuation of lift, affecting flight comfort and safety, easily causing fatigue damage to the wing structure, and limiting the flight envelope in the transonic section. In order to eliminate the fluctuating load of transonic buffet and improve the flow stability, various passive and active control schemes are used to modify the flow in the shock wave/boundary layer interaction region and trailing edge region [3]. The control devices mainly include vortex generator [4,5], shock control bump [6,7], streamwise slot [8], plasma actuator [9,10], and trailing edge flap [11,12]. Unfortunately, due to the complexity of shock buffet, a robust control scheme for eliminating shock buffet load has not been obtained in experiments. Researchers seek to analyze the advantages and disadvantages of different control schemes through numerical simulation, so as to provide some support for the realization of shock buffet control. The active control with the trailing edge flap as the actuator can effectively eliminate the fluctuating load, but it will occupy the control channel of the flap surface. Passive control represented by shock control bump (SCB) is widely studied because of its simple structure and effective improvement of transonic buffet boundary. However, the problem of SCB is that when the position and shape of the SCB are determined, it can only eliminate the buffeting load in the partial incoming flow state, and will reduce the aerodynamic performance in the non-buffeting state. Therefore, the smart skin technology of a deformable wing [13] can combine the advantages of both control schemes mentioned above. The shape of the bump can be dynamically adjusted after transonic buffet occurs, which has the potential to effectively eliminate the transonic buffeting load without affecting the aerodynamic performance of the non-buffeting state.

With the development of new intelligent materials and distributed sensing and control technology, the deformable wing technology is evolving towards the direction of intelligence and flexibility. The use of intelligent materials such as piezoelectric materials [14] and shape memory alloys [15] to drive the deformation of aircraft and achieve the integration of flexible skin materials, structure, and actuator is a current research hotspot and a development direction for future applications. Table 1 compares the actuation performance of two intelligent materials: piezoelectric materials and shape memory alloys. Among the advantages of the piezoelectric materials is the fast response speed, while the shape memory alloy has the advantage of large output force and output displacement, compared to conventional wings that improve the aerodynamic performance of an aircraft under only a few specific flight conditions through the simple rigid deformation of flaps, leading edge slats, and other mechanisms. The intelligent deformable wings composed of intelligent deformable materials combined with distributed sensors and microcontrollers do not produce discontinuous changes and abrupt curvature changes, or reduce the separation of flow, and can optimize deformation parameters for each flight condition, which has better flight performance [16,17]. The deformable wing includes folding wing [18], flexible trailing edge [19], adaptive wingtip [20], active camber wing [21] and smart skin [22]. To enhance the performance of the aircraft, researchers use wings with variable camber [23,24], variable thickness [25], variable trailing edge deflection [19,26], and deformable winglets [27] to improve aerodynamic performance; use adaptive wings with adjustable maximum height and maximum height position to improve stall aerodynamic performance [28]; use deformable wings to adapt to different flight environments [29]; use flexible trailing edges to reduce structural weight of swept-wing aircraft [30]; use deformable trailing edges to control the boundary layer flow to reduce noise [31]; use adaptive wingtips to reduce fuel consumption and gust loads through cant, twist, and camber deformation [20]; use smart skin to adjust the inlet shape to make the engine have good performance over a wide range of rotation speed range [32]; and use smart skin to suppress panel flutter [33] and wing flutter [34].

Table 1. Comparison of intelligent material actuation performance.

Intelligent Material	Response Speed	Output Force	Output Displacement	Frequency Band
Piezoelectric materials	fast	moderate	small	wide
Shape memory alloys	slow	high	large	narrow

In this study, smart skin technology is used to suppress the fluctuating load of transonic buffeting flow. Smart skin technology forms the neural network of aircraft by implanting intelligent structures within the aircraft skin and other components, including detection elements (sensors), microprocessor control systems (signal processors), drive elements (micro actuator), and interconnecting lines, etc. It can not only sense its own physical conditions (such as structural damage), but also be sensitive to the external environment (such as incoming flow condition). In this way, the components and even the whole aircraft body are endowed with self-monitoring, self-correction, self-adaptation, and memory, thinking, judgment, and response functions.

This paper is organized as follows: In Section 2, the composition of the smart skin system and the flow numerical simulation method to simulate the control performance are introduced. In Section 3, the adopted control method is introduced and simulations are performed to verify the effectiveness of the method in suppressing the transonic buffet fluctuating load. Section 4 analyzes the influence of smart skin system parameters on the control effect and tests the robustness of the system to different incoming flow states. Conclusion and discussion are presented in Section 5.

2. Smart Skin System and Simulation Method

2.1. Smart Skin System

The smart skin system includes sensors, microprocessor control systems, micro actuators, and interconnecting lines. The feedback signal used for the active control of transonic buffet is mainly the force coefficient and the wall pressure. Compared with the wall pressure, the lift coefficient (C_l) is obtained by the integration of the pressure distribution (C_p), which is smooth and beneficial to the stability of the system. Therefore, the lift coefficient is used as a feedback signal in the smart skin system. Since there is a direct proportional relationship between the lift force and the wing root bending moment, the lift coefficient is obtained by measuring the wing root bending moment using strain gauges. The design of the control law is described in Section 3.1.

The micro actuator uses the local deformation skin with adjustable height; the shape of deformation skin is a widely used Hicks–Henne shape function [7], as shown in Figure 1. The shape parameters include: (1) the length of deformation skin, l_b, (2) the starting position of deformation skin, x_0, (3) crest location, $x_{h\max}/l_b$, and (4) the height of deformation skin, h_b (defined as the ratio of the actual height to the chord length). The height of deformation skin is output by the control law and can be adjusted dynamically. The Hicks–Henne shape function is

$$\begin{aligned} f_b(x_b) &= h_b H(x_b) \\ H(x_b) &= \sin^4(\pi x_b^m), \quad m = \ln(0.5)/\ln(x_{h\max}/l_b) \end{aligned} \tag{1}$$

where, x_b is the dimensionless chordwise position of deformation skin, which is defined as $0 \le x_b = (x - x_0)/l_b \le 1$. In this study, the length of deformation skin is taken to be 0.2 times the chord length, $l_b = 0.2\,c$, and the shape of deformation skin is symmetric, and the crest location is $x_{h\max}/l_b = 0.5$.

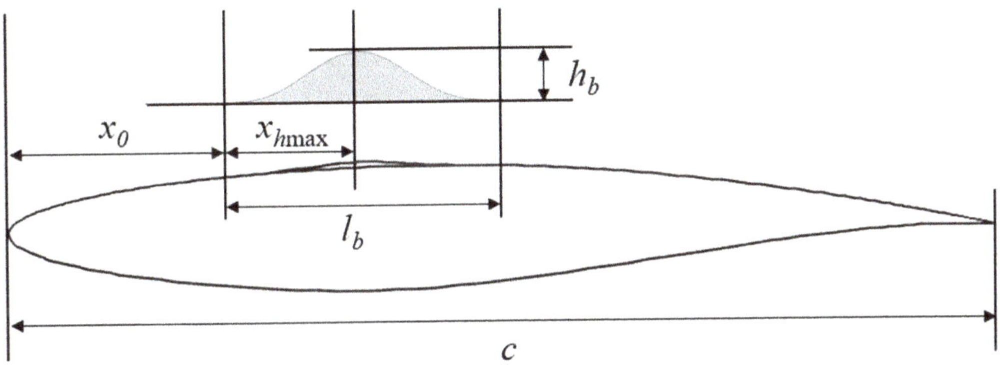

Figure 1. Diagram of smart skin micro actuator.

2.2. Flow Numerical Simulation Method

In this study, the control performance of the smart skin system is analyzed by flow numerical simulation, and the numerical simulation uses the unsteady Reynolds-averaged Navier–Stokes (URANS) solver based on the finite volume method. The two-dimensional compressible URANS governing equations can be expressed as:

$$\frac{\partial}{\partial t}\iiint_{\Omega} \mathbf{W} dV + \iint_{\partial\Omega} \mathbf{F}^i(\mathbf{W}, \mathbf{V}_{grid})\cdot \mathbf{n} dS = \iint_{\partial\Omega} \mathbf{F}^v(\mathbf{W})\cdot \mathbf{n} dS \tag{2}$$

where $\boldsymbol{W}$ is the flux variables, $\boldsymbol{F}^i(\boldsymbol{W}, V_{grid})$ is the inviscid flux, $\boldsymbol{F}^v(\boldsymbol{W})$ is the viscous flux, V_{grid} is the grid velocity, Ω is the control volume, $\partial\Omega$ is the control volume boundary, and $\boldsymbol{n}$ is the unit vector of normal direction outside the boundary.

The S-A turbulence model is introduced to close the Reynolds-averaged equation, and the spatial discretization and time integration of turbulent and mean flows are carried out in a loosely coupled way. The second-order advection upstream splitting method (AUSM) scheme is used to evaluate the inviscid flux with a reconstruction technique. The viscous flux term is discretized by the standard central scheme. In the turbulence model, the convective term is discretized by the second-order AUSM scheme and the destruction term by the second-order central scheme. For unsteady computations, the dual time stepping method is used to solve the governing equations. At the sub-iteration, the fourth-stage Runge–Kutta scheme is used with a local time stepping.

Due to the airfoil shape change during the numerical simulation, a grid deformation method is required to ensure the update and deformation of the airfoil surface and the solution domain grid. The grid deformation scheme uses the radial basis function (RBF) interpolation method [35].

The controlled object adopts the transonic buffet of RAE2822 airfoil, and the computational flow domain is discretized by a hybrid unstructured grid. The far field extends 20 chords away from the airfoil. There are 12,958 surface nodes and 40 layers of structured viscous grids around the airfoil. The distance between the first layer and the wall in the perpendicular direction is 5×10^{-6} chords ($y^+ < 1$). For a transonic flow: $M = 0.73$, $\alpha = 3.19°$, and $Re = 6.5 \times 10^6$. Figure 2 shows the comparison between the calculated pressure distribution and the experimental result [36], which is basically consistent with each other. By fixing the Mach number of the incoming flow and simulating different angles of attack (the interval of angles of attack is 0.1°) it can be determined whether the transonic buffet occurs. The buffet onset boundary obtained is shown in Figure 3, which is almost consistent with the onset boundary [7] calculated by Tian et al.

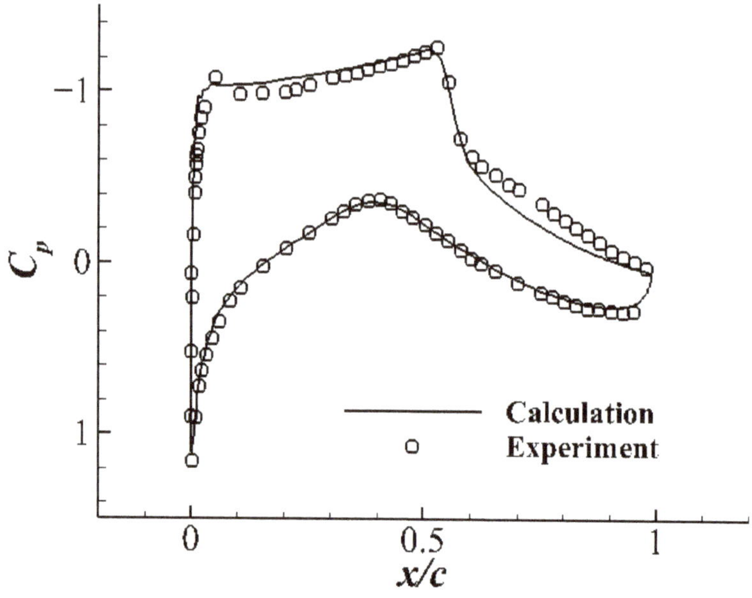

Figure 2. Comparison of pressure distribution on airfoil surface.

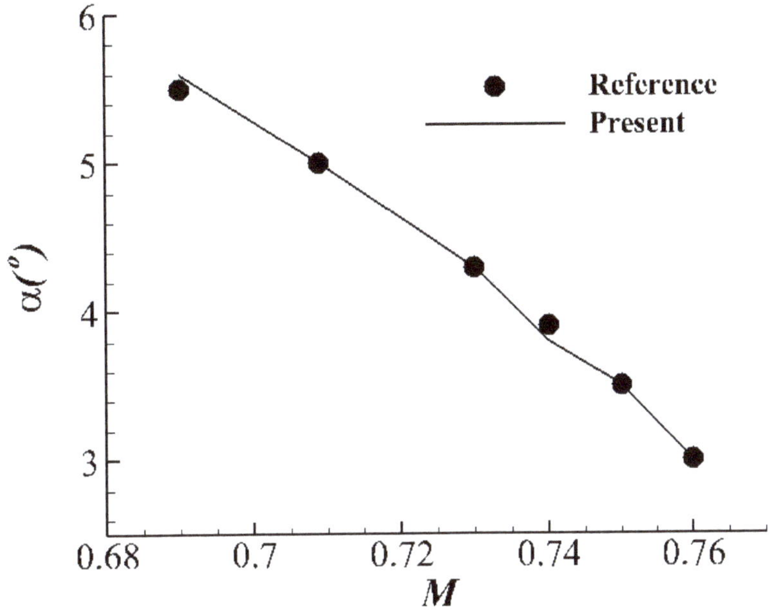

Figure 3. Comparison of transonic buffet onset boundary.

Choose the incoming flow state: M = 0.75, α = 4°, and $Re = 6.5 \times 10^6$ to simulate the time history of buffet. It can be seen from Figure 3 that this flow state is above the buffet onset boundary, and when the control is not turned on, the system will have a large lift fluctuation, as shown in Figure 4. Figure 5 shows the flow field pressure contours at different moments, corresponding to the four moments in Figure 4 (*a* and *c* are the middle positions when the shock wave moves upstream and downstream, respectively, and *b* and *d* are the most upstream and downstream positions of the shock wave, respectively). When the shock wave moves upstream, the length of the trailing edge separation bubble increases, and when the shock wave moves downstream, the separation bubble at the foot of the shock wave and the trailing edge separation bubble approach and merge gradually. From the pressure contours, it can be seen that the dimensionless chordwise range of the shock wave oscillation for this flow condition is 0.45~0.56, and the shock wave is at the most downstream position when the lift coefficient is maximum, and vice versa.

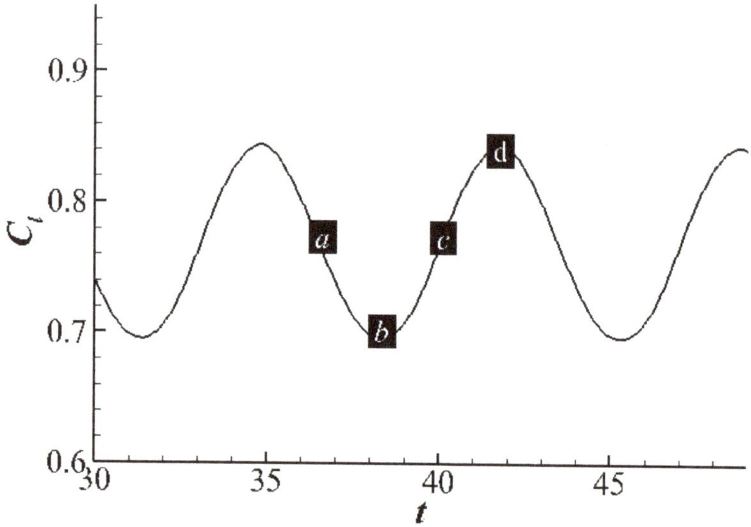

Figure 4. Lift coefficient time history.

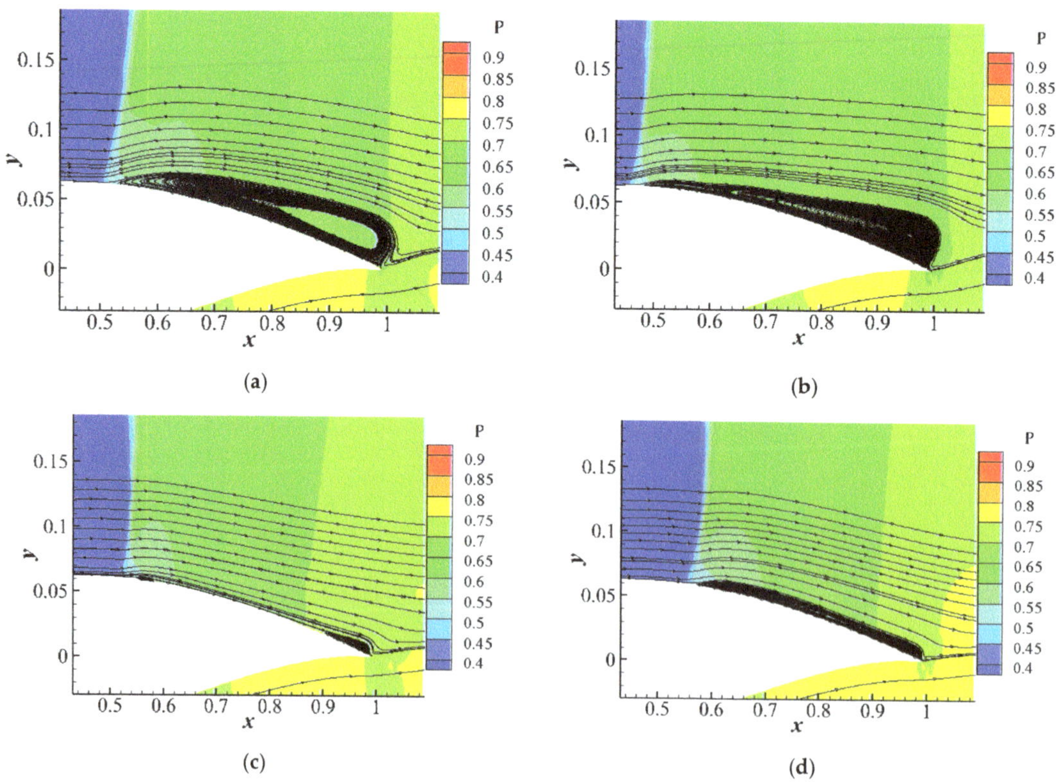

Figure 5. Local pressure contours and streamlines of flow field at different moments: (**a**) moment *a*; (**b**) moment *b*; (**c**) moment *c*; (**d**) moment *d*.

3. Transonic Buffet Control

3.1. Model-Free Adaptive Control

Model-free adaptive control [37] is a data-driven control method. Its basic idea is to equivalently transform the discrete-time nonlinear system into a dynamic linearized data model based on the input and output data of the system at each moment using a pseudo partial derivative (PPD). Without establishing an accurate mathematical model of the flow system, the data-driven adaptive control of the flow system is achieved by estimating the system PPD online using the input and output data of the flow system, and then minimizing a given cost function.

For the transonic buffet flow system in this study, the feedback signal is the lift coefficient, C_l, the control signal is the actuator height, h_b, and the controlled plant can be described by the following single-input, single-output, discrete-time nonlinear system:

$$C_l(k+1) = f(C_l(k), \cdots, C_l(k-n_o), h_b(k), \cdots, h_b(k-n_i)) \quad (3)$$

where n_i and n_o are unknown input and output delay orders, respectively, and $f(\cdot)$ is an unknown nonlinear function describing the relationship between lift coefficient and actuator height.

Based on PPD, the above flow plant is equivalently transformed into the following dynamic linearized data model:

$$C_l(k+1) = C_l(k) + \phi(k)\Delta h_b(k) \quad (4)$$

where $\phi(k)$ is the PPD of system (3), it is estimated from the historical input and output data of the system by minimizing the following cost function:

$$J(\phi(k)) = |C_l(k) - C_l(k-1) - \phi(k)\Delta h_b(k-1)|_2 + \mu|\phi(k) - \hat{\phi}(k-1)|^2 \quad (5)$$

The estimation algorithm of the PPD (where $\mu > 0$ is the weighting factor):

$$\hat{\phi}(k) = \hat{\phi}(k-1) + \frac{\eta \Delta h_b(k-1)}{\mu + \Delta h_b(k-1)^2}(\Delta C_l(k) - \hat{\phi}(k-1)\Delta h_b(k-1)) \quad (6)$$

$\eta \in (0, 1]$ is the update step of PPD, which can make the algorithm more flexible, and $\eta = 1$ is taken in all of this paper.

In order to eliminate fluctuating load of the transonic buffet flow and improve the flow stability, the following cost function are considered:

$$J(h_b(k)) = |C_l^*(k+1) - C_l(k+1)|_2 + \lambda_1|h_b(k) - h_b(k-1)|_2 + \lambda_2|h_b(k) - h_{b0}|^2 \quad (7)$$

where $C_l^*(k+1)$ is the desired lift coefficient, h_{b0} is the initial height of the actuator, and λ_1 and λ_2 are positive weighting factors. The first term of the cost function is used to limit the deviation between the lift coefficient and the desired lift coefficient, and the second term is used to limit the change of the actuator height. When the unstable steady flow of transonic buffet cannot be obtained in advance, the corresponding desired lift coefficient cannot be obtained, so the desired lift coefficient is taken as the time-averaged lift coefficient, and the deviation between the actuator height and the initial height of the actuator is limited by introducing the third term of the cost function.

Combined with the dynamic linearized data model (4) to minimize the cost function (7), the control law is:

$$h_b(k) = \frac{1}{\lambda_1 + \lambda_2 + \phi(k)^2}\left(\phi(k)(C_l^*(k+1) - C_l(k)) + (\lambda_1 + \phi(k)^2)\, h_b(k-1) + \lambda_2 h_{b0}\right) \quad (8)$$

The control system block diagram is shown in Figure 6. The loop in the upper part of the block diagram is an adaptive loop, indicating that the adaptive law uses the input and output data of the system to estimate the PPD online and adjust the controller parameters when the plant parameters change, so as to make the system work in an optimal or near-optimal state. The loop of the lower part is a general feedback loop, which is obtained by minimizing the cost function.

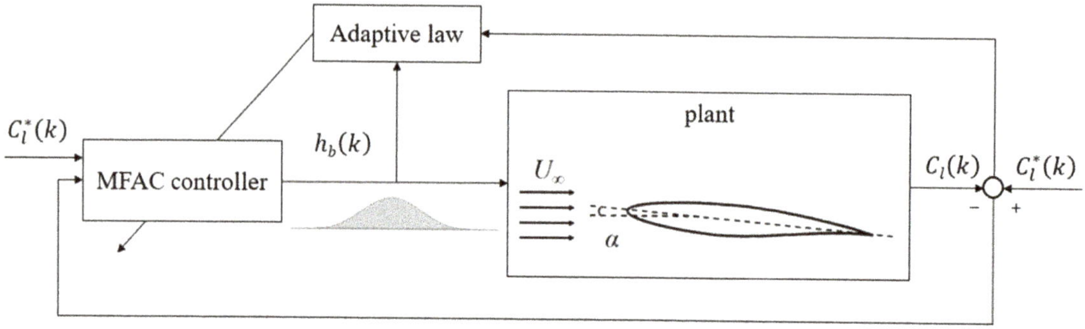

Figure 6. Block diagram of model-free adaptive control system.

3.2. Numerical Simulation and Mechanism Analysis of Control Law

For the incoming flow state of $M = 0.75$, $\alpha = 4°$, and $Re = 6.5 \times 10^6$, it can be seen from Section 2.2 that this state is above the buffet onset boundary. When the control is not turned on, the dimensionless chordwise range of the shock wave oscillation is 0.45~0.56. Smart skin system is used to eliminate the fluctuating load in this state. According to the analysis results by Tian et al., the fluctuating load can be effectively suppressed when the bump is located after the shock wave [7], so the range of actuator is chosen to be 0.55~0.75, that is, the starting position of deformation skin $x_0 = 0.55$, and the length of deformation skin l_b is 0.2 c.

The transonic buffet fluctuating load in this state is suppressed by the above control system configuration. The time history of control signal (actuator height, h_b), feedback signal (lift coefficient, C_l), and drag coefficient (C_d) are shown in Figure 7. It can be seen that when the control is turned on, the fluctuating load (C_l and C_d) has been completely eliminated after a few buffeting cycles, and the actuator height has returned to the initial 0, that is, the airfoil returns to the shape of RAE2822 airfoil. The maximum height of actuator during the control process is 0.00299; there is no additional increase in airfoil drag, and the final steady flow field pressure contour is shown in Figure 8.

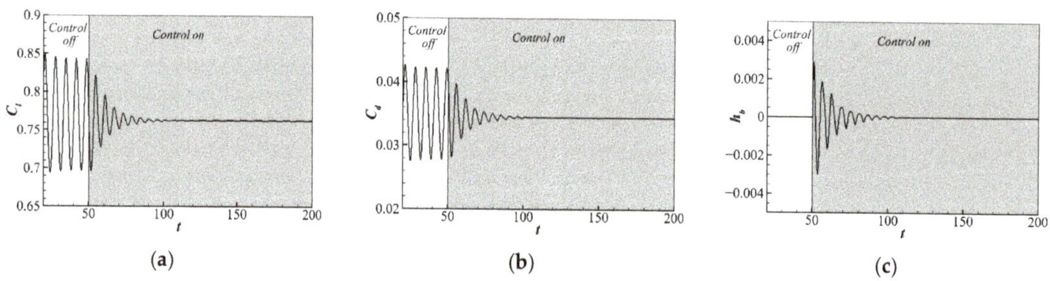

Figure 7. Time history of the system: (**a**) lift coefficient; (**b**) drag coefficient; (**c**) actuator height.

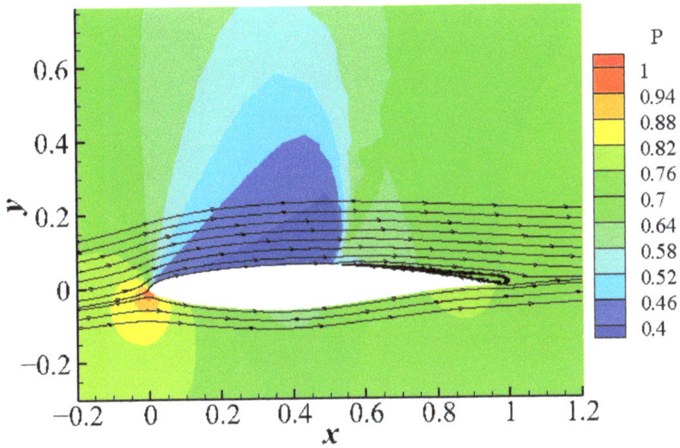

Figure 8. Flow field pressure contour and streamlines after control convergence.

Figure 9 shows the time history of the shock wave dimensionless position x_{sw}/c and the height of the actuator. It can be seen that the height of the actuator located at the downstream of the shock wave is close to the lowest when the shock wave is at the most downstream, and the height of the actuator is close to the highest when the shock wave is at the most upstream. By changing the height of the actuator, the oscillation of the shock wave is limited, thus eliminating the fluctuating load of the buffet. Figure 10 shows the pressure contour and streamlines for the four moments in Figure 9; b' and d' are the middle positions where the shock wave moves downstream and upstream, respectively, and a' and c' are the most upstream and downstream positions of the shock wave, respectively. When the shock wave moves from upstream to downstream, the height of the actuator decreases, which limits the movement of the separation bubble at the shock wave foot to downstream and prevents the merging of the separation bubble at the shock wave foot with the trailing edge separation bubble. When the shock wave moves upstream, the actuator height increases, which limits the growth of the trailing edge separation bubble upstream.

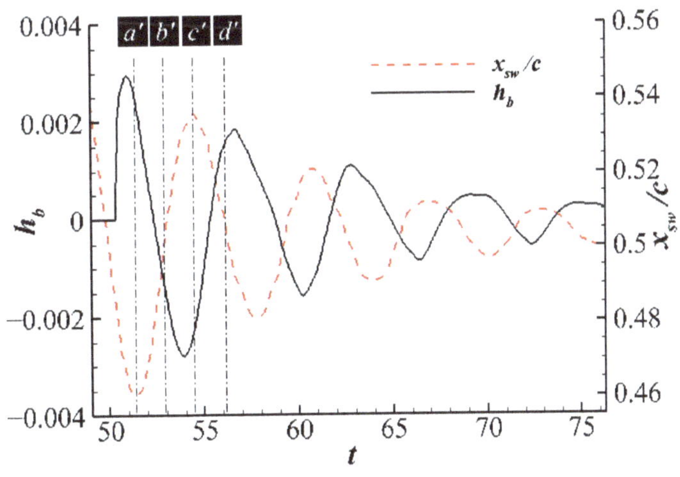

Figure 9. Time history of the shock wave position and the height of the actuator.

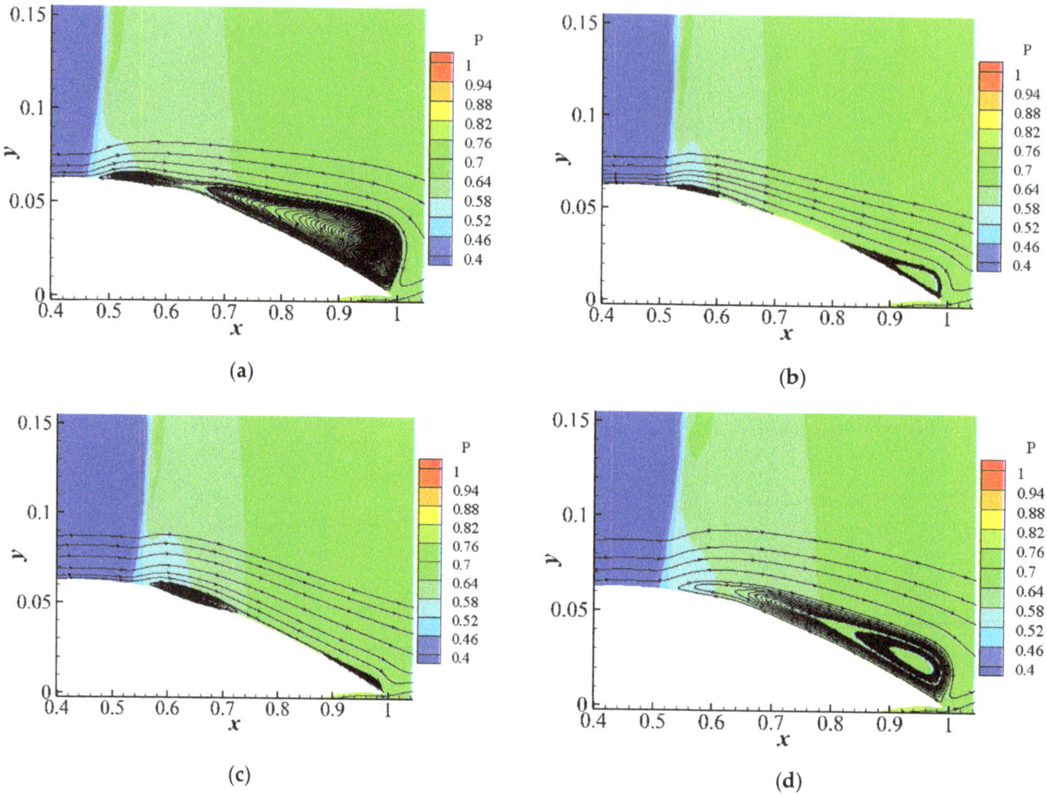

Figure 10. Local pressure contours and streamlines of flow field at different moments: (**a**) moment a'; (**b**) moment b'; (**c**) moment c'; (**d**) moment d'.

4. Smart Skin Performance Analysis

4.1. The Influence of Actuator Position

In the previous section, for the incoming flow state of $M = 0.75$, $\alpha = 4°$, and $Re = 6.5 \times 10^6$, the dimensionless chordwise range of the shock wave oscillation is 0.45~0.56, and the actuator range is chosen to be 0.55~0.75. In this section, the influence of actuator position on the buffeting load suppression is analyzed. Two actuator positions are selected: (1) the actuator position is upstream of the shock wave oscillation range (actuator range: 0.25~0.45); (2) the actuator position coincides with the shock wave oscillation range (actuator range: 0.4~0.6). The input and output time history of control system are shown in Figures 11 and 12, respectively. It should be noted that in Section 3.2, we analyzed that for the actuator located downstream of the shock wave, when the shock wave was at the most downstream, the height of the actuator was close to the highest. For the actuator located upstream of the shock wave (actuator range: 0.25~0.45), the height of the actuator should be close to the lowest when the shock wave is at the most downstream. That is, the initial value of the PPD in the control law should be opposite to the sign in Section 3.2.

(a) (b)

Figure 11. Time history of the system (actuator range: 0.25~0.45): (**a**) lift coefficient; (**b**) actuator height.

(a) (b)

Figure 12. Time history of the system (actuator range: 0.4~0.6): (**a**) lift coefficient; (**b**) actuator height.

From Figures 11 and 12, we find that the required actuator height is smaller when the actuator position is close to the oscillation range of shock wave. As shown in Figure 13, we compare the maximum height of the actuator for different crest chordwise positions under the premise of basically the same settling time. Because the function of the actuator is to suppress the oscillation of the shock wave and to prevent the merging of the separation bubble at the shock wave foot and the trailing edge separation bubble, the position of the actuator far away from the shock wave needs a larger height to prevent the merging of the two separation bubbles, and to generate enough pressure disturbance to suppress the oscillation of the shock wave.

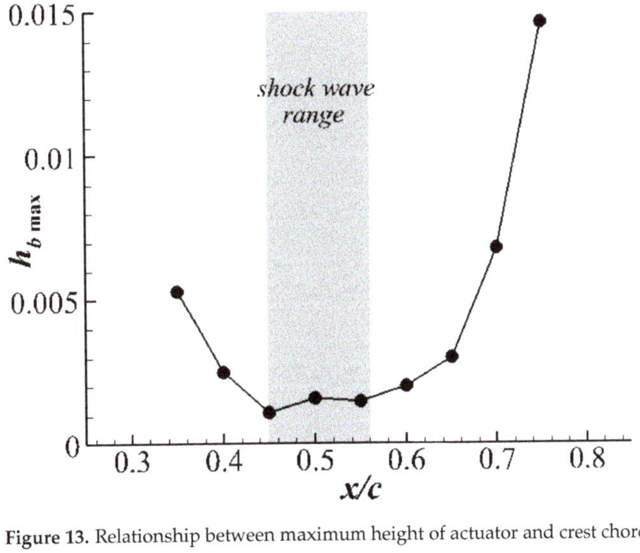

Figure 13. Relationship between maximum height of actuator and crest chordwise position.

4.2. The Influence of Actuator Height

Real wings design often has multiple constraints, such as the wing volume will limit the amplitude of the actuator subsidence, and actuator structure and drag constraints will limit the maximum height of the actuator. Therefore, this subsection will analyze the performance change of the smart skin when the actuator height has the maximum amplitude limit h_s for the incoming flow state of $M = 0.75$, $\alpha = 4°$, and $Re = 6.5 \times 10^6$, and the actuator range of 0.55~0.75. As can be seen in Figure 14, the smart skin system can eliminate the transonic buffet fluctuating load under different restrictions of maximum height of the actuator, but with the decrease of the maximum actuator height, the system settling time gradually increases. According to the previous smart skin control mechanism, the phase relationship between the actuator height and the shock wave position (or lift coefficient) plays a decisive role, while the height only affects the dynamic characteristics such as the settling time.

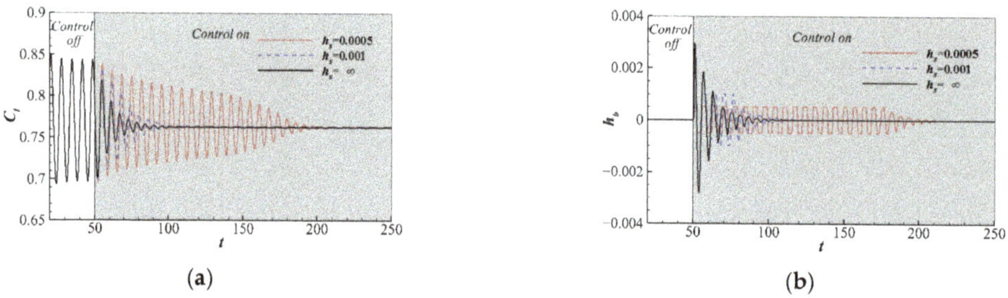

Figure 14. Influence of actuator height constraint h_s on dynamic characteristics of the control system: (**a**) time history of lift coefficient; (**b**) time history of actuator height.

As shown in Table 2, the required actuator height for active control can be smaller than that for a passive bump device located immediately downstream of the shock wave, which reduces the requirement for the actuator and is conducive to engineering implementation. Although compared with passive control, active control requires energy input from the time history of input and output of the control system; the output displacement of actuator and settling time are small, so the energy input required is acceptable.

Table 2. Maximum height of actuator required for different control strategy.

Control Strategy	Passive Control		Active Control
	Jones et al. [6]	**Tian et al. [7]**	**Present**
entry 2	0.012	0.008	less than 0.003

4.3. Robustness of The Control Law

Due to the existence of atmospheric turbulence, the incoming flow state may change, and the dynamics characteristics of the transonic buffet flow system will change, which is manifested in the change of the oscillation range of shock wave, the mean value, and fluctuation amplitude of lift, such as for the other two states above the buffet onset boundary: (1) $M = 0.73$, $\alpha = 4.8°$, and $Re = 6.5 \times 10^6$, and (2) $M = 0.71$, $\alpha = 5.5°$, and $Re = 6.5 \times 10^6$. In state (1), oscillation range of shock wave is from 0.35~0.55, and fluctuation range of lift coefficient is 0.723~0.994. In state (2), oscillation range of shock wave is from 0.24~0.48, and fluctuation range of lift coefficient is 0.673~1.042. Figures 15 and 16 show the lift coefficient, drag coefficient, and actuator height time history of the system using smart skin to suppress the fluctuating load in the above two incoming flow states, respectively, and the actuator range from 0.55 to 0.75. It can be seen from the Figures 15 and 16 that the control strategy in this study has strong robustness, and after turning on the control, by adjusting the height of smart skin, the large fluctuating load of transonic buffet can be completely eliminated, and there is no additional increase in airfoil drag. Compared with state (1), the angle of attack of state (2) increases, the shock wave position moves forward, and the fluctuating amplitude of lift coefficient increases, so the maximum height required to suppress the fluctuating load in state (2) increases.

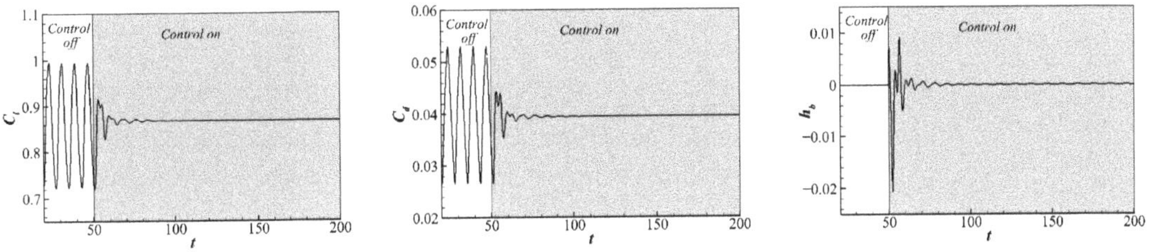

(a) (b) (c)

Figure 15. Time history of the system (M = 0.73, α = 4.8°): (**a**) lift coefficient; (**b**) drag coefficient; (**c**) actuator height.

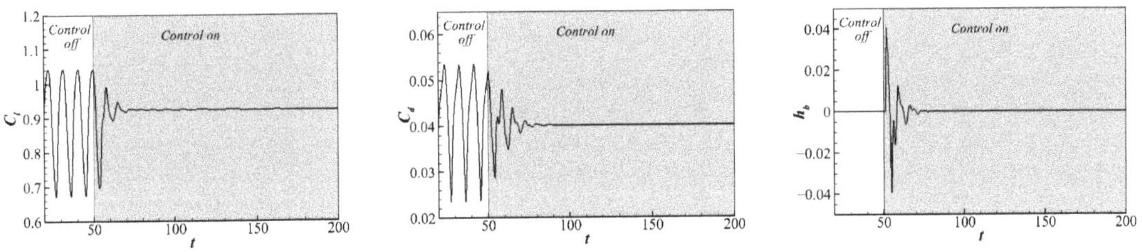

(a) (b) (c)

Figure 16. Time history of the system (M = 0.71, α = 5.5°): (**a**) lift coefficient; (**b**) drag coefficient; (**c**) actuator height.

5. Conclusions

In this study, the smart skin system is used to suppress the fluctuating load of transonic buffet flow for RAE2822 airfoil. The smart skin system uses the lift coefficient as feedback signal and adopts a model-free adaptive control to dynamically adjust the local skin height of airfoil. When the incoming flow state enters above the transonic buffet onset boundary, by preventing the merging of the separation bubble at the shock wave foot with the trailing edge separation bubble, the smart skin system can completely suppress the fluctuating load and does not affect the aerodynamic performance in the non-buffeting state. From the time history of the control system input and output, the system has a short settling time and does not need too large an output displacement, so the piezoelectric material with high precision of displacement control, fast response speed, low power consumption, and small size can be used as the actuator.

By analyzing the control performance of the smart skin system, the following conclusions were obtained:

(1) The smart skin system can completely suppress the fluctuating load regardless of whether the actuator position is upstream or downstream of the shock wave, but the maximum height of the actuator required increases when the actuator position is far away from the shock wave oscillation range.

(2) The fluctuating load of transonic buffet can be completely suppressed under different maximum height constraints, and the height constraint will affect the dynamic performance of the system, such as settling time.

(3) When the incoming flow state changes, although the oscillation range of shock wave and the fluctuation range of lift coefficient change greatly, the buffeting load in different incoming flow states can be suppressed by the model-free adaptive control with the mean value of the lift coefficient as the desired output of the control system.

The above research is carried out in a simulation environment in order to provide some theoretical support for the realization of shock buffeting control. To realize the control of shock buffet, more detailed parameters, such as the specification and distribution of sensors and the bandwidth and hysteresis of actuators, need to be further studied.

With the progress of material and control technology, new flexible controllers such as smart skin can have more flexible control methods without abrupt curvature changes on the surface of the aircraft, and have the ability to obtain the best control effect through adaptive and intelligent control for different flight conditions, which can greatly improve the aerodynamic performance of the aircraft.

Author Contributions: K.R. designed the controller and writing-original draft preparation.; C.G. conceived the original ideals, review; F.Z. proposed the comment for the paper; W.Z. provided research methods, supervision. All authors have read and agreed to the published version of the manuscript.

Funding: This research was funded by the National Natural Science Foundation of China (No. 11902269), the Aviation Science Foundation of China (No. 2019ZH053003), the Fundamental Research Funds for the Central Universities, and State Key Laboratory of Aerodynamics innovation Fund Project (JBKYC190102).

Data Availability Statement: Not applicable.

Acknowledgments: Fangqi Zhou is particularly grateful for the support of the State Key Laboratory of Aerodynamics innovation Fund Project.

Conflicts of Interest: The authors declare no conflict of interest.

References

1. Gao, C.; Zhang, W.; Kou, J.; Liu, Y.; Ye, Z. Active control of transonic buffet flow. *J. Fluid Mech.* **2017**, *824*, 312–351. [CrossRef]
2. Giannelis, N.F.; Vio, G.A.; Levinski, O. A review of recent developments in the understanding of transonic shock buffet. *Prog. Aerosp. Sci* **2017**, *92*, 39–84. [CrossRef]
3. Gao, C.; Zhang, W.; Ye, Z. Numerical study on closed-loop control of transonic buffet suppression by trailing edge flap. *Comput. Fluids* **2016**, *132*, 32–45. [CrossRef]
4. Titchener, N.; Babinsky, H. Shock wave/boundary-layer interaction control using a combination of vortex generators and bleed. *AIAA J.* **2013**, *51*, 1221–1233. [CrossRef]
5. Kouchi, T.; Yamaguchi, S.; Koike, S.; Nakajima, T.; Sato, M.; Kanda, H.; Yanase, S. Wavelet analysis of transonic buffet on a two-dimensional airfoil with vortex generators. *Exp. Fluids* **2016**, *57*, 166. [CrossRef]
6. Jones, N.R.; Jarrett, J.P. Designing a shock control bump array for a transonic wing–body model. *AIAA J.* **2018**, *56*, 4801–4814. [CrossRef]
7. Tian, Y.; Gao, S.; Liu, P.; Wang, J. Transonic buffet control research with two types of shock control bump based on RAE2822 airfoil. *Chin. J. Aeronaut.* **2017**, *30*, 1681–1696. [CrossRef]
8. Smith, A.; Babinsky, H.; Fulker, J.; Ashill, P. Shock wave/boundary-layer interaction control using streamwise slots in transonic flows. *J. Aircr.* **2004**, *41*, 540–546. [CrossRef]
9. Polivanov, P.A.; Sidorenko, A.A.; Maslov, A.A. Effective plasma buffet and drag control for laminar transonic aerofoil. *Proc. Inst. Mech. Eng. Part G* **2020**, *234*, 58–67. [CrossRef]
10. Sidorenko, A.; Budovsky, A.; Polivanov, P.; Vishnyakov, O.; Sudakov, V.; Ishchenko, V. Suppression of transonic buffet with plasma vortex generators. *Thermophys. Aeromechanics* **2019**, *26*, 465–480. [CrossRef]
11. Caruana, D.; Mignosi, A.; Corrège, M.; Le Pourhiet, A.; Rodde, A. Buffet and buffeting control in transonic flow. *Aerosp. Sci. Technol.* **2005**, *9*, 605–616. [CrossRef]
12. Ren, K.; Chen, Y.; Gao, C.; Zhang, W. Adaptive control of transonic buffet flows over an airfoil. *Phys. Fluids* **2020**, *32*, 096106. [CrossRef]
13. Li, D.; Zhao, S.; Da Ronch, A.; Xiang, J.; Drofelnik, J.; Li, Y.; Zhang, L.; Wu, Y.; Kintscher, M.; Monner, H.P. A review of modelling and analysis of morphing wings. *Prog. Aerosp. Sci.* **2018**, *100*, 46–62. [CrossRef]
14. Nir, A.; Abramovich, H. Design, analysis and testing of a smart fin. *Compos. Struct.* **2010**, *92*, 863–872. [CrossRef]

15. Sinn, T.; Barrett, R. Design, manufacturing and test of a high lift secondary flight control surface with shape memory alloy post-buckled precompressed actuators. *Actuators* **2015**, *4*, 156–171. [CrossRef]
16. Sun, J.; Guan, Q.; Liu, Y.; Leng, J. Morphing aircraft based on smart materials and structures: A state-of-the-art review. *J. Intell. Mater. Syst. Struct.* **2016**, *27*, 2289–2312. [CrossRef]
17. Fincham, J.; Friswell, M. Aerodynamic optimisation of a camber morphing aerofoil. *Aerosp. Sci. Technol.* **2015**, *43*, 245–255. [CrossRef]
18. Wang, I.; Gibbs, S.C.; Dowell, E.H. Aeroelastic model of multisegmented folding wings: Theory and experiment. *J. Aircr.* **2012**, *49*, 911–921. [CrossRef]
19. Zhao, S.; Li, D.; Zhou, J.; Sha, E. Numerical and experimental study of a flexible trailing edge driving by pneumatic muscle actuators. *Actuators* **2021**, *10*, 142. [CrossRef]
20. Cooper, J.; Chekkal, I.; Cheung, R.; Wales, C.; Allen, N.; Lawson, S.; Peace, A.; Cook, R.; Standen, P.; Hancock, S. Design of a morphing wingtip. *J. Aircr.* **2015**, *52*, 1394–1403. [CrossRef]
21. Woods, B.K.; Bilgen, O.; Friswell, M.I. Wind tunnel testing of the fish bone active camber morphing concept. *J. Intell. Mater. Syst. Struct.* **2014**, *25*, 772–785. [CrossRef]
22. Thill, C.; Etches, J.; Bond, I.; Potter, K.; Weaver, P. Morphing skins. *Aeronaut. J.* **2008**, *112*, 117–139. [CrossRef]
23. Bilgen, O.; Kochersberger, K.B.; Inman, D.J.; Ohanian, O.J., III. Novel, bidirectional, variable-camber airfoil via macro-fiber composite actuators. *J. Aircr.* **2010**, *47*, 303–314. [CrossRef]
24. Li, D.; Liu, Q.; Wu, Y.; Xiang, J. Design and analysis of a morphing drag rudder on the aerodynamics, structural deformation, and the required actuating moment. *J. Intell. Mater. Syst. Struct.* **2018**, *29*, 1038–1049. [CrossRef]
25. Grigorie, T.; Botez, R.; Popov, A. How the airfoil shape of a morphing wing is actuated and controlled in a smart way. *J. Aerosp. Eng.* **2015**, *28*, 04014043. [CrossRef]
26. Lv, B.; Wang, Y.; Lei, P. Effects of trailing edge deflections driven by shape memory alloy actuators on the transonic aerodynamic characteristics of a super critical airfoil. *Actuators* **2021**, *10*, 160. [CrossRef]
27. Dimino, I.; Gallorini, F.; Palmieri, M.; Pispola, G. Electromechanical actuation for morphing winglets. *Actuators* **2019**, *8*, 42. [CrossRef]
28. Munday, D.; Jacob, J. Active control of separation on a wing with conformal camber. In Proceedings of the 39th Aerospace Sciences Meeting and Exhibit, Reno, NV, USA, 8–11 January 2001; p. 293.
29. Baldelli, D.H.; Lee, D.-H.; Peña, R.S.S.; Cannon, B. Modeling and control of an aeroelastic morphing vehicle. *J. Guid. Control Dyn.* **2008**, *31*, 1687–1699. [CrossRef]
30. Perera, M.; Guo, S. Optimal design of an aeroelastic wing structure with seamless control surfaces. *Proc. Inst. Mech. Eng. Part G* **2009**, *223*, 1141–1151. [CrossRef]
31. Ai, Q.; Azarpeyvand, M.; Lachenal, X.; Weaver, P.M. Aerodynamic and aeroacoustic performance of airfoils with morphing structures. *Wind Energy* **2016**, *19*, 1325–1339. [CrossRef]
32. Zhao, C.; Zhou, L.; Qiu, T. Research on flexible skin technique for adaptive bump inlet structure. *Aircr. Eng. Aerosp. Technol.* **2022**, *94*, 981–993. [CrossRef]
33. Lee, C.Y.; Kim, J.H. Active futter suppression of smart-skin antenna structures with piezoelectric sensors and actuators. *Aerospace* **2021**, *8*, 257. [CrossRef]
34. Tsushima, N.; Su, W. Flutter suppression for highly flexible wings using passive and active piezoelectric effects. *Aerosp. Sci. Technol.* **2017**, *65*, 78–89. [CrossRef]
35. Wang, G.; Mian, H.H.; Ye, Z.-Y.; Lee, J.-D. Improved point selection method for hybrid-unstructured mesh deformation using radial basis functions. *AIAA J.* **2015**, *53*, 1016–1025. [CrossRef]
36. Cook, P.; Firmin, M.; McDonald, M. *Aerofoil RAE 2822: Pressure Distributions, and Boundary Layer and Wake Measurements*; AGARD Advisory Report; Advisory Group for Aerospace Research and Development: Neuilly sur Seine, France, 1979; p. A6.
37. Hou, Z.S.; Wang, Z. From model-based control to data-driven control: Survey, classification and perspective. *Inf. Sci.* **2013**, *235*, 3–35. [CrossRef]

actuators

Article

Influence of Synthetic Jets on Multiscale Features in Wall-Bounded Turbulence

Biaohui Li [1], Jinhao Zhang [1] and Nan Jiang [1,2,*]

[1] School of Mechanical Engineering, Tianjin University, 135 Yaguan Road, Tianjin 300350, China; libiaohui915@outlook.com (B.L.); zhangjh0625@outlook.com (J.Z.)
[2] Tianjin Key Laboratory of Modern Engineering Mechanics, Tianjin University, 135 Yaguan Road, Tianjin 300350, China
* Correspondence: nanj@tju.edu.cn

Abstract: This experimental research focuses on the impacts of submerged synthetic jets on a fully-developed turbulent boundary layer (TBL) under a drag reduction working case. Two-dimensional velocity vectors in the flow field are captured with the aid of a particle image velocimetry (PIV) system. Proper orthogonal decomposition (POD) analyses provide evidence that synthetic jets notably attenuate the induction effect of prograde vortex on the low-speed fluid in large-scale fluctuation velocity field, thereby weakening the bursting process of near-wall turbulent events. Furthermore, the introduced perturbance redistributes the turbulent kinetic energy (TKE) and concentrates the TKE onto small-scale coherent structures. Modal time coefficients in various orders of POD are divided into components of multiple frequency bands by virtue of complementary ensemble empirical mode decomposition (CEEMD). It is found that the turbulence signals are shifted from low-frequency to high-frequency bands thanks to synthetic jets, thus revealing the relationship between scales and frequency bands. One further method of scale decomposition is proposed, that is, the large-scale fluctuating flow field will be obtained after removing the high-frequency noise data with the help of continuous mean square error (CMSE) criterion.

Keywords: active control; multiscale; synthetic jet; turbulent boundary layer

Citation: Li, B.; Zhang, J.; Jiang, N. Influence of Synthetic Jets on Multiscale Features in Wall-Bounded Turbulence. *Actuators* **2022**, *11*, 199. https://doi.org/10.3390/act11070199

Academic Editor: Luigi de Luca

Received: 4 July 2022
Accepted: 17 July 2022
Published: 18 July 2022

1. Introduction

Wall turbulence drag reduction technologies have represented a dramatic topic in aerospace and other engineering applications over the past few decades as a result of the oil crisis in the early 1970s [1]. So far, various control schemes [2] have been put forward with remarkable success and are of great significance in improving the performance of vehicles and saving energy [3]. The consensus is that effective drag reduction is achieved by impairing coherent structures [4] in the turbulent boundary layer (TBL). Generally, passive approaches that only apply to a single design condition are reliable and well-studied. In contrast, active control techniques are too complicated to be attained, which has become a vital research direction to be conquered at present [5]. Among them, the accomplishment of skin friction drag reduction by employing blowing and suction at the wall is a widely studied strategy [6]. Fortunately, the synthetic jet actuator (SJA) has the ability to generate local blowing and suction without additional air supply. This facilitates interaction with vorticity structures in the TBL [7], which has tremendous potential.

Ricco et al. [8] reviewed sundry state-of-the-art techniques oriented toward turbulent drag reduction via spanwise actuation, including wall motion, plasma body forces and synthetic jets, pointing out the preponderance of synthetic jets for resistance reduction control. The interaction experiment of the inclined circular synthetic jets with turbulent channel flow was carried out by Iuso et al. [9]. Reductions in the average surface friction of up to 15% as well as in the turbulence pulsation of up to 12% were realized. Yao et al. [10] imposed a body force in the motion equation to simulate the spanwise motion induced

Table 1. Experimental parameters.

Parameter	Value
Field of view	99.97 mm × 62.48 mm
Sampling frequency	600 Hz
Resolution	1280 pixels × 800 pixels
Number of pictures	8216 × 5
Scale factor	0.0781 mm/pix
Interrogation window	32 pixels × 32 pixels, 75% overlap
Vector pitch	0.6248 mm

3. Results and Analyses

3.1. Proper Orthogonal Decomposition

For the two-dimensional two-component (2D-2C) velocity field, the snapshot POD [19] could be executed to availably separate the scale feature of flow. The specific idea is that the instantaneous field can be regarded as the superposition of the mean field and the fluctuation field:

$$x'(t_i) = x(\xi, t_i) - \bar{x}(\xi) \in \mathbb{R}^m, i = 1, 2, \ldots n, n \ll m \tag{1}$$

where $x = (u, v)$, m and n are space and time series, respectively, and $m = 92 \times 156 \times 2$. Since POD is not sensitive to time resolution [20], and in order to reduce the computational burden, one snapshot is extracted for every five snapshots, so the time series is $n = 8215$. In view of this, the time scale equivalent is to be amplified by a factor of 5, and as a result, the value of frequency needs to be reduced by 5 times, accordingly, in the power spectral density (PSD) analysis of the time coefficients in the following. The specific process is as follows:

(1) The time series (fluctuation velocity) matrix is expressed as:

$$X = \left[x'(t_1), x'(t_2), x'(t_3), \ldots, x'(t_{n-1}), x'(t_n)\right] \in \mathbb{R}^{m \times n} \tag{2}$$

(2) Solving the eigenvectors Ψ_j and eigenvalues λ_j of the correlation matrix:

$$X^T X \Psi_j = \lambda_j \Psi_j,\ \Psi_j \in \mathbb{R}^n \tag{3}$$

(3) The basis function for POD mode is written as:

$$\Phi_j = \frac{X\Psi_j}{\|X\Psi_j\|} \in \mathbb{R}^m, j = 1, 2, \ldots, n \tag{4}$$

(4) The time coefficient (expansion coefficient) of each mode will be realized as:

$$a_j = \Phi_j^T X \tag{5}$$

(5) Reconstructing the fluctuating velocity field:

$$x'(\xi, t_i) = \sum_{j=1}^{n} a_j(t_i) \Phi_j(\xi) \tag{6}$$

(6) Single-order mode turbulent kinetic energy ratio:

$$P(\lambda) = \frac{\lambda_j}{\Sigma \lambda}, j = 1, 2, \ldots, n \tag{7}$$

(7) The proportion of cumulative energy is defined as:

$$C(\lambda) = \frac{\Sigma_{j=1}^{k} \lambda_j}{\Sigma \lambda}, 1 \le k \le n \quad (8)$$

The contribution of each POD mode to total turbulent kinetic energy (TKE) is given in Figure 3a. In the control state, the energy proportion is lower than that without control in the first four modes, and starting with the fifth mode, energy distribution of the two conditions is opposite. Thus, it can be inferred that the synthetic jets restrain the large-scale (low-order) TKE and strengthen the small-scale (high-order) turbulent events. Figure 3b plots the cumulative energy ratio. For the perturbed condition, more modes are required to achieve the same levels of TKE as the uncontrolled flow field, which is because the periodic disturbance generated by synthetic jets accelerates the migration of turbulent energy from large-scale to small-scale.

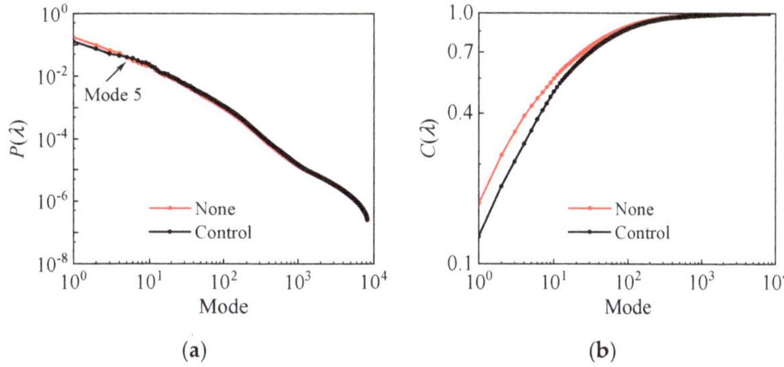

Figure 3. Flow field energy distributions: (**a**) Single energy; (**b**) Cumulative energy.

The levels of TKE of the aforementioned various modes are determined by normalized eigenvalues, so the amount of information of turbulent events contained will be attained by introducing information entropy [21]. The formula is as follows:

$$H(\lambda) = -\sum_{j=1}^{k} P(\lambda_j) \log P(\lambda_j), 1 \le k \le n \quad (9)$$

where the base of the logarithm is two and the unit is bit. Information entropy is the expectation of information content, as a result, in turbulent events, low-order modes have higher probability, smaller information entropy, less uncertainty and events are ordered; high-order modes have lower probability, larger information entropy, more uncertainty and events are disordered.

The accumulated curve of information entropy along with modes is reported in Figure 4a. The cumulative $H(\lambda)$ of the first seventeen modes in the forced case is smaller than that of the undisturbed state. Since the eighteenth mode, the $H(\lambda)$ of the former exceeds the latter, and the difference between the two is further expanded as the number of modes increases. On the whole, the $H(\lambda)$ of the flow field before and after the application of control is 6.13 and 6.53, respectively. In other words, the existence of the synthetic jet leads to disordered flow field. However, it should be pointed out that in the low-order modes range (1~17), the uncertainty of the turbulent events contained in the controlled condition is relatively low, and the flow field is more orderly, which is attributed to the perturbation exerting a positive modulation influence on the large-scale coherent structures.

Table 2. The number of modes with zero signal reconstructed at various turbulent energy.

	0.5	0.6	0.7	0.8	0.9	1.0
None	0	0	0	0	28	7992
Control	1	2	11	20	59	8003

Then, the global frequency information of the time series within the sampling period is solved using the Hilbert transform [31], which is defined as the convolution of the signal itself and the reciprocal of the time domain:

$$y(t) = x(t) * \frac{1}{\pi t} \tag{13}$$

where $x(t)$ is the real part of the complex signal, and $y(t)$ stands for the imaginary part.

Next, the instantaneous phase $\theta(t)$ and instantaneous frequency $f(t)$ of the analytical signal will be represented by the following equation:

$$\theta(t) = \arctan\frac{y(t)}{x(t)}, f(t) = \frac{1}{2\pi}\frac{d\theta(t)}{dt} \tag{14}$$

The instantaneous frequencies of the original signal and the reconstructed sequence in the time domain are indicated in Figure 14. The time coefficients of the two working states have abrupt changes in high- and low-frequency components at certain moments, which is a typical feature of non-stationary signals. In the disturbed state, the time proportion of high-frequency bands is much higher than that in the unforced case. In comparison to the original sequence, the instantaneous frequency mutation of the reconstructed signal is immensely shortened because of filtering out high-frequency noise data, and it is maintained at a lower level. Some useful high-frequency constituents are retained in the reconstructed signal of disturbed condition, embodying the characteristics of external disturbances.

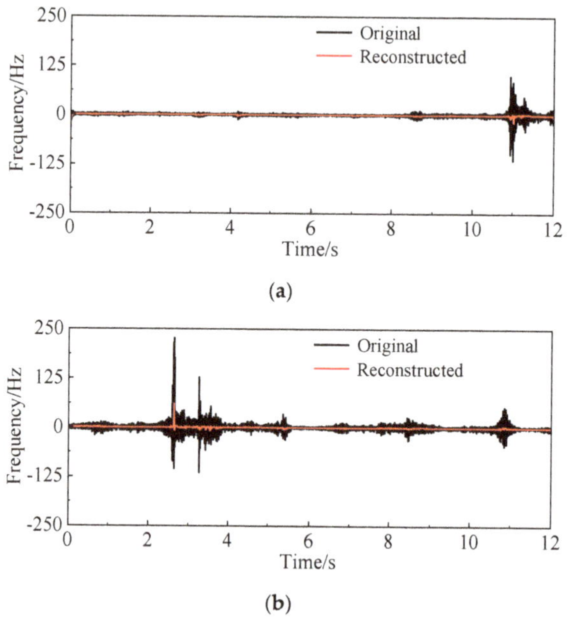

Figure 14. Instantaneous frequency distribution: (**a**) None; (**b**) Control.

Up to now, the time coefficient of each order is reconstructed, and then the pulsating velocity fields will be determined by formula 5. In this section, the original flow field and the reconstructed flow field at a certain instant are selected for comparative analysis (see Figure 15). In the original flow field, there are large-scale structures of red and blue alternating flow direction, and small-scale pulsations are distributed near the edge. As for the reconstructed flow field, on the basis of retaining the large-scale motion, the small-scale structures are filtered out, and the shape of the flow structures is relatively smooth.

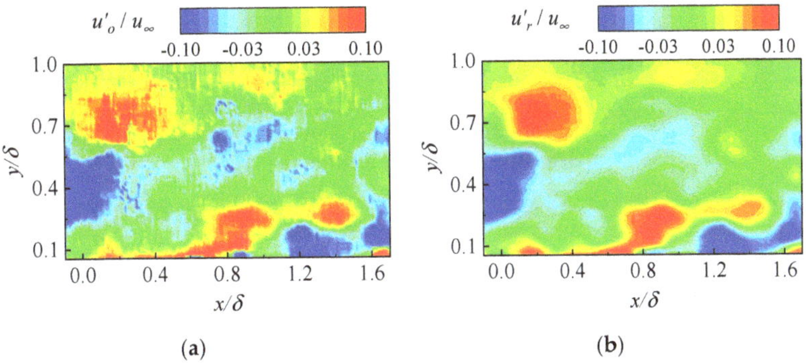

Figure 15. Comparison of (**a**) original and (**b**) reconstructed fluid field.

Figure 16 manifests the relative probability statistics of the dimensionless large-scale streamwise and wall-normal pulsation velocities over the whole viewing window. The ejection event in the second quadrant and the sweep event in the fourth quadrant are the main constituents of the turbulent events, and the intensity of the burst events is directly related to the Reynolds shear stress [32]. In contrast, the relative probability of large-scale streamwise direction pulsation in the disturbed flow field decreases, which can be confirmed by the decline in the flow direction length of the contour. The enhancement in the area proportion of the red region indicates that the flow direction and wall-normal fluctuation of the smaller scale are heightened. Because the synthetic jet has a weak impact on the large-scale wall-normal pulsation, the variation of the outermost contour of the cloud image in the wall-normal direction is not prominent.

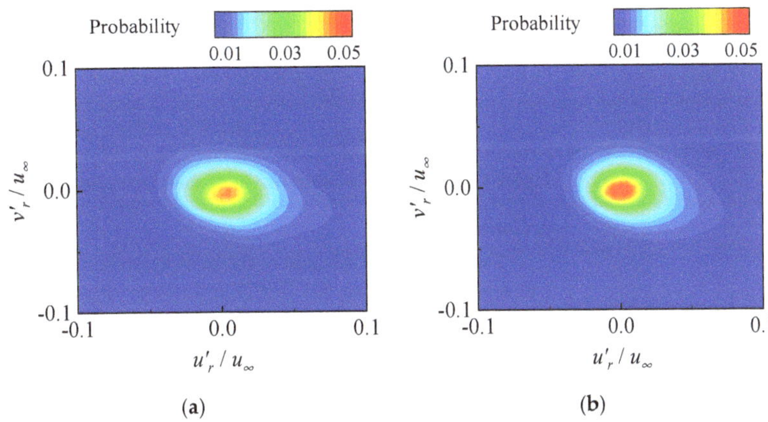

Figure 16. Relative probability of flow and wall-normal fluctuation velocity: (**a**) None; (**b**) Control.

The detection of burst events can also be implemented with the spatial local average velocity structure function, overcoming the limitation of traditional condition sampling

15. Segawa, T.; Li, F.-C.; Yoshida, H.; Murakami, K.; Mizunuma, H. Spanwise oscillating excitation for turbulence drag reduction using alternative suction and blowing. In Proceedings of the 43rd AIAA Aerospace Sciences Meeting and Exhibit, Reno, NV, USA, 10–13 January 2005. [CrossRef]
16. Spinosa, E.; Zhong, S. Reduction of skin friction drag in a turbulent boundary layer using circular synthetic jets. In Proceedings of the 55th AIAA Aerospace Sciences Meeting, Grapevine, Texas, 9–13 January 2017.
17. Marusic, I.; McKeon, B.J.; Monkewitz, P.A.; Nagib, H.M.; Smits, A.J.; Sreenivasan, K.R. Wall-bounded turbulent flows at high Reynolds numbers: Recent advances and key issues. *Phys. Fluids.* **2010**, *22*, 065103. [CrossRef]
18. Li, B.; Zhang, J.; Tian, H.; Ma, X.; Tang, Z.; Jiang, N. Effects of submerged synthetic jet on the coherent structures in turbulent boundary layer. *Acta Mech. Sin.* **2022**, *38*, 321590. [CrossRef]
19. Taira, K.; Hemati, M.S.; Brunton, S.L.; Sun, Y.Y.; Duraisamy, K.; Bagheri, S.; Dawson, S.T.M.; Yeh, C.A. Modal Analysis of Fluid Flows: Applications and Outlook. *AIAA J.* **2020**, *58*, 998–1022. [CrossRef]
20. Wang, X.; Wang, Y.; Tian, H.; Jiang, N. Effects of the slip wall on the drag and coherent structures of turbulent boundary layer. *Acta Mech. Sin.* **2021**, *37*, 1278–1290. [CrossRef]
21. Shannon, C.E. A mathematical theory of communication. *Bell Syst. Tech. J.* **1948**, *27*, 623–656. [CrossRef]
22. Kullback, S.; Leibler, R.A. On information and sufficiency. *Ann. Math. Stat.* **1951**, *22*, 79–86. [CrossRef]
23. Wu, Y.; Christensen, K.T. Spatial structure of a turbulent boundary layer with irregular surface roughness. *J. Fluid Mech.* **2010**, *655*, 380–418. [CrossRef]
24. Zhou, J.; Adrian, R.J.; Balachandar, S.; Kendall, T.M. Mechanisms for generating coherent packets of hairpin vortices in channel flow. *J. Fluid Mech.* **1999**, *387*, 353–396. [CrossRef]
25. Fukagata, K.; Iwamoto, K.; Kasagi, N. Contribution of Reynolds stress distribution to the skin friction in wall-bounded flows. *Phys. Fluids.* **2002**, *14*, L73–L76. [CrossRef]
26. Yang, S.Q.; Jiang, N. Tomographic TR-PIV measurement of coherent structure spatial topology utilizing an improved quadrant splitting method. *Sci. China Phys. Mech. Astron.* **2012**, *55*, 1863–1872. [CrossRef]
27. Huang, N.E.; Shen, Z.; Long, S.R.; Wu, M.C.; Shih, H.H.; Zheng, Q.; Yen, N.C.; Tung, C.C.; Liu, H.H. The empirical mode decomposition and the Hilbert spectrum for nonlinear and non-stationary time series analysis. *Proc. Math. Phys. Eng. Sci.* **1998**, *454*, 903–995. [CrossRef]
28. Huang, N.E.; Wu, M.L.; Qu, W.; Long, S.R.; Shen, S.S.P. Applications of Hilbert-Huang transform to non-stationary financial time series analysis. *Appl. Stoch. Models Bus. Ind.* **2010**, *19*, 245–268. [CrossRef]
29. Yeh, J.R.; Shieh, J.S.; Huang, N.E. Complementary ensemble empirical mode decomposition: A novel noise enhanced data analysis method. *Adv. Adapt. Data Anal.* **2010**, *2*, 135–156. [CrossRef]
30. Wu, H.; Fang, Q.; Peng, Y.; Gong, Z.M.; Kong, X.Z. Hard projectile perforation on the monolithic and segmented RC panels with a rear steel liner. *Int. J. Impact Eng.* **2015**, *76*, 232–250. [CrossRef]
31. Huang, N.E.; Shen, Z.; Long, S.R. A new view of nonlinear water waves: The Hilbert spectrum. *Annu. Rev. Fluid Mech.* **1999**, *31*, 417–457. [CrossRef]
32. Tang, Z.; Jiang, N.; Schröder, A.; Geisler, R. Tomographic PIV investigation of coherent structures in a turbulent boundary layer flow. *Acta Mech. Sin.* **2012**, *28*, 572–582. [CrossRef]
33. Yang, S.; Li, S.; Tian, H.; Wang, Q.; Jiang, N. Coherent spanwise structures in turbulent boundary layer over drag-reducing riblets. *Trans. Tianjin Univ.* **2015**, *21*, 317–323. [CrossRef]
34. Tian, H.; Zhang, J.; Wang, E.; Yao, Z.; Jiang, N. Experimental investigation on drag reduction in turbulent boundary layer over superhydrophobic surface by TRPIV. *Theor. Appl. Mech. Lett.* **2015**, *5*, 45–49. [CrossRef]

actuators

MDPI

Article

Dual Synthetic Jets Actuator and Its Applications—Part I: PIV Measurements and Comparison to Synthetic Jet Actuator

Zhenbing Luo *, Zhijie Zhao, Xiong Deng, Lin Wang and Zhixun Xia

College of Aerospace Science, National University of Defense Technology, Changsha 410073, China; 18041360895@163.com (Z.Z.); badi_arg@126.com (X.D.); wanglin-2007@126.com (L.W.); zxxia@nudt.edu.cn (Z.X.)
* Correspondence: luozhenbing@163.com

Abstract: In order to understand the differences between dual synthetic jets (DSJs) and synthetic jets (SJs), particle image velocimetry (PIV) technology is used to capture the basic flow field characteristics of a dual synthetic jet actuator (DSJA) and a synthetic jet actuator (SJA), and then a careful comparison between them is implemented. The results indicate that a cycle of the DSJ is divided into two stages. In the near-field downstream, a pair of synthetic jets entrain fluid around them and interact with each other, making the flow field complex, and the time-periodic diaphragm dominates them. There is an unfavorable phenomenon of "self-support" between the two jets. In the far-field downstream, the two jets merge into a single, more stable SJ with a higher velocity and a double characteristic frequency. The DSJs have also shown good vectoring characteristics, with the vectoring deflection angle (VDA) changing from about -46° to 46°. The above results demonstrate that the DSJA may replace the traditional SJA in all kinds of applications and extend the applying area of the SJ to more active flow control systems, which cannot be qualified by traditional SJA.

Keywords: dual synthetic jets; synthetic jet; vectoring characteristics; PIV

Citation: Luo, Z.; Zhao, Z.; Deng, X.; Wang, L.; Xia, Z. Dual Synthetic Jets Actuator and Its Applications—Part I: PIV Measurements and Comparison to Synthetic Jet Actuator. *Actuators* **2022**, *11*, 205. https://doi.org/10.3390/act11080205

Academic Editors: Luigi de Luca, Hui Tang and Xin Wen

Received: 23 June 2022
Accepted: 17 July 2022
Published: 22 July 2022

1. Introduction

Synthetic jets have been widely researched in the past twenty years, and flow control means based on synthetic jet actuators have become a critical tool for the aerodynamicist [1–5]. The SJA is a zero-mass-flux device that converts electrical energy into momentum. A schematic of a traditional SJA and the schlieren flow visualization are shown in Figure 1 [6]. The traditional SJA is composed of a slot and an enclosed cavity, bounded by one or two piezoelectric (PZT) diaphragms. A piezoelectric element can vibrate by applying an alternating voltage to it, and the outside air is periodically pushed out or sucked into the chamber through the orifice with a rapid response. The SJA is deemed to be a "zero-mass-flux" device, because the net mass flowing from the jet to the outside air is zero. The mass of the drawn air is equal to the mass of the ejected air. As an example: imagine that you are sitting on a frictionless surface. Make an "O" shape with your mouth, and then breathe in. You do not move. Breathe out through your mouth, keeping an "O" shape, and you will move in the opposite direction to which you are breathing, based on Newton's Third Law of Motion. If you repeat these processes several hundred or thousand times per second, you will be an SJ. In fact, the SJA, with its advantages of simple structure, low cost, easy operation, no need of an air source and pipes, could establish an SJ more rapidly than other actuators that generate a steady jet or a pulsed jet. The distinctive characteristics of the synthetic jet flow field and the good working attributes of actuators make the SJA suitable for a variety of applications, such as active flow control, flight control, and thermal management in microelectronic devices [6–17].

The traditional SJA, however, with one side of the PZT diaphragm exposed to the air drawn from the base flow (BF) and the other surface exposed to the environment flow (EF), will have two problems of pressure failure and inefficient energy utilization, as shown

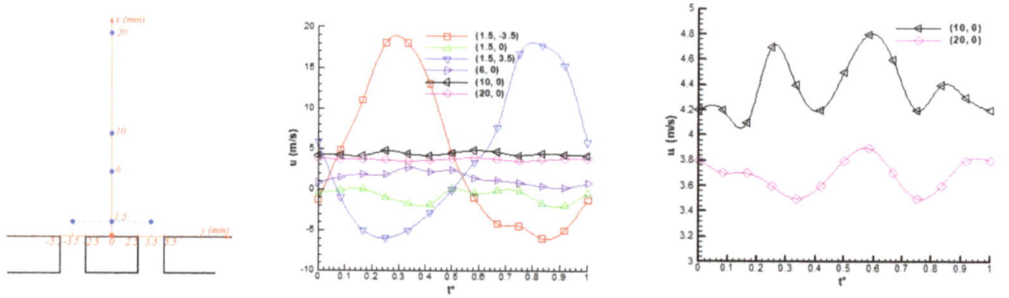

(**a**) Locations of six points (**b**) Results of the six measured points (**c**) Detailed results on two points

Figure 7. Time-locked velocity traces on six measured points per unit circle.

Figure 6 also shows the vorticity magnitude and two counter-rotating vortex pairs produced by the DSJA. The vorticity of the vortex pairs contributes to the pressure difference. An "induced" velocity V, a translation velocity with which the vortex ring travels, is given by Hill and Saffman [15] as follows:

$$V = \frac{\Gamma}{2\pi a}\left[\ln\frac{8a}{(4\nu t)^{1/2}} - 0.558 + O\left(\frac{\nu t}{a^2}\right)^{1/2}\right] \quad (5)$$

where Γ is the circulation and a is the radius of the vortex ring. The vortices decelerate not only by losing energy due to friction, but also through entrainment of fluid from the environment. The circulation $\Gamma \propto \omega$ and the "induced" velocity V contributes to the pressure difference. The pressure in the area of the main vortex pairs is lower than in other flow domains. The stronger the vortex ω, the stronger the circulation Γ, the higher the velocity V, and the lower the pressure will get. Equation (5) also shows that the velocity V is strongly associated with the time t, indicating that the phase angle also contributes to the "induced" velocity V and the pressure difference. The results show that in the far field downstream, the fore vortex pairs entrain the aft vortex pairs, and the aft vortex pairs are deflected to the fore vortex pairs due to the lower pressure. Meanwhile, the fore vortex pairs will decelerate gradually by losing energy due to friction and entrainment of fluid from the environment, and the aft vortex pairs will also impact the fore vortex pairs. Lastly, a single, more stable synthetic jet could be realized by the merging of the two synthetic jets. Figure 7 also indicates that the two jets of the DSJA could merge into a single, more stable synthetic jet in the far field. Furthermore, Figure 7b,c shows that the DSJ downstream has a double characteristic frequency, which broadens the applying frequency domain.

The detailed comparisons between the DSJ and the SJ are shown in Figures 8–10. In the near-field downstream, the flow fields controlled by the DSJ and the SJ similarly emerge into four stages (accelerating blow, decelerating blow, accelerating suction and decelerating suction), while the flow characteristic of the DSJ is more complex than that of the SJ. Moreover, the blowing or sucking peak velocity of the DSJ is larger, showing a higher control ability. It is worth noting that in the mode of the DSJ, the left peak velocity is slightly lower than the peak velocity of the right exit, which may result from the preload differences of both sides. Considering that the PZT diaphragm is fixed by bolts, the clamping conditions on both sides of the diaphragm may be different, resulting in the inconsistent vibrating amplitude in both directions of the PZT diaphragm, thus showing the different velocity evolution. Furthermore, the two synthetic jets of the DSJ could merge into a single, more stable jet with a higher velocity in the far field downstream.

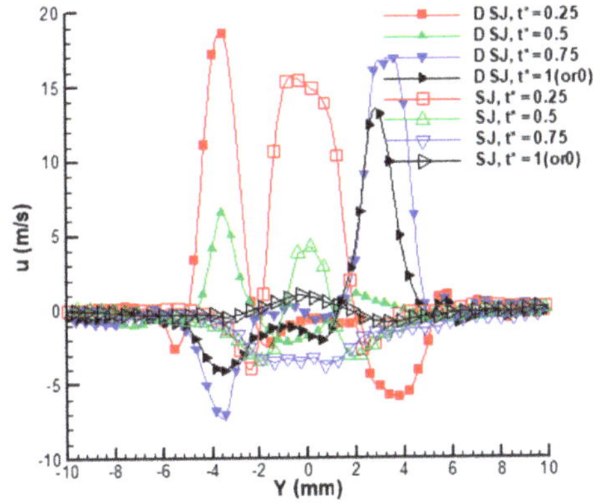

Figure 8. Streamline velocity of DSJ and SJ on x = 1 mm at different times.

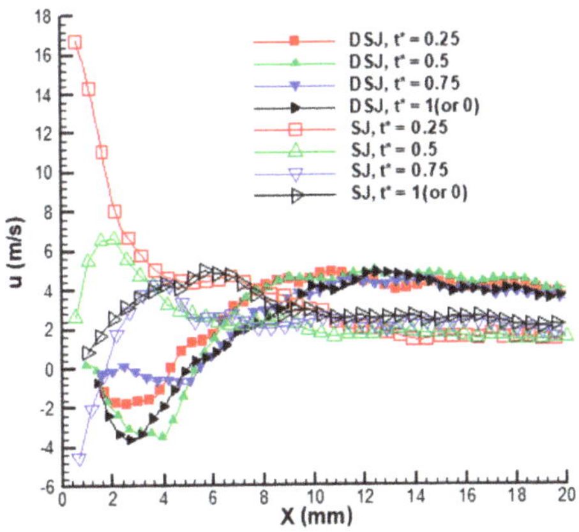

Figure 9. Streamline velocity of DSJ and SJ on y = 0 at different times.

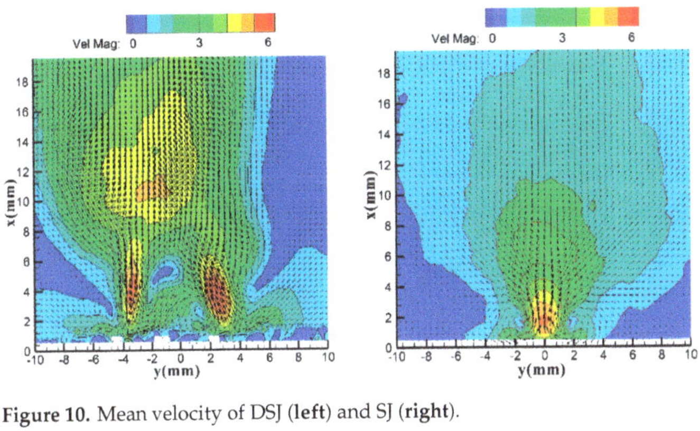

Figure 10. Mean velocity of DSJ (**left**) and SJ (**right**).

Based on the above analysis, the structure of the two cavities sharing a single diaphragm ensures that the DSJA not only has the unique property of zero mass flux, but also solves the problems of pressure failure and inefficient energy utilization existing in a traditional SJA, which extends the applications of the DSJA to more flow control systems, such as thrust vectoring control and flight control [23,24].

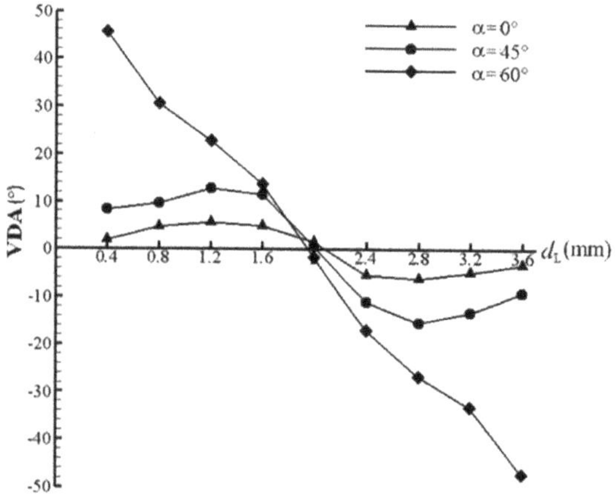

Figure 13. VDA measurements of DSJA with different α and d_L.

4. Conclusions

To understand the detailed discrepancy between the DSJ and the SJ in quiescent surroundings, a novel DSJA was investigated using PIV, and its flow-field characteristics were compared with the traditional SJA. A transfer-phase and sub-frequency technique was introduced to capture the arbitrary phase of the DSJ, and a transfer-phase-to-equal technique was provided to determine the phase of the DSJ. In addition, the unique vectoring characteristics of the DSJA with a variable α and d_L were also explored.

The results indicate that a cycle of the DSJ can be divided into a "left" stage and a "right" stage. In the "left" stage, the flow field is dominated by the left jet, while the flow field is dominated by the right jet in the "right" stage. In the near-field downstream of the DSJA, the two jets entrain air around them and interact with each other, and the flow field is more complex than that of the SJA. There is an unfavorable phenomenon of "self-support" between the two synthetic jets, weakening the strength of the DSJ downstream. In the-far field downstream of the DSJA, the two vortex pairs interact with each other and entrain fluid from the surroundings, and the DSJs merge into a single, more stable synthetic jet (similar to a steady jet) with a higher velocity and a double characteristic frequency. In addition, the DSJA has shown great vectoring characteristics, with the VDA changing from approximately $-46°$ to $46°$, suggesting a broader applying prospect. These results indicate that the traditional SJA may be replaced by the DSJA in many applications, and the applying area of the synthetic jets could be extended to more active flow control systems that cannot be qualified by the traditional SJA.

In this paper, the comparison of the DSJ and the SJ in quiescent environments is explored, but its evolution process under the condition of incoming flow was not shown, which may have more engineering significance. Therefore, further work will focus on the comparison of the evolution process with the incoming flow, and wing tunnel tests will also be implemented.

Author Contributions: Conceptualization, Z.L. and Z.X.; methodology, Z.Z. and L.W.; formal analysis, Z.Z. and Z.L.; writing—original draft preparation, X.D. and Z.Z.; writing—review and editing, Z.L. and Z.Z. All authors have read and agreed to the published version of the manuscript.

Funding: The authors are grateful to the National Nature Science Foundation of China for its financial support for this project (Grants: U2141252, 11972369, 11872374).

Conflicts of Interest: The authors declare no conflict of interest.

Nomenclature

D	Diameter of cylindrical chambers
H	Height of cylindrical chambers
U_A	Driving voltage amplitude
f	Driving frequency
h_t	Thickness of rigid wall
l	Length of slots
h	Width of slots
d	Distance between the two slots
Re_u	Reynolds number based on jet exit
St	Strouhal number
w	Vorticity
α	Exit chamfer
ρ	Air density
μ	Air viscosity
u_{amp}	Velocity amplitude of SJ
L_0	Stroke length of SJ
Re_{I0}	Re based on the blowing phase per unit width
$u_0(t)$	Instantaneous centerline velocity at the actuator slot exit
T	Time of a DSJ period
t^*	Dimensionless time
V	"Induced" velocity
Γ	Circulation
a	Radius of the vortex ring
d_L	Length of left exit

References

1. Glezer, A.; Amitay, M. Synthetic jets. *Annu. Rev. Fluid Mech.* **2002**, *34*, 503–529. [CrossRef]
2. Luo, Z.B.; Xia, Z.X. Advances in synthetic jet technology and applications in flow control. *Adv. Mech.* **2005**, *35*, 221–234. (In Chinese)
3. Zhang, P.F.; Wang, J.J.; Feng, L.H. Review of zero-net-mass-flux jet and its application in separation flow control. *Sci. China Ser. E Technol. Sci.* **2008**, *51*, 1315–1344. [CrossRef]
4. Wen, X.; Tang, H.; Liu, Y.Z. Interaction of twin synthetic jets in attached and separated boundary layers: Effects of yaw angle and phase difference. *J. Vis.* **2018**, *21*, 949–963. [CrossRef]
5. Wen, X.; Tang, H.; Duan, F. Interaction of in-line twin synthetic jets with a separated flow. *Phys. Fluids* **2016**, *28*, 043602. [CrossRef]
6. Smith, B.L.; Glezer, A. The formation and evolution of synthetic jets. *Phys. Fluids* **1998**, *10*, 2281–2297. [CrossRef]
7. Singh, P.K.; Renganathan, M.; Yadav, H.; Sahu, S.K.; Upadhyay, P.K.; Agrawal, A. An experimental investigation of the flow-field and thermal characteristics of synthetic jet impingement with different waveforms. *Int. J. Heat Mass Transf.* **2022**, *187*, 122534. [CrossRef]
8. Kim, M.; Essel, E.E.; Sullivan, P.E. Effect of varying frequency of a synthetic jet on flow separation over an airfoil. *Phys. Fluids* **2022**, *34*, 015122. [CrossRef]
9. Xu, X.P.; Zhou, Z. Analytical study on the synthetic jet control of asymmetric flow field of flying wing unmanned aerial vehicle. *Aerosp. Sci. Technol.* **2016**, *56*, 90–99. [CrossRef]
10. Trávnícek, Z.; Tesar, V. Annular synthetic jet used for impinging flow mass-transfer. *Int. J. Heat Mass Transf.* **2003**, *46*, 3291–3297. [CrossRef]
11. Pitital, I.; Rajnish, N.S. Large Eddy simulation of a NACA0015 circulation control airfoil using synthetic jets. *Aerosp. Sci. Technol.* **2018**, *82–83*, 545–556.
12. Wen, X.; Tang, H.; Duan, F. Vortex dynamics of in-line twin synthetic jets in a laminar boundary layer. *Phys. Fluids* **2015**, *27*, 083601. [CrossRef]
13. Rice, T.T.; Amitay, M. Forced Flow Reattachment Process and the Effect of Pressure Gradient. *AIAA J.* **2019**, *57*, 2795–2807. [CrossRef]
14. Xia, H.; Qin, N. Detached-eddy simulation for synthetic jets with moving boundaries. *Mod. Phys. Lett. B* **2005**, *28*, 1429–1434. [CrossRef]
15. Hill, D.J.; Saffman, P.G. Counter-rotating vortex patches in shear: A model of the effect of wind shear on aircraft trailing vortices. *Proc. R. Soc. Lond. A* **2002**, *458*, 1527–1553. [CrossRef]
16. Wang, Q.T.; Cheng, K.M.; Gu, Y.S.; Li, Z.Q. Continuous control of asymmetric forebody vortices in a bi-stable state. *Phys. Fluids* **2018**, *30*, 024102. [CrossRef]

and the upstream convergent section instead of air supply. This improvement reduces the layout difficulty of using a dual-throat vectoring nozzle and achieves an equivalent or even better performance than the conventional dual-throat vectoring nozzle. In this study, the realizable $k-\varepsilon$ turbulent model with the standard wall function is selected.

Meanwhile, predecessors have developed some FTV methods with no dependence on air source, called passive FTV methods. Using an ejecting-mixing fluidic vectoring nozzle [10,11] takes advantage of the entrainment effect of the primary jet on the surrounding air. In detail, the deflection angle of the primary jet could be controlled by adjusting the mass rate of the entrainment secondary jet through the control valves. The results of a supersonic primary jet numerical simulation [10], a flight test powered by electric duct fan [12], and a 6 kg turbojet engine ground experiment [13] show that this method is suitable to control the direction of subsonic and supersonic primary jets.

As one of the lightest active-flow control means, synthetic jet (SJ) actuators have been widely applied in many fields, such as separation control [14], vortex-induced vibration control [15], and flight control [16]. Researchers have found that with the help of zero-mass-flux synthetic jets, the direction of the primary jet can be deflected even without any moving components. SJ actuators can be driven by piezoelectric (PZT) diaphragms or electrodes fixed in a cavity, capable of switching work state rapidly in response to different electrical signals.

Smith [17,18] found that a synthetic jet actuator installed close to the primary jet outlet could deflect the primary jet with a low momentum coefficient. However, the deflection becomes undetectable when the velocity of the primary jet increases to 30 m/s. Pack [19] carried out an experimental study with a diffuser attached to the jet exit, and the excitation was introduced at the junction between the jet exit and the diffuser inlet, demonstrating that the presence of the wide-angle diffuser increases the effectiveness of the added periodic momentum caused by a favorable interaction among the excitation, the jet shear layer, and the diffuser wall. However, Pack [19] and Chiekh [20] found that the direct deployment of the diffuser wall will lead to no-control primary flow into a bistable situation, which is intolerable in an FTV system as it will result in an unstable control moment.

The PZT diaphragm on the conventional SJ actuators is set between different pressure conditions of basic flow (the flow field to be controlled) and environment flow (usually the equipment cabin), which indicates that half of the radiation energy of the vibrating diaphragm is injected into somewhere unwanted. It also means that the PZT must overcome the pressure difference on two sides of the diaphragm to vibrate. When the pressure differential is large, the driving vibration will be difficult or even unachievable. In short, the structural design of conventional PZT-driven SJ actuators limits their energy efficiency and environmental adaptability. Subsequently, Luo [21] invented a new-generation synthetic jet actuator, known as a dual synthetic jet actuator (DSJA). Different from conventional SJ actuators, DSJ actuators have two cavities on both sides of the PZT diaphragm. Both cavities are connected to the controlled flow field, which guarantees that the pressure difference between the two sides of the PZT diaphragm will not be too large. Therefore, the energy efficiency and environmental adaptability of DSJ actuators are improved. Moreover, DSJ actuators have greater flexibility and stronger control ability. For example, DSJ could perform both functions of "push" and "pull" to vector a primary jet [22]. However, according to the latest explorations, DSJ actuators or SJ actuators are only capable of deflecting a primary jet with low velocity and small height.

To sum up, plenty of significant studies have investigated the feasibility of synthetic jet application to FTV. As a control method without introducing an air source, FTV by DSJ is a technology worthy of further in-depth research and development. However, existing studies have shown that the simple deployment of DSJ is insufficient to provide effective FTV control. This work explores the potential of dual synthetic jets for thrust vector control by proposing a novel method and its configuration.

2. Materials and Methods

2.1. Method and Configuration

The configuration of this novel FTV method is shown in Figure 1. The new method was created from the reference and improvement of existing DSJ actuators jet-deflecting method and co-flow FTV nozzle. The height of the primary jet inlet is 50 mm. Avoiding the problems of pressure loading and energy inefficiency, DSJ actuators were selected (comparing to SJ actuators) as the provider of zero-mass jet to adapt to the changing external environment in flight. In this configuration, the DSJ actuators are installed on the edge of the primary nozzle outlet, with a total thickness of 10 mm. To prevent the unexpected deflection from damaging the practicability of the nozzle, passive secondary jet inlets are arranged as gaps between the primary jet inlet wall and the nozzle wall. The gaps are connected to the outside atmosphere and are approximately 8 mm high at the downstream edge. The exits of DSJ actuators are arranged at the two downstream corners of the edge and are set 45 degrees up and down from the mainstream direction, respectively.

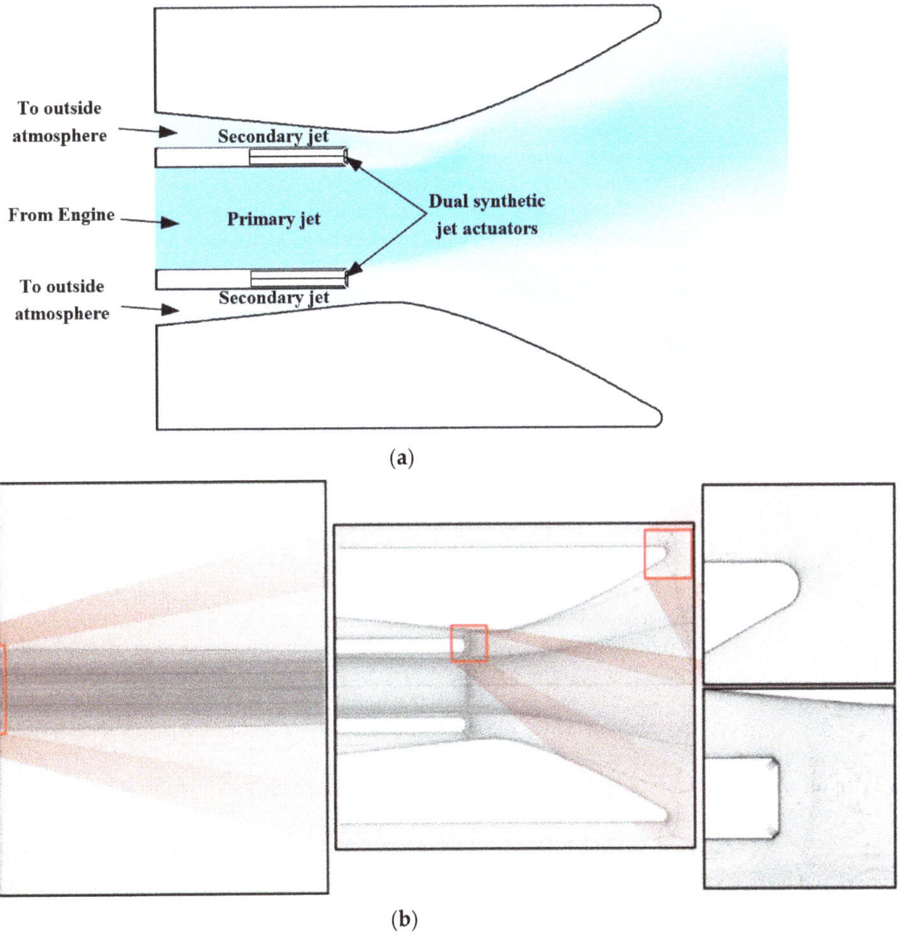

Figure 1. Configuration design (**a**) and numerical calculation mesh configuration (**b**).

(including DSJ), considering air as an incompressible gas, the momentum coefficient can be defined as:

$$C_\mu = \frac{n \times V_j^2 h_j}{V_p^2 h_p} \tag{6}$$

where V_j is the peak velocity of synthetic jets, h_j is the height of dual synthetic jet exits, V_p is the velocity of the primary jet and h_p denotes the height of the primary jet outlet. n denotes the number of zero-mass synthetic jets activated for control and n = 2 for a single DSJ actuator.

3. Results

3.1. Control Characteristics of Thrust Vectoring

To illustrate the effectiveness of DSJ in this configuration, Figure 4 shows the velocity contours of the flow field under different conditions. The primary jet velocity V_p is 100 m/s (based on the height of primary jet inlet, corresponding to Re = 3.9 × 10^4). The DSJ peak velocity V_j in Figure 5b is set as 70 m/s and its frequency f is 500 Hz (based on the height of primary jet inlet, corresponding to St = 0.25).

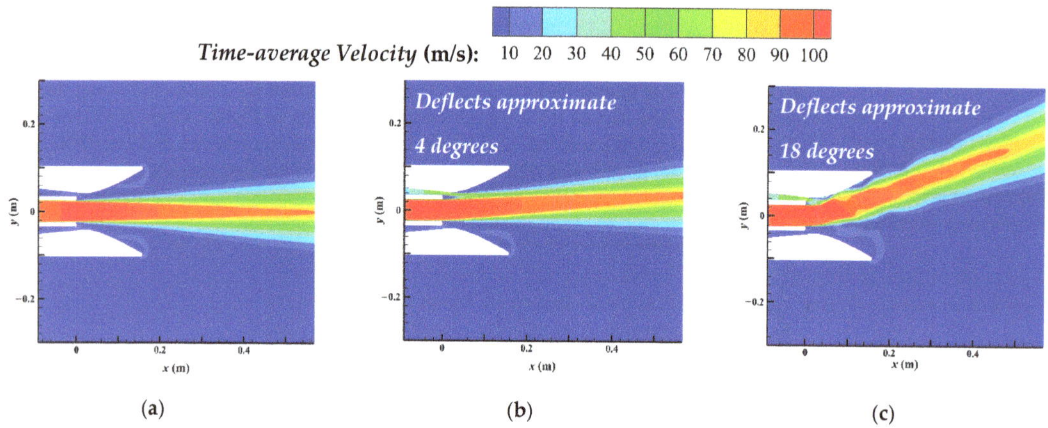

Figure 4. Velocity contours of flow field with no control (**a**), co-flow effect (**b**) and DSJ control (**c**) (V_p = 100 m/s).

The activation of DSJ makes the primary jet deflect upward; meanwhile, the secondary jet is accelerated. The primary jet finally gets a stable vector downstream. The primary jet is vectored upside with an eventual mean angle of approximately 18 degrees. The secondary jet will attain a mass flow of 0.541 kg/s by the ejection of the vectored primary jet and DSJ. As a contrast, injecting a secondary jet with the same mass flow can lead to a deflection of 4 degrees alone by the co-flow effect, as shown in Figure 4b. In this configuration, the properties of DSJ are critical in the deflection of the primary jet. Meanwhile, through the ejection effect enhanced by DSJ, the deflected primary jet activates the secondary jet to be a co-flow, offering an improvement in deflection performance.

To investigate the control characteristics of this configuration, simulations are implemented with different DSJ peak velocities (10 m/s, 20 m/s, ... , 100 m/s). The velocity distribution, deflection angles, and velocity distribution, are shown and analyzed as follows.

Figure 5 shows the time-average velocity distribution of controlled flow with different DSJ peak velocities. The DSJ frequency f is maintained at 500 Hz in this work. As shown in the velocity contours, the vectoring becomes observable when peak velocity increases to 30 m/s and increases as DSJ peak velocity increases.

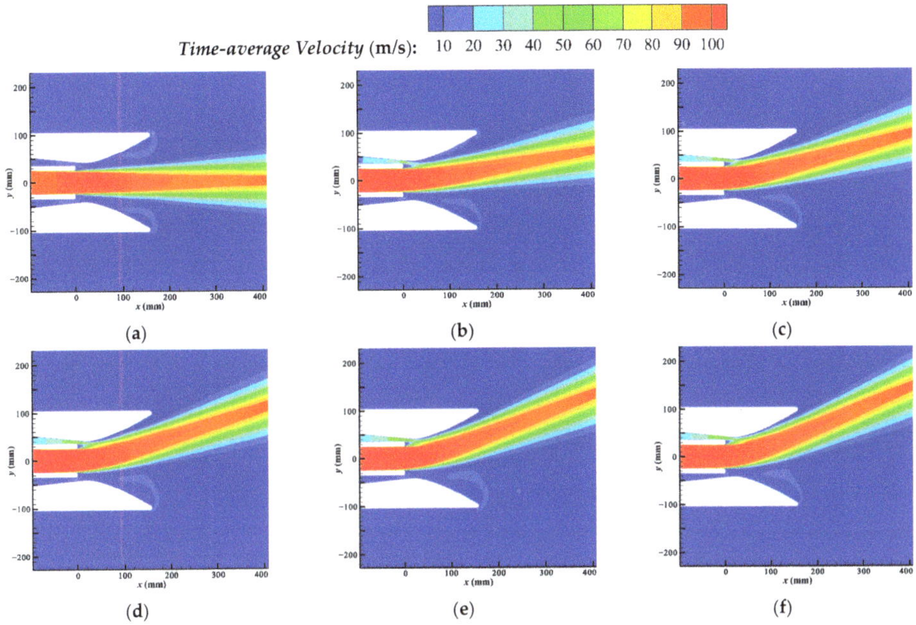

Figure 5. Time-average velocity contours of controlled flow with DSJ of different intensities. (**a**) V_j = 10 m/s (C_μ =0.04%), (**b**) V_j = 30 m/s (C_μ =0.36%), (**c**) V_j = 50 m/s (C_μ =1.00%), (**d**) V_j = 70 m/s (C_μ =1.96%), (**e**) V_j = 90 m/s (C_μ =3.24%), (**f**) V_j = 100 m/s (C_μ =4.00%).

Figure 6 shows the variation of the deflection angle with DSJ peak velocity and the momentum coefficient. To most FTV means, the changing law between the deflection angle and DSJ peak velocity was supposed to be characterized by a dead zone, a linear control region, and saturation (according to Mason [5]). In this configuration, the establishment of deflection occurs when DSJ peak velocity is less than 20 m/s, and the control characteristic remains linear with peak velocity until the deflection angle reaches 23 degrees. Considering that the geometric exit angle of the nozzle is 28 degrees (which limits the maximum of the deflection angle), it shows a comparatively linear control characteristics in almost all control ranges. The highest efficiency occurs at C_μ, ranging from 0 to 0.5%, and the deflection angle reaches 18 degrees at C_μ of 1.96%. This indicates that this configuration has good efficiency performance in FTV control.

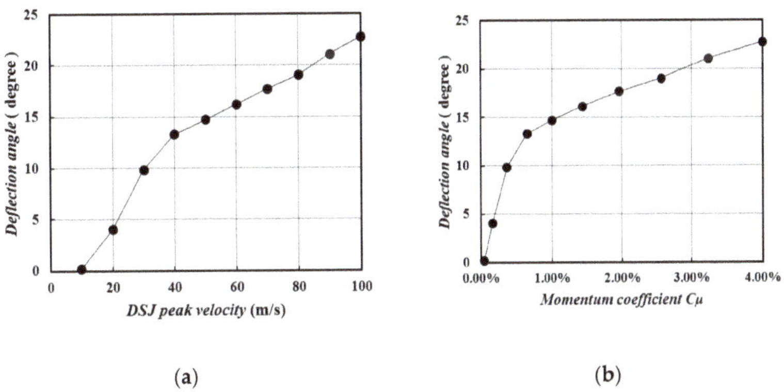

Figure 6. Variation of deflection angle with DSJ peak velocity (**a**) and DSJ momentum coefficient (**b**).

3.2. Pressure Characteristics of Flow Field

This section investigates the controlled flow in pressure characteristics. The time-average pressure distributions of the controlled flow with different DSJ peak velocities are shown in Figure 7.

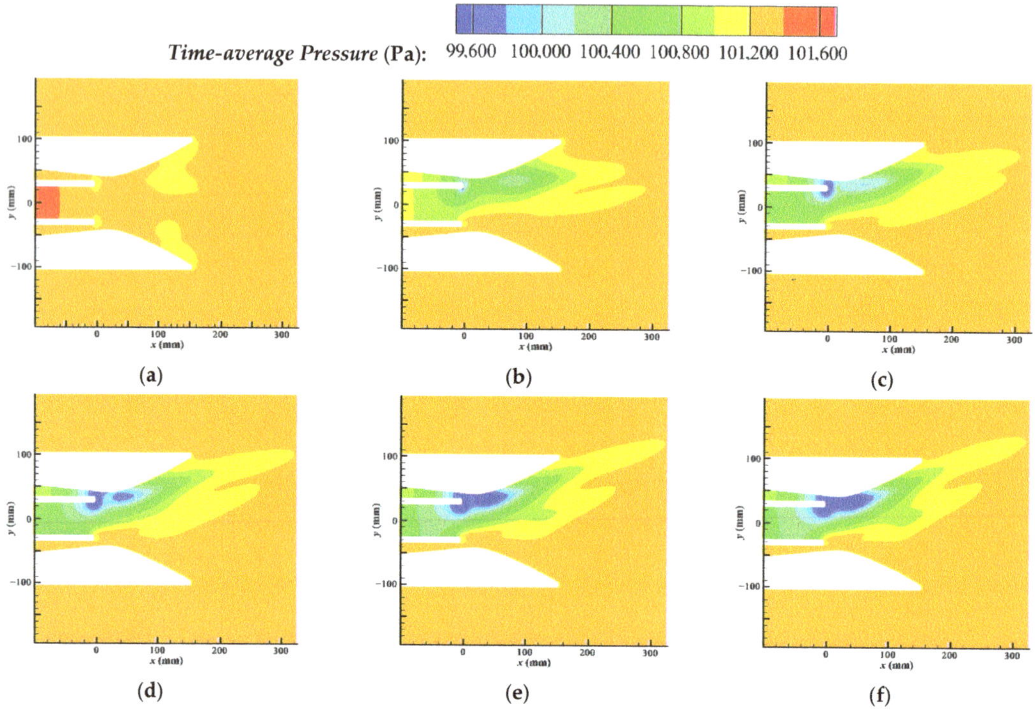

Figure 7. Time-average pressure contours of the controlled flow with DSJ of different intensities. (**a**) $V_j = 10$ m/s (C_μ =0.04%), (**b**) $V_j = 30$ m/s (C_μ =0.36%), (**c**) $V_j = 50$ m/s (C_μ =1.00%), (**d**) $V_j =$ 70 m/s (C_μ =1.96%), (**e**) $V_j = 90$ m/s (C_μ =3.24%), (**f**) $V_j = 100$ m/s (C_μ =4.00%).

Based on time-average pressure, the iconic flow field structure is the low-pressure zone formed by DSJ. There are two cores of the low-pressure zone, one near the DSJ actuator exits and the other located at the wall of the throat. The greater the DSJ peak velocity, the larger the low-pressure zone. In addition, DSJ also reduces the pressure in the primary jet outlet. In Figure 8, the upper surface and outlet section are selected to visually compare the pressure change under different peak velocities.

Corresponding to two low-pressure core areas, two peaks of low-pressure zones along the nozzle wall are shown in Figure 7. Peak 1 is close to the exits of DSJ, directly caused by DSJ. With the augmentation of DSJ peak velocity, the growth rate of the low pressure in peak 1 decreases gradually, and the position of peak 1 moves downstream slightly. Peak 2 of the low-pressure zone appears on the wall of the throat, further downstream of peak 1.

With small DSJ velocity (below 50 m/s), the augmentation of peak 1 is significant, while the primary jet is far from the wall and peak 2 is not yet established. When the DSJ peak velocity is above 60 m/s, peak 2 begins to be established with an obvious augmentation while the augmentation of peak 1 is descending. This shows the significant effect of the wall on deflection when the primary jet is close enough to the wall. Eventually, the establishment of a low-pressure zone maintains a relatively stable level in all deflection angles. This indicates that the direct effects of DSJ (corresponding to peak 1) and the evolution of flow of the controlled primary jet (corresponding to peak 2) provide comparable contributions in deflection.

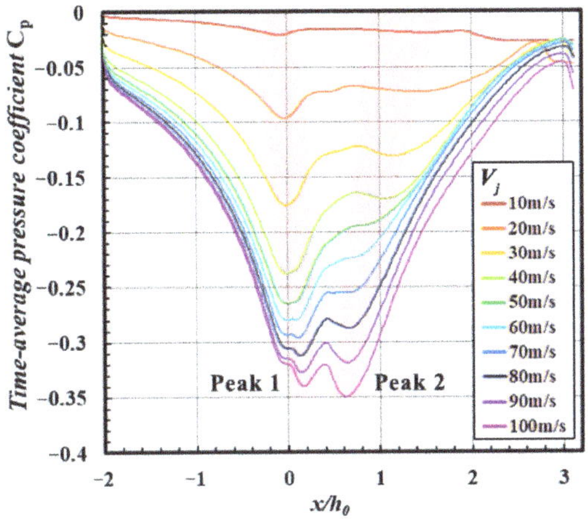

Figure 8. Time-average pressure coefficient C_p distributed along the upper wall.

3.3. Periodic Evolution of Flow Field

To investigate the evolution of controlled flow, the characteristics of flow field in a DSJ cycle are explored. Figure 9 shows the velocity and pressure distribution of the flow field near the outlet at four typical moments (0T, T/4, T/2, 3T/4) of the DSJ cycle, starting from 0.1 s. To control flow, at the T/4 moment, the upper exit of DSJ is blowing at peak velocity, while the lower exit is sucking at peak velocity. Parts of the primary jet are absorbed in the DSJ cavity, and a strong low-pressure zone is formed downstream, which causes a substantial deflection of the primary jet. The blowing at the upper exit does not have a significant impact at this stage. At the T/2 moment, the upper exit finishes blowing and is going to absorb, while the lower exit is in the opposite situation, and the velocity of both exit inlets is almost zero. The primary jet is deflected at the previous moment develop to attach to the wall and the low-pressure zone generated at previous moment extends in the primary jet tunnel. At the 3T/4 moment, the upper exit of DSJ is absorbing at peak velocity, while the lower exit is blowing at peak velocity. The primary jet deflected at the moment of T/4 begins to attach to the wall. The pressure rises at the attachment area as the wall blocks the y-direction velocity of the primary jet, indicating that the primary jet is continuously deflecting when passing through the low-pressure zone. The low-pressure zone extends and moves downstream. The secondary jet is accelerated by the ejecting of the attached primary jet and the absorbing of the upper exit. An adverse pressure gradient occurs between the low-pressure zone and high-pressure attachment, and a separation vortex occurs at the throat. At the 0T moment, the upper exit finishes absorbing and is going to blow, while the lower exit is in the opposite situation. The separation vortex moves downstream. Throughout the cycle of flow-field evolution, the low-pressure generated by DSJ, the ejecting enhanced by DSJ, and the co-flow effect produced by the accelerated secondary jet all seem to play roles in the deflection of the primary jet.

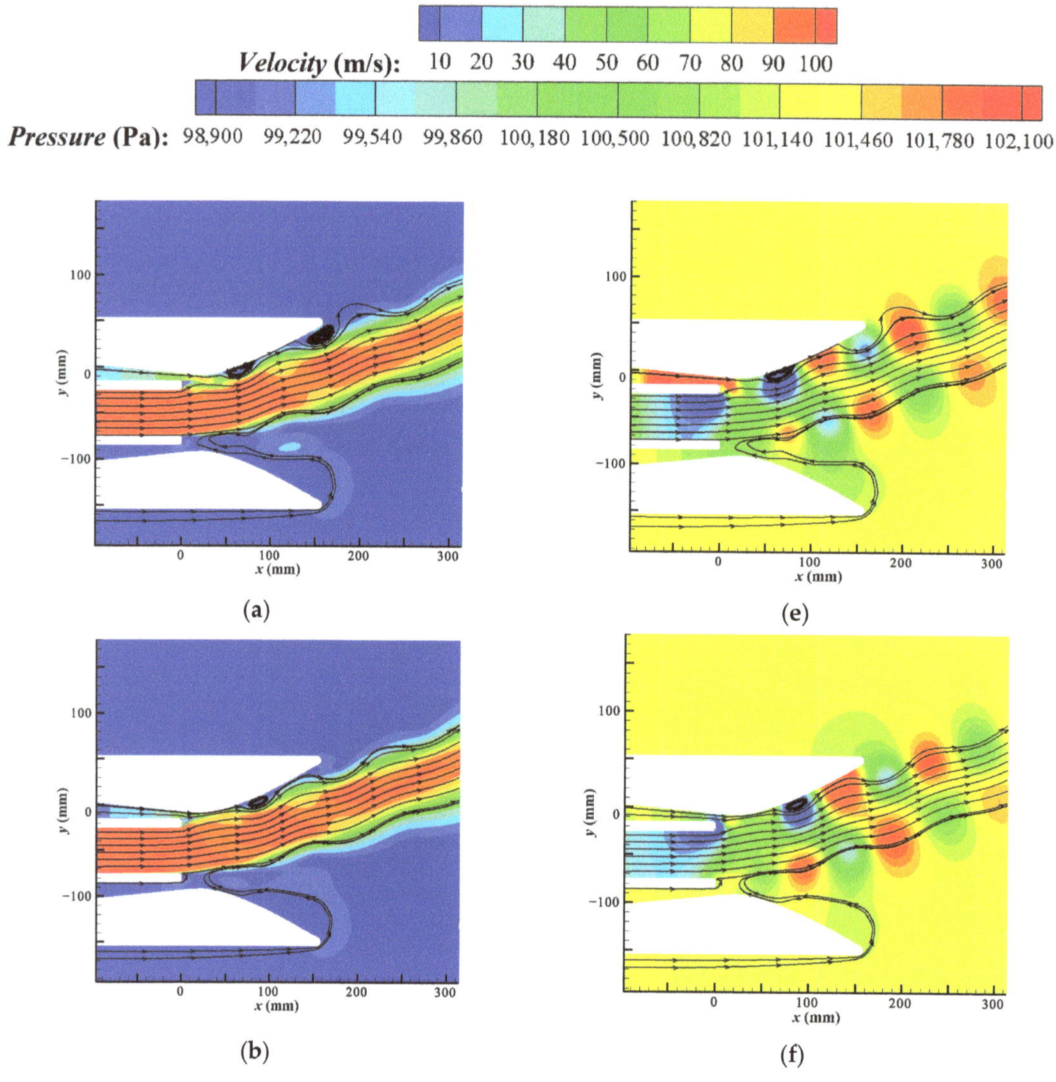

Figure 9. *Cont.*

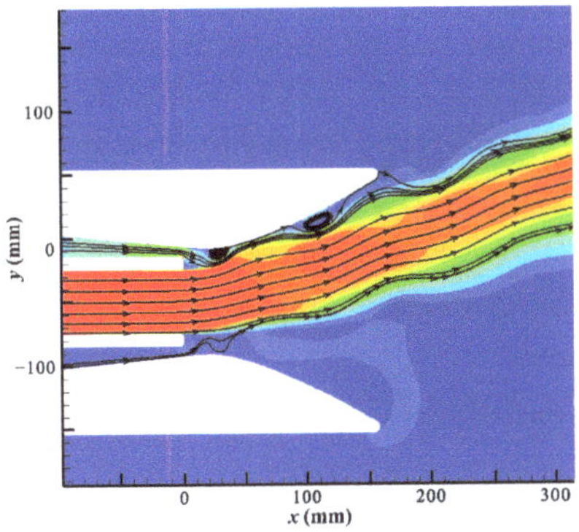

(c)

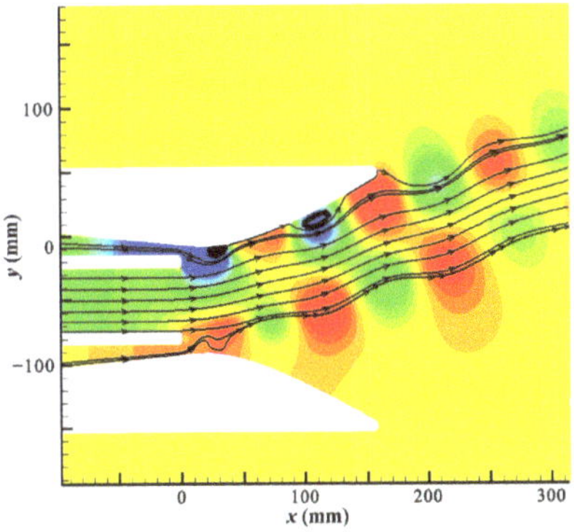

(g)

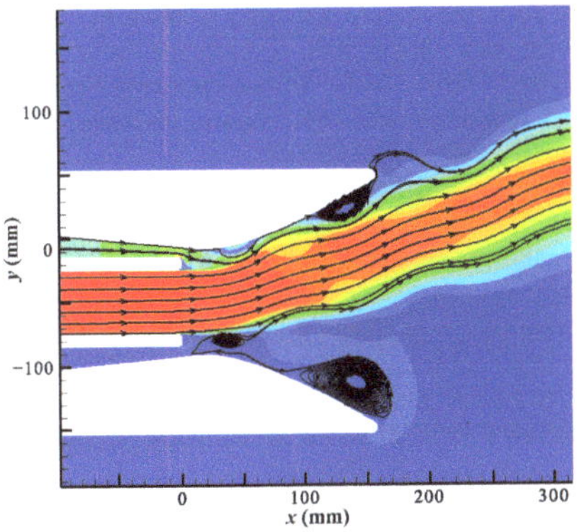

(d)

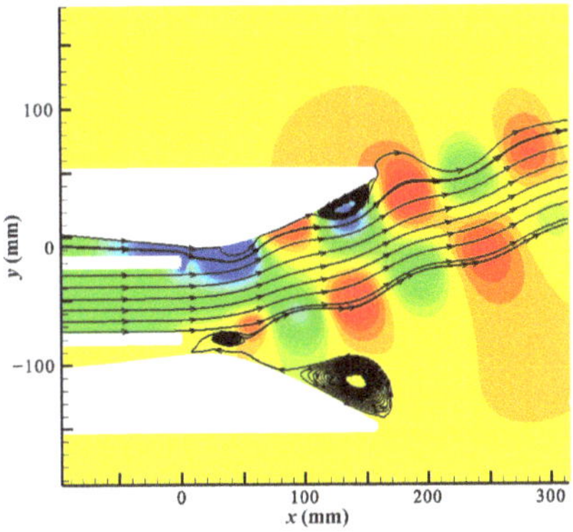

(h)

Figure 9. Periodic variation of velocity and pressure near the DSJ (V_j = 70 m/s). (**a**) T/4 velocity, (**b**) T/2 velocity, (**c**) 3T/4 velocity, (**d**) 0T velocity, (**e**) T/4 pressure, (**f**) T/2 pressure, (**g**) 3T/4 pressure, (**h**) 0T pressure.

4. Conclusions

A novel method of thrust vectoring based on DSJ and its configuration are proposed and evaluated by simulations. Different from previous schemes, it has two passive secondary flow outlets to stabilize the primary jet, and a finely designed nozzle wall geometric shape to optimize deflection capability. Numerical results reveal that this configuration has a high efficiency in deflecting a primary jet with a velocity of 100 m/s and a height of 50 mm and shows comparatively linear control characteristics in almost all control ranges.

For the momentum coefficient C_μ, the highest efficiency occurs at C_μ, ranging from 0 to 0.5%, and the deflection angle reaches approximate 18 degrees at a C_μ of 1.96%. The low pressure generated by DSJ, the ejecting enhanced by DSJ, and the co-flow effect produced by accelerated secondary jet all seem to play roles in the deflection of the primary jet.

Since the primary jet is strong enough to drive a small aircraft, the potential of dual synthetic jets to provide FTV control is initially revealed. The demonstrated control capability of DSJ makes it possible to provide thrust vector control without any moving parts or air supply. The detailed structure of flow field and the mechanism of this method need to be further studied.

In the future, relative experimental observations will be conducted to further investigate this mechanism, and flight tests will be implemented to further verify its effectiveness and control characteristics.

discovered that a secondary vortex with an opposite circulation was generated after the vortex pair arrived at the heated wall. Greco et al. [25] investigated the combined effect of impingement distance and stroke length on the cooling performance of an impinging synthetic jet. Both inner and outer ring-shaped regions occurred at a short impingement distance under the highest dimensionless stroke length. As the impingement spacing increased, the outer ring-shaped regions disappeared owing to the weakening of secondary vortex rings. Qiu et al. [26] analyzed the effect of the interactions between a circular synthetic jet and a cross flow in a microchannel on the heat transfer process. They pointed out that the heat transfer enhancement could be divided into two parts of an impinging region and an entraining region. Lyu et al. [27] investigated the heat transfer characteristics of a single synthetic jet with two planar-lobed orifices impinging on flat and concave surfaces. It was demonstrated that the N = 6 petal-shaped orifice was a most promising orifice configuration. For the baseline round orifice, the spatially-averaged Nu was reduced up to 25% on the concave surface with respect to the flat surface. Lau et al. [28] analyzed the effects of nanoparticle types and volume concentrations on the heat transfer in a three-dimensional microchannel with a synthetic jet based on a Eulerian approach. With the increase in the volume concentration, the viscosity became higher, resulting in a poorer cooling performance. The overall cooling performance of Al_2O_3-water nanofluid was the best at the volume concentration of 5%. Wang et al. [29] investigated the flow characteristics and unsteady heat transfer of noncircular synthetic jets impinging on a heated plate. Compared with the circular orifice, the jet penetration from the square orifice into the wall shear layer was deeper, and the maximum stagnation cooling coefficient increased 42%. The axis-switching phenomenon of elliptic and rectangular orifices enhanced the near-wall mixing and the turbulent kinetic energy. The heat transfer was improved compared with the circular orifice. The flow and thermal behavior of impingement synthetic jet driven by sinusoidal and square signals were analyzed by Singh et al. [30]. It was found that the cooling performance in the stagnation region with square signal is 18.94% higher than that with sinusoidal signal, but the averaged Nusselt number with the sinusoidal signal is 20.86% higher.

As above mentioned, the impingement cooling based on a traditional synthetic jet cannot vary the jet direction, and thus the cooling area is certain and limited. It is not applicable to the large-area cooling and the removal of dynamic hotspots. Considering this, vectoring dual synthetic jets (DSJ) was developed realized by controlling the movement of a slider based on a synthetic jet [31–34]. It has been found that DSJ has exhibited an excellent performance in impingement cooling and spray cooling [35–38]. Moreover, vectoring DSJ has been proven to be a useful cooling technology in space-constrained and large-area electronic cooling [39].

The vectoring function and the doubled energy utilization efficiency of vectoring DSJ make it potentially useful for cooling applications. The aim of the present work is to investigate the effects of slider location, driving voltage and driving frequency on the impingement flow field and cooling characteristics of vectoring DSJ. A compact thermal management strategy is provided for space-restricted electronic cooling based on a vectoring DSJ.

2. Experimental Device and Methods

2.1. Experimental Device

The experiment set up, as sketched in Figure 1, included a vectoring DSJ actuator, an impinged plate, a particle image velocimetry (PIV) system, and an infrared camera. The vectoring DSJ actuator driven by a sine signal was installed under the plate with a constant distance of 40 mm. The configurations of two slots are shown in Figure 1a, and each width of two slots could be adjusted by changing the slider location through a step motor (seeing Figure 1b). A stainless-steel foil (196 mm × 100 mm × 0.08 mm) was selected as the impinged plate.

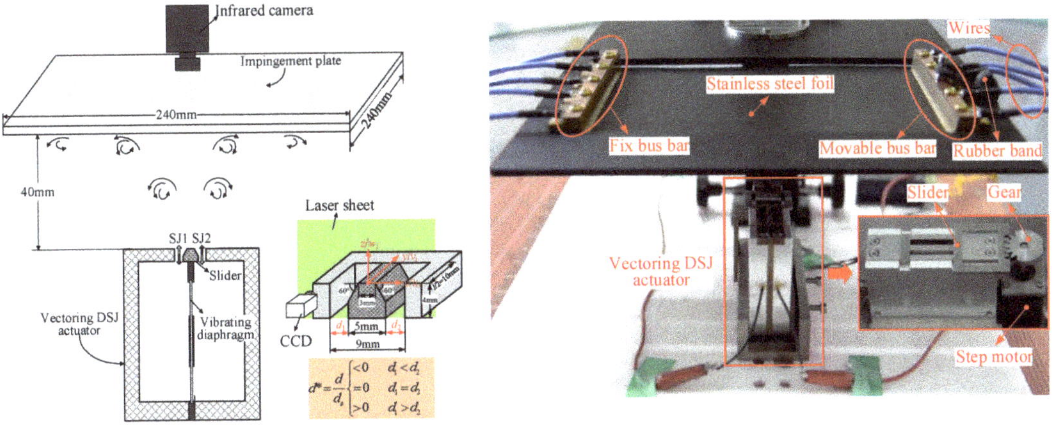

(a) schematic diagram (b) physical photo

Figure 1. Experimental device.

A 2D PIV system of MicroVec Corp. was used to measure the impingement flow field of vectoring DSJ. A sheet light with a thickness of 1 mm was generated from a twin-cavity Nd: YAG laser with a separation of 12 μs. The output energy of each laser pulse was about 200 mJ with a duration of 7 ns. The impingement model of vectoring DSJ was covered in a Plexiglas® container. The *xoz* plane was brightened by a laser sheet of 1 mm thickness at a frequency of 10 Hz. The particle images with a resolution of 0.035 mm/pixel were recorded by a charge-coupled device (CCD) through an optical filter. The cross-frame time was set according to the DSJ velocity. Phase-locked and phase-shifting technologies were used to capture the images at 16 equal-divided phases of the DSJ period. A velocity field has been processed using the MicroVec 3.0 software package based on a two-frame cross-correlation algorithm with an interrogation window of 32 × 16 pixels and 50% overlap. A phase-averaged flow field was calculated according to 24 pairs of velocity fields at each phase point. Then the period-averaged flow field of impingement DSJ was achieved by averaging the 16 phase-averaged flow fields.

A FLIR infrared camera was installed above to measure the temperature distribution on the top surface of the impinged plate. The measurement range was from 5 °C to 2500 °C with an accuracy of ±1 °C. The impinged plate was steadily and uniformly heated by a direct-current source. Since the maximum Biot number was 0.0013, which was much less than the critical value of 0.1 along the thickness direction, the temperature of the top surface and impinged surface at the same location could be deemed to be consistent. To improve the radiation feature and the measurement accuracy, a thin and uniform layer of a black matted paint with an emissivity of 0.95 was painted on both sides of the plate. When the plate reached its thermal equilibrium, fifty thermal images were recorded with a sampling frequency of 20 Hz. The steady-status temperature distribution was obtained by averaging the fifty thermal images.

2.2. Vectoring Evaluation Method of Impinging DSJ

In order to explicitly quantify the vectoring angle of DSJ, an evaluation method was proposed based on the potential core. Firstly, a potential core was selected to cover the vast bulk of jet momentum in an evaluated flow field. Secondly, we considered a line and projected the velocity vectors in the potential core onto the line. An objective function was defined as the sum of the projection values of the velocity vectors in the potential core. Thus, the objective function could be deemed as the impinging strength of the jet momentum in the potential core along the line. Finally, the maximum of the objective function could be achieved for a specific projected line. The inclination angle of the specific line was prescribed as the vectoring angle of flow field. The detailed algorithm is explained as follows.

In this paper, the critical velocity V_c is defined as half of the maximum velocity magnitude in the period-averaged flow field of the vectoring DSJ. The region with velocity magnitude beyond the critical velocity is regarded as the potential core.

$$V_c = \frac{V_{\max}}{2} = \frac{\max(|\boldsymbol{V}_i|)}{2} = \frac{\max(\sqrt{u_i^2 + w_i^2})}{2} \tag{1}$$

where $\boldsymbol{V}_i$ denotes the velocity vectors with components of u_i and w_i along x-axis and z-axis, respectively.

According to the definition above, the objective function S_o is expressed as

$$S_o = \sum_{i=1}^{n} (\boldsymbol{V}_i \cdot \boldsymbol{e}_\theta) \Big|_{|\boldsymbol{V}_i| \geq V_c} \tag{2}$$

where $\boldsymbol{e}_\theta$ is the unit vector along an assuming line with an inclination angle θ with respect to z-axis, depicted as

$$\boldsymbol{e}_\theta = \left(\frac{\tan\theta}{\sqrt{1+\tan^2\theta}}, \frac{1}{\sqrt{1+\tan^2\theta}}\right) \tag{3}$$

Then, the objective function can be written as

$$\begin{aligned} S_o &= \sum_{i=1}^{n} (\boldsymbol{V}_i \cdot \boldsymbol{e}_\theta) \Big|_{|\boldsymbol{V}_i| \geq V_c} \\ &= \left(\sum_{i=1}^{n} \frac{u_i \tan\theta}{\sqrt{1+\tan^2\theta}} + \sum_{i=1}^{n} \frac{w_i}{\sqrt{1+\tan^2\theta}} \right) \Big|_{|\boldsymbol{V}_i| \geq V_c} \\ &= \frac{\tan\theta \sum_{i=1}^{n} u_i + \sum_{i=1}^{n} w_i}{\sqrt{1+\tan^2\theta}} \Bigg|_{|\boldsymbol{V}_i| \geq V_c} \end{aligned} \tag{4}$$

It can be concluded that there must be a specific angle θ_0, at which the objective function reaches its maximum. The mathematical relation is given by

$$(S_o)_{\max} = \lim_{\theta \to \theta_0} S_o \tag{5}$$

As mentioned above, θ_0 represents the direction of the jet momentum in the potential core, and also the vectoring angle of flow field.

2.3. Cooling Performance Evaluation Method of Impinging DSJ

The local cooling performance of impingement DSJ, namely the Nusselt number (Nu), is evaluated by

$$Nu = \frac{h d_e}{\lambda} \tag{6}$$

where λ is the thermal conductivity of air at the average temperature of the impinged surface. The slider width of 5 mm is chosen as the characteristic length d_e. h is the local heat transfer coefficient of impingement DSJ expressed as

$$h = \frac{q_{net}}{T_s - T_j} \tag{7}$$

where q_{net} is the net heat flux removed by impingement jet, T_s and T_j are the local temperature of impingement surface and jet temperature, respectively.

The net heat flux is estimated as follows

$$q_{net} = \frac{Q_{ele} - Q_{loss}}{A} \tag{8}$$

where Q_{ele} is the heating power, Q_{loss} is the total heat loss, A is the effective heating area of impinged surface.

The area-averaged Nusselt number (Nu_{avg}) of the whole impingement surface is calculated by:

$$Nu_{avg} = \frac{\int_{\Omega} Nudxdy}{A_{\Omega}} \tag{9}$$

where A_{Ω} is the image area recorded by the infrared camera.

In the experiments, the heating power is kept a constant of 40 W with a measurement error less than 1.4%. The total loss heat resulted from the radiation and the lateral is estimated to be within 6.4% of the heating power. The uncertainty of Nu is less than 8% through the evaluation method [40].

3. Results and Discussion

3.1. Effect of Slider Location

The slider location determines the area ratio of two slots which influences the momentum ratio of two synthetic jets and the strength of the low-pressure region. Thus, different momentums of two synthetic jets result in the deflection of DSJ. Comparing to a traditional synthetic jet, vectoring DSJ can greatly extend the impingement area. The PIV results with different dimensionless slider location d^*, defined as d/d_e, are exhibited in Figure 2. The driving voltage and driving frequency are set to ±150 V and 450 Hz, respectively.

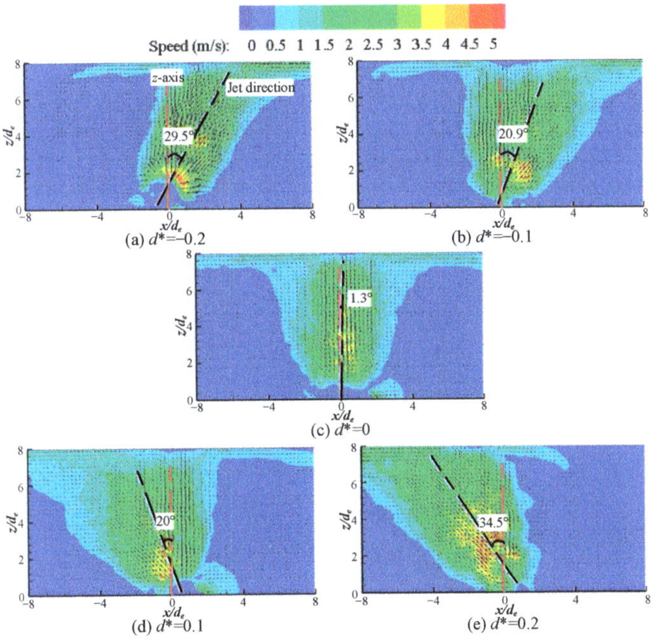

Figure 2. Velocity vector maps, contours, and vectoring angles at different slider locations.

As demonstrated in Figure 2, the DSJ deflects toward the side of the slot with a greater cross section. The reason is that the jet momentum from a smaller width slot is higher, which can be deemed as a primary jet. Conversely, the other jet with a lower momentum is regarded as a disturbing jet. As the primary jet issues from the slot, the other jet is in the suction phase. A low-pressure region is generated near the slot, which causes the primary jet deflects. As the disturbing jet is formed in the blowing phase, the primary jet has moved downstream. The entrainment of vortex structures in the disturbing jet accelerates the deflection of the downstream primary jet. So, the DSJ deflects toward the slot with a larger cross section. With d^* varying from -0.2 to 0.2, the vectoring DSJ can deflect from 34.5° toward the left to 29.5° toward the right. Since the momentums of two synthetic jets are equal at $d^* = 0$, there is no deflection. The vectoring angle of 1.3° at $d^* = 0$ (in Figure 2c) may be attributed to the asymmetry of two slots resulting from machining errors. It is observed that the stagnation region length is about $8d_e$ (from $-4d_e$ to $4d_e$) for the normal DSJ (seeing Figure 2c), while for the vectoring DSJ it increases to $15d_e$ (from more than $-8d_e$ in Figure 2a to $7d_e$ in Figure 2e). Additionally, the jet velocity becomes a little higher at an off-center slider location. It has been proved that the heat transfer coefficient decreases with the increase in the inclination angle of synthetic jet, and reaches the maximum for the normal impingement [41]. Considering the vectoring characteristic, the vectoring DSJ can be obliquely installed to make its impingement normal. This feature will be greatly useful in spaced-constrained cooling applications.

Figure 3 shows the Nu distributions on the impinged surface at different slider locations. The driving voltage and the driving frequency are kept at ±150 V and 500 Hz, respectively. It is indicated that the Nu distribution at $d^* = 0$ is self-symmetrical and two Nu distributions at a same $|d^*|$ are symmetrical about the line $x/d_e = 0$. The high Nu area gradually moves away from the center with the increase in $|d^*|$. The reason is that with the increase in $|d^*|$, the vectoring angle of DSJ becomes larger which causes the stagnation region moving away from the center.

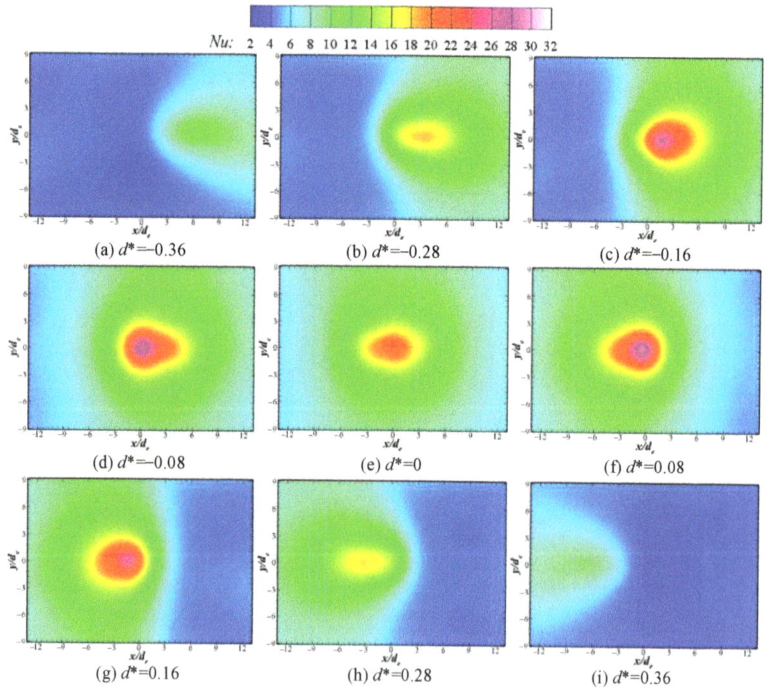

Figure 3. Nu contours on the impingement surface at different slider locations.

Figure 4a draws the Nu on the centerline $y/d_e = 0$ along the x-axis. The solid dot represents the peak Nu on the curves. It can be seen that the curves at the same $|d^*|$ are symmetrical with the line $x/d_e = 0$. With the increase in $|d^*|$, the curves and peak point of

Nu gradually move away from the center, which is consistent with the variation trend of the vectoring angle (shown in Figure 2). The peak Nu reaches its maximum at $|d^*| = 0.08$ rather than at the center location. The reason is that the DSJ velocity increases with the increase in $|d^*|$, which improves the local cooling performance. As $|d^*|$ continues to increase, the thickness of the viscous layer in the slot neck gradually grows and eventually fills the entire neck. As a result, the choking effect occurs in the slot neck and the DSJ velocity reduces. Moreover, the vectoring angle increases with the increase in $|d^*|$ (shown in Figure 2) and the impingement DSJ will gradually transform to a cross flow, resulting in the weakness of the cooling performance. There are three local peaks of Nu_{avg} at $d^* = 0$ and ± 0.28 (shown in Figure 4b). The reason is that at the beginning of $|d^*|$ increase, the DSJ velocity increases, which improves the whole cooling performance. As $|d^*|$ continues to increase, the choking effect impairs the DSJ velocity, and meanwhile the increase in vectoring angle enhances the cross-flow effect of the DSJ. Thus, the whole cooling performance reduces rapidly. Although the maximum values of Nu at $d^* = \pm 0.08$ and ± 0.16 are larger than that at $d^* = 0$ (shown in Figure 4a), the values of Nu_{avg} are smaller (shown in Figure 4b). Since the DSJ velocity $d^* = \pm 0.08$ and ± 0.16 are higher than that at $d^* = 0$ owing to the smaller cross sections, and the local cooling performance. However, the vectoring angles at $d^* = \pm 0.08$ and ± 0.16 enhance the cross-flow effect and reduce the impingement stagnation region of the DSJ, which causes the decrease in overall cooling performance.

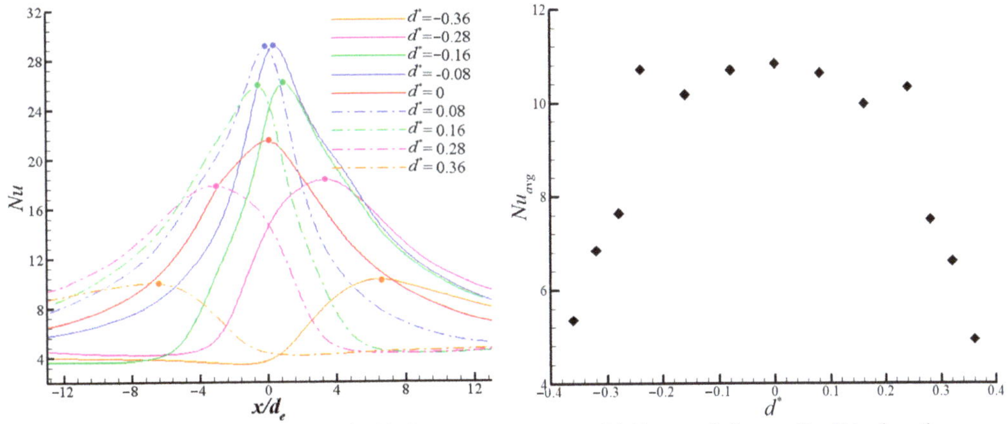

(a) Nu on the centerline of y/d_e=0 (b) Nu_{avg} variations with slider location

Figure 4. Variations of Nu and Nu_{avg}.

3.2. Effect of Driving Frequency

Previous work [42] has shown that the frequency response of a piezoelectric-driven synthetic jet actuator is consistent with that of a damped fourth order system governed by two key resonance frequencies—the natural frequency of the vibrating diaphragm and the Helmholtz resonance frequency. There are two local peaks of jet velocity and energy efficiency at the two resonance frequencies. In this section, experiments on the effect of driving frequency are carried out at a constant driving voltage of ±150 V and $d^* = 0.2$. The PIV results are illustrated in Figure 5.

It is observed from Figure 5 that the vectoring angle shows a non-monotonic variation trend with the increase in driving frequency. The minimum vectoring angle is 9.7° at the driving frequency of 350 Hz. The natural frequency of vibrating diaphragm is about 360 Hz. Since the amplitude of the vibrating diaphragm will increases significantly near the natural frequency, the DSJ velocity at 350 Hz is much higher than that at the other two frequencies. A higher DSJ velocity requires a greater force to deflect. Thus, the vectoring angle and the impingement stagnation region become small. To prevent electrical and mechanical failure of the vibrating diaphragm, the driving frequency should not be set at the resonant frequency. The driving frequency should be chosen by balancing the cooling performance and the service life of the DSJ actuator.

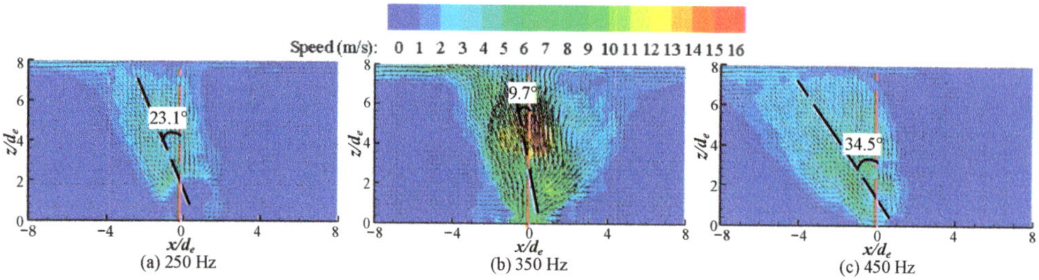

Figure 5. Velocity vector maps, contours and vectoring angles at different driving frequencies.

Nu contours on the impinged surface at different driving frequencies and slider locations are displayed in Figure 6. It can be seen that the peak *Nu* and the high *Nu* area of 350 Hz are obviously larger than those of 250 Hz and 450 Hz for each slider location. It is attributed to the frequency of 350 Hz close to the resonant frequency where the jet velocity is very higher (shown in Figure 5). As a result, the local heat transfer coefficients are higher than those of other two frequencies which are far away from the resonant frequency. For the driving frequency of 350 Hz, the peak *Nu* at $d^* = 0$ is greater than those at $d^* = \pm 0.24$. The tendencies are contrary at driving frequency of 250 Hz and 450 Hz. The reason is that the DSJ velocity of 350 Hz is high enough at $d^* = 0$, and the choking effect occurs in the narrow slot neck during the slider moving away from the center. It means although the cross section of the slot becomes smaller, the jet velocity may decrease, such as the slider locations of $d^* = \pm 0.24$. For the driving frequency of 250 Hz and 450 Hz, the DSJ velocities are low at $d^* = 0$ and increase with the decrease in the slot cross section at $d^* = \pm 0.24$. A higher DSJ velocity produces a better local cooling performance and a higher local *Nu*.

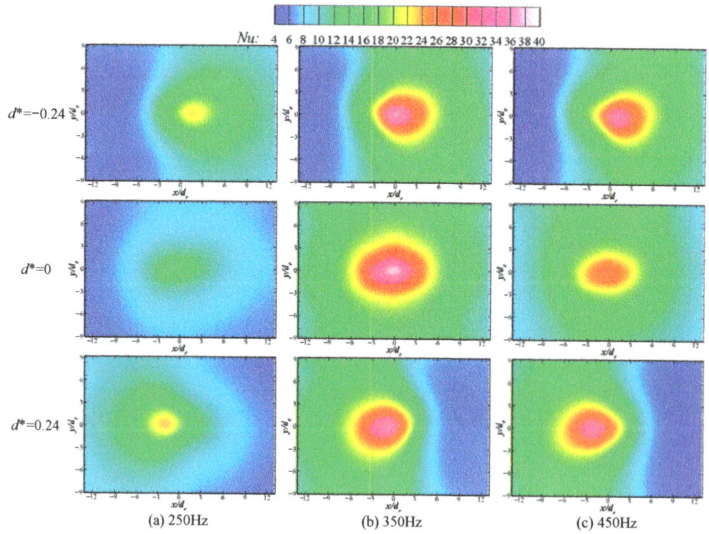

Figure 6. *Nu* contours at different driving frequencies and slider locations.

Nu_{avg} with different drive frequencies and slider locations are shown in Figure 7. Results in Figure 7 indicate that there are three local peaks at $d^* = 0$ and $d^* = \pm 0.24$ both for 250 Hz and 450 Hz, which is similar to Figure 4b. The distinction is that the maximum Nu_{avg} of 450 Hz occurs at $d^* = 0$ rather than that of 250 Hz at $d^* = \pm 0.24$. The reason is that at

$d^* = 0$ the DSJ velocity of 250 Hz is smaller than that of 450 Hz and increases more obviously at $d^* = \pm 0.24$. In addition, the impingement area is larger at $d^* = \pm 0.24$ than that at $d^* = 0$. As a result, the cooling performance at $d^* = \pm 0.24$ is better than that at $d^* = 0$. There is only one local peak with the drive frequency of 350 Hz, which is different from the drive frequency of 250 Hz and 450 Hz. It is attributed to that the DSJ velocity at 350 Hz becomes small owing to the choking effect in the slot neck when the slider moves away from the center. The decrease in DSJ velocity is more remarkable than the increase in impingement stagnation region, and thus the whole cooling performance gradually becomes weak. For the driving frequency of 250 Hz, although the DSJ velocity at $d^* = \pm 0.08$ is a little higher than that at $d^* = 0$, the cross-flow effect is more obvious due to the vectoring deflection. Overall, the whole cooling performance decreases a bit. As the slider location moves to $d^* = \pm 0.24$, the increase in the DSJ velocity is more remarkable than the enhancement of the cross-flow effect. Thus, the whole cooling performance gradually improves. As $|d^*|$ continues to increase, the choking effect occurs and impairs the DSJ velocity. The whole cooling performance reduces. The explanation for the variation trend of Nu_{avg} at 350 Hz is similar to the above mentioned about 250 Hz. The differences are that the maximum Nu_{avg} of 450 Hz occurs at $d^* = 0$ rather than that of 250 Hz at $d^* = \pm 0.24$. The reason is that the DSJ velocity of 250 Hz can be obviously enhanced with the increase in $|d^*|$, which improves the whole cooling performance.

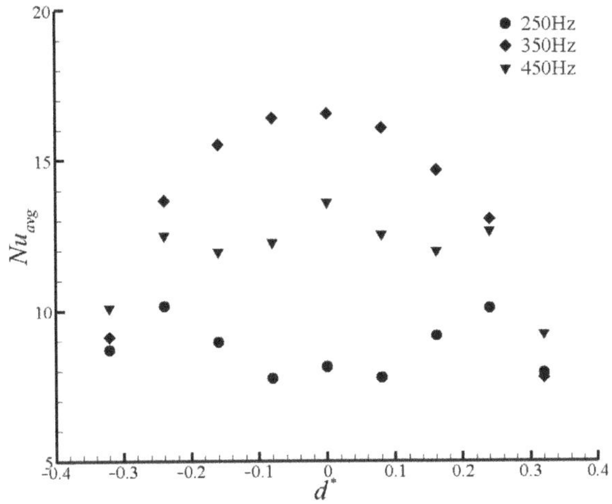

Figure 7. Nu_{avg} variations with the driving frequency and the slider location.

3.3. Effect of Driving Voltage

The driving voltage influences the input power imposed on the vibrating diaphragm, and then regulates the velocities of the primary jet and the disturbing jet. To investigate the effect of driving voltage experiments under three different driving voltages are executed. The corresponding velocity vector maps and contours at the driving frequency of 450 Hz and $d^* = 0.2$ are shown in Figure 8.

It can be seen that the velocity becomes high with the increase in the driving voltage, but the vectoring angle of DSJ decreases from 46.9° to 22.2°. It is attributed to that a high driving voltage can improve the amplitude of vibrating diaphragm, as well as the jet velocity. As the driving voltage increases, both of the primary jet and the disturbing jet are enhanced. Thus, the primary jet requires more force to deflect. However, the deflection force generated by the suction phase of the disturbing jet increases a little. The reason is that the fluid away from the primary jet is easily inhaled due to the three-dimensional effect of the suction process. Thus, the vectoring angle of DSJ becomes small. It is also observed that in view of the deflection symmetry, the impingement region of vectoring DSJ can be obviously extended. This feature is useful for improving the impingement cooling uniformity and removing dynamic hotspots.

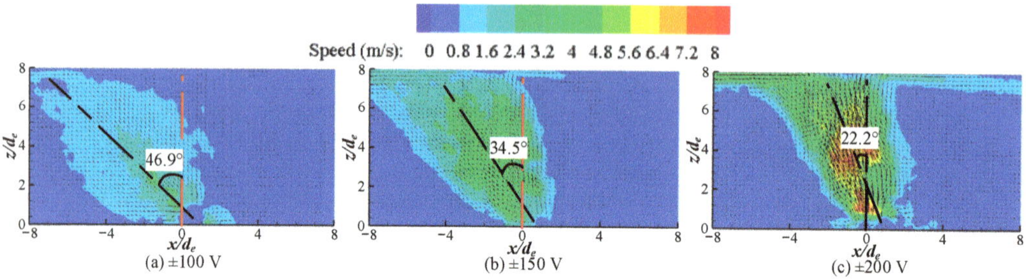

Figure 8. Velocity vector maps, contours and vectoring angles at different driving voltages.

Figure 9 shows Nu distributions on impinged surface at different driving voltages and slider locations. It is indicated that for each slider location, the peak Nu and the high Nu region gradually increase with the increase in the driving voltage. The reason is that the primary jet velocity becomes high with the increase in the driving voltage, and thus the local heat transfer coefficient increases. The core of the high Nu region can move toward the center location, which is agreement with the variation trend of the vectoring angle with the increase in the driving voltage.

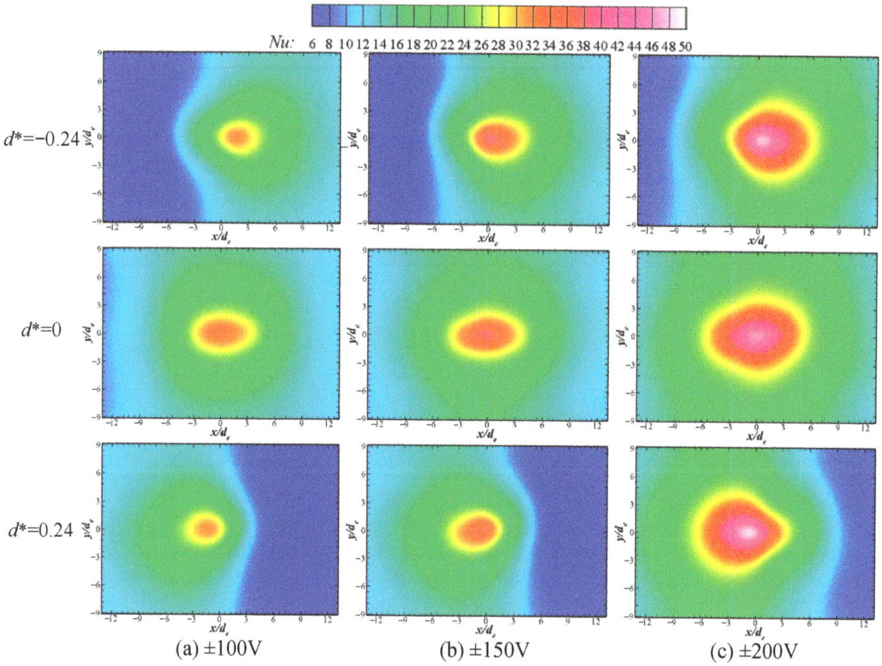

Figure 9. Nu contours at different driving voltages and slider locations.

Nu_{avg} at different driving voltages and slider locations are plotted in Figure 10. The curves indicate that Nu_{avg} improves with the increase in the driving voltage at each slider location. It is attributed to the obvious increases in the peak Nu and the high Nu region (shown in Figure 9). For each driving voltage, there is one peak of Nu_{avg} at the center location of the slider and decreases when the slider moves away from the center. The reason is that the primary jet velocity is so high at the center location of the slider that the choking effect occurs when the slider moves away from the center. As a result, the DSJ velocity reduces, as well as the heat transfer coefficient. Moreover, the impingement DSJ can transform to a cross flow owing to the vectoring deflection. Above two reasons make the whole cooling performance weak. Although a high driving can improve the cooling performance, the energy consumption also increases. The driving voltage should be selected by considering the energy utilization efficiency in the cooling applications.

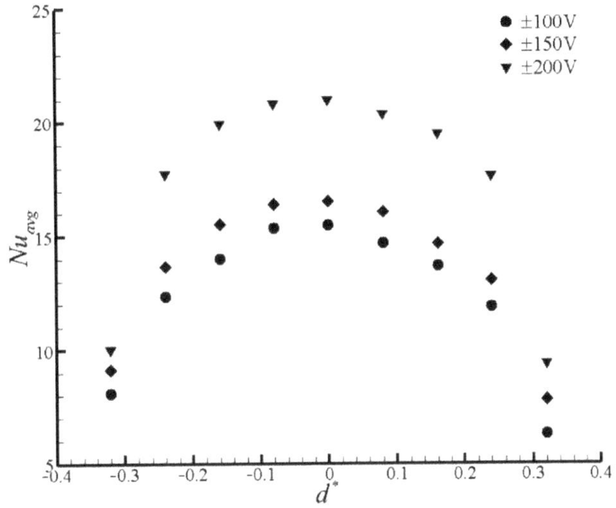

Figure 10. Nu_{avg} variations with driving voltage and slider location.

4. Conclusions

The effects of slot location, driving voltage, and driving frequency on impingement flow field and cooling characteristics of vectoring DSJ are experimentally investigated. Conclusions are summarized as follows:

(1) The DSJ deflects toward the side of the slot with a greater cross section. The vectoring angle can be adjusted from 34.5° toward left to 29.5° toward right. The length of the impingement region is over two times as large as that of the normal DSJ. With the increase in $|d^*|$, the maximum Nu increases first and then decreases. There are three local peaks of Nu_{avg} at $d^* = 0$ and ± 0.28. Although maximum Nu at $d^* = \pm 0.08$ and ± 0.16 are larger than that at $d^* = 0$, the Nu_{avg} are smaller.

(2) The vectoring angle is lowest of 9.7° at the driving frequency of 350 Hz, but the peak Nu and high Nu area are obviously augmented. There is one local peak of Nu_{avg} at 350Hz rather than three local peaks of Nu_{avg} at 250 Hz and 450 Hz. The difference is that the maximum Nu_{avg} of 250 Hz and 450 Hz, respectively, occurs at $d^* = 0$ and at $d^* = \pm 0.24$.

(3) With the increase in the driving voltage, the vectoring angle decreases from 46.9° to 22.2°, but the length of the impingement area is extended. There is one local peak of Nu_{avg} at $d^* = 0$ for each driving voltage.

Author Contributions: Conceptualization, X.D. and Z.L.; methodology, X.D. and W.H.; formal analysis, X.D. and Z.D.; investigation, X.D. and Q.L.; writing—original draft, X.D.; writing—review and editing, C.P. and Z.D. All authors have read and agreed to the published version of the manuscript.

Funding: This research was funded by the National Natural Science Foundation of China (grant numbers 11972369, 11872374) and the Science and Technology Innovation Program of Hunan Province (2021RC3075).

Data Availability Statement: Data available on request from the authors.

Conflicts of Interest: The authors declare no conflict of interest.

Nomenclature

A	effectively heated area (mm^2)
A_Ω	integration area (mm^2)
d	slot location (mm)
d_1, d_2	width of two slots (mm)
d_e	characteristic length (mm)
d^*	slot dimensionless location
e_θ	unit vector along the projected line
h	convective heat transfer $W/(m^2K)$
I	current through the foil (A)
Ma	Mach number
Nu	local Nusselt number
Nu_{avg}	area-averaged Nusselt number
Q_{ele}, Q_{loss}	input power and heat loss (W)
q_{net}	net removed heat flux (W/m^2)
S_o	objection function (m/s)
u_i, w_i	velocity components along x, z axis
T_s, T_j	impingement surface and jet temperature (°C)
V	voltage across the foil (V)
V_c	critical velocity (m/s)
Greek symbols	
θ	oblique angle of projected line (°)
θ_0	vectoring angle (°)
λ	thermal conductivity (W/(m·K))
Subscripts	
avg	spatially-averaged
net	net
elc	electric
j	jet
loss	heat radiation loss
max	maximum value
s	surface
Ω	integral domain
Acronyms	
CCD	charge-coupled device
DSJ	dual synthetic jet
PIV	particle image velocimetry

References

1. Glezer, A. Some aspects of aerodynamic flow control using synthetic-jet actuation. *Philos. Trans. R. Soc. A Math. Phys. Eng. Sci.* **2011**, *369*, 1476–1494. [CrossRef] [PubMed]
2. Taylor, K.; Amitay, M. Dynamic stall process on a finite span model and its control via synthetic jet actuators. *Phys. Fluids* **2015**, *27*, 077104. [CrossRef]
3. Wu, J.; Shen, M.; Jiang, L. Role of synthetic jet control in energy harvesting capability of a semi-active flapping airfoil. *Energy* **2020**, *208*, 118389. [CrossRef]
4. Farrell, G.; Gibbons, M.; Persoons, T. Combined Passive/Active Flow Control of Drag and Lift Forces on a Cylinder in Crossflow Using a Synthetic Jet Actuator and Porous Coatings. *Actuators* **2022**, *11*, 201. [CrossRef]
5. Luo, Z.; Zhao, Z.; Liu, J.; Deng, X.; Zheng, M.; Yang, H.; Chen, Q.; Li, S. Novel roll effector based on zero-mass-flux dual synthetic jets and its flight test. *Chin. J. Aeronaut.* **2022**, *35*, 1–6. [CrossRef]
6. Tang, H.; Salunkhe, P.; Zheng, Y.; Du, J.; Wu, Y. On the use of synthetic jet actuator arrays for active flow separation control. *Exp. Therm. Fluid Sci.* **2014**, *57*, 1–10. [CrossRef]
7. Zhao, G.; Zhao, Q. Parametric analyses for synthetic jet control on separation and stall over rotor airfoil. *Chin. J. Aeronaut.* **2014**, *27*, 1051–1061. [CrossRef]
8. Wen, X.; Tang, H.; Liu, Y. Interaction of twin synthetic jets in attached and separated boundary layers: Effects of yaw angle and phase difference. *J. Vis.* **2018**, *21*, 949–963. [CrossRef]
9. Rathay, N.; Amitay, M. Interaction of synthetic jets with a massively separated three-dimensional flow field. *Phys. Rev. Fluids* **2022**, *7*, 034702. [CrossRef]

10. Xia, Q.; Zhong, S. Enhancement of inline mixing with lateral synthetic jet pairs at low Reynolds numbers: The effect of fluid viscosity. *Flow Meas. Instrum.* **2017**, *53*, 308–316. [CrossRef]
11. Wang, L.; Feng, L.-H.; Wang, J.-J.; Li, T. Characteristics and mechanism of mixing enhancement for noncircular synthetic jets at low Reynolds number. *Exp. Therm. Fluid Sci.* **2018**, *98*, 731–743. [CrossRef]
12. Zhang, B.; Liu, H.; Li, Y.; Liu, H.; Dong, J. Experimental Study of Coaxial Jets Mixing Enhancement Using Synthetic Jets. *Appl. Sci.* **2021**, *11*, 803. [CrossRef]
13. Smith, B.L.; Glezer, A. Jet vectoring using synthetic jets. *J. Fluid Mech.* **2002**, *458*, 1–34. [CrossRef]
14. Luo, Z.-B.; Xia, Z.-X.; Xie, Y.-G. Jet Vectoring Control Using a Novel Synthetic Jet Actuator. *Chin. J. Aeronaut.* **2007**, *20*, 193–201. [CrossRef]
15. Ryota, K.; Yuki, W.; Yu, T.; Koichi, N.; Donghyuk, K.; Kotaro, S. Jet vectoring using secondary Coanda synthetic jets. *Mech. Eng. J.* **2020**, *7*, 20-00215.
16. Yu, Y.; Simon, T.W.; Zhang, M.; Yeom, T.; North, M.T.; Cui, T. Enhancing heat transfer in air-cooled heat sinks using piezoelectrically-driven agitators and synthetic jets. *Int. J. Heat Mass Transf.* **2014**, *68*, 184–193. [CrossRef]
17. Yeom, T.; Simon, T.; Zhang, M.; Yu, Y.; Cui, T. Active heat sink with piezoelectric translational agitators, piezoelectric synthetic jets, and micro pin fin arrays. *Exp. Therm. Fluid Sci.* **2018**, *99*, 190–199. [CrossRef]
18. Gil, P.; Smyk, E.; Gałek, R.; Przeszłowski, Ł. Thermal, flow and acoustic characteristics of the heat sink integrated inside the synthetic jet actuator cavity. *Int. J. Therm. Sci.* **2021**, *170*, 107171. [CrossRef]
19. Mohammadpour, J.; Salehi, F.; Lee, A.; Brandt, L. Nanofluid heat transfer in a microchannel heat sink with multiple synthetic jets and protrusions. *Int. J. Therm. Sci.* **2022**, *179*, 107642. [CrossRef]
20. Maghrabie, H.M. Heat transfer intensification of jet impingement using exciting jets—A comprehensive review. *Renew. Sustain. Energy Rev.* **2021**, *139*, 110684. [CrossRef]
21. Glezer, A.; Amitay, M. Synthetic jets. *Annu. Rev. Fluid Mech.* **2002**, *34*, 503–529. [CrossRef]
22. Greco, C.S.; Cardone, G.; Soria, J. On the behaviour of impinging zero-net-mass-flux jets. *J. Fluid Mech.* **2016**, *810*, 25–59. [CrossRef]
23. Pavlova, A.; Amitay, M. Electronic Cooling Using Synthetic Jet Impingement. *J. Heat Transf.* **2006**, *128*, 897–907. [CrossRef]
24. Luis, S.L.; Alfonso, O. Vortex dynamics and mechanisms of heat transfer enhancement in synthetic jet impingement. *Int. J. Therm. Sci.* **2017**, *112*, 153–164.
25. Greco, C.S.; Paolillo, G.; Ianiro, A.; Cardone, G.; de Luca, L. Effects of the stroke length and nozzle-to-plate distance on synthetic jet impingement heat transfer. *Int. J. Heat Mass Transf.* **2018**, *117*, 1019–1031. [CrossRef]
26. Qiu, Y.-L.; Hu, W.-J.; Wu, C.-J.; Chen, W.-F. Flow and heat transfer characteristics in a microchannel with a circular synthetic jet. *Int. J. Therm. Sci.* **2021**, *164*, 106911. [CrossRef]
27. Lyu, Y.-W.; Zhang, J.-Z.; Tan, J.-W.; Shan, Y. Impingement heat transfer on flat and concave surfaces by piston-driven synthetic jet from planar lobed orifice. *Int. J. Heat Mass Transf.* **2021**, *167*, 120832. [CrossRef]
28. Lau, G.; Mohammadpour, J.; Lee, A. Cooling performance of an impinging synthetic jet in a microchannel with nanofluids: An Eulerian approach. *Appl. Therm. Eng.* **2021**, *188*, 116624. [CrossRef]
29. Wang, L.; Feng, L.H.; Xu, Y.; Xu, Y.; Wang, J.J. Experimental investigation on flow characteristics and unsteady heat transfer of noncircular impinging synthetic jets. *Int. J. Heat Mass Tran.* **2022**, *190*, 122760. [CrossRef]
30. Singh, P.K.; Renganathan, M.; Yadav, H.; Sahu, S.K.; Upadhyay, P.K.; Agrawal, A. An experimental investigation of the flow-field and thermal characteristics of synthetic jet impingement with different waveforms. *Int. J. Heat Mass Transf.* **2022**, *187*, 122534. [CrossRef]
31. Smith, B.L.; Glezer, A. The formation and evolution of synthetic jets. *Phys. Fluids* **1998**, *10*, 2281–2297. [CrossRef]
32. Luo, Z.-B.; Xia, Z.-X.; Liu, B. New Generation of Synthetic Jet Actuators. *AIAA J.* **2006**, *44*, 2418–2420. [CrossRef]
33. Deng, X.; Xia, Z.-X.; Luo, Z.-B.; Li, Y.-J. Vector-Adjusting Characteristic of Dual-Synthetic-Jet Actuator. *AIAA J.* **2015**, *53*, 794–797. [CrossRef]
34. Luo, Z.; Zhao, Z.; Deng, X.; Wangw4, L.; Xia, Z. Dual Synthetic Jets Actuator and Its Applications—Part I: PIV Measurements and Comparison to Synthetic Jet Actuator. *Actuators* **2022**, *11*, 205. [CrossRef]
35. Deng, X.; Luo, Z.; Xia, Z.; Gong, W.; Wang, L. Active-passive combined and closed-loop control for the thermal management of high-power LED based on a dual synthetic jet actuator. *Energy Convers. Manag.* **2017**, *132*, 207–212. [CrossRef]
36. Deng, X.; Luo, Z.-B.; Xia, Z.-X.; Gong, W.-J. Experimental investigation on the flow regime and impingement heat transfer of dual synthetic jet. *Int. J. Therm. Sci.* **2019**, *145*, 105864. [CrossRef]
37. He, W.; Luo, Z.-B.; Deng, X.; Xia, Z.-X. Experimental investigation on the performance of a novel dual synthetic jet actuator-based atomization device. *Int. J. Heat Mass Transf.* **2019**, *142*, 118406. [CrossRef]
38. He, W.; Luo, Z.-B.; Deng, X.; Peng, W.-Q.; Zhao, Z.-J. Experimental investigation on the vectoring spray based on a novel synthetic jet actuator. *Appl. Therm. Eng.* **2020**, *179*, 115677. [CrossRef]
39. Luo, Z.-B.; Deng, X.; Xia, Z.-X.; Wang, L.; Gong, W.-J. Flow field and heat transfer characteristics of impingement based on a vectoring dual synthetic jet actuator. *Int. J. Heat Mass Transf.* **2016**, *102*, 18–25. [CrossRef]
40. Moffat, R.J. Describing the uncertainties in experimental results. *Exp. Therm. Fluid Sci.* **1988**, *1*, 3–17. [CrossRef]

41. Luo, Z.-B.; Xia, Z.-X. The mechanism of jet vectoring using synthetic jet actuators. *Mod. Phys. Lett. B* **2005**, *19*, 1619–1622. [CrossRef]
42. Gallas, Q.; Holman, R.; Nishida, T.; Carroll, B.; Sheplak, M.; Cattafesta, L. Lumped Element Modeling of Piezoelectric-Driven Synthetic Jet Actuators. *AIAA J.* **2003**, *41*, 240–247. [CrossRef]

actuators

Article

Dual Synthetic Jet Actuator and Its Applications—Part IV: Analysis of Heat Dissipation and Entropy Generation of Liquid Cooling with Dual Synthetic Jet Actuator

Ying Kang, Zhenbing Luo *, Xiong Deng, Yinxin Zhu and Zhixun Xia

College of Aerospace Science and Engineering, National University of Defense Technology, Changsha 410073, China
* Correspondence: luozhenbing@163.com

Abstract: Increasing heat flux restricts the development of the miniaturization of electronic devices. There is an urgent need for a heat dissipation method that will efficiently cool the chip. This paper presents a novel liquid cooling device based on dual synthetic jets actuator (DSJA) technology. The characteristics of the temperature and velocity field of the device are numerically studied by a three-dimensional coupled heat transfer model. The entropy generation rate caused by heat transfer and fluid friction was studied to analyze the effective work loss and irreversibility of the heat transfer process. When the DSJA is turned on, the temperature of the heat source with a heat flux of 200 W/cm^2 is 73.07 °C, and the maximum velocity is 24.32 m/s. Compared with the condition when the the DSJA is closed, the temperature decreases by 25.15 °C, and the velocity increases by nearly 20 m/s. At this time, the total inlet flow is 1.26 L/min. The larger frictional entropy generation is mainly distributed near the inlet and outlet of the channel and the jet orifice. The higher the velocity is, the more obvious the frictional entropy generation is. Due to the large temperature gradient, there is a large thermal entropy generation rate at the fluid–solid interface.

Keywords: liquid cooling; dual synthetic jets actuator; entropy generation

Citation: Kang, Y.; Luo, Z.; Deng, X.; Zhu, Y.; Xia, Z. Dual Synthetic Jet Actuator and Its Applications—Part IV: Analysis of Heat Dissipation and Entropy Generation of Liquid Cooling with Dual Synthetic Jet Actuator. *Actuators* **2022**, *11*, 382. https://doi.org/10.3390/act11120382

Academic Editor: Luigi de Luca

Received: 8 November 2022
Accepted: 16 December 2022
Published: 19 December 2022

1. Introduction

The continuous development of micro-electronic devices will depend on efficient heat dissipation technology. At present, it is a great challenge to design a heat sink with a heat flux exceeding 1 MW/m^2 [1]. Many electronic devices are damaged by overheating. When the temperature of electronic equipment increases by 10 °C, the reliability decreases by 50% [2]. Therefore, it is urgent that we seek an efficient heat dissipation technology.

Heat dissipation technology is divided into natural convection technology and forced convection technology. Forced convection technology can rapidly reduce the surface temperature of electronic devices, including spray cooling technology, microchannel cooling technology, synthetic jet technology, etc. The spray cooling device has high heat dissipation efficiency but the system is complex. The microchannel technology has the advantage of small volume and high heat transfer. Many researchers have studied the structural parameters of microchannels [3–13]. Rhombus fractal-like units [5] and spider-netted microchannels [7] have excellent heat dissipation performance. Rectangular microchannels are widely used in our lives due to their good heat dissipation and hydraulic performance. Many studies have shown that microchannels have a strong heat dissipation capability. However, to meet the heat dissipation demand, they often need to provide more pump power [14].

Synthetic jet technology [15] is an active flow control technology, which can be used in the field of heat transfer enhancement. The synthetic jet actuator (SJA) comprises a diaphragm, a cavity, and a jet orifice. The diaphragm vibrates periodically, changing the volume of the cavity. When the cavity volume is compressed, the fluid in the cavity is ejected through the jet orifice, forming a jet at the outlet of the actuator, and developing continuously downstream. When the cavity volume is expanded, the external fluid is sucked into the cavity through both sides of the jet orifice. So far, the SJ has completed a periodic movement [15,16]. Different parameters have different effects on the heat transfer of the SJ. There is an optimal frequency to realize the maximum Nusselt number of SJs impinging on a constant heat flux disk. In addition, when the jet-to-surface distance is small, the Nusselt number increases with the jet-to-surface distance [17]. SJs can significantly destroy the thermal boundary layer to enhance heat transfer [18,19]. However, one side of the diaphragm of the SJA is located at the air side, which can easily cause ballast failure and low energy utilization.

Based on SJ technology, Luo et al. [20] invented the dual synthetic jet (DSJ) technology. The DSJ has superior performance and overcomes the problems of SJs [21]. The DSJ technology consists of two cavities, one diaphragm, and two jet outlets. The diaphragm separates the cavity. When the diaphragm vibrates, the cavity on one side expands and the cavity on the other side compresses, the jet outlet on the expanding side inhales the surrounding fluid, and the compressed side ejects the fluid in the cavity. The alternately ejected/inhaled jets interact near the outlet and merge into the DSJ, which continues to develop downstream. Research on SJ and SJA technology mostly focuses on the field of air cooling. The SJA technology has been proven to be well applied to heat dissipation in confined spaces [22].

For the heat dissipation of higher heat flux, the ability of air cooling is not enough. Therefore, some scholars have preliminarily studied the heat transfer mechanism of SJs in the liquid cooling field. Combining SJs with microchannels, the influence of the interaction between the jet and incoming flow on heat transfer is studied [23–28]. The PIV technique was used to observe the flow field of SJs underwater. Compared with the condition without SJs, the effect of an SJ makes the heat transfer increase by about 4.3 times [23]. The SJ promotes fluid mixing in the channel and enhances heat transfer [29]. It is worth noting that the SJ contributes little to the pressure drop in the channel. With a low flow rate, heat transfer can be enhanced by 130% [30]. Multiple synthetic jets can be combined with microchannels at the same time for better heat dissipation [26]. The convective heat transfer capacity of two SJs is more effective when 180° is out of phase [31,32]. The higher the frequency and amplitude of the diaphragm, the higher the convective heat transfer capacity. There is an optimal fluid inlet temperature of 297.15 K for achieving maximum heat transfer. The overall performance of the heat sink increases significantly with the Reynolds number at the inlet [32]. The underwater PIV experiment of DSJ technology shows that the jet strength is the highest when the driving frequency is 30 Hz [33]. The heat transfer characteristics and physical properties of the SJ flow field in the channel are very complex, so a comprehensive parameter study of SJAs is required [34].

Some other scholars have carried out entropy generation analysis on the studied device. Entropy generation analysis is helpful to understand the source of irreversible loss. Luis et al. [35] analyzed local entropy generation and global entropy generation of different parts of a hybrid microjet heat sink. It is found that the irreversibility of the system mainly comes from the heat transfer of the copper plate. The total entropy generation consists of thermal entropy generation and friction entropy generation. Mehdi et al. [36] found that the latter has little influence on the irreversible loss of the heat sink. Omid et al. [37] reviewed the entropy generation theory of nanofluids and introduced two calculation methods for entropy generation.

In our previous research, a side-mounted liquid cooling device was proposed [38]. Good heat transfer and flow performance are obtained. To further simplify the fluid circuit and reduce the flow resistance, this paper proposes a front-placement liquid cooling device. In this paper, the DSJ is combined with a channel for liquid cooling. The entropy generation of the cooling device is analyzed from the point of view of the second law of thermodynamics. The novelty of this study is that a new type of forwarding compound DSJ liquid cooling device is proposed. It is of great significance to solve the heat dissipation of electronic devices in a high heat flux confined space. At the same time, this study fills in the blanks that remain from the study of DSJs by the entropy generation analysis method. The research contents of this paper are as follows: Section 2 mainly introduces the relevant models and numerical methods; Section 3 is the result discussion. The temperature field, velocity field, entropy generation, and the effect of diaphragm frequency have been studied. Section 4 is the conclusion.

2. Numerical Methods

2.1. Physical Model

In this paper, the DSJA cooling device consists of a DSJA, a channel, and an aluminum plate, as shown in Figures 1 and 2. The jet orifices are directly arranged on the upper side of the cavity to connect the channel and the actuator cavity. The channel consists of an inlet and an outlet. Jet orifices set to 5 × 5 arrays. The fluid enters the cooling device in three ways. One fluid enters the channel, and the other two fluids enter the channel through the jet orifices, through the actuator. The three converge in the channel and move downstream. The yellow arrow in Figure 1 indicates the path of fluid flow. The jet orifices are row 1, row 2, row 3, row 4, and row 5 along the positive direction of the y-axis.

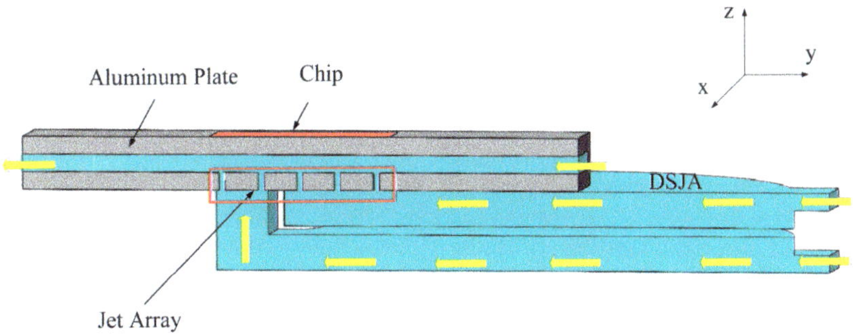

Figure 1. Schematic diagram of cooling device.

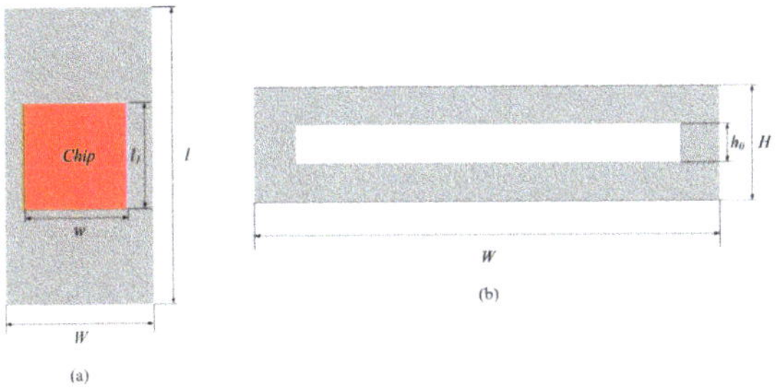

Figure 2. Schematic diagram of channel: (**a**) Vertical view; (**b**) Side view.

When the DSJA diaphragm is not vibrating, the actuator only acts as a fluid circuit. At this time, jets from the jet orifices can be regarded as steady jets. When the DSJA diaphragm vibrates upward, the cavity on the upper side of the actuator shrinks, and fluid is ejected from the third, fourth, and fifth-row jet orifices. At the same time, the cavity at the lower side of the actuator expands, and the fluid in the channel flows into the cavity at the lower side through the 1st and 2nd discharge orifices. When the DSJA diaphragm vibrates downward, the flow situation is opposite to the above process. Different jet outlets alternately eject/inhale fluid, and the formed jets fuse and develop with each other, and interacts with the incoming flow of the channel to form a violent disturbance.

2.2. Governing Equations

In this paper, a three-dimensional fluid–solid coupling model is established. The thermophysical properties of fluids and solids are considered constant. The fluid is assumed to be incompressible. The working fluid is water, and the solid is aluminum. The mass, momentum, and energy equations of the problem [25] can be written as

$$\frac{\partial \rho}{\partial t} + \nabla \cdot \rho U = 0 \tag{1}$$

$$\frac{\partial}{\partial t}(\rho U) + \nabla \cdot (\rho U U) = -\nabla p + \nabla \cdot (\overline{\tau}) \tag{2}$$

$$\frac{\partial}{\partial t}(\rho E) + \nabla \cdot (U(\rho E + p)) = \nabla \cdot \left(\lambda_{\text{eff}} \nabla \mathrm{T} + \left(\overline{\overline{\tau}}_{\text{eff}} \cdot U\right)\right) \tag{3}$$

2.3. Boundary Conditions and Numerical Solver Setting

The simulation software used in this research is Fluent. Pressure-based solver is selected for calculation. The energy equation is opened. Considering the need to solve the jet impact problem and large-scale flow changes, the SST k-ω model is selected. The calculation includes a fluid domain and a solid domain. The material of the fluid domain is water, and the material of the solid domain is aluminum.

The boundary conditions of the cooling device are set as shown in Figure 3. Considering that the cooling device is completely symmetrical, only half of the model is calculated. Symmetrical boundary conditions are adopted for the model section. The entrance of the channel and the entrance of the actuator adopt the velocity inlet boundary. The inlet velocity is 1.5 m/s and 0.3 m/s, respectively. The channel outlet is set as the pressure outlet. The chip uses a constant heat flux boundary. The inlet water temperature is 25 °C. The fluid–solid interface is set as the coupling surface. The rest are set as non-sliding walls.

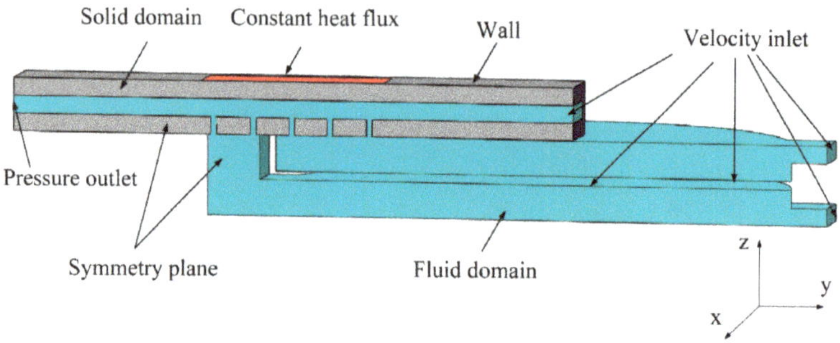

Figure 3. Boundary conditions of the calculation model.

The actuator diaphragm uses velocity inlet boundary conditions of user-defined functions (UDF) as

$$\begin{cases} u(t)_l = 0.02\sin(2\pi f t + 0^\circ) \\ u(t)_r = 0.02\sin(2\pi f t + 180^\circ) \end{cases} \tag{4}$$

The pressure and velocity are coupled by a SIMPLE algorithm. The relaxation factors for pressure, density, and momentum are set by default to 0.3, 1, and 0.7, respectively. The criterion of the energy equation is 10^{-6}, and the convergence criterion of mass and momentum is 10^{-3}. In this paper, the frequency is selected as f = 30 Hz [33]. The time step is set to 2.778×10^{-4} s, and the total time is 6.67 s.

2.4. Numerical Method Validation

Grid independence verification is performed on the model, as shown in Table 1. Because the model is completely symmetrical, only half of the model is used for structural mesh generation. Three sets of grids are divided, and the $y+$ value is less than 1. The error between the number of grids (8,555,843) and the number of grids (11,713,823) is less than 8%. To ensure the calculation accuracy and velocity, 8,555,843 grids are selected for calculation, as shown in Figure 4.

Table 1. Grid independence verification.

	Grid Number	Pressure Drop (kPa)	Convergence Criteria	Base Temperature (°C)	Convergence Criteria
DSJ off	4,374,057	4.35	0.685%	97.82	1.082%
	8,555,843	4.38	0.000%	98.22	0.677%
	11,713,823	4.38	-	98.89	-
DSJ on(T)	4,374,057	5.93	18.767%	76.31	0.883%
	8,555,843	6.78	7.123%	76.61	0.493%
	11,713,823	7.30	-	76.99	-

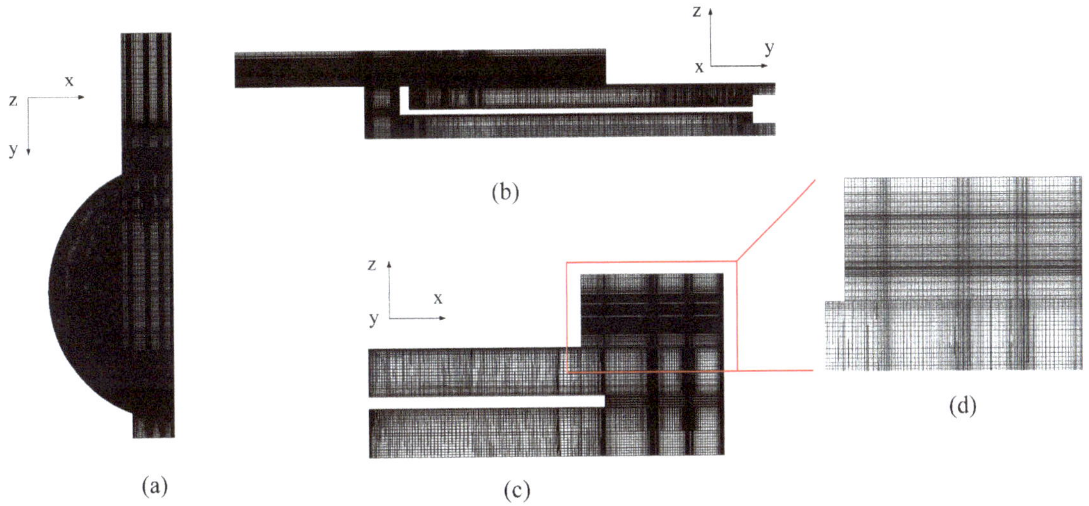

Figure 4. Structural grid of the model: (**a**) Vertical view; (**b**) Front view; (**c**) Side view; (**d**) Local details of (**c**).

The experimental model in reference [33] is selected to verify the numerical method in this paper. As shown in Figure 5, the RNG k-ε model and SST k-ω model are selected. The velocity distribution at 2 mm away from the DSJ orifice was studied. The results show that the SST k-ω model is in better agreement with the experiment.

2.5. Relevant Definitions

The hydraulic diameter of the channel is

$$D_h = \frac{2w\, h_0}{w + h_0} \tag{5}$$

where w, and h_0 are the channel width and height, respectively. In this paper, D_h = 1.8182 mm.

The channel Reynolds number Re_c [28] and the DSJ Reynolds number Re_{dsj} are defined as follows:

$$Re_c = \frac{\rho u_{in} D_h}{\mu} \tag{6}$$

$$Re_{dsj} = \frac{\rho u_{dsj} d}{\mu} \tag{7}$$

where u_{in} = 1.5 m/s, u_{dsj} = 24.32 m/s, and d = 0.36 mm are the flow velocity of the channel inlet, the maximum flow velocity at the exit of the actuator, and the diameter of the jet orifice, respectively. The value of the channel Reynolds number is 2714. The range of jet Reynolds numbers Re_{dsj} is 43,998.192.

The Nusselt number of a channel is defined as

$$Nu = \frac{D_h h}{\lambda} \tag{8}$$

where $h = \frac{q}{(T_w - T_f)}$ is the convective heat transfer coefficient of DSJA, where T_w and T_f are the solid surface temperature and the fluid temperature, respectively.

The volume entropy production rate includes the entropy production rate caused by irreversible heat transfer and the irreversible flow loss caused by fluid friction. All formulas [36] are as follows:

$$\dot{S}'''_{gen,total} = \dot{S}'''_{gen,thermal} + \dot{S}'''_{gen,frictional} \tag{9}$$

$$\dot{S}'''_{gen,thermal} = \frac{\lambda_f}{T_f^2}\left[\left(\frac{\partial T}{\partial x}\right)^2 + \left(\frac{\partial T}{\partial y}\right)^2 + \left(\frac{\partial T}{\partial z}\right)^2\right] \tag{10}$$

$$\dot{S}'''_{gen,frictional} = \frac{\mu}{T_f}\left\{2\left[\left(\frac{\partial u}{\partial x}\right)^2 + \left(\frac{\partial v}{\partial y}\right)^2 + \left(\frac{\partial w}{\partial z}\right)^2\right] + \left(\frac{\partial u}{\partial y} + \frac{\partial v}{\partial x}\right)^2 + \left(\frac{\partial u}{\partial z} + \frac{\partial w}{\partial x}\right)^2 + \left(\frac{\partial v}{\partial z} + \frac{\partial w}{\partial y}\right)^2\right\} \tag{11}$$

u, v, w are velocity components in x, y, z directions respectively.

The total entropy generation is the integral sum of thermal entropy generation and frictional entropy generation in the entire cooling device.

$$\int\int\int_\Omega \dot{S}_{gen} dV = \int\int\int_\Omega \dot{S}_{gen,thermal} dV + \int\int\int_\Omega \dot{S}_{gen,frictional} dV \tag{12}$$

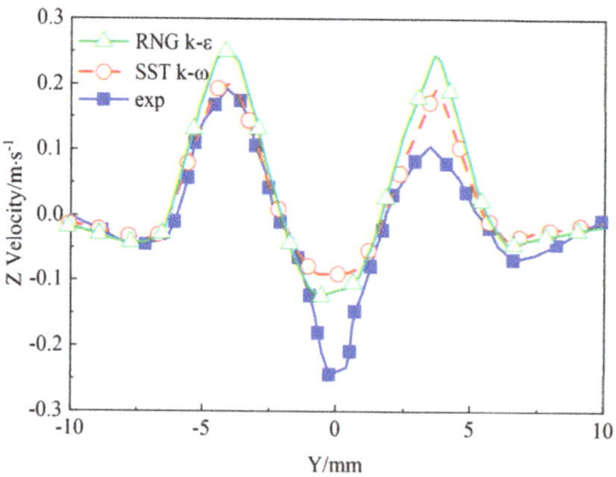

Figure 5. Numerical simulation and experimental [33] verification of DSJ underwater (z = 2 mm).

3. Results

3.1. Temperature Field

To study the enhanced heat transfer mechanism of the DSJ, the central section of the channel (section x = 28.5 mm) is taken as an example for analysis. The inlet velocity of the channel is 1.5 m/s, and the inlet velocity of DSJA is 0.3 m/s. The heat flux of the heat source is 200 W/cm^2. The heat source is located at -5.25 mm < y < 5.25 mm.

The temperature field of the channel is shown in Figure 6. The average temperature of the chip is 120.35 °C. It is observed that the substrate temperature near the outlet is higher than that of the inlet. This is because the fluid takes away heat along the flow direction, increasing the fluid temperature. A thermal boundary layer attached to the substrate is gradually thickening along the flow direction, weakening the heat transfer capability. There are partitions between adjacent channels to enhance the heat conduction effect.

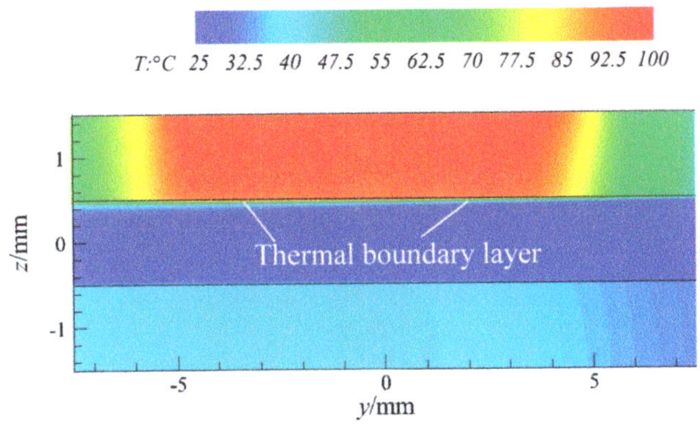

Figure 6. Temperature distribution of channel's center section (x = 28.5 mm).

The temperature fields of the DSJ when off and the DSJ when on are shown in Figures 7 and 8, respectively. In Figure 7, the maximum temperature of the heat source is 98.22 °C. The thermal boundary layer gradually develops and thickens along the channel's top wall from the channel inlet. It is destroyed in the jet impact area. Compared with other areas, a part of the thermal boundary layer is observed above the fourth and fifth row of jet orifices, considering the jet deflection caused by the weak jet of the upper cavity itself and the incoming flow. The impact damage ability to the thermal boundary layer is weakened. The existence of the thermal boundary layer will deteriorate the heat transfer, which will help to reduce the temperature of the actuator.

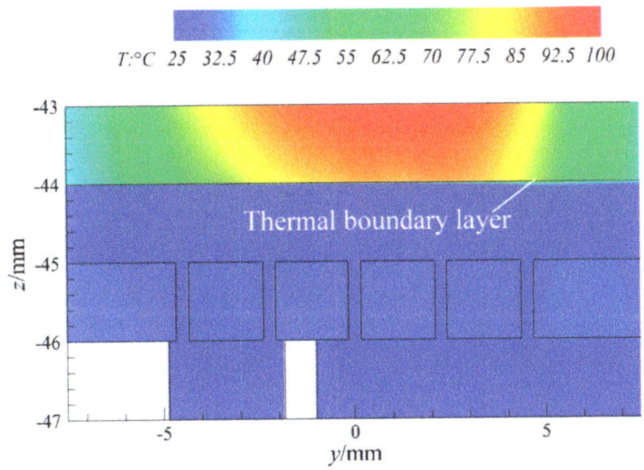

Figure 7. Temperature distribution (DSJ off) of channel's center section (x = 28.5 mm).

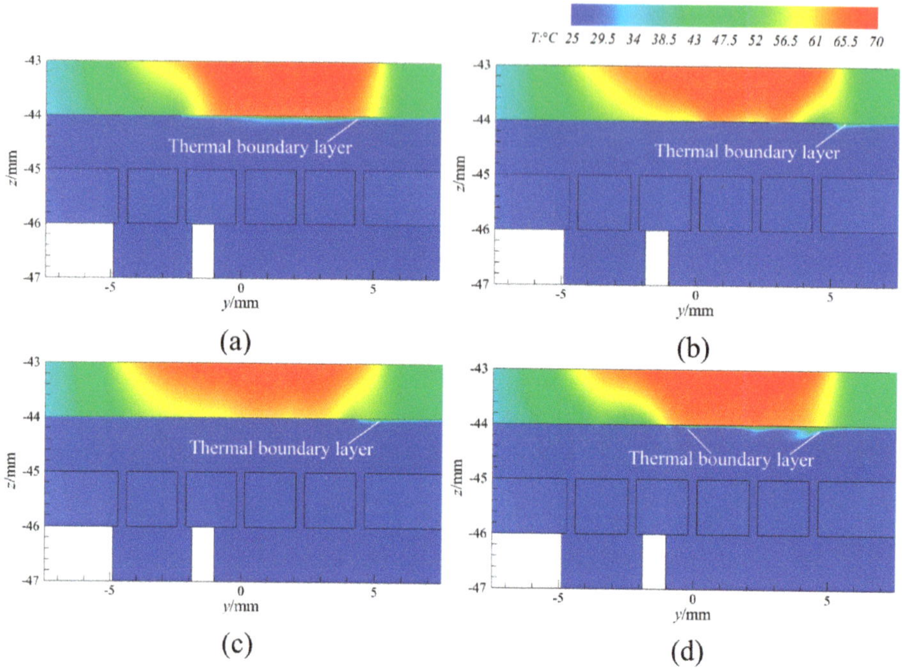

Figure 8. Temperature distribution of channel's center section (x = 28.5 mm) (DSJ on): (**a**) 0 T (1 T); (**b**) 0.25 T; (**c**) 0. 5 T; (**d**) 0.75 T.

Compared with the DSJ when off, the maximum temperature drop of the heat source is 25.15 °C when the DSJ is on, as shown in Figure 8. The alternating vibration of the diaphragm of the DSJA has a periodic effect on the temperature distribution of the aluminum plate. At 0 T (1 T), the thermal boundary layer was observed above the third, fourth, and fifth-row jet orifice. Consider that the upper cavity is in the suction stage at this time. The attachment of the thermal boundary layer results in an obvious high-temperature region. The temperature of the aluminum plate above the first and second rows of jet orifices is obviously lower, which is caused by jet impingement cooling. At 0.25 T, the temperature contour shows an obvious low-temperature area, and the thermal boundary layer has been destroyed. At this time, the upper cavity is in the ejection stage, and the jet ability is strong. At 0.5 T, the temperature distribution near the impact surface is relatively uniform. Consider the weakening stage of the jet from the upper cavity. At 0.75 T, the temperature of the corresponding impact area of the first and second row of jet orifices decreased significantly. This is due to the impact of the jet at the outlet of the lower cavity. Influenced by the flow of the jets and incoming flow in the suction channel of the cavity, a partially disordered thermal boundary layer was observed. Then, the cooling device enters the stage of 0 T (1 T) again, and the above process is repeated. The periodic vibration of the diaphragm brings about the periodic change of the temperature of the aluminum plate, and the reciprocating disturbance of the jet and its interaction with the incoming flow strengthen the convection heat transfer.

The distribution of Nu on the impact surface (z = 44 mm) along the y direction is shown in Figure 9. When the DSJ is off, the distribution of Nu is low in the middle and high on both sides. This is because the middle area is close to the heat source and has a higher temperature. Both sides have low-temperature regions because they are far away from heat sources and close to the inlet and outlet of the channel. When the DSJ is on, the distribution of Nu varies at different times. The distribution of Nu is opposite to the temperature distribution of the DSJA in Figure 8. The area where the thermal boundary layer is damaged has a large Nu number. The importance of breaking the thermal boundary layer for enhancing heat transfer is further illustrated. In addition, at the moment of 0.75 T~0 T (1 T), a larger Nu number is observed in the area of $y < 0$. It indicates that the lower cavity is in the injection stage at this time. Since there are only two rows of jet outlets in the lower cavity, the jet velocity is higher. At 0.75 T, the impact area is wider, corresponding to the wider Nu distribution in Figure 9. In a word, the effect of the DSJ increases the Nu number and presents a "wavy" Nu distribution curve. It reflects the effective disturbance of the DSJ to the flow field.

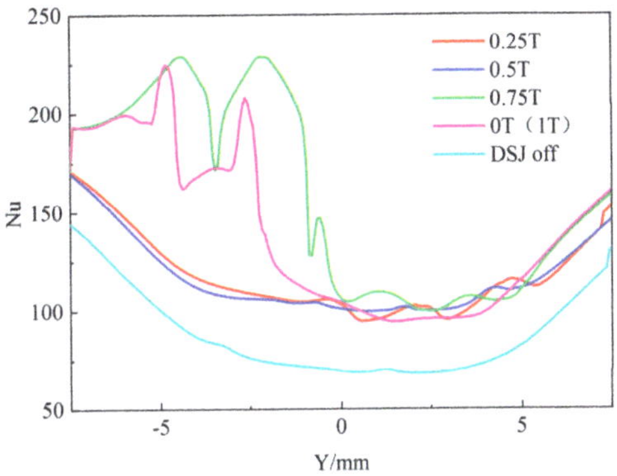

Figure 9. The distribution of *Nu* on the impact surface (z = 44 mm) along the *y* direction.

3.2. Flow Field Characteristics

This section studies the flow field characteristics of cooling devices, as shown in Figures 10–13. It can be observed from Figure 10 that when there is only a channel, the streamlines in the channel are parallel to each other along the streamlines. The fluid in the channel does not affect each other. Thick velocity boundary layers were observed near the upper and lower walls of the channel.

When the DSJ is off, the maximum velocity in the flow field reaches 4.5 m/s. The fluid entering the channel flows parallel to the channel at the beginning, and its flow path is disturbed when it passes near the jet orifice under the influence of the jet. The fluid becomes disordered. The channel flow is "squeezed" upward by the jet, and a low-velocity area (the blue part in Figure 11) is observed between adjacent jets. At the same time, the jet deflects along the flow direction under the action of channel flow. The fourth and fifth rows of jets deflect most obviously. It is consistent with the phenomenon described in Section 3.1.

When the DSJ is on, the flow field distribution of one cycle of the DSJA is shown in Figures 12 and 13. At 0 T, the actuator diaphragm vibrates downward. The upper cavity volume expands while the lower cavity volume compresses. An obvious vortex structure was observed in the upper cavity. Since the vibration of the diaphragm is small at this time, the flow from the actuator will inhibit the suction of fluid in the channel. In addition, when the jet interacts with the incoming flow, the flow in the channel is disordered, while when the jet is far away, the flow in the channel remains stable. The region of a steady fluid flow corresponds to the gradually thickened thermal boundary layer in Figure 8a. It further shows that the enhancement of fluid kinetic energy destroys the thermal boundary layer and enhances heat transfer. At 0.25 T, the upper cavity ejects fluid, and the jet impinges on the aluminum plate surface to form an obvious low-temperature area. At 0.5 T, the diaphragm is in a balanced position. The ability of the upper cavity to eject the jet is weakened. The jet deflects under the influence of incoming flow. At 0.75 T, the lower cavity ejects a strong jet. An obvious vortex structure is formed in the channel, and the temperature of the jet impingement zone is obviously reduced. At the same time, it was observed that the fluid in some channels was sucked into the upper cavity. The large vortex structure was observed above the third, fourth, and fifth jet orifices, because it was affected by the three actions of ejected fluid, inhaled fluid, and incoming flow. The resulting temperature distribution is consistent with that observed in Figure 8.

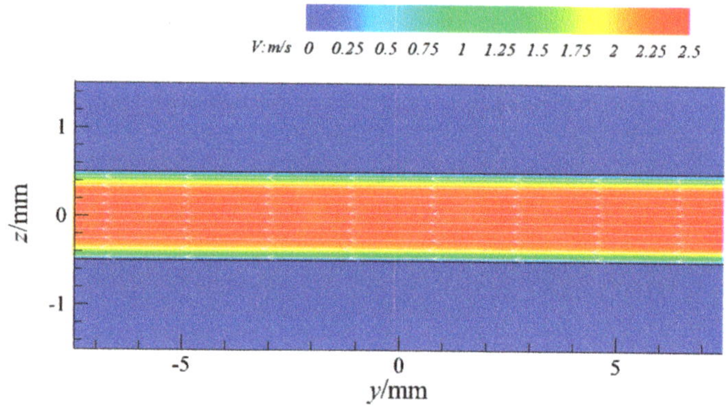

Figure 10. Velocity distribution of channel's center section (x = 28.5 mm).

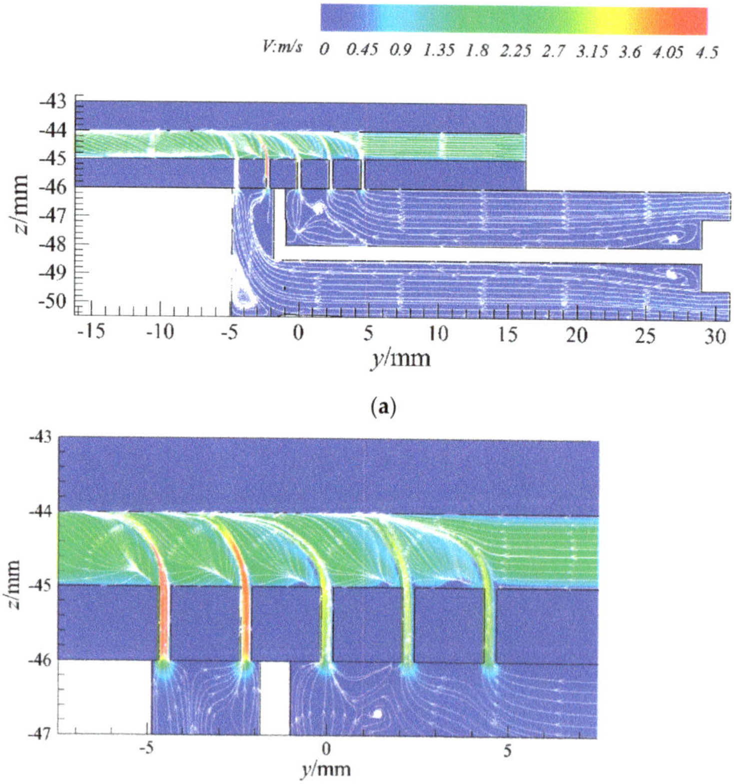

Figure 11. Velocity distribution (DSJ off) of channel's center section (x = 28.5 mm): (**a**) overall view; (**b**) partial enlarged view of jet orifice area.

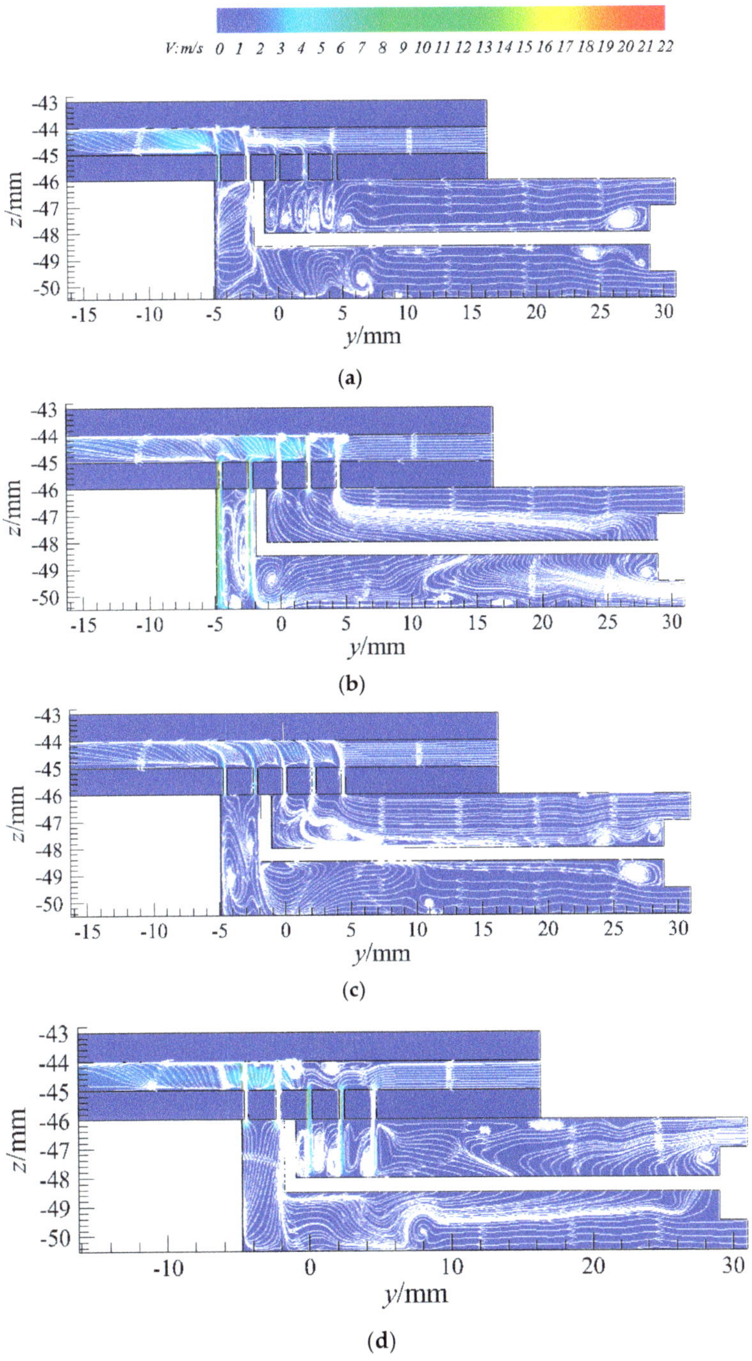

Figure 12. Velocity distribution of channel's center section (x = 28.5 mm, overall view) (DSJ on): (**a**) 0 T (1 T); (**b**) 0.25 T; (**c**) 0.5 T; (**d**) 0.75 T.

To further analyze the flow field characteristics of the DSJA, the vorticity distribution of the DSJA is shown in Figure 14. When the DSJ is off, only large vortices are observed in the channel, especially near the jet orifices, as shown in Figure 14a. The vorticity distribution of one cycle after the DSJ is turned on is shown in Figure 14b–e. It can be found that the vorticity increases significantly, and the vorticity distribution also changes significantly. Under the action of the DSJ, the fluid in the channel is affected by alternate blowing and suction, forming a periodic disturbance. A larger range of large vorticity distribution was observed in the channel. In addition, the reciprocating up and down the vibration of the diaphragm increases the vorticity in the cavity. It further shows that the fluid mixing under the action of the DSJ is enhanced, which enhances the convective heat transfer capability of the fluid.

3.3. Entropy Generation Analysis

The entropy generation rate is used to reflect the irreversibility of the heat transfer and flow process. This section studies the entropy generation characteristics of the DSJA from the perspective of the second law of thermodynamics. The total entropy production rate includes the entropy production rate caused by irreversible heat transfer and the irreversible flow loss caused by fluid friction.

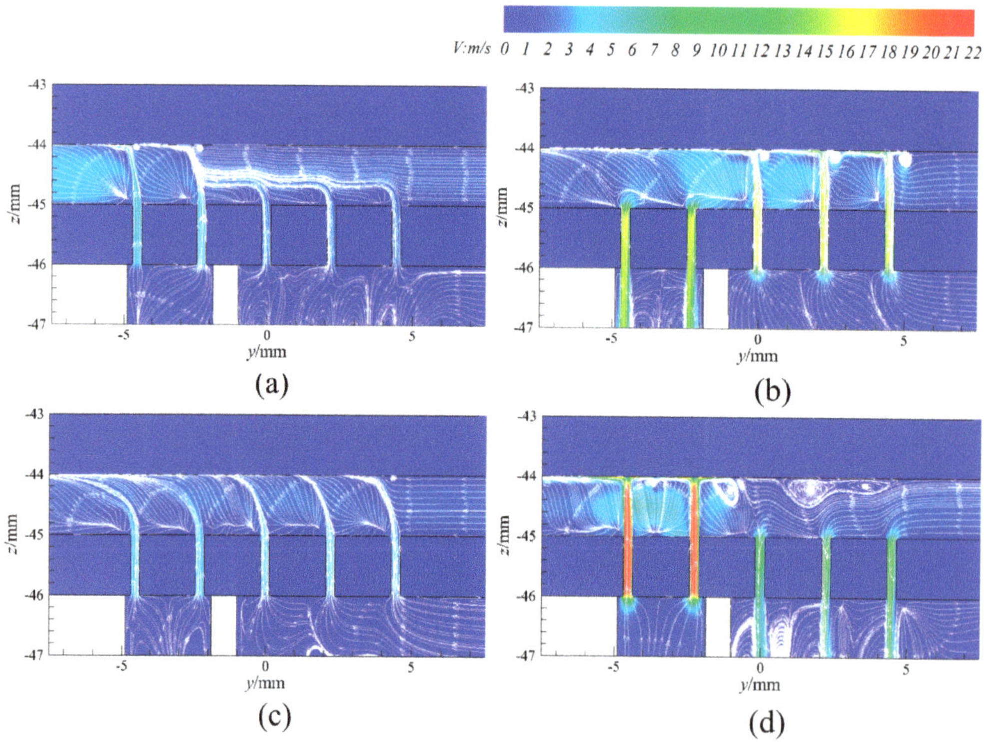

Figure 13. Velocity distribution of channel's center section (x = 28.5 mm, partial enlarged view of jet orifice area) (DSJ on): (**a**) 0 T (1 T); (**b**) 0.25 T; (**c**) 0.5 T; (**d**) 0.75 T.

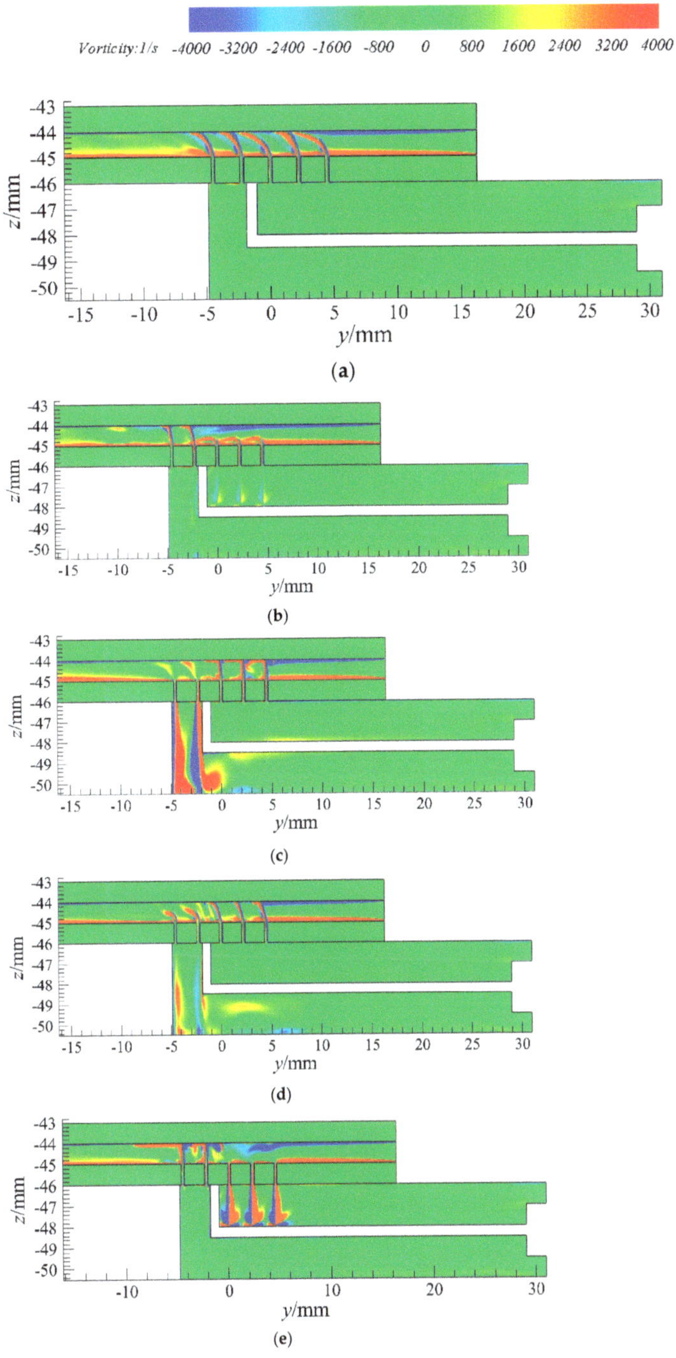

Figure 14. X vorticity distribution of channel's center section (x = 28.5 mm, overall view): (**a**) DSJ off; (**b**) 0 T (1 T); (**c**) 0.25 T; (**d**) 0.5 T; (**e**) 0.75 T (**d**,**e**, DSJ on).

The thermal entropy generation rate of the DSJ when off and the DSJ when on are shown in Figures 15 and 16. When the DSJ is off, the higher thermal entropy generation rate is mainly located at the interface between the fluid domain and the solid domain. This is mainly due to the large temperature gradient in these places. The water temperature at the inlet and outlet of the channel is low, the temperature gradient is large, and the thermal entropy generation rate is higher. When the DSJ is on, a high thermal entropy generation rate is also observed at the fluid–solid interface. The heat transfer in the region where the thermal boundary layer exists is poor and the thermal entropy generation rate is high. Under the disturbance of the DSJ, the distribution of the thermal entropy generation rate near the jet orifice is obviously disordered. At the same time, the impingement of the jet makes the thermal boundary layer thinner and the thermal entropy generation rate lower.

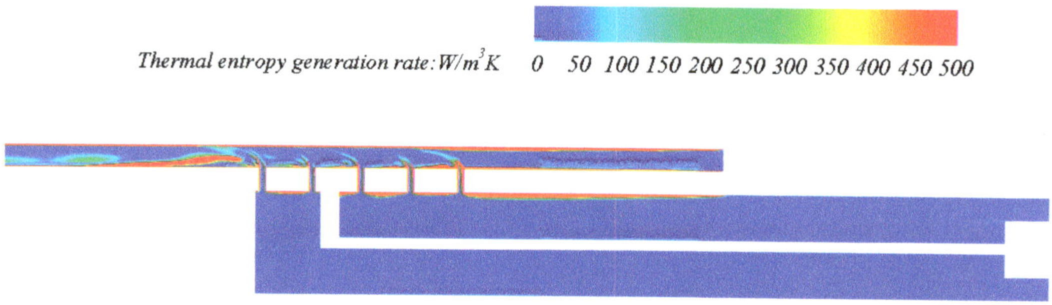

Figure 15. Thermal entropy generation rate in the fluid domain (x = 28.5 mm, DSJ off).

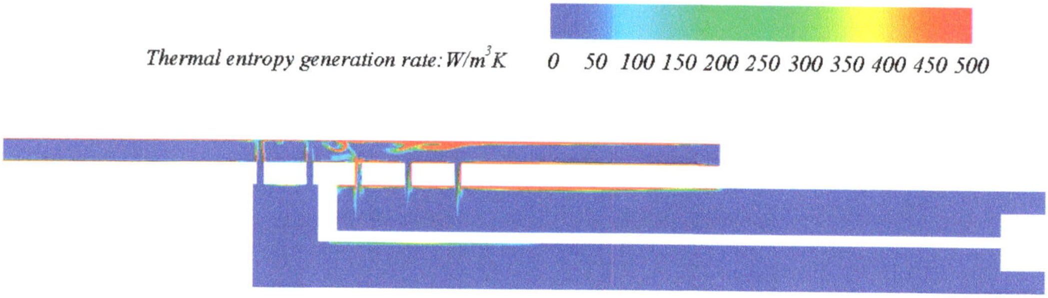

Figure 16. Thermal entropy generation rate in the fluid domain (x = 28.5 mm, 0.75 T of DSJ on).

The frictional entropy generation rate of the DSJ when off and the DSJ when on are shown in Figures 17 and 18. The higher frictional entropy generation rate is mainly located near the inlet and outlet of the channel and the jet orifice, especially in the jet impingement area. This is mainly due to the large velocity gradient at the inlet and outlet of the channel. The jet orifice diameter is small, and the velocity will change suddenly. When the DSJ is off, the jet deflects under the influence of incoming flow. The deflection area has a low-velocity area. Therefore, the frictional entropy generation rate is large. This is consistent with the phenomenon observed in Figure 11.

When the DSJ is on, Figure 18 observed the frictional entropy output value in a larger area than when the DSJ is off. This is because the diaphragm is at a limited position of downward vibration at this moment. The lower cavity ejects the jet and the upper cavity inhales the jet. The high strength of the jet results in a large velocity gradient. Observing the DSJ-off and DSJ-on cases, it can be found that the frictional entropy generation increases with the increase of the velocity of the DSJA. This is due to the increase in velocity gradient and the decrease in temperature. In the whole heat dissipation device, the greater the velocity, the more obvious the frictional entropy generation. Small channels and jet orifices favor greater velocity gradients.

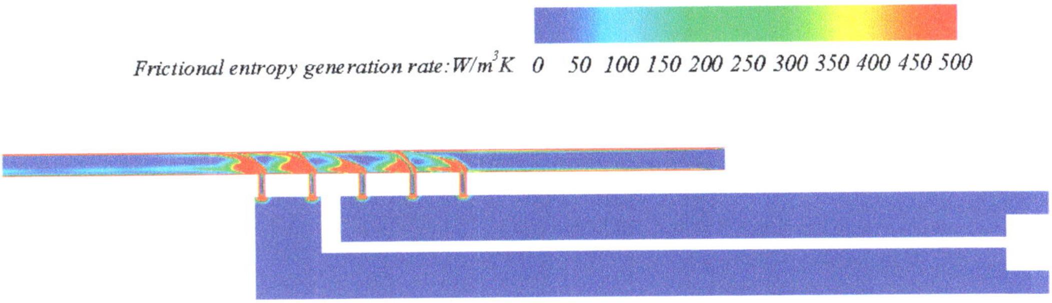

Figure 17. Frictional entropy generation rate in the fluid domain (x = 28.5 mm, DSJ off).

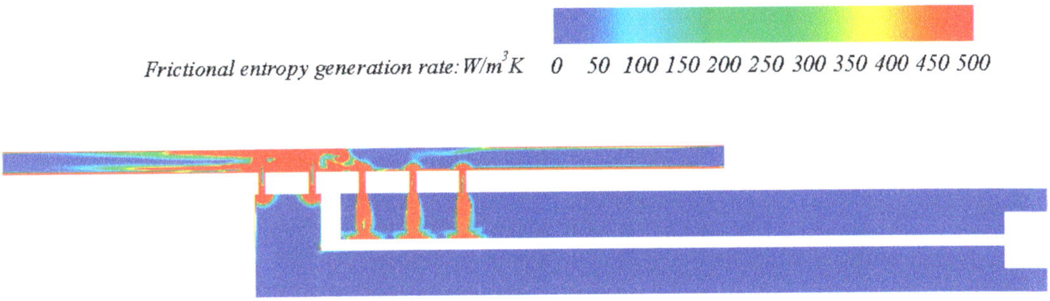

Figure 18. Frictional entropy generation rate in the fluid domain (x = 28.5 mm, 0.75 T of DSJ on).

The total entropy generation rate of the DSJ when off and the DSJ when on are shown in Figures 19 and 20. It can be seen that the high total entropy generation rate is mainly located near the jet orifice and the inlet and outlet of the channel. We use Formula (12) to calculate the overall value of entropy, as shown in Table 2. It can be seen that the total entropy generation is dominated by the value of thermal entropy generation because the contribution of frictional entropy generation is very small. Since there is no velocity in the solid domain, its frictional entropy generation rate is 0. The total entropy generation in the fluid domain is much greater than that in the solid domain. Moreover, the total entropy generation of the DSJ when off is 0.01309, while the value of the DSJ when on is only 0.00414. The entropy production is reduced, and the latter is more than three times lower than the former. It indicates that the irreversible loss of the DSJA is greatly reduced.

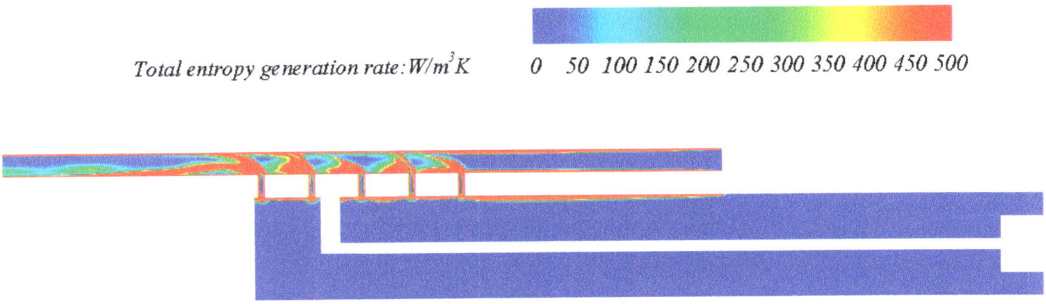

Figure 19. Total entropy generation rate in the fluid domain (x = 28.5 mm, DSJ off).

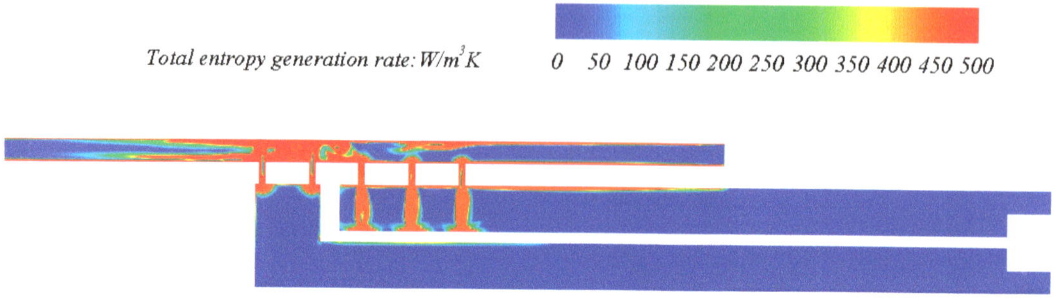

Figure 20. Total entropy generation rate in the fluid domain (x = 28.5 mm, 0.75 T of DSJ on).

Table 2. Thermal entropy generation, frictional entropy generation, and total entropy generation in different areas of the cooling device.

Case	Fluid			Solid		Cooling Device		
	Thermal	Frictional	Total	Thermal	Total	Thermal	Frictional	Total
DSJ off	0.01298	0.00005	0.01303	0.00006	0.00006	0.01304	0.00005	0.01309
DSJ on (0.75 T)	0.00380	0.00030	0.00410	0.00004	0.00004	0.00384	0.00030	0.00414

3.4. Effect of Diaphragm Frequency

The influence of the diaphragm frequency on the performance of the cooling device was studied, keeping other settings unchanged.

The change of the periodic average temperature of the wall along the y direction along the channel with the diaphragm frequency is shown in Figure 21. The temperature distribution of the impact surface is not monotonic with the frequency. The impact surface temperature is the highest when the frequency is 10 Hz. When the frequency is 30 Hz, the impact surface has the lowest temperature and the temperature change is more obvious. The temperature near the jet impingement point is lower than that in other areas. This is because the effect of the dual synthetic jets strengthens the fluid-mixing and improves the convection heat transfer capacity of the fluid.

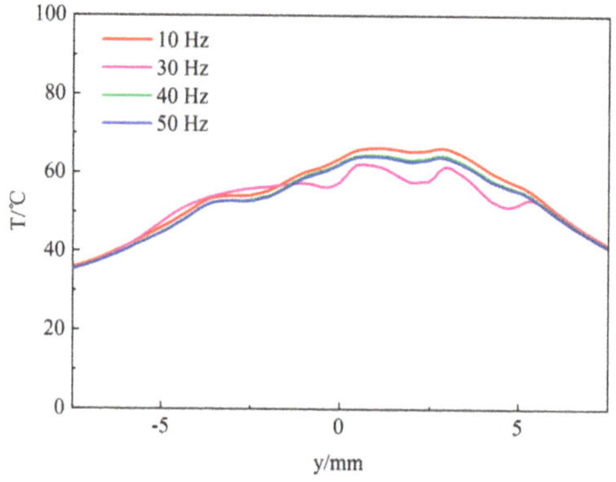

Figure 21. The change of the periodic average temperature of the wall along the y direction along the channel with the diaphragm frequency (x = 28.5 mm).

The channel inlet and outlet pressure drop changes with the diaphragm frequency as shown in Figure 22. The channel inlet and outlet pressure drop is the lowest when the diaphragm frequency is 10 Hz and the highest when the diaphragm frequency is 40 Hz.

It shows that the pressure drop at the inlet and outlet of the channel is not monotonous with the change in the diaphragm frequency. It can be seen from Figures 21 and 22 that the cooling device has the best performance due to the optimal frequency. In the frequency range studied in this paper, the best comprehensive performance of the heat sink is obtained when the frequency is 30 Hz.

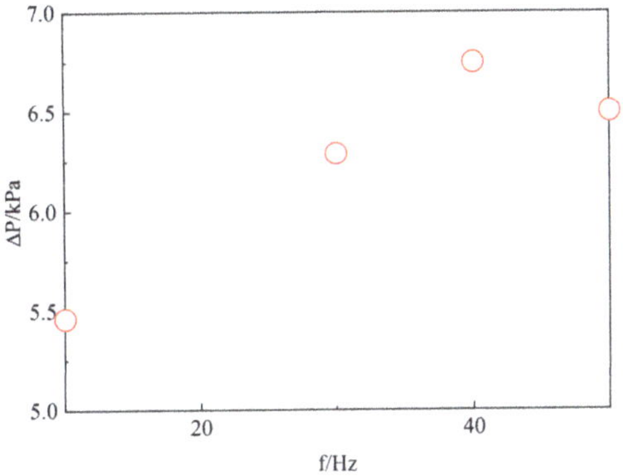

Figure 22. The channel inlet and outlet pressure drop changes with the diaphragm frequency (x = 28.5 mm).

4. Conclusions

In this paper, combined with the DSJ technology, a forward liquid cooling device is designed. The heat transfer and flow characteristics of the device are analyzed. The entropy generation of the DSJA is analyzed from the point of view of the second law of thermodynamics. The main conclusions are as follows

- When the DSJ is off, the cooling device observes an asymmetric temperature distribution. The maximum temperature reaches 98.22 °C. When the DSJ is on, the maximum temperature is only 73.07 °C. A 25.60% heat transfer enhancement was achieved. The temperature field of the cooling device changes periodically with the DSJ. The thermal boundary layer attached to the impact surface was periodically destroyed by the DSJ. It is helpful to enhance convection heat transfer.
- When the DSJ is off, the jet deflects under the influence of a flow velocity of 1.5 m/s in the channel. It is difficult to achieve full-impact heat dissipation on the wall. The DSJ on solves this problem. At this time, the DSJ can directly impact the wall and interacts with the incoming flow to enhance mixing. The rich vortex structure is helpful for convection heat transfer.
- The results of entropy generation analysis show that total entropy generation mainly comprises the contribution of thermal entropy generation, and the influence of frictional entropy generation is small. The entropy generation of the DSJ when on is much lower than that of the DSJ when off.
- The optimal frequency results in the best overall performance of the cooling device. This section studies the performance of the cooling device when the diaphragm frequency is 10 Hz, 30 Hz, 40 Hz, and 50 Hz. It is found that the device has the best comprehensive performance when the frequency is 30 Hz.

Author Contributions: Writing: original draft, Y.K.; Writing: review and editing, Z.L., X.D., Y.Z. and Z.X. All authors have read and agreed to the published version of the manuscript.

Funding: This research was funded by the National Natural Science Foundation of China, grant numbers 11872374, 11972369; the Science and Technology Innovation Program of Hunan Province, grant number 2021RC3075; and the Natural Science Foundation of Hunan Province of China, grant number 2021JJ40672.

Institutional Review Board Statement: Not applicable.

Informed Consent Statement: Not applicable.

Data Availability Statement: Not applicable.

Conflicts of Interest: The authors declare no conflict of interest.

References

1. Hao, X.H.; Peng, B.; Xie, G.N.; Chen, Y. Efficient on-chip hotspot removal combined solution of thermoelectric cooler and mini-channel heat sink. *Appl. Therm. Eng.* **2016**, *100*, 170–178. [CrossRef]
2. Viswanath, R.; Wakharkar, V.; Watwe, A.; Lebonheur, V. Thermal performance challenges from silicon to systems. *Int. Technol. J.* **2000**, *3*, 1–16.
3. Tuckerman, D.B.; Pease, R.F.W. High-performance heat sinking for VLSI. *IEEE Electron Device Lett.* **1981**, *2*, 126–129. [CrossRef]
4. Zeng, S.; Lee, P.S. Topology optimization of liquid-cooled microchannel heat sinks: An experimental and numerical study. *Int. J. Heat Mass Transf.* **2019**, *142*, 118401. [CrossRef]
5. Zhuang, D.W.; Yang, Y.F.; Ding, G.L.; Du, X.Y.; Hu, Z.T. Optimization of microchannel heat sink with rhombus fractal-like units for electronic chip cooling. *Int. J. Refrig.* **2020**, *116*, 108–118. [CrossRef]
6. Alugoju, U.K.; Dubey, S.K.; Javed, A. Optimization of converging and diverging microchannel heat sink for electronic chip cooling, IEEE Transactions on Com-ponents. *Packag. Manuf. Technol.* **2020**, *10*, 817–827. [CrossRef]
7. Tan, H.; Wu, L.W.; Wang, M.Y.; Yang, Z.H.; Du, P.A. Heat transfer improvement in microchannel heat sink by topology design and optimization for high heat flux chip cooling. *Int. J. Heat Mass Transf.* **2019**, *129*, 681–689. [CrossRef]
8. Wang, R.; Qian, J.Y.; Wei, T.; Huang, H.J. Integrated closed cooling system for high-power chips. *Case Stud. Therm. Eng.* **2021**, *26*, 100954. [CrossRef]
9. Ma, D.D.; Xia, G.D.; Wang, J.; Yang, Y.C.; Jia, Y.T.; Zong, L.X. An experimental study on hydrothermal performance of microchannel heat sinks with 4-ports and offset zigzag channels. *Energy Convers. Manag.* **2017**, *152*, 157–165. [CrossRef]
10. Zhang, J.R.; Zhang, T.T.; Prakash, S.; Jaluria, Y. Experimental and numerical study of transient electronic chip cooling by liquid flow in microchannel heat sinks. *Numer. Heat Transf. Part A Appl.* **2014**, *65*, 627–643. [CrossRef]
11. Xie, G.N.; Liu, J.; Zhang, W.H.; Sunden, B. Analysis of flow and thermal performance of a water-cooled transversal wavy microchannel heat sink for chip cooling. *J. Electron. Packag.* **2012**, *134*, 041010. [CrossRef]
12. Tseng, P.H.; Tsai, K.T.; Chen, A.L.; Wang, C.C. Performance of novel liquid-cooled porous heat sink via 3-D laser additive manufacturing. *Int. J. Heat Mass Transf.* **2019**, *137*, 558–564. [CrossRef]
13. Chiu, H.C.; Jang, J.H.; Yeh, H.W.; Wu, M.S. The heat transfer characteristics of liquid cooling heatsink containing microchannels. *Int. J. Heat Mass Transf.* **2011**, *54*, 34–42. [CrossRef]
14. Robinson, A.J. A thermal–hydraulic comparison of liquid microchannel and impinging liquid jet array heat sinks for high-power electronics cooling. *IEEE Trans. Compon. Packag. Technol.* **2009**, *32*, 347–357. [CrossRef]
15. Smith, B.L.; Glezer, A. The formation and evolution of synthetic jets. *Phys. Fluids* **1998**, *10*, 2281–2297. [CrossRef]
16. van Buren, T.; Whalen, E.; Amitay, M. Vortex formation of a finite-span synthetic jet: Effect of rectangular orifice geometry. *J. Fluid Mech.* **2014**, *745*, 180–207. [CrossRef]
17. Bazdidi-Tehrani, F.; Eghbali, A.; Karami, M. Influence of jet-to-surface distance and frequency on unsteady heat transfer and mass flowrates in an impingement synthetic jet. *J. Enhanc. Heat Transf.* **2013**, *20*, 115–136. [CrossRef]
18. Zhang, J.Z.; Tan, X.M. Experimental study on flow and heat transfer characteristics of synthetic jet driven by piezoelectric actuator. *Sci. China Ser. E Technol. Sci.* **2007**, *50*, 221–229. [CrossRef]
19. Jagannatha, D.; Narayanaswamy, R.; Chandratilleke, T.T. Analysis of a synthetic jet-based electronic cooling module. *Numer. Heat Transf. Part A Appl.* **2009**, *56*, 211–229. [CrossRef]
20. Luo, Z.B.; Xia, Z.X.; Liu, B. New Generation of Synthetic Jet Actuators. *AIAA J.* **2006**, *44*, 2418–2419. [CrossRef]
21. Luo, Z.B.; Zhao, Z.J.; Deng, X.; Wang, L.; Xia, Z.X. Dual Synthetic Jets Actuator and Its Applications—Part I: PIV Measurements and Comparison to Synthetic Jet Actuator. *Actuators* **2022**, *11*, 205. [CrossRef]
22. Deng, X.; Luo, Z.B.; Xia, Z.X.; Gong, W.J. Experimental investigation on the flow regime and impingement heat transfer of dual synthetic jet. *Int. J. Therm. Sci.* **2019**, *145*, 105864. [CrossRef]
23. Jagannatha, D.; Chandratilleke, T.T.; Narayanaswamy, R. Interactive flow behaviour and heat transfer enhancement in a microchannel with cross flow synthetic jet. *Int. J. Emerg. Multidiscip. Fluid Sci.* **2010**, *2*, 27–43. [CrossRef]
24. Lee, A.; Yeoh, G.H.; Timchenko, V.; Reizes, J. Numerical computation and investigation of the characteristics of microscale synthetic jets. *Model. Simul. Eng.* **2011**, *2011*, 12. [CrossRef]
25. Pokharel, P. Convective Heat Transfer Enhancement of a Channel-Flow Using Synthetic Jet. Doctoral Dissertation, The University of Mississippi, Oxford, MS, USA, May 2021.
26. Lee, A.; Yeoh, G.H.; Timchenko, V.; Reizes, J.A. Heat transfer enhancement in micro-channel with multiple synthetic jets. *Appl. Therm. Eng.* **2012**, *48*, 275–288. [CrossRef]
27. Lau, G.E.; Mohammadpour, J.; Lee, A. Cooling performance of an impinging synthetic jet in a microchannel with nanofluids: An Eulerian approach. *Appl. Therm. Eng.* **2021**, *188*, 116624. [CrossRef]

28. Qiu, Y.L.; Hu, W.J.; Wu, C.J.; Chen, W.F. Flow and heat transfer characteristics in a microchannel with a circular synthetic jet. *Int. J. Therm. Sci.* **2021**, *164*, 106911. [CrossRef]
29. Giachetti, B.; Fénot, M.; Couton, D.; Plourde, F. Influence of Reynolds number synthetic jet dynamic in crossflow configuration on heat transfer enhancement. *Int. J. Heat Mass Transf.* **2018**, *118*, 1–13. [CrossRef]
30. Fang, R.X.; Khan, J.A. Active heat transfer enhancement in single-phase microchannels by using synthetic jets. *J. Therm. Sci. Eng. Appl.* **2013**, *5*, 011006. [CrossRef]
31. Mohammadpour, J.; Salehi, F.; Lee, A.; Brandt, L. Nanofluid heat transfer in a microchannel heat sink with multiple synthetic jets and protrusions. *Int. J. Therm. Sci.* **2022**, *179*, 107642. [CrossRef]
32. Mohammadpour, J.; Salehi, F.; Lee, A. Performance of nano encapsulated phase change material slurry heat transfer in a microchannel heat sink with dual-circular synthetic jets. *Int. J. Heat Mass Transf.* **2022**, *184*, 122265. [CrossRef]
33. Peng, L.; Luo, Z.B.; Deng, X.; Yang, S.K. Experimental investigation on characteristics of flow field and propulsion of dual synthetic jets in water. *Acta Aerodyn. Sin.* **2017**, *35*, 290–298.
34. Ikhlaq, M.; Yasir, M.; Demiroğlu, M.; Arik, M. Synthetic Jet Cooling Technology for Electronics Thermal Management—A Critical Review. IEEE Transactions on Components. *Packag. Manuf. Technol.* **2021**, *11*, 1156–1170. [CrossRef]
35. Paniagua-Guerra, L.E.; Ramos-Alvarado, B. Efficient hybrid microjet liquid cooled heat sinks made of photopolymer resin: Thermo-fluid characteristics and entropy generation analysis. *Int. J. Heat Mass Transf.* **2020**, *146*, 118844. [CrossRef]
36. Bahiraei, M.; Heshmatian, S.; Keshavarzi, M. A decision-making based method to optimize energy efficiency of ecofriendly nanofluid flow inside a new heat sink enhanced with flow distributor. *Powder Technol.* **2019**, *342*, 85–98. [CrossRef]
37. Mahian, O.; Kianifar, A.; Kleinstreuer, C.; Al-Nimr, M.A.; Pop, I.; Ahmet, Z.; Wongwises, S.S. A review of entropy generation in nanofluid flow. *Int. J. Heat Mass Transf.* **2013**, *65*, 514–532. [CrossRef]
38. Kang, Y.; Luo, Z.B.; Deng, X.; Cheng, P.; Peng, C.; He, W.; Xia, Z.X. Numerical study of a liquid cooling device based on dual synthetic jets actuator. *Appl. Therm. Eng.* **2023**, *219*, 119691. [CrossRef]

actuators

MDPI

Article

Evolution and Near-Wall Effect of the Vortex Structures Induced by In-Line Twin Synthetic Jets in a Crossflow

Hongxin Wang [1,2], Degang Xu [1,*], Linwen Li [2], Kaiwen Zhou [3], Xin Wen [3,*] and Hui Tang [4]

1 School of Automation, Central South University, Changsha 410083, China
2 Shanghai Aircraft Design and Research Institute, Shanghai 201210, China
3 Key Lab of Education Ministry for Power Machinery and Engineering, School of Mechanical Engineering, Shanghai Jiao Tong University, 800 Dongchuan Road, Shanghai 200240, China
4 Department of Mechanical Engineering, The Hong Kong Polytechnic University, Kowloon, Hong Kong, China
* Correspondence: dgxu@csu.edu.cn (D.X.); wenxin84@sjtu.edu.cn (X.W.)

Abstract: This paper aims to further the understanding of the mixing process of in-line twin synthetic jets (SJs) and their impact in the near-wall region in a flat-plate laminar boundary layer. A numerical study has been carried out, in which colored fluid particles and the Q criterion are used to track the SJ-induced vortex structures at the early stage of the evolution. Interacting vortex structures at four selected phase differences are presented and analyzed. It is found that the fluid injected at the early stage of the blowing stroke mainly contributes to the formation of the hairpin legs, the fluid injected near the maximum blowing mainly contributes to the formation of the hairpin head, and the fluid injected at the late stage of the blowing stroke contributes very little to the formation of the hairpin vortex. It is also confirmed that, irrespective of the phase difference, the hairpin vortex issued from the upstream actuator is more capable of maintaining its coherence than its counterpart issued from the downstream actuator. The influence of the interacting vortex structures on the boundary layer is also studied through investigating excess wall shear stress. In all cases, a pair of streaks of high wall shear stress can be observed with similar size. Among them, the streaks have the strongest wall shear stress, with the largest gap at phase difference 0 when partially interacting vortex structures are produced. The findings can provide valuable guiding information for the applications of synthetic jets in heat transfer, mixing control, and flow control in a crossflow.

Keywords: twin synthetic jets; laminar boundary layer; vortex interaction; simulation

Citation: Wang, H.; Xu, D.; Li, L.; Zhou, K.; Wen, X.; Tang, H. Evolution and Near-Wall Effect of the Vortex Structures Induced by In-Line Twin Synthetic Jets in a Crossflow. *Actuators* **2022**, *11*, 234. https://doi.org/10.3390/act11080234

Academic Editor: André Preumont

Received: 22 July 2022
Accepted: 12 August 2022
Published: 16 August 2022

1. Introduction

A synthetic jet (SJ) actuator can produce a train of vortical structures by the periodical motion of an oscillatory diaphragm on a cavity [1–3]. These vortical structures can effectively enhance fluid mixing in the crossflow, thereby providing an effective mechanism of boundary layer flow control [4–9].

Previous works were carried out to investigate the impact of the SJ-induced vortical structures in the near-wall region of a boundary layer, where effective flow separation control and wall heat transfer are ultimately desired. Using dye visualization, Zhong et al. [10] and Jabbal and Zhong [11] looked at more detailed interactions and identified three key SJ-induced vortex structures, i.e., hairpin vortices, stretched vortex rings, and tilted vortex rings, under different jet operation and boundary layer flow conditions. Different thermal footprints on the wall are correlated to the vortical structures induced by SJs. Using high-frame-rate particle image velocimetry (PIV) experiments, Jabbal and Zhong [12] confirmed that the wall shear stress shares similar patterns with the thermal footprints, and found that hairpin vortices and stretched vortex rings are able to offer enhanced heat transfer in the near-wall region. Using numerical methods, Lee et al. [13] confirmed the effective wall cooling of single SJs in micro-channels. They also found that the flow structures induced by stronger SJs in the micro-channel can exert larger mixing regions and stronger convection at the heated wall surface.

To increase the impact region, SJ arrays are usually applied. For example, in the application of fluid cooling, Lee et al. [14] found that twin SJs in out-of-phase operation can significantly enhance wall surface cooling due to the enhanced fluid mixing. However, unlike single SJs, the interactions between multiple circular SJs and cross-flows, as well as among multiple SJs, have not been fully studied and understood. Very limited investigations have been reported on this topic in the literature, especially on the near-wall effect. Based on oil visualization, Watson et al. [15] examined the in-teraction of two neighboring SJs in a boundary layer flow, and pointed out that the twin SJs can perform in either a constructive (enhancing the strength of the SJs) or a destructive (weakening the SJs) manner under different configurations. Using hot-wire measurements, Liddle and Wood [16] and Liddle et al. [17] confirmed that the phase difference plays a very important role in the interaction of twin SJs in a crossflow. At a 2700 phase difference, the SJ actuators in the upstream and downstream locations yielded a more concentrated movement of near-wall fluid towards the vortex core than at a 900 phase difference. With PIV measurements of in-line twin SJs in a channel flow, Honami and Motosuke [18] found that the twin SJs can have different vortex strengths with different configurations. Although these studies revealed useful physical insights, they are not enough to give a holistic picture of the flow structures induced by multiple circular SJs in a crossflow and their near-wall effect. Therefore, more detailed studies are highly desired.

Although these works have revealed some useful physical insights into the interaction of in-line twin SJs in boundary layers, the detailed vortex interaction and evolution of twin SJs are not well understood. Limited by the spatial and temporal resolution, it is difficult for current experimental methods to capture the three-dimensional flow structures of twin SJs. In addition, it is also difficult to examine the evolution of the interacting flow structure issued from different SJ actuators in Cartesian coordinates. Therefore, two research questions still remain: First, how does the mixing process of the twin-SJ-induced flow structures occur in detail? Second, what is the influence of these flow structures on the boundary layer flow, especially in the near-wall region? By examining the flow structures and their impact in the boundary layer, we can obtain results which are valuable for flow control and heat transfer applications. For example, strong vortices and high wall shear stress can produce high flow mixing in the boundary layer, delaying flow separation and enhancing heat transfer on the wall. Therefore, to address these questions in the present research, a numerical study is conducted. A cluster of colored fluid particles are used to track the evolution and mixing process of the single and twin SJs at the early stage. In the Lagrangian perspective, fluid particles are applied to examine the flow dynamical system rather than a continuum [19]. It is widely used to present high-quality visible details of time-dependent vortex structures, which may not be apparent in the classical Eulerian fields [20,21]. In addition, to capture the associated vortex structures, Q iso-surfaces are defined [22]. Finally, the excess wall shear stress is eval-uated and compared.

2. Methodology

2.1. Problem Description

The twin SJ actuators used in the present numerical study are identical to those used in our previous experiment [23,24]. The SJ cavity has an orifice with a diameter of D_o = 5 mm. The centers of the two SJ actuator orifices are at a distance of d = 10 mm. Similar to previous experiments [23], the oscillation diaphragm of each actuator has a diameter of D_d = 45 mm, which oscillates like a piston, with velocity varying in a sinusoidal manner.

$$u_d(t) = \pi \Delta f \sin(2\pi f t) \quad (1)$$

where Δ is the peak-to-peak displacement of the diaphragm and f the oscillation frequency. During an actuation cycle, the beginning of blowing is defined as t/T = 0 (and 1). In the present study, the working flow is water, which is incompressible. According to Tang and Zhong [25], the jet velocity $\overline{U}_o$ averaged over an entire actuation cycle can be calculated as

$$\overline{U}_o = \frac{1}{T}\int_0^{T/2} u_o(t)dt = \Delta f\left(\frac{D_d}{D_o}\right)^2 \quad (2)$$

where the instantaneous jet velocity is $u_o(t)$.

In current study, the bulk flow has a velocity of U_∞ = 0.055 m/s. The bulk flow has a boundary layer thickness of 15 mm, i.e., $\delta = 3D_o$, at the actuator orifices. Based on Reynolds number, the momentum thickness is $Re_\theta \approx 106$. The actuators have a diaphragm displacement of Δ = 0.158 mm with a frequency of f = 1 Hz. To characterize the performance of SJs in a laminar boundary layer, the dimensionless stroke length L and velocity ratio VR are also defined according to Wen and Tang [24]. The stroke length L is defined as

$$L = \frac{L_o}{D_o} = \frac{\overline{U}_o}{fD_o} \quad (3)$$

where L_o is the jet column length during the SJ blowing stroke. The jet strength is characterized by the velocity ratio VR as

$$VR = \frac{\overline{U}_o}{U_\infty} \quad (4)$$

From Equations (2)–(4), the settings in the present study yield L = 2.6 and VR = 0.23, to produce hairpin vortices according to the parametric map revealed by Jabbal & Zhong [11]. These hairpin vortices can stay in the boundary layer and hence are very desired in SJ-based flow separation delay.

In the present study, the two SJ actuators are arranged in line with the crossflow and are separated by a fixed distance $d = 2D_o$. Phase difference between the two SJs are investigated by $\Delta\phi = 0, \pi/2, \pi$ and $3\pi/2$ (or $-\pi/2$). Here $\Delta\phi = 0$ means that the twin SJs are in phase, and the $\Delta\phi = \pi/2$ means the downstream actuator is $\pi/2$ behind, and so on.

2.2. Numerical Approaches

Figure 1 shows the computational geometry and boundary conditions, which are similar to those used in Wen and Tang [26] and Zhou and Zhong [27]. In the x (streamwise) direction, the simulation domain has a length of 55D$_o$. In the y (wall-normal) and z (spanwise) directions, the domain is $30D_o$ and $20D_o$, respectively. The origin of the coordinate system is set at the midpoint of two actuator orifice centers, which are $15D_o$ downstream from the crossflow inlet. Although not presented in Figure 1, the cavities of the two SJ actuators are included in the simulation. The crossflow in-let on the left of the computational domain is set as the velocity inlet with a laminar Blasius velocity profile. The wall of flat plate in the boundary layer and the SJ actuators are set as no-slip conditions. Both lateral sides of the crossflow domain are set as periodic conditions. The upper face and the downstream boundary on the right of the crossflow domain are set as the pressure outlet. On the diaphragms of the actuators, a time-dependent velocity condition is set according to Equation (1).

To activate the SJ actuator, the motion of the diaphragm in the cavity is controlled by setting a time-dependent velocity boundary condition according to Equation (1). In this study, water is used, which is incompressible. The fluid is governed by mass con-servation and momentum conservation. The flow is laminar; therefore, no turbulent model is used. The second-order implicit scheme is used in time, and the second-order upwind scheme is used in space. For the pressure–velocity coupling, the pressure-implicit with splitting of operators method is used. The unsteady, three-dimensional incompressible Navier–Stokes equations are solved by a commercial CFD code, ANSYS FLUENT 2020 R2.

$$\frac{\partial u_i}{\partial x_i} = 0 \quad (5)$$

$$\frac{\partial u_i}{\partial t} + u_j \frac{\partial u_i}{\partial x_j} = -\frac{1}{\rho}\frac{\partial p}{\partial x_i} + v\frac{\partial^2 u_i}{\partial x_j \partial x_j} \tag{6}$$

where ρ is the fluid density and ν is viscosity, and the summation notion is applied. Sensitivity studies have been conducted to ensure the mesh number and time-step size are optimum. Three different mesh numbers, i.e. 2.1 × 106, 2.8 × 106, and 3.7 × 106, and three different time-step sizes, i.e. 1/240 s, 1/360 s, and 1/480 s, are examined. Mesh and time sensitivity studies have been conducted for the case of twin SJs operating in phase (i.e., $\Delta\phi = 0$). With different mesh sizes and time steps, the time-averaged velocity profiles along a spanwise, wall-parallel line (i.e., $x/D_o = 5$ and $y/D_o = 2$) within the boundary layer are compared. As shown in Figure 2, the simulation result seems independent of the selected mesh size and time step, confirming the mesh and time convergence in the current study. The total number of meshes used in the simulation is about 2.8 million. In one diaphragm oscillation cycle, there are 240 time steps as a compromise between computational accuracy and time.

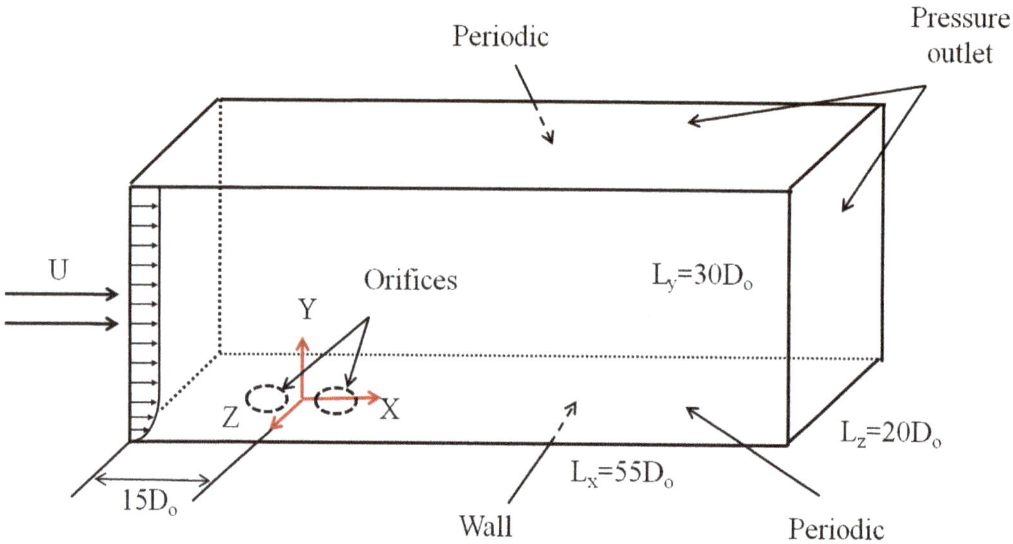

Figure 1. Computational geometry and boundary conditions for current CFD simulations. The actuator cavities are included in the simulation (although the cavities are not presented). The orifices of the two actuators are presented as dash circles located close to the origin of the coordination.

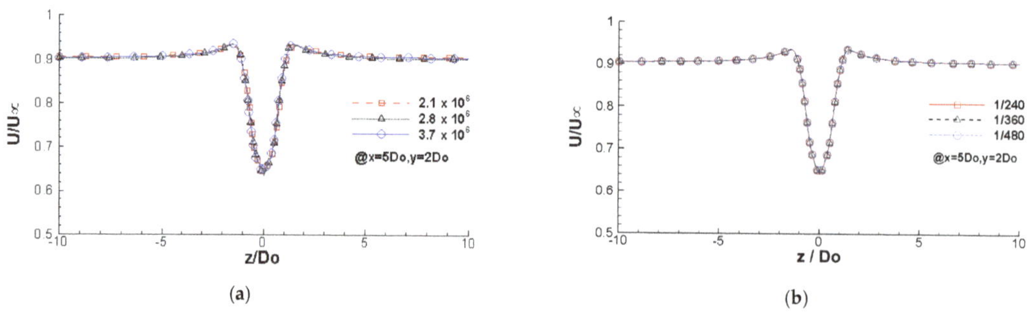

Figure 2. Comparison of the time-averaged velocity profiles along a spanwise, wall-parallel line (i.e., $x/D_o = 5$ and $y/D_o = 2$) within the boundary layer for the (**a**) mesh size (chosen value 2.8 × 10^6) and (**b**) time step (chosen value 1/240) sensitivity studies for the twin SJs operating in phase.

In the present study, the SJ-induced vortex structures are identified using the Q criterion [22]. A vortex structure can be captured by an iso-surface of a positive Q value. To differentiate individual SJs in the present simulation, particles of two different colors are continuously and uniformly released from the two actuator orifices during the blowing cycle once the flow field reaches its steady state. The density of these particles is set to that of water, and the discrete particle model (DPM) is used to track them. Similar to our dye visualization experiment [23,24], the particles issued from the upstream actuator are colored in red, whereas those from the downstream actuator are in green.

To reflect the capability of the SJ-induced vortex structures in flow separation control, wall shear stress is calculated based on the least squares cell-based gradient discretization proposed by Anderson and Bonhaus [28] to ensure a second-order interpolation, with a spatial resolution of 0.5% of the boundary layer thickness or $y+ = 0.2$.

2.3. Validation

The present CFD framework has been well validated with a single SJ in our previous work [26]. To further validate this framework for the current twin-SJ configuration, experiments are conducted using color dye visualization and PIV measurement. In dye visualization, food dye with two different colors fills the two actuator cavities, i.e. red color for the upstream actuator and green color for the down-stream. Methanol is premixed with the dye to achieve a similar density to water. Two cameras are used to capture images from the side and bottom views. In the PIV measurement, the Dantec system is used. Polyamide seeding particles are used to track the flow, and a FlowSense 2M charge-coupled-device camera is used to capture the particle images. A double pulsed Nd:YAG laser is used to illuminate the seeding particles. The sampling frequency of PIV measurement is 10 Hz. For each case, 300 image pairs are taken to achieve convergence. The PIV measurement results are verified with Blasius solution. More details of the dye visualization and PIV measurement can be found in our previous work [23]. To validate the CFD framework, flow patterns represented by colored fluid particles in the simulation are first compared to those in the color dye visualization experiment. As shown in Figure 3, the simulated particle patterns match with the dye visualization in a good manner. As shown in Figure 4, instantaneous and time-averaged velocity profiles are also compared at $x = 4Do$, with a good agreement between the simulation results and the PIV data.

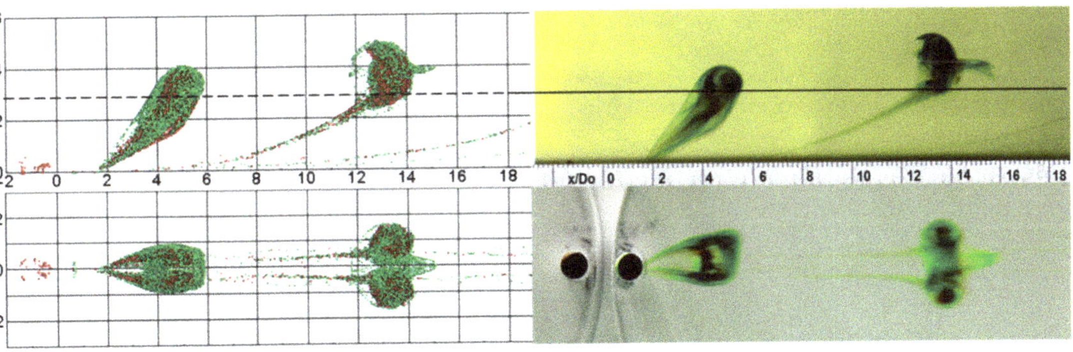

Figure 3. CFD Particle patterns (**left**) and experimental dye visualization (**right**) for the twin SJs at $\Delta\phi = \pi/2$ and $t/T = 0.25$ (The lines indicate the edge of the boundary layer).

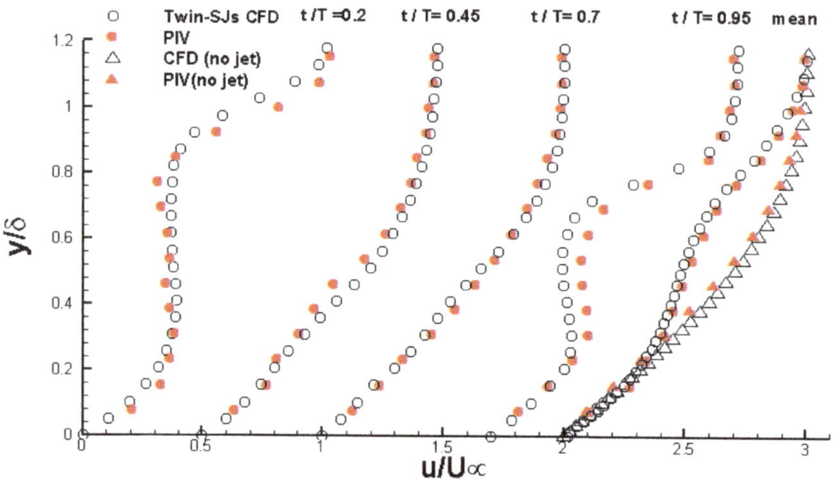

Figure 4. Comparison of instantaneous and time-averaged streamwise velocity profiles taken at $x = 4D_o$ from the CFD and PIV data for twin SJs operating at $\Delta\phi = 0$. The $t/T = 0.45, 0.7, 0.95$ profiles and the mean profile are shifted by $u/U_\infty = 0.5, 1.0, 1.7$ and 2, respectively.

3. Evolution and Interaction of In-Line Twin SJ-Induced Flow Structures

3.1. Flow Structures of a Single SJ in a Crossflow

The single-SJ-induced flow structures visualized by the fluid particles, and the Q iso-surfaces are presented first to serve as the baseline case. As shown in Figure 5a, iso-surfaces of $Q = 5$ are superimposed over the particles, showing that the flow structures represented by the fluid particles coincide with the primary hairpin vortices captured by the Q iso-surfaces in terms of the shape, size, and location. This confirms that, in addition to the Q iso-surfaces, the SJ-induced hairpin vortices can also be examined through the pattern and motion of fluid particles. As shown in Figure 5b, where only the fluid particles are presented, each hairpin vortex is well represented by a strong head and a pair of slim, stretched legs. The hairpin head also carries a spiral of fluid particles when seen from the side cross-section view, indicating very strong rotation inside. However, no particles are observed in the induced secondary vortices that are captured by the Q iso-surfaces as shown in Figure 5a. This is not surprising because the particles are all released from the actuator orifice, while the secondary vortices form with ambient fluid that does not carry any particles. These secondary vortices are very close to the wall and hence play an important role in SJ-related flow separation control [26].

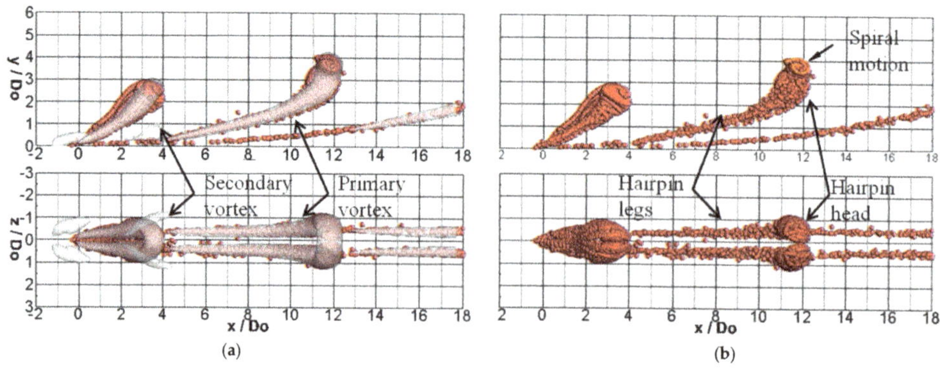

Figure 5. (**a**) Fluid particles plus $Q = 5$ iso-surfaces and (**b**) fluid particles for a single SJ at $t/T = 0$.

3.2. Evolution of Single and In-Phase Twin SJs

By using a sequence of snapshots of fluid particles, the evolution of the single SJ is presented in the upper two rows of Figure 6. It can be seen that, after expulsion from the actuator, the SJ-induced vortex ring gradually tilts downstream because of the velocity gradient in the boundary layer. Its upstream part is weakened by the boundary layer. As time advances, a hairpin vortex is formed by the shear stress in the boundary layer. It should be noted that the hairpin vortex is formed in a boundary layer with a flat plate. The wall-surface shape and obstacles in the boundary layer can significantly affect the evolution of hairpin vortices and horseshoe vortices, which is comprehensively discussed in previous works [29–31]. To further reveal the contribution of various SJ portions to the resulting hairpin vortex, three groups of fluid particles are marked in the snapshots, which are released from the orifice exit at three consecutive instants, i.e., $t = 0.0625T$ (group one), $0.25T$ (group two), and $0.4375T$ (group three). Each group consists of four selected particles, representing the upstream, central, downstream, and side portions of the SJ cross-section, respectively. By tracking these three groups of particles, it is found that the hairpin head contains fluid particles mainly from group two, while the hairpin legs contain fluid particles mainly from group one. As for the fluid particles in group three, most of them are inhaled back into the actuator during the SJ ingestion process ($t = 0.5T$~T). This result indicates that the fluid injected at the early stage of the blowing stroke ($t \approx 0$) mainly contributes to the formation of hairpin legs, the fluid injected near the maximum blowing ($t \approx 0.25T$) mainly contributes to the formation of the hairpin head, and the fluid injected at the late stage of the blowing stroke ($t \approx 0.5T$) contributes very little to the formation of the hairpin vortex.

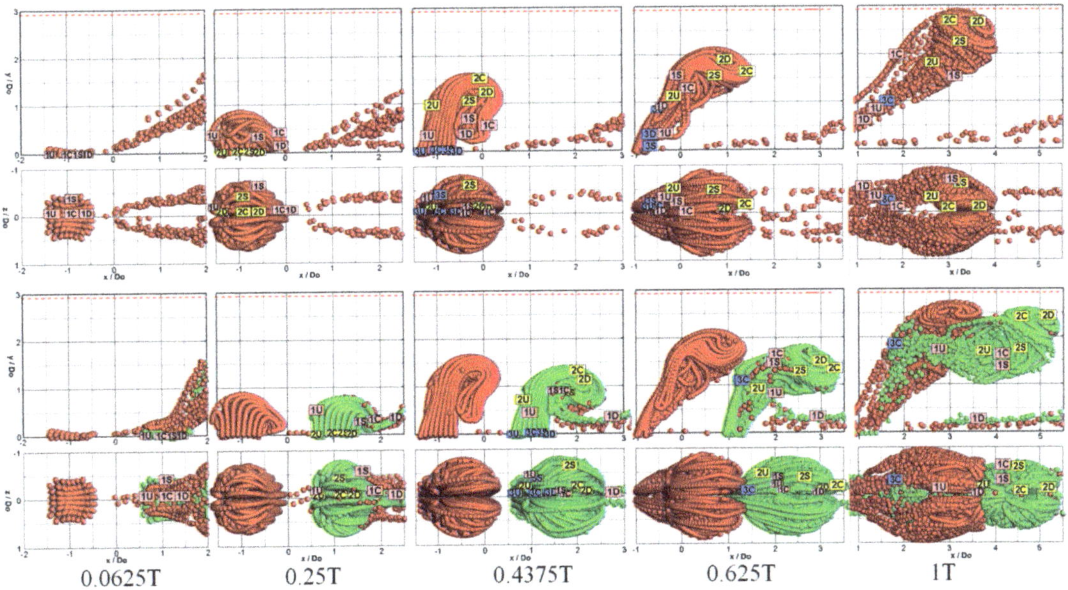

Figure 6. Evolution of fluid particles of the single SJ (**upper** two rows) and in-phase twin SJs (**lower** two rows). Group 1 particles (in pink) are released at $t/T = 0.0625$, Group 2 particles (in yellow) are released at $t/T = 0.25$, and Group 3 particles (in blue) are released at $t/T = 0.4375$. All particles are released at the actuator orifice exit.

If another SJ is introduced near the single SJ, the interaction between these two SJs will result in the formation of new flow structures. Take the in-phase, in-line twin SJs as an example, whose evolution is presented in the lower two rows of Figure 6. The SJ issued from the upstream actuator (in red, also called the trailing vortex in this case) is similar to the single SJ before it entrains its downstream counterpart, while the SJ issued from the downstream actuator (in green, also called the leading vortex) is significantly affected by its upstream counterpart. The interaction between these two SJs is evidenced by the lift-up of the leading vortex legs and their entrainment into the trailing vortex. Furthermore, since it is the leading vortex that is significantly affected by the interaction, the same three groups of fluid particles as in the single-SJ case are released from the downstream orifice exit. By tracking these particles and comparing their traces with those in the single SJ, it can be seen that during the blowing stroke (t = 0~0.5T), the evolution of these particles is similar to that in the single-SJ case, except that the particles released from the downstream portion of the SJ cross-section at the early stage of the blowing stroke (marked as "1D") are stripped out from the main vortex structure under the influence of the preceding vortices. When the twin SJs start interacting with each other in the ingestion period (t = 0.5T~T), particles originally in the lower portion of the vortex leg, e.g. the particles "1U" and "3C", are significantly lifted up by the trailing vortex, while particles originally in the vortex head, e.g. the particles in group two, still remain close in the highly rotating vortex head, indicating a well-maintained coherence in the head even under the strong interaction.

3.3. *Impact of Phase Differences*

With the simulated colored fluid particles, three different vortical structures are observed at the four selected phase differences. At $\Delta\phi = \pi/2$ (Figure 7b), the two SJs merge together to form one combined vortex, which is much stronger than single hairpin vortices and can pass through the boundary layer quickly. At $\Delta\phi = 3\pi/2$ (Figure 7d), the two hairpin vortices are separate from each other without obvious interaction, forming two completely separated vortices. At $\Delta\phi = 0$ (Figure 7a) and π (Figure 7c), the legs of one hairpin vortex are entrained into the head of the other, forming partially interacting vortex structures. As such, the twin SJs have distinct vortex strengths at various phase differences, as revealed in the authors' previous PIV experiments [23,24]. Furthermore, it is seen that, no matter which type of interaction occurs, the green particles issued from the downstream actuator are mainly scattered in the outer layer of the interacting structures, whereas the red particles is-sued from the upstream actuator are mainly enclosed inside. The different particle distributions can be caused by the different strengths of the vortical structures issued from the twin SJ actuators. The flow structures issued from the upstream actuator can hold vortical structures better than those from the downstream. Therefore, the parti-cles issued from the upstream actuator are more concentrated inside of the interacting flow structures. This is especially so when the interacting structures develop into the second actuation period, as shown in the plane view of Figure 7a–c. It is also seen that at $\Delta\phi = 3\pi/2$ (Figure 7d), unlike its counterpart, the head of the SJ issued from the upstream actuator is not affected at all by the near-wall particles. All these observations support our previous statement, i.e., in any case, the SJ issued from the up-stream actuator can better maintain its vortical structure than the SJ issued from the downstream actuator [23,24], and hence it dominates the interaction. The reason is obvious: the vortex produced by the upstream actuator has sufficient time to develop itself before the interaction, while the SJ issued from the downstream actuator is disturbed from the very beginning of its emergence.

The different capability of the two hairpin vortices in maintaining their own coherence can be further demonstrated in snapshots from the streamwise cross-section view. Figure 8 shows the snapshots from the $\Delta\phi = 0$, $\pi/2$ and π cases, at the same instant $t/T = 0$ as in Figure 7. In all three snapshots, the boundaries between the particles from different actuators are clear. At $\Delta\phi = \pi/2$ (Figure 8b), the head of the combined vortex includes two cores of fluid particles issued from the upstream actuator (in red) which are surrounded by scattered fluid particles issued from the downstream actua-tor (in green). At $\Delta\phi = 0$, the twin SJs produce partially interacting vortex structures (Figure 8a). The fluid particles issued from the upstream actuator still appear in two cores, forming the head of the trailing vortex, which is surrounded by the fluid particles issued from the downstream actuator. Moreover, two legs of the leading vortex with green particles appear near the mid-span plane, whose induced flow pushes the upper portion of the trailing vortex heads (two red cores) away from each other, which is prominent if compared with the combined vortex head shown in Figure 8b. In the partially interacting vortex structures at $\Delta\phi = \pi$ (Figure 8c), the head of the trailing vortex (in green) no longer appears as two solid cores. Instead, it appears as two hollow tubes enclosing the two counter-rotating legs of the leading vortex with red particles, and its shape differs greatly from those in Figure 8a,b.

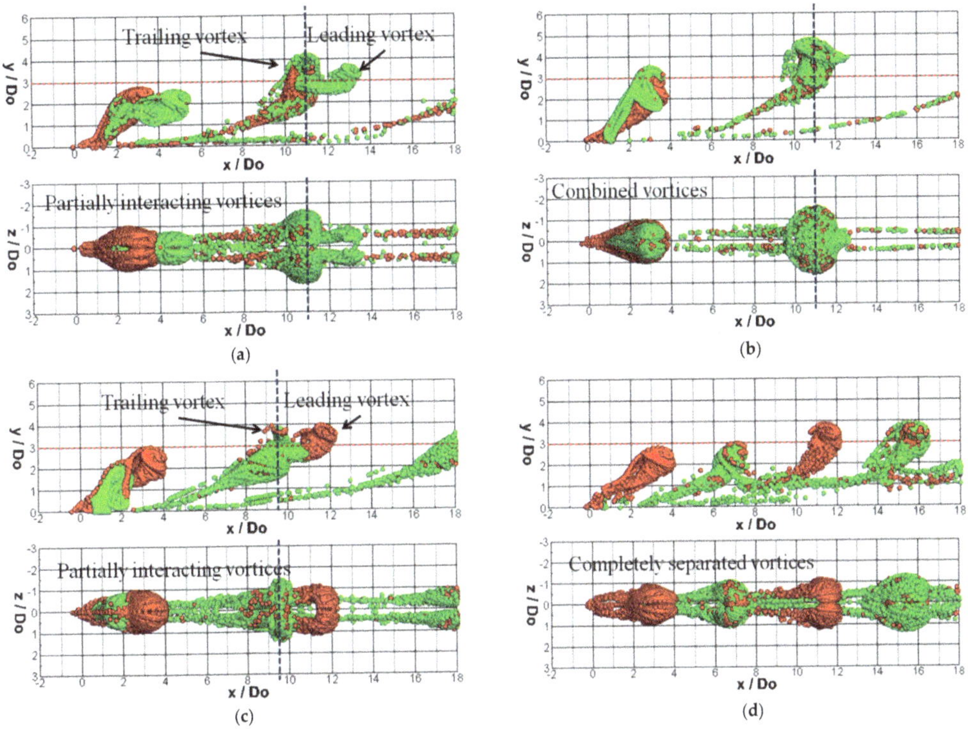

Figure 7. Colored fluid particles of the in-line twin SJs ($t/T = 0$), (**a**) $\Delta\phi = 0$, (**b**) $\Delta\phi = \pi/2$, (**c**) $\Delta\phi = \pi$, (**d**) $\Delta\phi = 3\pi/2$.

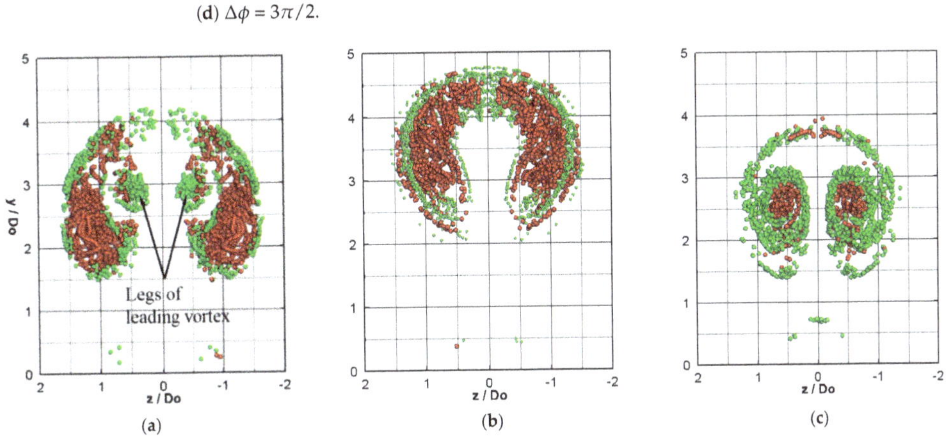

Figure 8. Snapshots of colored fluid particles from the streamwise cross-section view at (**a**) $x = 11D_o$ for the $\Delta\phi = 0$ case, (**b**) $x = 11D_o$ for the $\Delta\phi = \pi/2$ case and (**c**) $x = 9.5D_o$ for $\Delta\phi = \pi$ case ($t/T = 0$). These streamwise cross-section planes are indicated with dashed lines in Figure 7.

Particles issued from selected positions of the SJ actuator orifices are also tracked to study the interactions of SJ vortex structures, as shown in Figure 9. The single SJ and the twin SJs at $\Delta\phi = \pi/2$ and $3\pi/2$ are selected for presentation. Compared to those in the single SJ, the particles issued at the center of the orifices at $\Delta\phi = \pi/2$ are lifted to a much higher position at x = 3Do outside of the boundary layer. This is because a stronger vortex is merged from the twin SJs. The two particles issued from different actuators have strong interactions and rotate around the merged vortex head. In addition, the particles issued at $z/Do = 0.4$ have a slightly faster rotation than that in a single SJ with a shorter traveling distance. On the other hand, the trajectories of particles at $\Delta\phi = 3\pi/2$ are similar to those in the single SJ because the two separated hairpin vortices are produced, with the exception that the particle issued at z/Do = 0.4 from the downstream orifice is trapped in the tip of the hairpin leg with a straighter trajectory.

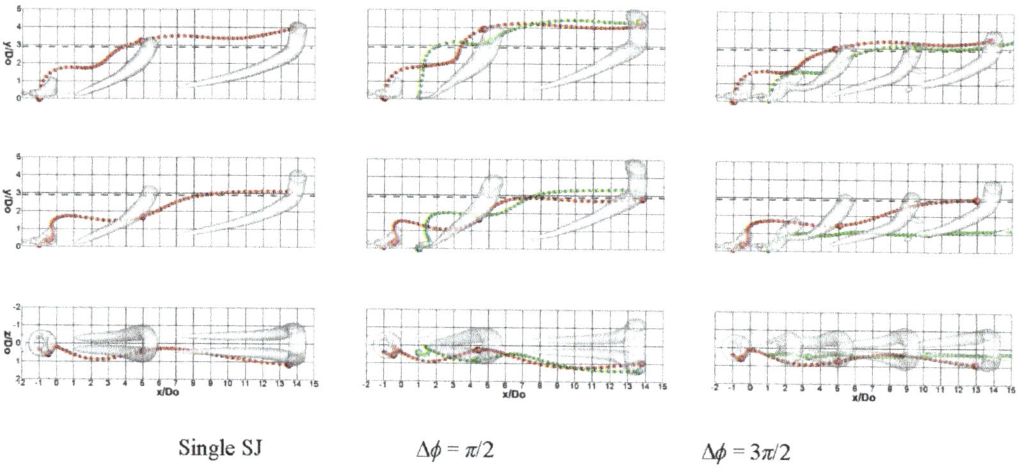

Figure 9. Tracking of particles that are released from different positions of the actuator orifices, top row: at center, middle & buttom rows: at $z/D_o = 0.4$ (side and plan views)). Vortex structures are also added at $t/T = 0.25$ indicated by the particles with larger diameters.

Figure 10 shows the correlations of the streamwise and wall-normal velocity components at three selected stations in the mid-span plane. In the single-SJ case, the upwash flow of hairpin vortices can increase the velocity in the wall-normal direction and decrease the velocity in the streamwise direction in the boundary layer [23,24]. For the twin SJs at $\Delta\phi = \pi/2$, the disturbance of the boundary layer is much stronger, indicated by the larger correlation loop of the streamwise and wall-normal velocity components. The shape of correlation loop is similar to that in the single SJ, because only the stronger hairpin vortices are merged with similar vortex structures. On the other hand, the shape of the correlation loop becomes very different at $\Delta\phi = 3\pi/2$. At a phase difference of $\Delta\phi = 3\pi/2$, the SJ flow structures are produced alternately from the twin actuators with an almost equal time interval. Therefore, the twin SJs have a weak interaction but with a doubled frequency compared to single SJs. On the other hand, the twin SJs merge into a combined vortex at $\Delta\phi = \pi/2$ with a high strength but the same frequency to the single SJs. Differing from the single loop in the single-SJ case, two loops are observed with substantial differences. This indicates that the two separated hairpin vortices at $\Delta\phi = 3\pi/2$ also have notable differences in vortex strength. In addition, the smaller loops also imply that the hairpin vortices at $\Delta\phi = 3\pi/2$ are slightly weaker than that in the single-SJ case.

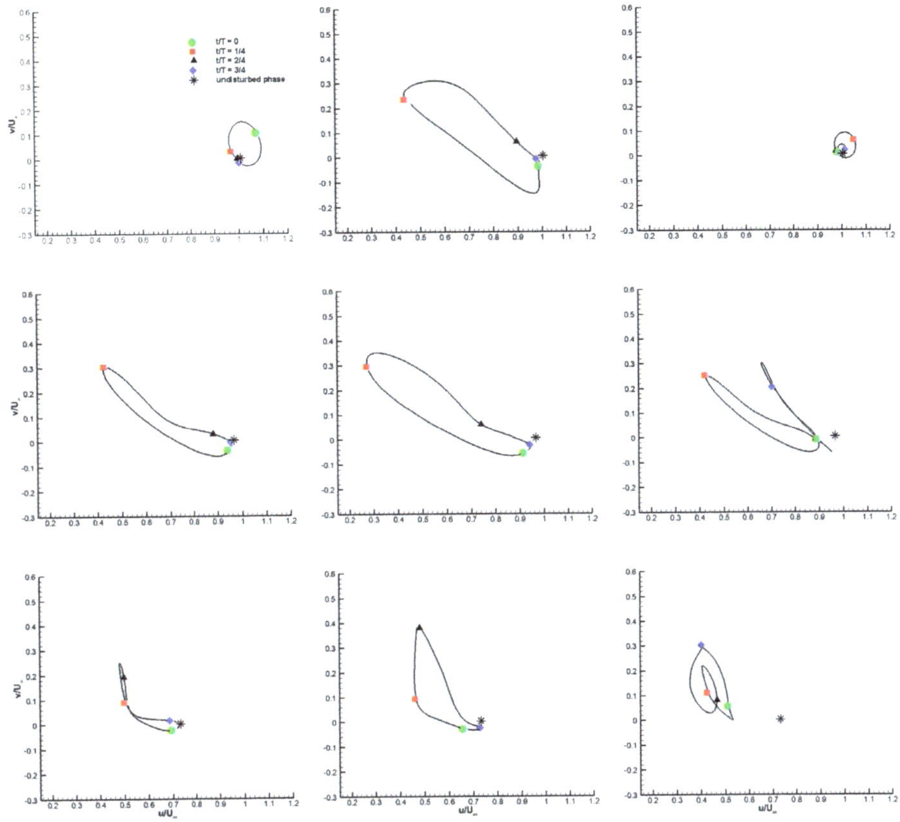

Single SJ $\Delta\phi = \pi/2$ $\Delta\phi = 3\pi/2$

Figure 10. Correlations of streamwise and wall-normal velocity components in the mid-span plane at three selected positions, top: ($x = 5D_o$, $y = 3.4D_o$), middle: ($x = 5D_o$, $y = 2.4D_o$), bottum: ($x = 5D_o$, $y = 1.4D_o$).

4. Impact of SJ Vortices on Wall Shear Stress

Previous studies have revealed that the SJ-induced near-wall streamwise vortices, i.e., the hairpin legs and induced secondary vortices, play an important role in flow separation control or enhancement of heat and mass transfer [12]. In this section, in order to reveal details of the evolution of these streamwise vortices, iso-surface of a smaller value of $Q = 1$ are colored by streamwise vorticity. To examine the impact of the SJ vortices on the boundary layer, the excess wall shear stress is presented and calculated as $(\tau_w - \tau_{w,nojet})/\tau_{w,nojet}$, which is a useful indicator of the SJ effectiveness in heat transfer enhancement [12].

4.1. Instantaneous Influence

Figure 11 shows the single-SJ-induced vortices and the corresponding contours of excess wall shear stress at the instant $t/T = 0$. Evolution of the primary hairpin vortices and secondary vortices has been discussed in our previous study [26]. These vortex structures influence the boundary layer as well as the wall shear stress. A pair of streaks of positive excess wall shear stress is observed outside of the SJs. This streak pair becomes wider near the regions between the legs of hairpin vortices and the secondary vortices (at about $8 < x/D_o < 10$), where the downwash flow is strengthened by the interaction of the vortices.

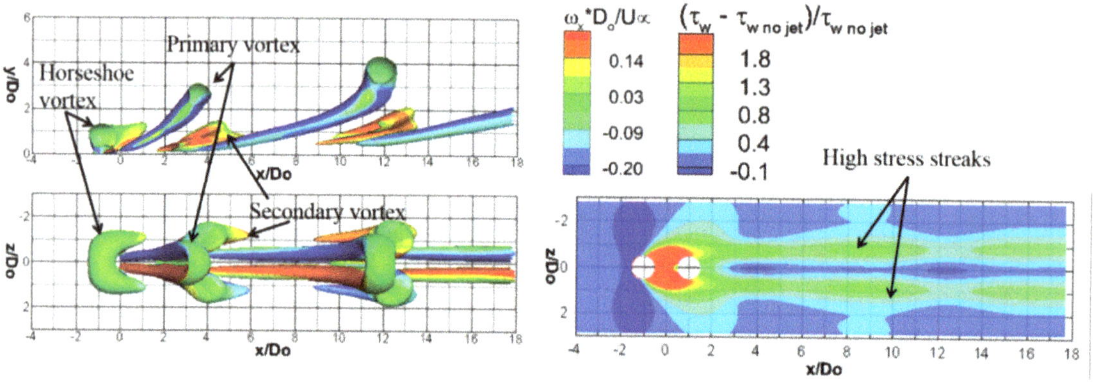

(a) (b)

Figure 11. Vortex structures in the single-SJ case, captured using $Q = 1$ iso-surfaces (**a**) and excess wall shear stress (**b**) at $t/T = 0$.

The two SJs merge into a stronger vortex when their phase difference is $\Delta\phi = \pi/2$. As shown in Figure 12a, both the hairpin vortices and the induced vortices, including the secondary vortices and horseshoe vortex, become stronger compared to those induced by the single SJ. The evolution of these vortices presented in the upper two rows of Figure 12b reveals that, at $t/T = 0.25$ when the upstream SJ is at its maximum blowing, a vortex ring (labeled as A) is issued the upstream actuator. It lifts up the horseshoe vortex produced in the previous cycle. In the downstream part of this horseshoe vortex a pair of secondary vortices begin to form. At $t/T = 0.5$, the vortex ring A begins to evolve into a hairpin vortex due to its interaction with the crossflow. Meanwhile a new vortex ring (labeled as B) emerges at the downstream orifice, breaking the horseshoe vortex. In addition, the downstream secondary vortices become more obvious. At $t/T = 0.75$ & 1, the twin SJs combine into a stronger hairpin vortex, which is accompanied by stronger secondary vortices and a larger horseshoe vortex.

The enhancement of wall shear stress is more significant, especially near to the orifice ($x/D_o < 12$) in Figure 12a. However, the combined vortices are so strong that they quickly penetrate the boundary layer, resulting in significant decay of the excess wall shear stress in the far downstream. In addition, some deficit of the wall shear stress is found outboard of the secondary vortices, which is associated with the induced upwash flow [26]. Detailed examination in the near-orifice region shows that changing of the excess wall shear stress is obviously periodic during one actuation cycle, as shown in the bottom row of Figure 12b. At $t/T = 0.25$, besides the area around the upstream orifice, high excess wall shear stress is also found near the downstream orifice, which is induced by the legs of the preceding combined vortex and the associated secondary vortices. At $t/T = 0.5$, high excess wall shear stress is induced around the downstream orifice due to the emergence of the vortex ring B. The excess wall shear stress downstream of this orifice becomes weaker. At $t/T = 0.75$, as the two SJs begin to combine, the regions of high excess wall shear stress around the two orifices connect. At downstream of the orifices, a large area of negative excess wall shear stress forms between the new forming vortex and the preceding one. At $t/T = 1$, as the newly formed combined vortex convects downstream, positive excess wall shear stress recovers near the orifices.

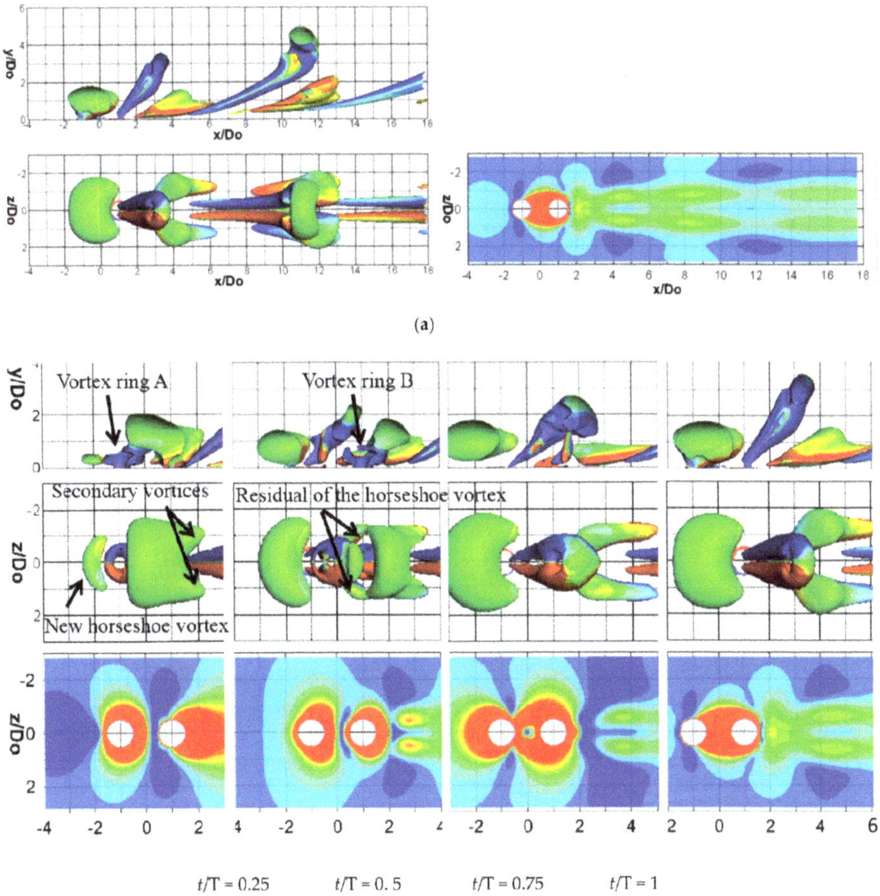

(b)

Figure 12. Vortex structures and excess wall shear stress for the twin SJs operating with $\Delta\phi = \pi/2$. (**a**) Vortex structures (**left**) and excess wall shear stress (**right**) at $t/T = 0$. (**b**) Evolution of vortices (**upper** two rows) and corresponding excess wall shear stress (**bottom** row). The color map can be referred to Figure 11.

Two completely separated vortices are produced at $\Delta\phi = 3\pi/2$. Primary hairpin vortex and outboard secondary vortices are induced as shown in left subplot of Figure 13a. However, their evolution processes start at different locations and instants. As shown in the upper two rows of Figure 13b, a pair of secondary vortices (labeled as #1) begins to form between the two orifices at $t/T = 0.25$. At $t/T = 0.5$, these secondary vortices become more obvious, as the horseshoe vortex is eliminated by the suction of the downstream actuator. They further grow at $t/T = 0.75$ and 1, and are sandwiched by the head of the SJ issued from the upstream actuator and the legs of the SJ issued from the downstream actuator. At $t/T = 1$, a vortex ring emerges from the downstream orifice and begins to evolve into a hairpin vortex at $t/T = 1.25$ (equivalently $t/T = 0.25$). By interacting with the preceding hairpin legs, it induces a pair of secondary vortices (labeled as #2) downstream. This pair of secondary vortices grows at $t/T = 1.5$ (equivalently $t/T = 0.5$), and finally becomes comparable to its counterpart (i.e., #1) at $t/T = 1.75$ (equivalently $t/T = 0.75$). Therefore, both the primary hairpin vortices and the induced secondary vortices double their appearance frequency (in other words, double their number) at far downstream. On the other hand, they are all shorter than those in single SJ case. The strip pair of enhanced wall shear stress also reflects this doubled frequency, as shown in the right subplot of Figure 13a. The strips are narrower compared to single-SJ case. Also, due to the doubled frequency effect, the strips in the near-orifice region can maintain the high excess wall shear stress, as revealed in the bottom row of Figure 13b.

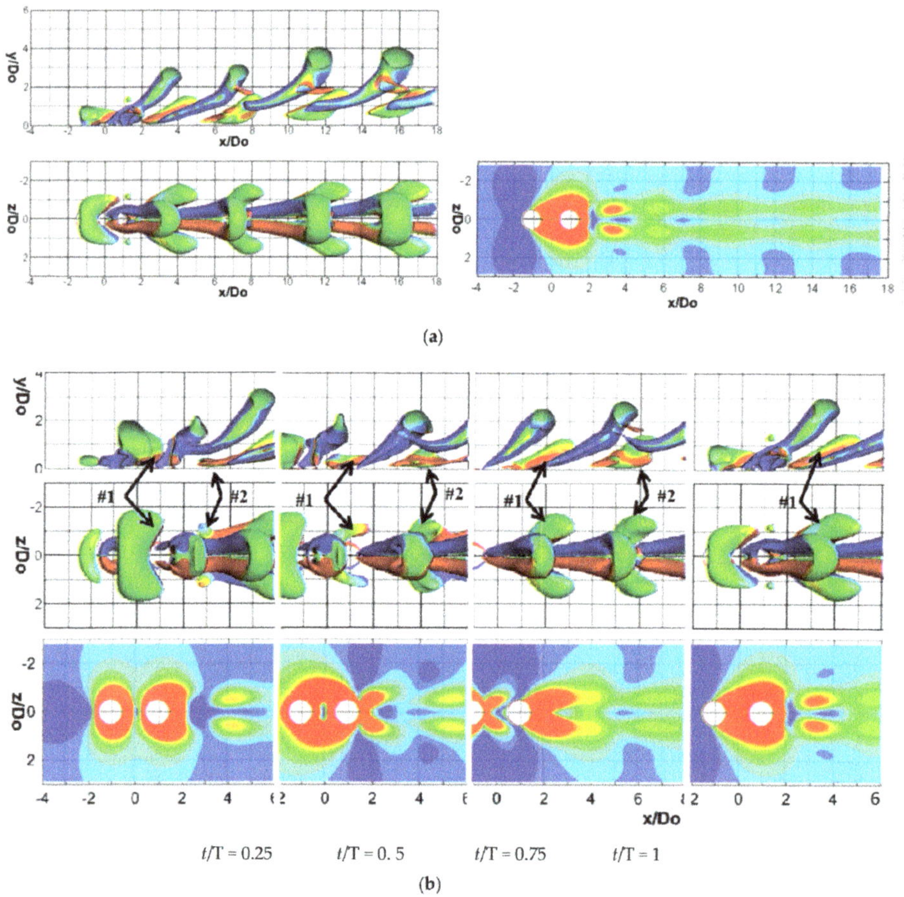

Figure 13. Vortex structures and excess wall shear stress for the twin SJs operating with $\Delta\phi = 3\pi/2$. (**a**) Vortex structures (**left**) and excess wall shear stress (**right**) at $t/T = 0$. (**b**) Evolution of vortices (**upper** two rows) and corresponding excess wall shear stress (**bottom** row). The color map can be referred to Figure 11.

Partially interacting vortex structures are produced at $\Delta\phi = 0$ and π. The Q iso-surfaces reveal that, no matter whether it is the trailing vortex (left subplot of Figure 14a) or leading vortex (left subplot of Figure 15a), the vortex produced from the upstream actuator dominates the interacting vortex structure in terms of vortex coherence. This is consistent to what has been revealed with the colored fluid particles in Section 3.3. For this reason, the vortex produced from the upstream actuator is also called the *dominant hairpin vortex* hereinafter, despite the phase difference. In both cases, a pair of much longer secondary vortices is induced by the interacting vortex structure, compared to the single-SJ case. Although appearing similar, these two partially interacting vortex structures show an interesting difference. That is, the lateral distance between the two legs of the partially interacting SJs at $\Delta\phi = 0$ is much larger than that at $\Delta\phi = \pi$. This is because in both cases the legs of the interacting vortex structure are from the dominant hairpin vortex. At $\Delta\phi = 0$, the leading vortex is entrained through the gap between the legs of trailing dominant vortex. Therefore, the gap is enlarged during the process. On the other hand, at $\Delta\phi = \pi$, the legs of dominant leading vortex is pushed together by the trailing vortex, resulting in a smaller gap. This difference also affects the pattern of excess wall shear stress. Although the patterns in both cases are similar to that in the $\Delta\phi = \pi/2$ case in far downstream, the deficit in the centerline low stress region in the $\Delta\phi = 0$ case is significantly larger (right subplot of Figure 14a). Furthermore, at $\Delta\phi = 0$, the side view reveals that the legs of the dominant hairpin vortex are gradually "pulled" down toward the wall as the reaction to their entrainment of the leading hairpin legs, which are then able to exert more influence on the wall shear stress, especially in the far-downstream region. As such, the two high stress streaks in the mid- and far-downstream regions are stronger than those in the near-downstream region, which is quite different from what is observed in the other cases.

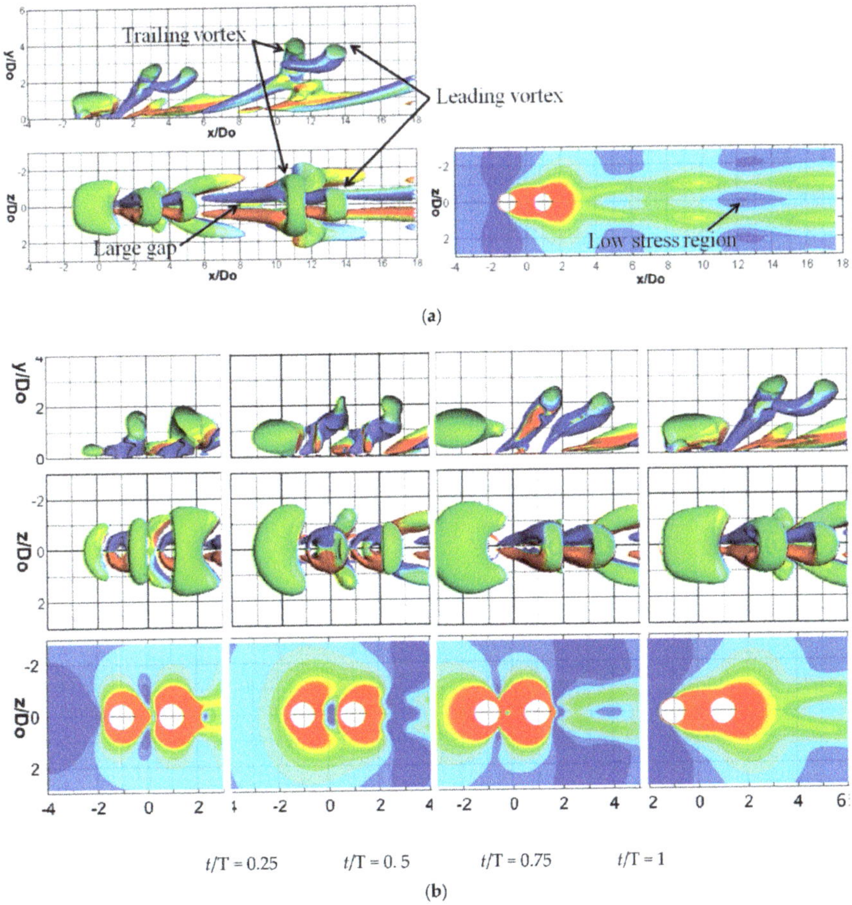

Figure 14. Vortex structures and excess wall shear stress for the twin SJs operating with $\Delta\phi = 0$. (**a**) Vortex structures (**left**) and excess wall shear stress (**right**) at $t/T = 0$. (**b**) Evolution of vortices (**upper** two rows) and corresponding excess wall shear stress (**bottom** row). The color map can be referred to Figure 11.

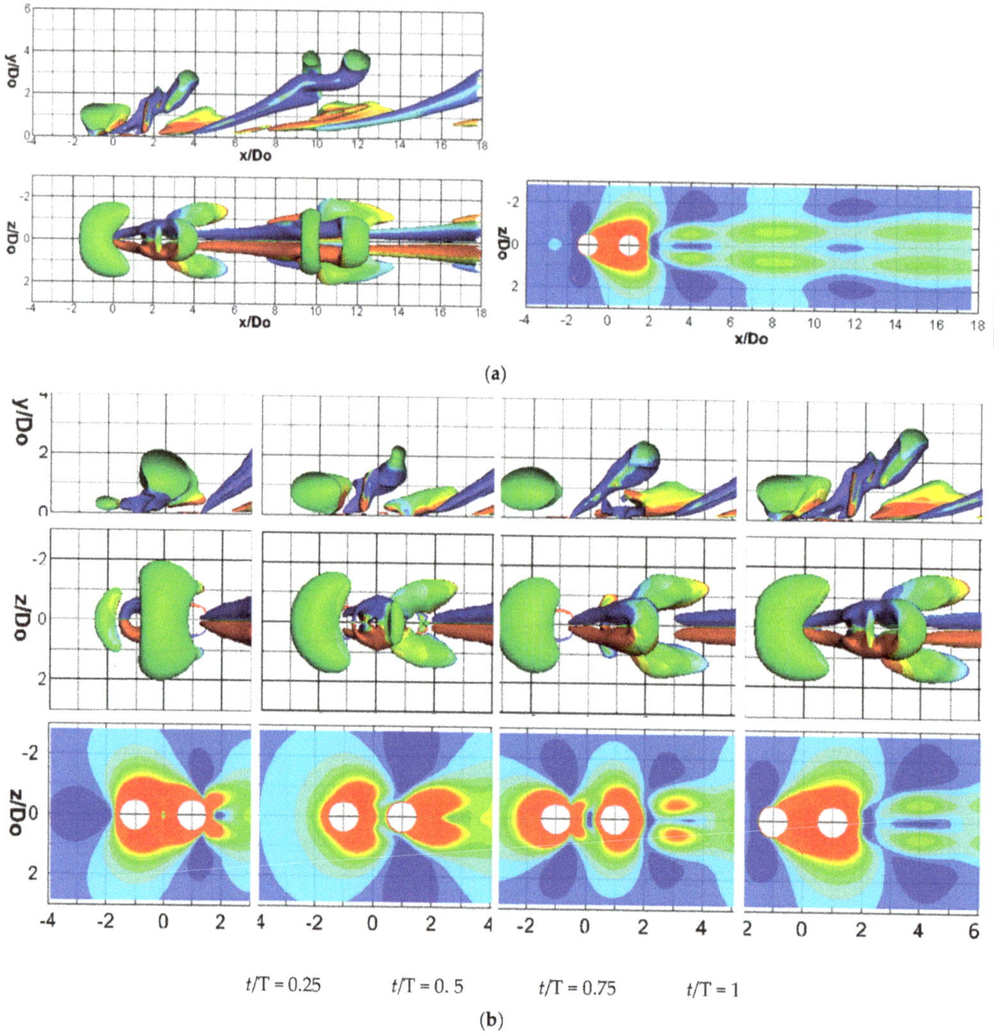

Figure 15. Vortex structures and excess wall shear stress for the twin SJs operating with $\Delta\phi = \pi$. (**a**) Vortex structures (**left**) and excess wall shear stress (**right**) at $t/T = 0$. (**b**) Evolution of vortices (**upper** two rows) and corresponding excess wall shear stress (**bottom** row). The color map can be referred to Figure 11.

4.2. Time-Averaged Influence

The time-averaged influence of the single and twin SJs on the wall shear stress is compared in Figure 16. It is seen that about $1D_o$ downstream of each operating actuator there exists an obvious region of negative excess stress, which is caused by the blockage of the SJs and the roll-up of the downstream branch of the SJ-induced vortex rings. Further downstream, the averaging removes the periodic patterns of the two streaks in all cases. Regardless the phase difference, the twin SJs can generate significantly more excess stress than the single SJ. But the persistency of the influence varies with phase difference. At $\Delta\phi = \pi/2$ the combined vortex structures induce a pair of wall shear stress streaks that is strong only within a short distance downstream from the orifices and starts decaying further downstream (see Figure 16c). This is because the combined vortices can penetrate the boundary layer deeply. The vortices move away from the wall as they travel in the crossflow, resulting in a fast decaying influence on the wall in the far-downstream region. This is evidenced by the particle tracking (Figure 7b) and instant flow structures (Figure 12) On the contrary, at $\Delta\phi = 3\pi/2$ the two completely separated vortices can stay close to the wall (Figures 7d and 13), and produce a pair of high stress streaks that sustain their strength for a much longer distance. The partially interacting vortices at $\Delta\phi = 0$ induce two streaks of high wall shear stress with gradually increased strength as shown in Figure 11b, between which a much wider strip of low stress value is observed. However, the time averaged high stress streaks produced by the other partially interacting vortices at $\Delta\phi = \pi$ (Figure 16d) appears similar to that in the $\Delta\phi = \pi/2$ case.

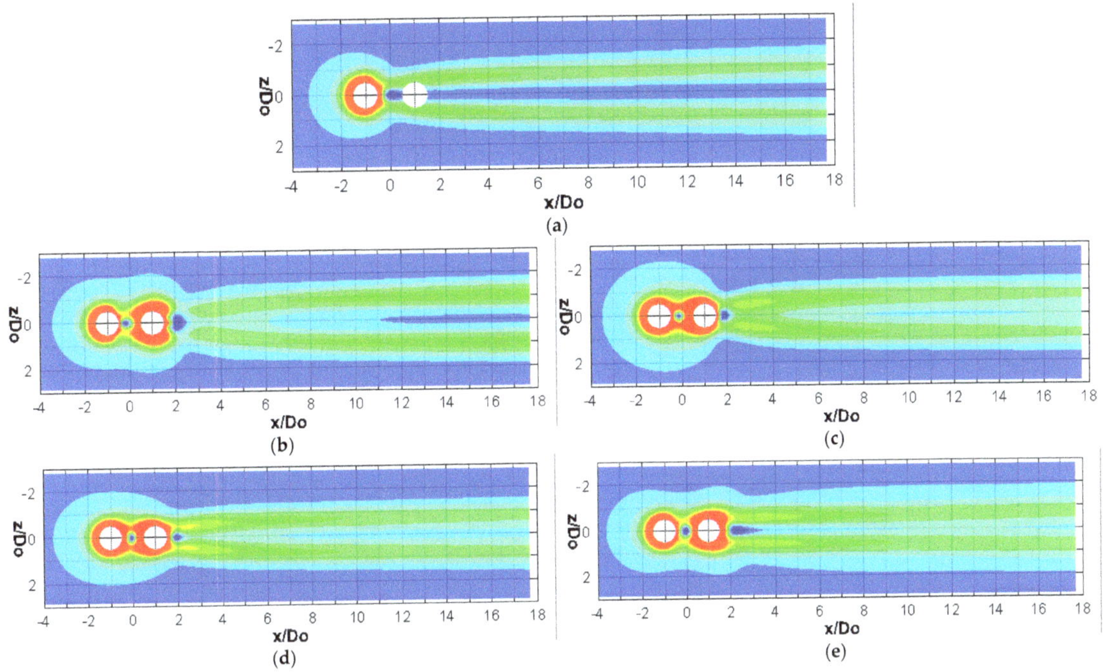

Figure 16. Time-averaged excess wall shear stress, (**a**) single SJ, (**b**) $\Delta\phi = 0$, (**c**) $\Delta\phi = \pi/2$, (**d**) $\Delta\phi = \pi$, (**e**) $\Delta\phi = 3\pi/2$.The color map can be referred to Figure 11.

In Figure 17, the excess wall shear stress is time-averaged and compared at two positions, i.e., $x = 4D_0$ and $18D_0$. In all cases, the two high stress streaks are represented by two peaks. At $x = 4D_0$, the two peaks induced by the single SJ are symmetrically located about $0.8D_0$ away from the mid-span plane, whereas the two peaks induced by all the twin SJs are slightly closer. It is also seen that the $\Delta\phi = 0$ case has a peak value close to that in the single SJ case. On the other hand, the other three twin SJ cases all have a much larger peak value (at least 20% higher). This is because in the $\Delta\phi = 0$ case the legs of the leading vortex (issued from the downstream actuator) are lifted up immediately after their emergence (as shown in Figures 6 and 7a), and hence the near-field wall shear stress is only affected by the legs of the trailing vortex (issued from the upstream actuator), while in the other three twin-SJ cases, i.e., $\Delta\phi = \pi/2$, π and $3\pi/2$, the near-field wall shear stress is affected by the legs of both vortices.

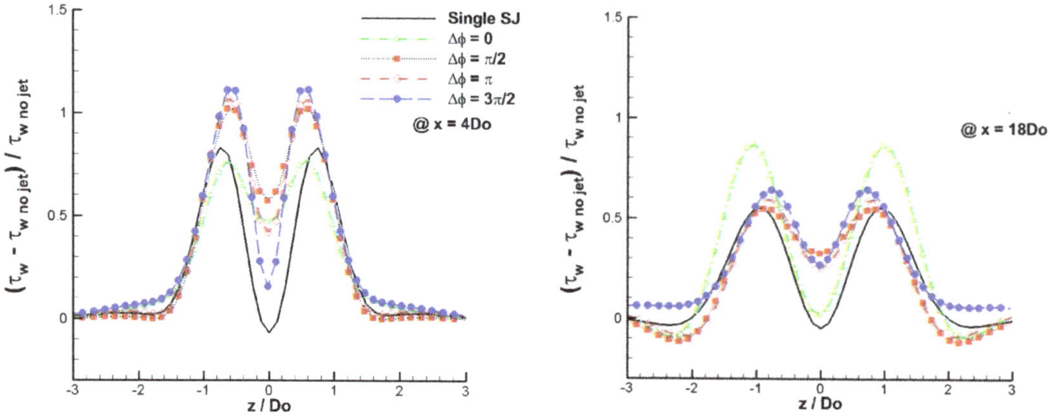

(**a**) (**b**)

Figure 17. Time-averaged excess wall shear stress along spanwise direction at (**a**) $x = 4D_o$ and (**b**) $x = 18D_o$.

At $x = 18D_o$ that is far from the two orifices, due to the spreading of the vortex structures and their penetrating in the boundary layer, the two stress peaks in all cases are separated from each other with a larger distance, and, except for the $\Delta\phi = 0$ case, their peak values reduce quite significantly if compared with those at $x = 4D_o$. In the $\Delta\phi = 0$ case, wall shear stress gets larger as the flow structures traveling to downstream. This may be caused by the closer distance of the hairpin vortex to the wall and the increasing strength of the hairpin vortex during the merging process of the twin SJs. In addition, since the two legs of the dominant vortex is pushed away from each other by its counterpart vortex, the distance of the corresponding peaks becomes much larger than that in the other cases including the single SJ case. This is also consistent with what is observed in Figure 16. In addition, negative values of excess wall shear stress are also found at $x = 18D_o$ in $|z/D_o| > 2$ regions, except at $\Delta\phi = 3\pi/2$. From the instantaneous patterns of excess wall shear stress (Figures 12a, 14a and 15a), these negative values are ascribed to the deficit of wall shear stress outboard of the secondary vortices.

The excess wall shear stress is time-averaged, and then spanwise averaged within a range of $z = -1.6D_o$~$1.6D_o$ where most of the non-zero excess stress is covered as revealed in Figure 17, and the streamwise distribution of the newly calculated quantity is then plotted and compared in Figure 18. It is seen that this time- and spatial-averaged excess stress in the single SJ case reaches a gentle peak at about $x = 3D_o$, and then decreases along the streamwise direction. In all the twin SJ cases, the curve begins with a significant rise, which corresponds to the region of negative excess stress right next to the downstream orifice as shown in Figure 16. Compared to the single SJ, in the $\Delta\phi = \pi/2$ case the combined vortex structures produce a peak of more than 40% larger at about $x = 3.5D_o$, after which the mean excess stress decreases with a higher rate, reaching a value of 26% larger at $x = 22D_o$. The trend is similar in the $\Delta\phi = 3\pi/2$ case where the two completely separated hairpin vortices are generated. Because of the doubling frequency effect in this case, the mean excess stress becomes almost unchanged after $x = 12D_o$, which turns to be larger than that in the $\Delta\phi = \pi/2$ case.

Although some difference is expected between the two cases that generate partially interacting vortex structures, it is still surprising to see a huge difference. The curve in the $\Delta\phi = \pi$ case follows a similar trend to that in the $\Delta\phi = \pi/2$ case, but with a slightly higher peak. But the curve in the $\Delta\phi = 0$ case varies quite differently: it reaches a plateau first at about $x = 4D_o$ where the excess stress is only 15% larger than that in the single SJ, and then increases further and reaches a gentle peak at about $x = 19D_o$ where the excess stress is about 50% greater. This huge discrepancy stems from the difference in the evolution of the dominant vortex structure in these two cases. At $\Delta\phi = 0$, the legs of the dominant hairpin vortex are gradually "pulled" down toward the wall, which can exert stronger influence on the wall shear stress, especially in the far-downstream region.

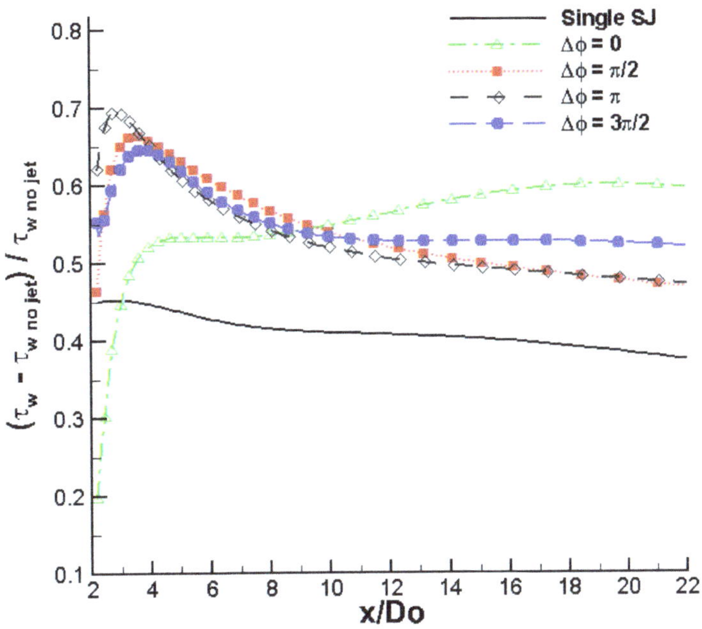

Figure 18. Mean excess wall shear stress along streamwise direction between $z = -1.6D_o$~$1.6D_o$.

5. Conclusions

The interaction of in-line twin SJs is examined in a laminar boundary layer numerically, that operate at various phase differences along a flat plate. Fluid particles of different colors released from the actuator orifices are tracked so as to show the mixing process of SJs in the boundary layer. In addition, the Q criterion is used to identify the twin SJ-induced vortex structures including both the primary and secondary vortices. The impact of these vortex structures on the boundary layer is also investigated through studying the excess wall shear stress. The major findings are as follows:

(1) The fluid injected at the early stage of the blowing stroke mainly contributes to the formation of hairpin legs, the fluid injected near the maximum blowing mainly contributes to the formation of hairpin head, and the fluid injected at the late stage of the blowing stroke contributes very little to the formation of hairpin vortex because most of them are inhaled back into the actuator during the ingestion process.

(2) When the twin SJs generate *a combined vortex* at $\Delta\phi = \pi/2$ or *partially interacting vortex structures* at $\Delta\phi = 0$ and π, fluid particles issued from the downstream actuator are mainly scattered in the outer layer of the interacting structures, whereas those issued from the upstream actuator are mainly enclosed inside. When the twin SJs generate *two completely separated hairpin vortices* at $\Delta\phi = 3\pi/2$, unlike its counterpart, the head of hairpin vortex issued from the upstream actuator is not affected at all.

(3) The trajectories of selected particles reveal that the fluid particle near the hairpin head rotates about the center of the vortex head, whereas the fluid particle near the hairpin leg is trapped in the leg and travels downstream with a helix path.

(4) The SJ-induced vortex structures make great impacts on the wall shear stress. In all cases a pair of streaks of positive excess wall shear stress can be observed. Although

both the $\Delta\phi = 0$ and $\Delta\phi = \pi$ cases produce partially interacting vortex structures, the induced two excess stress streaks are significantly different, which stems from the difference in the evolution of the vortex structures, especially of the dominant hairpin vortex, in these two cases.

Author Contributions: Writing—original draft preparation, H.W.; writing—review and editing, X.W.; supervision and project administration, D.X. and H.T.; methodology, L.L. and K.Z.; funding acquisition, X.W. All authors have read and agreed to the published version of the manuscript.

Funding: The authors gratefully acknowledge the financial support provided by the National Natural Science Foundation of China, China (grant Nos. 12072196) and the Science and Technology Commission of Shanghai Municipality, China (grant No. 19JC1412900), Advanced Jet Propulsion Innovation Center, China (AEAC HKCX2020-02-028), Aeronautical Science Foundation of China (2020Z006057001) and State Key Laboratory of Aerodynamics, China (SKLA-20200303).

Institutional Review Board Statement: Not applicable.

Informed Consent Statement: Not applicable.

Data Availability Statement: The data presented in this study are available on request from the corresponding author.

Conflicts of Interest: The authors declare no conflict of interest.

Nomenclature

D_o	orifice diameter: mm
D_d	diaphragm diameter, mm
d	actuator distance, mm
f	oscillation frequency, Hz
L	stroke length, mm
L_o	jet column length
p	pressure, Pa
Re_θ	momentum thickness
T	oscillation period, s
t	time, s
$\bar{U}_o$	jet velocity averaged over an entire actuation cycle, m/s
U_∞	bulk flow velocity, m/s
$u_o(t)$	instantaneous jet velocity, m/s
u_d	Diaphragm velocity, m/s
u_i	velocity, m/s
VR	velocity ratio
ν	viscosity, m^2s^{-1}
x_i	space location, mm
Δ	peak-to-peak displacement of the diaphragm, mm
δ	boundary layer thickness, mm
$\Delta\phi$	phase difference
ρ	fluid density, kg/m^3
τ_w	wall shear stress with control, Pa
$\tau_{w,nojet}$	wall shear stress without control, Pa
Acronyms	
SJ	synthetic jet
PIV	particle image velocimetry
CFD	computational fluid dynamics

References

1. Smith, B.; Glezer, A. The formation and evolution of synthetic jets. *Phys. Fluids* **1998**, *10*, 2281. [CrossRef]
2. Amitay, M.; Smith, D.R.; Kibens, V.; Parekh, D.E.; Glezer, A. Aerodynamic flow control over an unconventional airfoil using synthetic jet actuators. *AIAA J.* **2001**, *39*, 361–370. [CrossRef]

3. Amitay, M.; Glezer, A. Role of actuation frequency in controlled flow reattachment over a stalled airfoil. *AIAA J.* **2002**, *40*, 209–216. [CrossRef]
4. Liu, Y.H.; Tsai, S.Y.; Wang, C.C. Effect of driven frequency on flow and heat transfer of an impinging synthetic air jet. *Appl. Therm. Eng.* **2015**, *75*, 289–297. [CrossRef]
5. McGuinn, A.; Rylatt, D.I.; O'Donovan, T.S. Heat transfer enhancement to an array of synthetic air jets by an induced crossflow. *Appl. Therm. Eng.* **2016**, *103*, 996–1003. [CrossRef]
6. Luo, Z.B.; Deng, X.; Xia, Z.X.; Wang, L.; Gong, W.J. Flow field and heat transfer characteristics of impingement based on a vectoring dual synthetic jet actuator. *Int. J. Heat Mass Transfer* **2016**, *102*, 18–25. [CrossRef]
7. Xu, Y.; He, G.S.; Kulkarni, V.; Wang, J.J. Experimental investigation of influence of Reynolds number on synthetic jet vortex rings impinging onto a solid wall. *Exp. Fluids* **2017**, *58*, 6. [CrossRef]
8. Li, B.; Zhang, J.; Jiang, N. Influence of Synthetic Jets on Multiscale Features in Wall-Bounded Turbulence. *Actuators* **2022**, *11*, 199. [CrossRef]
9. Liu, J.-F.; Luo, Z.-B.; Deng, X.; Zhao, Z.-J.; Li, S.-Q.; Liu, Q.; Zhu, Y.-X. Dual Synthetic Jets Actuator and Its Applications—Part II: Novel Fluidic Thrust-Vectoring Method Based on Dual Synthetic Jets Actuator. *Actuators* **2022**, *11*, 209. [CrossRef]
10. Zhong, S.; Millet, F.; Wood, N. The behaviour of circular synthetic jets in a laminar boundary layer. *Aeronaut. J.* **2005**, *109*, 461–470. [CrossRef]
11. Jabbal, M.; Zhong, S. The near wall effect of synthetic jets in a boundary layer. *Int. J. Heat Fluid Flow* **2008**, *29*, 119–130. [CrossRef]
12. Jabbal, M.; Zhong, S. Particle image velocimetry measurements of the interaction of synthetic jets with a zero-pressure gradient laminar boundary layer. *Phys. Fluids* **2010**, *22*, 063603. [CrossRef]
13. Lee, A.; Timchenko, V.; Yeoh, G.H.; Reizes, J.A. Three-dimensional modelling of fluid flow and heat transfer in micro-channels with synthetic jet. *Int. J. Heat Mass Transfer* **2012**, *55*, 198–213. [CrossRef]
14. Lee, A.; Yeoh, G.H.; Timchenko, V.; Reizes, J.A. Heat transfer enhancement in micro-channel with multiple synthetic jets. *Appl. Therm. Eng.* **2012**, *48*, 275–288. [CrossRef]
15. Watson, M.; Jaworski, A.; Wood, N. Contribution to the understanding of flow interactions between multiple synthetic jets. *AIAA J.* **2003**, *41*, 747–749. [CrossRef]
16. Liddle, S.; Crowther, W.; Wood, N. Investigation of phase and spacing effects in synthetic jet actuator arrays. In Proceedings of the 43rd AIAA Aerospace Sciences Meeting and Exhibit, Reno, NV, USA, 10–13 January 2005; p. 107.
17. Liddld, S.; Wood, N. Investigation into clustering of synthetic jet actuators for flow separation control application. *Aeronaut. J.* **2005**, *109*, 35–44. [CrossRef]
18. Honami, S.; Motosuke, M. Vortex interaction of in-line synthetic jets injected at difference phase in low Reynolds number cross flow. In Proceedings of the 6th AIAA Flow Control Conference, New Orleans, LA, USA, 25–28 June 2012; p. 243.
19. Haller, G.; Yuan, G. Lagrangian coherent structures and mixing in twodimensional turbulence. *Phys. D Nonlinear Phenom.* **2000**, *147*, 352–370. [CrossRef]
20. Shadden, S.C.; Lekien, F.; Marsden, J.E. Definition and properties of Lagrangian coherent structures from finite-time Lyapunov exponents in two-dimensional aperiodic flows. *Phys. D* **2005**, *212*, 271–304. [CrossRef]
21. Franco, E.; Pekarek, D.N.; Peng, J.; Dabiri, J.O. Geometry of unsteady fluid transport during fluid–structure interactions. *J. Fluid Mech.* **2007**, *589*, 125–145. [CrossRef]
22. Hunt, J.; Wray, A.; Moin, P. Eddies, streams, and convergence zones in turbulent flows. In Proceedings of the 1988 Summer Program, Center for Turbulence Research, Stanford University, Stanford, CA, USA, 1 December 1988; pp. 193–208.
23. Wen, X.; Tang, H.; Duan, F. Vortex dynamics of in-line twin synthetic jets in a laminar boundary layer. *Phys. Fluids* **2015**, *27*, 083601. [CrossRef]
24. Wen, X.; Tang, H. Dye visualization of in-line twin synthetic jets in crossflows—A parametric study. *ASME J. Fluids Eng.* **2017**, *139*, 091203. [CrossRef]
25. Tang, H.; Zhong, S. Incompressible flow model of synthetic jet actuators. *AIAA J.* **2006**, *44*, 908–912. [CrossRef]
26. Wen, X.; Tang, H. On hairpin vortices induced by circular synthetic jets in laminar and turbulent boundary layers. *Comput. Fluids* **2014**, *95*, 1–18. [CrossRef]
27. Zhou, J.; Zhong, S. Numerical simulation of the interaction of a circular synthetic jet with a boundary layer. *Comput. Fluids* **2009**, *38*, 393–405. [CrossRef]
28. Anderson, W.K.; Bonhaus, D.L. An implicit upwind algorithm for computing turbulent flows on unstructured grids. *Comput. Fluids* **1994**, *23*, 1–21. [CrossRef]
29. Tafarojnoruz, A.; Gaudio, R. Vortex scouring process around bridge pier with a caisson. *J. Hydraul. Res.* **2012**, *50*, 443–444. [CrossRef]
30. Calomino, F.; Tafarojnoruz, A.; De Marchis, M.; Gaudio, R.; Napoli, E. Experimental and Numerical Study on the Flow Field and Friction Factor in a Pressurized Corrugated Pipe. *J. Hydraul. Eng.* **2015**, *141*, 04015027. [CrossRef]
31. Tafarojnoruz, A.; Lauria, A. Large eddy simulation of the turbulent flow field around a submerged pile within a scour hole under current condition. *Coast. Eng. J.* **2020**, *62*, 489–503. [CrossRef]

actuators

MDPI

Article

Characteristics of a Fluidic Oscillator with Low Frequency and Low Speed and Its Application to Stall Margin Improvement

Zhuoqi Liu [1], Tianyu Pan [1,*], Shiqi Wang [2,*] and Zhaoqi Yan [3]

1 Research Institute of Aero-Engine, Beihang University, Beijing 100191, China
2 Aero-Engine Academy of China, Aero-Engine Corporation of China, Beijing 101304, China
3 School of Energy and Power Engineering, Beihang University, Beijing 100191, China
* Correspondence: pantianyu@buaa.edu.cn (T.P.); wangsq6@126.com (S.W.)

Abstract: Active flow control methods are commonly used in expanding the operating range of compressors. Indeed, unsteady active control methods are the main focus of researchers due to their effectiveness. For constructing an unsteady active control system, reliable actuators are significant. To compare with conventional actuators such as synthetic jet actuators and rotating valves, fluidic oscillators have structurally robust characteristics and can generate self-excited and self-sustained oscillating jets, which leads to its higher applicability in compressors under severe working conditions. Thus, to explore the feasibility of unsteady active control systems by the usage of fluidic oscillators, a low-frequency and low-speed oscillator is first designed and experimentally studied for improving the stability of a low-speed axial flow compressor. During the experiments, a special casing is designed to install 15 uniformly distributed oscillators in the tip region of compressor. Based on the unsteady micro injections of the rotor tip with rotor rotation frequency, the results indicate that the frequency/period of oscillators are flexible, in which the values are decoupled with the variation of inlet pressure. When the inlet-to-outlet pressure ratio of the oscillator is in the range of 1.1~2.0, the maximum velocity ranges from 30 m/s to 80 m/s. Moreover, the mass flow rate of the single oscillator only varies from 0.017‰ to 0.059‰ from the designed compressor mass flow rate. For the improvement of the compressor stall margin, the value is 3.45% when the total mass flow of oscillators is 0.08% of the designed compressor mass flow.

Keywords: fluidic oscillator; pulsating jets; axial compressor; unsteady stall margin improvement

Citation: Liu, Z.; Pan, T.; Wang, S.; Yan, Z. Characteristics of a Fluidic Oscillator with Low Frequency and Low Speed and Its Application to Stall Margin Improvement. *Actuators* **2022**, *11*, 341. https://doi.org/10.3390/act11120341

Academic Editor: Luigi de Luca

Received: 26 October 2022
Accepted: 21 November 2022
Published: 22 November 2022

1. Introduction

The compressor is one of the key components of aero engines; it is expected to achieve not only high efficiency and a high-pressure ratio but also high stability. There are two ways to improve the stable operating margin, which are passive flow control and active flow control methods. [1] Passive flow control methods can remarkably extend the operating range of compressors. By changing the structure of the casing or blades, they reallocate the flow rate and blade loading in the tip region and extend the operating range of compressors. Typical passive flow control methods include casing treatment [2–4], blade tip winglets [5,6], and vortex generators [7,8]. However, passive control methods will inevitably cause a significant penalty in the efficiency and the pressure ratio [9]. Thus, the concept of active flow control was put forward by Epstein et al. [10] and Ffowcs Williams et al. [11]; by monitoring the flow field parameters and using the actuating mechanisms, signals can be added to the flow field as needed to inhibit the development of pre-stall disturbances. In this way, active flow control methods can not only delay the instability phenomenon but also reduce the efficiency loss.

Active flow control methods, such as tip injection [12,13], variable guide vanes [14,15], and synthetic jets [16,17], can be divided into steady and unsteady active control methods according to the temporal characteristics of the perturbation signals. Steady tip injection is a widely studied steady active control method. It has many technical advantages. By changing the jet parameters [18] or using the over-tip recirculation system [19], steady tip injection can act on the tip region and expand the operating range. Based on the total inject momentum, Lin et al. [20] and Li et al. [21] categorized two kinds of stall margin improvement mechanisms. According to the range of spanwise blade load changes caused by air injection from the steady aspect, these two kinds of mechanisms are micro and macro tip injection. Unsteady active flow control methods have higher control efficiency than steady control methods because they can obtain control effects that are several orders of magnitude higher than the percentage of excitation energy of the mainstream [22]. Li et al. [23] used acoustic excitation, Kefalakis et al. [24] used unsteady jets, and Saddoughi et al. [25] used plasma excitation to conduct unsteady active control of compressors, all of which achieved certain stall margin expansion with small energy inputs. However, there has not been a common understanding of the unsteady stall margin expansion mechanism caused by the unsteady nature of excitation.

Due to the harsh working conditions inside compressors, the difficulty of unsteady active flow control lies in the lack of reliable actuators. The wall-attachment oscillator [26] can generate self-excited and self-sustained unsteady jets based on the Coanda effect. It has a simple internal structure and contains no moving parts. Compared with traditional unsteady actuators, such as synthetic jet actuators, rotating valves, or plasma actuators, fluidic oscillators are structurally robust and resistant to electromagnetic interference. In addition, the wall-attachment oscillator has a large jet velocity range and less added mass flow [27]. Thus, it is a good choice to promote unsteady excitation engineering applications, especially when high frequencies are required. In recent years, wall-attachment oscillators have been widely studied in airfoil lift augmentation [28], cavity noise suppression [29], and heat conduction enhancement [30,31]. In the field of aero-engines, wall-attachment oscillators have been applied to adjust the thrust vector [32]; control flow separation at the s-shaped inlet [33], corner of compressor stator [34,35], and outlet guide vane of high-load turbine [36]; and reduce loss caused by turbine tip clearance leakage vortex [37].

Through careful design, wall-attachment oscillators can generate unsteady micro jets with very little additional flow rate; thus, they have the potential to realize effective unsteady active control while further reducing the efficiency penalty. Moreover, wall-attachment oscillators are more robust and reliable than traditional unsteady actuators. Therefore, when facing severe working environments inside aeroengines, it is essential to explore the feasibility of unsteady stall margin expansion based on fluidic oscillators. Thus, a low-frequency and low-speed oscillator configuration for a low-speed compressor was redesigned based on the models of Wang et al. [38] and Zhou et al. [39,40]. A series of experiments were conducted to investigate the characteristics of this configuration in detail by using hot-wire anemometer and other measurement methods. A specially designed casing was used to anchor 15 uniformly distributed oscillators in the circumferential direction in the tip region of a low-speed axial compressor rotor. Then, the micro tip injection experiments with the unsteady excitation frequency of rotor rotation were conducted.

This paper is organized as follows. First, a brief introduction to the research subjects and methodology is presented in Section 2. The oscillator characteristics and experimental results in active stall control are presented and analyzed in Section 3, which is followed by a discussion in Section 4. Finally, the main findings are concluded in Section 5.

2. Research Subjects and Methodology

2.1. Research Subjects

2.1.1. Low-Speed Axial Compressor Test Rig TA 36

This paper expects unsteady micro tip injection experiments based on fluidic oscillators on the low-speed compressor TA 36, so the test rig will be introduced first.

The schematic of TA 36 and the measurement sections are shown in Figure 1. At Plane A (compressor inlet) and Plane C (50 mm downstream of stator outlet), 4 pressure sensors are embedded in the casing wall at the position shown in Figure 1 to measure the static pressure of the inlet and outlet. In the meantime, 2 pressure rakes are respectively placed in the inlet and outlet ducts to measure total pressure. At Plane B (0.5 chord length upstream of the rotor inlet), 8 high-frequency response pressure sensors are used on TA 36 in order to capture the pressure perturbations near the rotor tip. All of the pressure sensors are equally spaced around the circumference.

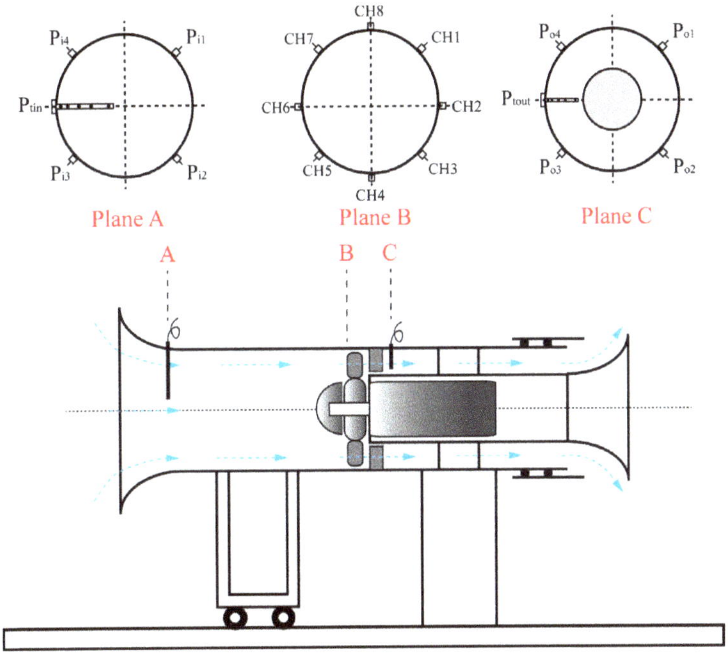

Figure 1. Schematic of TA 36 and measurement sections.

To design a fluidic oscillator suitable for TA 36, three compressor parameters are highlighted in the present paper, which are the design flow rate, the tip rotational velocity, and the characteristic frequencies associated with unstable phenomena.

Tables 1 and 2 shows the main design parameters of the TA 36 compressor [41]. The design mass flow is 6.5 kg/s. According to the design rotational speed (2930 rpm) and the tip diameter (600 mm), the tip rotational velocity is approximately 92 m/s. Figure 2 shows the static pressure signals of Plane B at 100% design rotational speed. From the change in the propagation velocity of pressure disturbances, it can be judged that TA36 is a spike-type stall. The calculated stall inception frequency and stall cell propagation frequency are about 38 Hz and 27 Hz, respectively. The rotor rotation frequency is 49 Hz since the rotating speed is 2930 rpm (the design speed).

Table 1. The aerodynamic design parameters of TA36 [41].

Aerodynamic Parameter	Value
Design speed (rpm)	2930
Design mass flow (kg/s)	6.5
Design efficiency	92%
Pressure ratio	1.0267

Table 2. The structural design parameters of TA36 [41].

Geometrical Parameter	Rotor	Stator
Number of blades	20	27
Aspect ratio	1.18	1.40
Tip clearance (mm)	0.6	0.5
Tip diameter (mm)	600	600
Hub clearance (mm)	346	401
Stagger angle at hub (°)	45	0

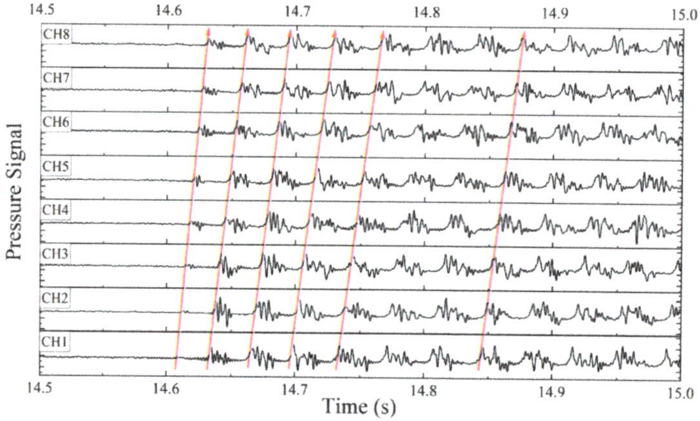

Figure 2. Time-resolved static pressure disturbance of Plane B at 100% design working speed.

2.1.2. Oscillator Configuration

Combined with the three key parameters of TA36 described above, a low-speed and low-frequency fluidic oscillator was redesigned based on models of Wang et al. [38] and Zhou et al. [39,40] in order to achieve unsteady micro tip injection on TA36.

This oscillator is a wall-attachment pulsed oscillator, and its interior and exterior structure are shown in Figure 3. The corresponding key parameters are listed in Table 3; all of the parameters are nondimensionalized by the depth of the interior structure, which is 2 mm. In Figure 3a, the blue arrows mark the planar core channel, and the orange tubes in Figure 3c are feedback channels. Together these two parts are named as the feedback loop. To meet the requirements of subsequent experiments, the oscillator's response frequency needs to be adjustable. Studies have shown that the response frequency of the wall-attachment pulsed oscillator is mainly controlled by the total length of the feedback loop [38,42]. Therefore, this configuration uses plastic tubes as feedback channels, which makes it easy to adjust its length by changing tubes. The interior structures are processed by the high-precision computerized numerical control method in order to ensure the consistency of the oscillators. In addition, feedback channels connect to the oscillator's main body by threaded pneumatic connectors.

Table 3. The key dimensions of the fluidic oscillator.

Symbol	Value	Parameter
t	0.2	Throat width and depth ratio
h	0.35	Control port height and depth ratio
w_1	1	Outlet width and depth ratio
w_2	12	Outlet distance and depth ratio
L_1	40	Oscillator height and depth ratio
L_2	20	Oscillator width and depth ratio

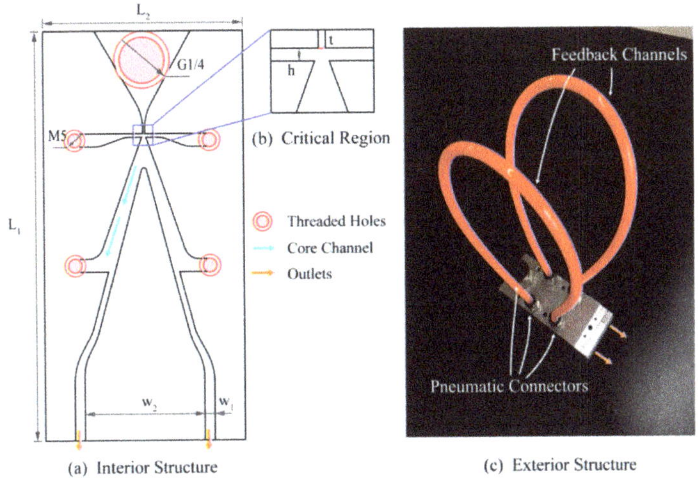

Figure 3. The interior and exterior structure of the fluidic oscillator: (**a**) interior structure; (**b**) critical region; and (**c**) exterior structure.

2.2. *Experiment Methods*

2.2.1. Oscillator Test Bench

As shown in Figure 4, the oscillator test bench includes an air pump (JUBA, TJ1600X6-3201, Wenling, Zhejiang, China) providing high-pressure air, a pressure regulator valve (SMC, AW20-02BG, Tokyo, Japan), a flow meter (Sevenstar, CS100, Tokyo, China), and a pressure gauge (Asmik, MIK-Y290, Hangzhou, Zhejiang, China). The test system is the StreamLine Pro CTA system. The hotwire (Dantec, 55P11, Skovlunde, Danmark) is placed at the center of the left outlet, less than 1 mm away from the outlet section along the depth direction. The hotwire calibration velocity range is 5~100 m/s and the sampling rate is 10 KHz.

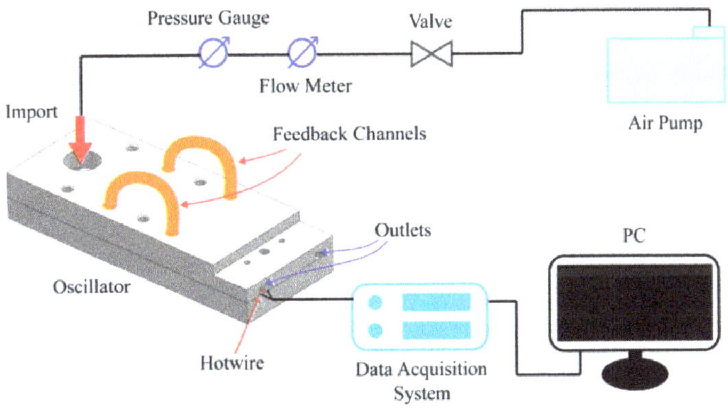

Figure 4. The oscillator test bench.

The oscillator outlet pressure is equal to the atmospheric pressure P_{atm}. The dimensionless parameter inlet-to-outlet pressure ratio is P_i/P_{atm}, where P_i is the total inlet pressure of the oscillator. The feedback channel length is L_f and the feedback channel inner diameter is d_f. During the test, d_f is 4 mm and L_f ranges from 2.5 m to 7 m. At each length, P_i/P_{atm} gradually increases from 1.1 to 2.0 in order to obtain thorough frequency and velocity characteristics of this oscillator.

2.2.2. Test Casing

Restricted by the outlet size, pulsating jets from a single oscillator can be quite limited. It is essential to arrange a certain number of oscillators on the casing to achieve the desired control effect. Figure 5 presents the test casing specially designed to fix oscillators. Thirty oscillators are anchored on the test casing, and the number of oscillators can be varied by disconnecting the air supply tubes of some of them. The axial position of oscillators is adjusted by switching the relative position between the inner ring and axial positioning rings. As shown in Figure 5, unsteady jets are 25° away from the mainstream and the axial position of actuation is 0.2 cord length downstream of the rotor tip leading edge.

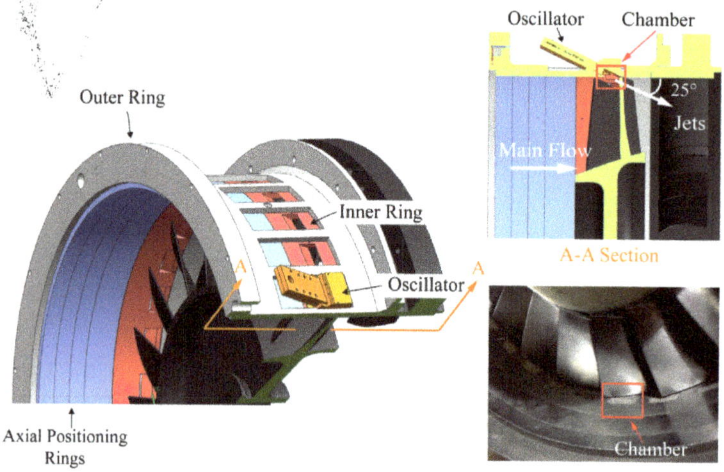

Figure 5. The test casing.

Meanwhile, the height of the oscillator is relatively short, so after the oscillator array is installed on the test casing, slot-like chambers still remain in the casing, as marked by the red boxes in Figure 5, in order to protect the rotor.

2.2.3. Oscillator–Compressor Experimental System

The composition of the oscillator–compressor experimental system is shown in Figure 6. It includes an air pump (JUBA, TJ1600X6-3201, China) providing high-pressure air, an air tank (JUBA, JB1.0-8, Wenling, Zhejiang, China) storing the compressed air, a strainer (JUBA, JB-30, Wenling, Zhejiang, China) filtering imported air, a manifold dividing the air 30 ways, a pressure regulator valve (SMC, AW20-02BG, Tokyo, Japan), a flow meter (SMC, PFMB7501, Tokyo, Japan), and a pressure gauge (Asmik, MIK-Y290, Hangzhou, Zhejiang, China). The data acquisition system is a NI-PXI system, which contains 48 channels, and its maximum sampling rate is 100 KHz. TA 36 has a controllable bleed valve to reach the pre-stall region as slowly as possible while recovering from the stall as quickly as possible. This setup ensures that all the steady operating characteristics of TA 36 can be recorded by the standard time–mean instruments.

During the experiment, the throttling process was stopped when the working condition of TA 36 reached the near-stall condition of baseline, the pressure regular valve was opened, and after the oscillators worked stably, we continued the throttling process. Pressure data was continuously collected during the whole process.

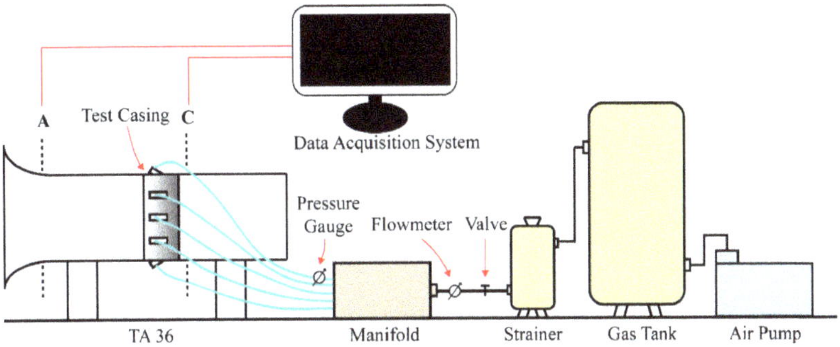

Figure 6. The oscillator–compressor experimental system.

3. Results

3.1. Oscillator Characteristics

3.1.1. Frequency Response

Hotwire tests were conducted under the condition that P_i/P_{atm} ranged from 1.1 to 2.0. The feedback channel's inner diameter d_f was 4 mm, and the length L_f range was 2.5–7 m.

The oscillator characteristic test results of some feedback channel lengths had very little difference, which is difficult to present. Therefore, in order to show the frequency characteristics that changed with L_f as clearly as possible, Figure 7 shows the experimental results with the feedback channel lengths L_f of 2.8 m, 3.0 m, 3.2 m, 3.8 m, 4.6 m, 5.2 m, and 6.2 m. With the decrease of L_f, the response frequency of the oscillator increased. All frequency–pressure response curves of different L_f show similar trends with the increase of P_i/P_{atm}. When P_i/P_{atm} increased from 1.1 to 1.25, the response frequency increased quickly, and after P_i/P_{atm} reached 1.25, the frequency basically did not change but floated around a certain value. According to the research of Cerretelli et al. [43], wall-attachment pulsed fluidic oscillators have the potential to realize frequency–pressure decoupling by designing interior configurations rationally. When P_i/P_{atm} was 1.25 to 2.0, the oscillator in this paper had response frequencies insensitive to the pressure change, that is, it had the frequency–pressure decoupling characteristics, which can effectively reduce the difficulty of frequency regulation in future experimental work.

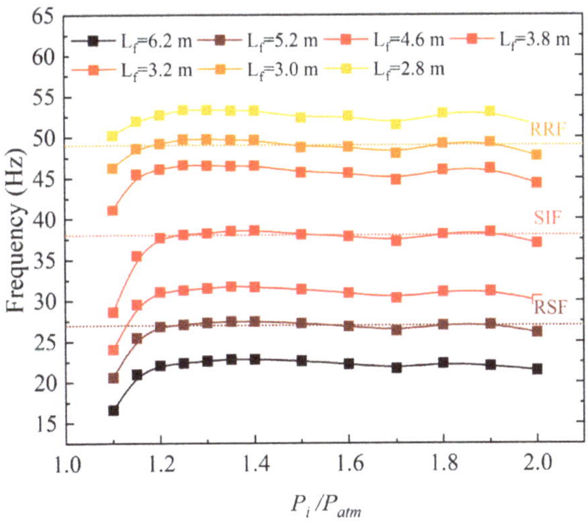

Figure 7. The oscillator frequency response spectrum.

In addition, three characteristic frequencies related to unstable phenomena are also marked in Figure 7. This configuration can achieve these frequencies by selecting the feedback channel length L_f. L_f corresponding to the rotating stall frequency (RSF, 27 Hz), the stall inception frequency (SIF, 38 Hz), and the rotor rotating frequency (RRF, 49 Hz) were respectively 5.2 m, 3.8 m, and 3.0 m. For each L_f, frequency deviations to the target frequency were less than 1 Hz, which means the error was no more than 3.70% and is in line with the experimental expectations.

When P_i/P_{atm} is from 1.25 to 2.0, the average frequency on each curve is defined as the average decoupled frequency. Further revealing the frequency characteristics of this oscillator, the average decoupled frequency and the average decoupled period T related to L_f of all test results are plotted in Figure 8. An empirical formula was obtained by linear fitting, where c represents the local speed of sound. Compared with the formulas proposed by Wang et al. [38] and Zhou et al. [39], the T of three curves are linear with L_f, but the slopes and the intercepts are different. The slope of the formula of this configuration is between Wang's and Zhou's formulas, and the intercept is negative while in the other two formulas, it is zero. These differences indicate that the slope and intercept of an oscillator are sensitive to its internal structure. In future work, if target frequencies are known, the feedback channel length L_f can be approximately predicted by this empirical formula.

$$T = 1/f \approx \frac{2.6L_f}{c} - \left(3.14 \times 10^{-3}\right) \quad (1)$$

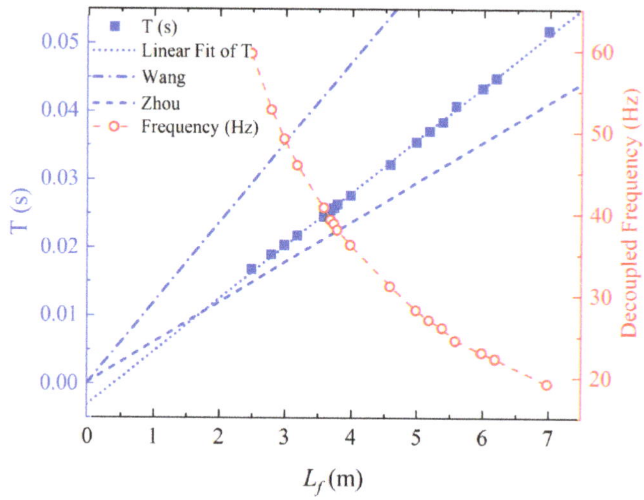

Figure 8. Changes of decoupled frequency and period T with L_f [38,39].

3.1.2. Velocity Response

Figure 9 shows the typical velocity response when L_f is 3 m and P_i/P_{atm} is 1.3. In each period T, the time of high pulsed speed is T_{on} and the time of low pulsed speed is T_{off}. The duty cycle, which is the ratio of T_{on} and T, was about 56%. In addition, the low pulsed speed was close to 0 and barely fluctuated, while the high pulsed frequency had obvious fluctuation. For the convenience of the following description, the average value of the high pulse velocity of each waveform is defined as V_{max}.

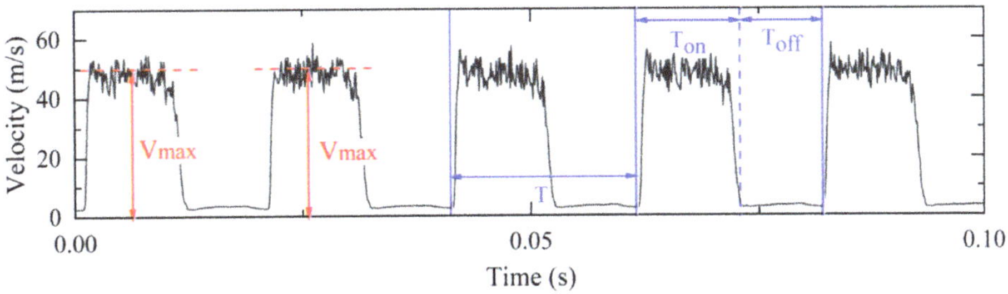

Figure 9. P_i/P_{atm} = 1.3, the typical velocity response of the oscillator.

Figure 10 shows the comparison of velocity responses with P_i/P_{atm} of 1.3, 1.5, 1.7, and 1.9, respectively. As the inlet pressure rose, the waveform of the velocity response was almost unchanged, and the V_{max} achieved increased as well as the fluctuating amplitude of the high pulsed velocity. Even if P_i/P_{atm} was the same, there were still some differences in neighboring waveforms due to velocity fluctuations. Therefore, the maximum velocity response of the oscillator was estimated by the arithmetic average V_{max} of 10 continuous waveforms, which was 30 m/s to 80 m/s, as shown by the blue curve in Figure 11. The unsteady jet speed generated by this oscillator configuration was lower than the tangential velocity of the rotor tip of TA36, which was larger than 92 m/s, so this configuration meets the requirements of the unsteady micro tip injection used in this paper.

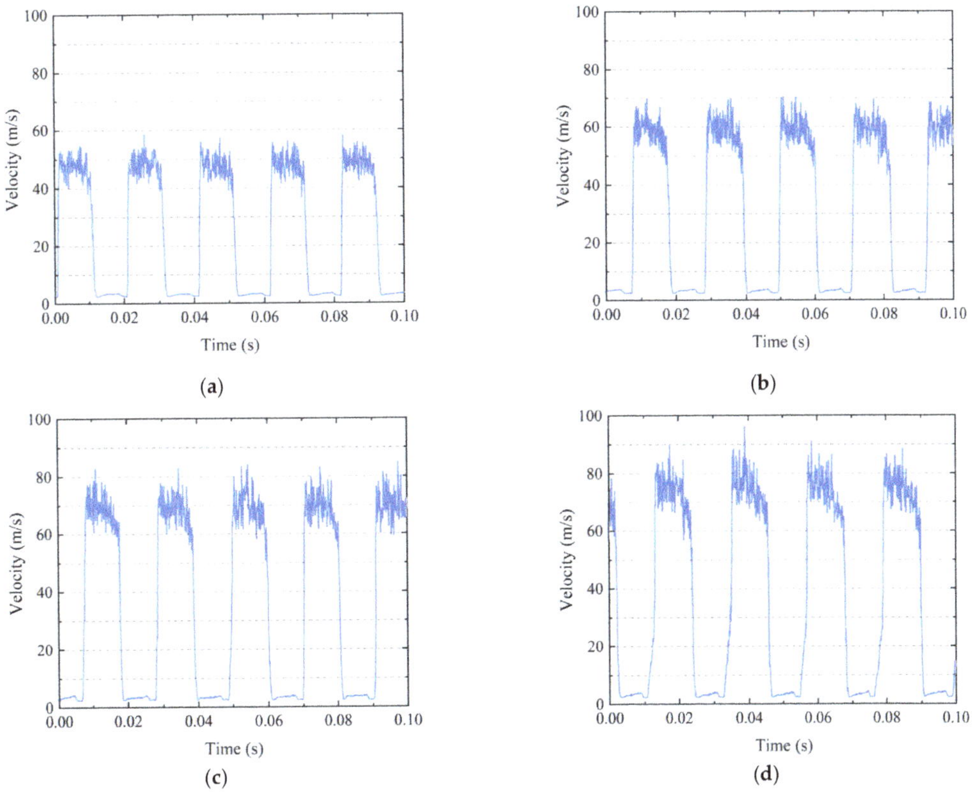

Figure 10. The oscillator velocity response at: (**a**) P_i/P_{atm} = 1.3; (**b**) P_i/P_{atm} = 1.5; (**c**) P_i/P_{atm} = 1.7; and (**d**) P_i/P_{atm} = 1.9.

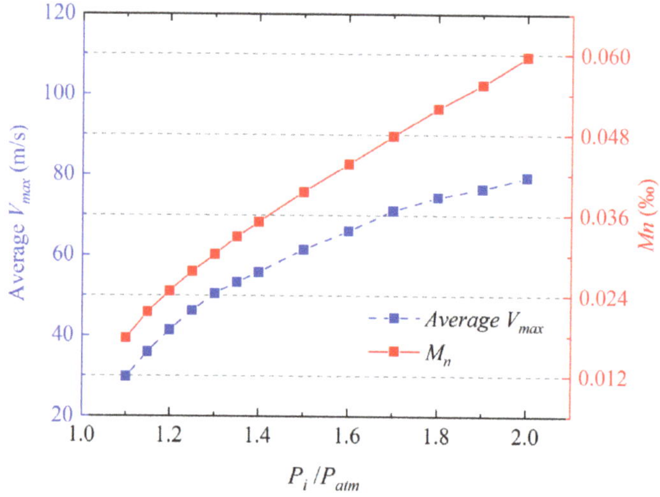

Figure 11. The oscillator's average V_{max} and mass flow rate response.

3.1.3. Mass Flow Response

The mass flow response of this oscillator is shown by the red curve in Figure 11. Restricted by the throat area, the flow rate increases monotonically with the increase of the inlet pressure. When the range of P_i/P_{atm} was 1.1 to 2.0, a single oscillator mass flow rate was from 0.115 g/s to 0.388 g/s. The dimensionless parameter M_n is the per millage of the oscillator flow rate to the design flow rate of the compressor TA36. In Figure 11, the M_n of a single oscillator ranges from 0.017 to 0.059, which is far less than the design flow rate of the compressor and meets the micro-injection requirement of the experiments.

In summary, this new oscillator configuration can realize the three frequencies related to the unstable phenomena of TA 36 by adjusting the feedback channel length. It can also achieve response frequency/period-pressure decoupling as an empirical formula between frequency/period and the feedback channel length was established, which is helpful to predict and select the corresponding target frequency. The P_i/P_{atm} range was between 1.1 and 2.0, and the maximum jet velocity range was 30 m/s to 80 m/s, which was smaller than the flow velocity at the rotor tip region. The mass flow range of the single oscillator was only 0.017~0.059‰ of the designed compressor flow rate, which ensures that the excitations were micro-injections, and the jet energy basically does not cause efficiency penalty.

3.2. *Stall Control Application*

In this study, 15 oscillators distributed evenly above the compressor rotor tip along the circumference were used, with an axial position of 0.2 times the tip cord length downstream of the leading edge. The total inlet pressure of all oscillators was 1.7 times atmospheric pressure, the jet frequency is rotor rotating frequency (RRF), and the maximum actuate velocity is 70 m/s. With a jet flow of only 0.08% of the compressor design mass flow and the maximum jet momentum of 0.27% of the compressor momentum corresponding to design mass flow, the active stall control experiments were conducted on TA 36 to explore the influence on the stability of the low-speed compressor.

3.2.1. Parameter Definition and Experimental Repeatability Verification

In order to quantitatively describe the effect of the unsteady excitation generated by oscillators on compressor stability, the following several parameters are defined.

The total-to-static pressure rise coefficient ψ is usually used to describe the low-speed compressor's pressure rise characteristics, which is defined as Formula (2). The flow coefficient ϕ is defined as Formula (3):

$$\psi = \frac{P_2 - P_{t1}}{0.5\rho U_m^2} \tag{2}$$

$$\phi = \frac{V_x}{U_m} = \frac{q_v}{U_m A_0} \tag{3}$$

where P_2 is the static pressure of the stator exit, P_{t1} is the total inlet pressure, U_m is the tangential velocity of the mid-span, q_v is volume flow, and A_0 is the inlet area.

Efficiency is expressed as Formula (4), P_{t2} is the total pressure at stator exit, and τ is the torque.

$$\eta = \frac{q_v(P_{t2} - P_{t1})}{\tau} \tag{4}$$

The comprehensive stall margin is adopted in this study to evaluate the compressor stability. The stall margin (SM) and stall margin improvement (SMI) are described as follows:

$$SM = \left(\frac{\psi_s/\phi_s}{\psi_d/\phi_d} - 1\right) \times 100\% \tag{5}$$

$$SMI = SM_{exp} - SM_{ref} \tag{6}$$

where subscripts s and d refer to stall and design operation points, respectively, and the working conditions corresponding to the subscripts *exp* and *ref* are selected as needed.

In Figure 12, a repetitive result of the baseline characteristics tested at the design speed is shown. The baseline cases are tested with the unmodified TA 36 casing. The results show good consistency, and the subsequent experimental results can be considered to be true and credible.

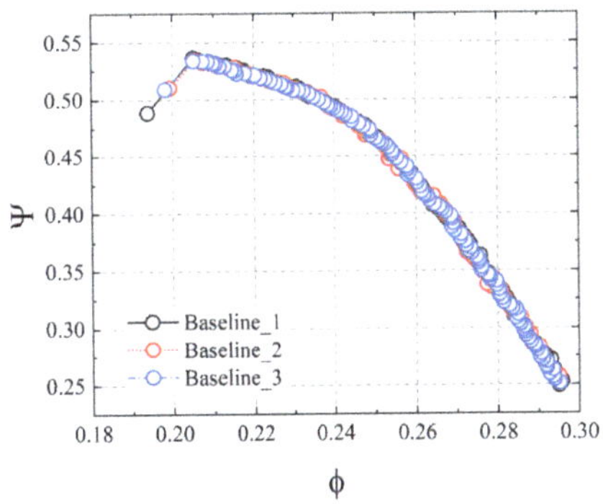

(a)

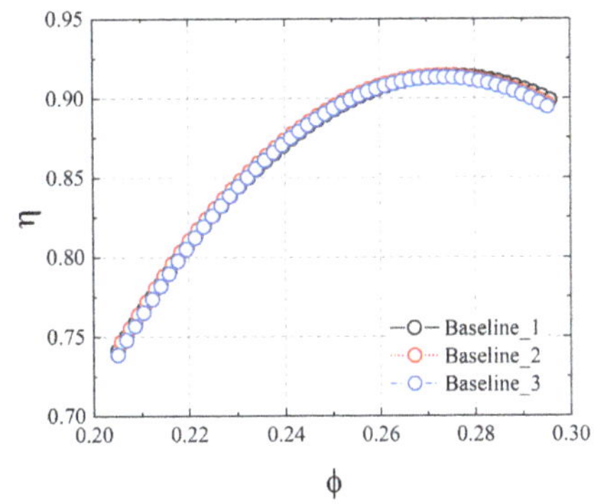

(b)

Figure 12. The repeated experiment results of TA36 under baseline condition: (**a**) pressure rise characteristics; and (**b**) efficiency characteristics.

3.2.2. Effect of Oscillators on Compressor Stability

Results of the unsteady actuation using oscillators are shown in Figure 13. In legends, EXP refers to the usage of test casing only, due to the assembly with oscillators; it has 30 slot-like chambers over the rotor tip region compared with the baseline case. EXP_RRF refers to the usage of test casing and RRF unsteady excitation, and SP refers to stall point.

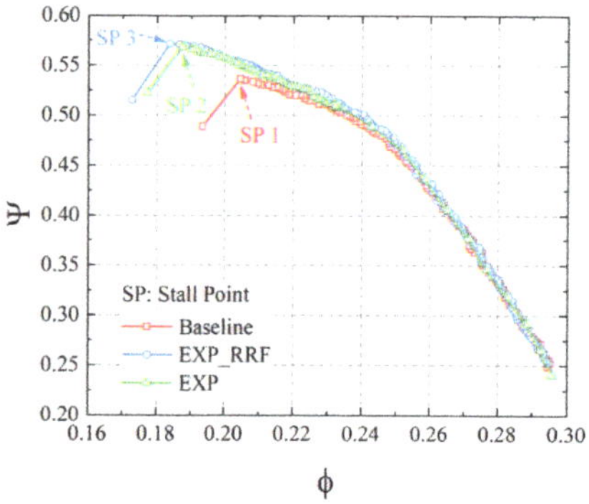

(a)

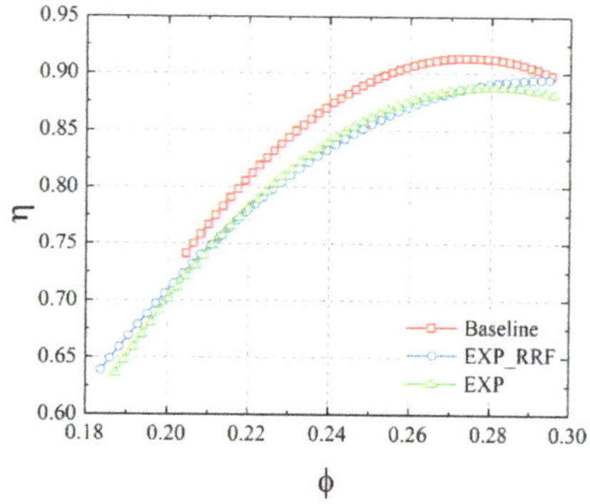

(b)

Figure 13. The results of the oscillator–compressor experiments: (**a**) pressure rise characteristics; and (**b**) efficiency characteristics.

In Figure 13a, compared with the baseline (SP 1), test casing (SP 2) itself can obviously improve the stall margin; meanwhile, it also has a significant efficiency loss of 2.77%. This is because of chambers in the rotor tip region after the casing and oscillators are assembled, as shown in Figure 5. The special structure of the test casing is designed by considering the safety and the related parameters of tip injection and casing treatment, which mainly include circumferential coverage and porosity. Studies [13,44] have shown that with the increase of the circumferential coverage of the tip injection, the SMI also increases, and the trend is basically first fast and then slow. The oscillators used in this experiment have two outlets, and the width of each is only 2 mm. When 15 oscillators are applied, the circumferential coverage of air injection is only 1.592%, and considering that the flow rate of air injection is only 0.08% of the compressor's designed flow rate, chambers are left on the casing. In these chambers, the momentum of the injected air is averaged, thereby expanding the circumferential coverage of the excitation to 20 times the original. At the same time, chambers on the casing are similar to the axially slotted casing treatment. According to previous research experience [45], with the increase of the porosity, the SMI and efficiency loss of the slotted casing treatment increase significantly. As mentioned above, 30 oscillators are always installed on the test casing, and the number of oscillators is adjusted by adjusting the air supply. Therefore, from the perspective of casing treatment, the porosity of the casing is independent of the number of oscillators used, which is always 63.67%, and the efficiency loss may be large. Therefore, it is necessary to reasonably design the depth of the oscillator insertion into the casing to reduce the efficiency loss as much as possible. At present, this paper only explores the feasibility of the application of fluidic oscillators on the stall margin control. This form of test casing was selected after considering both safety and its possible stall margin improvement ability. In future work, the influence of unsteady excitation of the oscillator directly acting on the rotor tip will be explored.

Moreover, in Figure 13a, after performing an unsteady excitation with the frequency of RRF, the flow coefficient of SP 3 was slightly smaller than that of SP 2, which was significantly less than the flow coefficient difference between SP 1 and SP 2. This illustrates that unsteady excitation generated by oscillators can have a certain effect. However, it is weaker than the test casing itself. Although the efficiency of case EXP_RRF is obviously lower than the baseline case, there is barely any efficiency loss compared with the case of EXP, and even some efficiency gain in near-choke and near-stall areas.

In this study, the combination of casing treatment and unsteady tip injection is technically realizing the decoupling of steady and unsteady stall margin improvement effects. When only using the test casing, the stall margin expansion (SP 1 to SP 2) completely reflected the steady mechanism, and adding the unsteady excitation on this basis, the extra stall margin expansion gain (SP 2 to SP 3) fully reflected the unsteady mechanism. To quantify the steady and unsteady effects in this work, the three stall points in Figure 13a are individually plotted in Figure 14. According to Formula (5), the stall margin of case Baseline was 32.68%, the stall margin of case EXP was 51.65%, and the stall margin of case EXP_RRF was 55.10%. According to Formula (6), if you select SM_{ref} as the stall margin of case Baseline and SM_{exp} as the stall margin of case EXP, the steady stall margin improvement of the case EXP is 18.97%. Select SM_{ref} as the stall margin of case EXP and SM_{exp} as the stall margin of case EXP_RRF, and the unsteady stall margin improvement of case EXP_RRF is 3.45%. Select SM_{ref} as the stall margin of case Baseline and SM_{exp} as the stall margin of case EXP_RRF, the steady and unsteady coupling improvement is 22.42%.

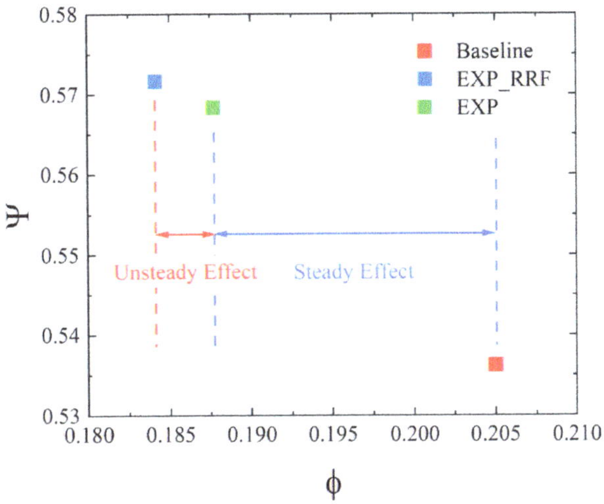

Figure 14. The summary of stall points of TA36.

4. Discussion

This oscillator configuration has the following advantages when applied to the low-speed compressor TA 36:

- The response frequency is adjustable and decoupled with pressure. An empirical formula between frequency/period and the feedback channel length L_f is established, which is helpful to predict and select the L_f of corresponding target frequency.
- The oscillators generate micro tip injections, which have almost no impact on the compressor characteristics compared with cases only using test casing, while it can obtain an unsteady stall margin expansion effect of two magnitudes greater than the excitation itself. In this study, the injected mass flow is only 0.08% of the compressor's design mass flow, and the unsteady stall margin expansion is 3.45%.

At the same time, because of the special structure of the test casing, the unsteady stall margin expanded by oscillators can be compared with that of the chambers. In Figure 13, the chambers are similar to the conventional passive flow control method and can produce up to 18.97% stall margin expansion, but at a cost of 2.77% efficiency loss. Unsteady micro jets have a weaker effect on stall margin than that of steady methods, but they produce basically no efficiency loss, perhaps because the influence mechanism is changed. Based on previous research experience, [20,21,23], the unsteady micro jets in this study input very small total energy, which can only act on the tip leakage vortex. Therefore, on the one hand, the jets themself suppress the advance and development of the leakage vortex. On the other hand, the unsteady frequency in the jet may interact with some frequencies in the leakage flow, stimulating the low-energy fluid in the tip region and delaying the leakage vortex generation, thus improving the stable working margin of the compressor.

This paper only verifies the feasibility of the innovative application of the oscillator in the compressor's active stall control field, and there are still many shortcomings. First, the unsteady excitation was not directly acting on the rotor tip flow field, and the stall margin expansion limit was not reached. Second, only the RRF unsteady excitation experiment was carried out, and the influence of other frequencies has not been clarified. Third, the data tested in the flow field were very limited, so it is difficult to fully explore and reveal the unsteady stall margin expansion mechanism. Lastly, compared with synthetic jet actuators or rotating valve actuators, oscillators have more application advantages in generating high excitation frequencies in the harsh operating environment of aeroengines, and their effects on the stability of high-speed compressors have not been explored. On the one hand, future work will optimize the appearance design of the oscillator and explore the influence of other excitation parameters on stability. Numerical simulation will be used to explore the unsteady stall margin mechanism combined with experimental results. On the other hand, oscillators will be applied on a high-speed compressor rig to explore the application value.

5. Conclusions

In this paper, the characteristics of a new pulsed fluidic oscillator with low frequency and low speed are experimentally studied. Fifteen unsteady oscillators are applied at the rotor tip of a low-speed axial flow compressor (TA36). Based on the usage of low-response steady transducers and high-response dynamic transducers, the influence of the unsteady excitation of oscillators on compressor stability is explored. The concluding remarks can be summarized as follows:

1. The oscillator configuration meets the requirements of unsteady micro tip injection experiments, which can cover the instability characteristic frequencies of TA36 by changing the feedback channel length. When P_i/P_{atm} ranges from 1.1 to 2.0, the maximum velocity ranges from 30 m/s to 80 m/s, which is smaller than the flow velocity in the rotor tip region. The flow rate of a single oscillator only varies from 0.017‰ to 0.059‰ of the compressor design flow rate, which causes low flow loss in the compressor.
2. The frequency/period of oscillator configuration is decoupled with the variation of inlet pressure. Furthermore, an empirical formula between frequency/period and the feedback channel length L_f is summarized, which is helpful to select the L_f corresponding target frequency in actual application.
3. The unsteady active stall margin control with the usage of fluid oscillators is feasible. When the excitation frequency of oscillators is RRF, the improvement of the compressor stall margin is 3.45%, in which the excitation flow and excitation momentum of oscillators are 0.08% and 0.27% of the compressor designed flow rate, respectively.

Author Contributions: Writing—original draft preparation, Z.L.; writing—review and editing, T.P., S.W., and Z.Y.; supervision and project administration, T.P. and S.W.; methodology, Z.L. and Z.Y.; funding acquisition, T.P. and S.W. All authors have read and agreed to the published version of the manuscript.

Funding: The authors acknowledge the support of the National Natural Science Foundation of China (No. 51976005 and No. 52006002), Science Center for Gas Turbine Project (P2022-B-II-004-001), National Science and Technology Major Project (2017-II-0005-0018), Advanced Jet Propulsion Innovation Center (AEAC HKCX2019-01-016), the Fundamental Research Funds for the Central Universities, and Beijing Nova Program.

Data Availability Statement: The data presented in this study are available on request from the corresponding author.

Conflicts of Interest: The authors declare no conflict of interest.

References

1. Day, I.J. Stall, surge, and 75 years of research. *J. Turbomach.* **2016**, *138*, 011001. [CrossRef]
2. Hathaway, M.D. Passive Endwall Treatments for Enhancing Stability. NASA/TM-2007-214409. 1 July 2007. Available online: https://ntrs.nasa.gov/citations/20070025023 (accessed on 25 October 2022).
3. Sun, X.F.; Sun, D.K.; Liu, X.H.; Yu, W.W.; Wang, X.Y. Theory of compressor stability enhancement using novel casing treatment, part I: Methodology. *J. Propul. Power* **2014**, *30*, 1224–1235. [CrossRef]
4. Sun, X.F.; Sun, D.K.; Liu, X.H.; Yu, W.W.; Wang, X.Y. Theory of compressor stability enhancement using novel casing treatment, part II: Experiment. *J. Propul. Power* **2014**, *30*, 1236–1247. [CrossRef]
5. Shivaramaiah, S.; Nagpurwala, Q.H.; Varpe, M.K.; Narahari, H.K. Effect of Rotor Tip Winglet on the Performance and Stability of a Transonic Axial Compressor. In Proceedings of the ASME 2017 Gas Turbine India Conference, Bangalore, India, 7–8 December 2017.
6. Han, S.B.; Zhong, J.J. Effect of blade tip winglet on the performance of a highly loaded transonic compressor rotor. *Chin. J. Aeronaut.* **2016**, *29*, 653–661. [CrossRef]
7. Bons, J.P.; Sondergaard, R.; Rivir, R.B. Turbine Separation Control Using Pulsed Vortex Generator Jets. *J. Turbomach.* **2001**, *123*, 198–206. [CrossRef]
8. Hergt, A.; Meyer, R.; Engel, K. Effects of Vortex Generator Application on the Performance of a Compressor Cascade. *J. Turbomach.* **2013**, *135*, 021026. [CrossRef]
9. Lu, X.G.; Chu, W.L.; Zhu, J.Q. A review of studies on casing treatment of axial flow compressor. *Adv. Appl. Mech.* **2006**, *2*, 222–232.
10. Epstein, A.H.; Ffowkes Williams, J.E.; Greitzer, E.M. Active suppression of aerodynamic instabilities in turbomachines. *J. Propul. Power* **1989**, *5*, 204–211. [CrossRef]
11. Ffowkes Williams, J.E. The aerodynamic potential of anti-Sound. In Proceedings of the 15th ICAS, Congress, London, UK, 7–12 September 1986.
12. Weigl, H.J.; Paduano, J.D.; Frechette, L.G.; Epstein, A.H.; Greitzer, E.M.; Bright, M.M.; Strazisar, A.J. Active stabilization of rotating stall and surge in a transonic single stage axial compressor. *J. Turbomach.* **1998**, *120*, 625–636. [CrossRef]
13. Suder, K.L.; Hathaway, M.D.; Thorp, S.A.; Strazisar, A.J.; Bright, M.B. Compressor stability enhancement using discrete tip injection. *J. Turbomach.* **2001**, *123*, 14–23. [CrossRef]
14. Paduano, J.D.; Epstein, A.H.; Valavani, L.; Longley, J.P.; Greitzer, E.M.; Guenette, G.R. Active control of rotating stall in a low speed axial Compressor. *J. Turbomach.* **1993**, *115*, 48–56. [CrossRef]
15. Haynes, J.M. Active Control of Rotating Stall in a Three-Stage Axial Compressor. Master's Thesis, Massachusetts Institute of Technology, Cambridge, MA, USA, 1993. GTL Report #218.
16. Glezer, A.; Amitay, M. Synthetic jets. *Annu. Rev. Fluid. Mech.* **2002**, *34*, 503–529. [CrossRef]
17. You, D.; Moin, P. Active control of flow separation over an airfoil using synthetic jets. *J. Fluid Struct.* **2008**, *24*, 1349–1357. [CrossRef]
18. Li, J.; Du, J.; Nie, C.; Zhang, H. Review of tip air injection to improve stall margin in axial compressors. *Prog. Aerosp. Sci.* **2019**, *106*, 15–31. [CrossRef]
19. Guinet, C.; Streit, J.A.; Kau, H.; Gümmer, V. Tip Gap Variation on a Transonic Rotor in the Presence of Tip Blowing. In Proceedings of the ASME Turbo Expo 2014: Turbine Technical Conference and Exposition, Düsseldorf, Germany, 16–20 June 2014; Volume 2A: Turbomachinery.
20. Lin, F.; Tong, Z.T.; Geng, S.J.; Zhang, J.; Chen, J.; Nie, C. A Summary of Stall Warning and Suppression Research with Micro Tip Injection. In Proceedings of the ASME 2011 Turbo Expo: Turbine Technical Conference and Exposition, Vancouver, BC, Canada, 6–10 June 2011; Volume 7: Turbomachinery, Parts A, B, and C.
21. Li, J.; Lin, F.; Tong, Z.; Nie, C.; Chen, J. The Dual Mechanisms and Implementations of Stability Enhancement with Discrete Tip Injection in Axial Flow Compressors. *J. Turbomach.* **2015**, *137*, 031010. [CrossRef]
22. Greenblatt, D.; Wygnanski, I. The control of flow separation by periodic excitation. *Prog. Aerosp. Sci.* **2000**, *36*, 487–545. [CrossRef]
23. Li, Z.; Gong, Z.; Liu, Y.; Lu, Y. The Experiment Research on the Performance of Low Speed Axial-Compressor by External Acoustic Excitation. In Proceedings of the ASME Turbo Expo 2004: Power for Land, Sea, and Air, Vienna, Austria, 14–17 June 2004; Volume 5: Turbo Expo 2004, Parts A and B.
24. Kefalakis, M.; Papailiou, K.D. Active Flow Control for Increasing the Surge Margin of an Axial Flow Compressor. In Proceedings of the ASME Turbo Expo 2006: Power for Land, Sea, and Air, Barcelona, Spain, 8–11 May 2006; Volume 6: Turbomachinery, Parts A and B.
25. Saddoughi, S.; Bennett, G.; Boespflug, M.; Puterbaugh, S.L.; Wadia, A.R. Experimental Investigation of Tip Clearance Flow in a Transonic Compressor with and Without Plasma Actuators. *J. Turbomach.* **2015**, *137*, 041008. [CrossRef]
26. Tomac, M.; Gregory, J. Internal jet interactions in a fluidic oscillator at low flow rate. *Exp. Fluids* **2014**, *55*, 1730. [CrossRef]
27. Gregory, J.; Tomac, M. A review of fluidic oscillator development and application for flow control. In Proceedings of the 43rd Fluid Dynamics Conference, San Diego, CA, USA, 24–27 June 2013.
28. Spens, A.; Pisano, A.P.; Bons, J.P. The application of curved fluidic oscillators for leading edge flow control. In Proceedings of the AIAA SCITECH 2022 Forum, San Diego, CA, USA, Virtual, 3–7 January 2022.
29. Raman, G.; Raghu, S. Cavity resonance suppression using miniature fluidic oscillators. *AIAA J.* **2004**, *42*, 2608–2612. [CrossRef]
30. Mohammadshahi, S.; Samsam-Khayani, H.; Nematollahi, O.; Kim, K.C. Flow characteristics of a wall attaching oscillating jet over single-wall and double-wall geometries. *Exp. Therm. Fluid Sci.* **2020**, *112*, 110009. [CrossRef]

31. Kim, S.; Kim, H.; Kim, K. Measurement of two-dimensional heat transfer and flow characteristics of an impinging sweeping jet. *Int. J. Heat Mass Transf.* **2019**, *136*, 415–426. [CrossRef]
32. Raman, G.; Packiarajan, S.; Papadopoulos, G.; Weissman, C.; Raghu, S. Jet thrust vectoring using a miniature fluidic oscillator. *Aeronaut. J.* **2005**, *109*, 129–138. [CrossRef]
33. Song, J.; Wang, S.; Wen, X.; Li, Z.; Lu, H.; Kong, X.; Liu, Y. Active flow control in an S-shaped duct at Mach 0.4 using sweeping jet actuators. *Exp. Therm. Fluid Sci.* **2022**, *138*, 110699. [CrossRef]
34. Staats, M.; Nitsche, W. Active control of the corner separation on a highly loaded compressor cascade with periodic nonsteady boundary conditions by means of fluidic actuators. *J. Turbomach.* **2016**, *138*, 031004. [CrossRef]
35. Meng, Q.; Du, X.; Chen, S.; Wang, S. Numerical study of dual sweeping jet actuators for corner separation control in compressor cascade. *J. Therm. Sci.* **2021**, *30*, 201–209. [CrossRef]
36. Kurz, J.; Hoeger, M.; Niehuis, R. Active Boundary Layer Control on a Highly Loaded Turbine Exit Case Profile. *Int. J. Turbomach. Propuls. Power* **2018**, *3*, 8. [CrossRef]
37. Chen, S.; Li, W.; Meng, Q.; Zhou, Z.; Wang, S. Effects of a sweeping jet actuator on aerodynamic performance in a linear turbine cascade with tip clearance. Proceedings of the Institution of Mechanical Engineers, Part G. *J. Aerosp. Eng.* **2019**, *233*, 4468–4481.
38. Wang, S.; Batikh, A.; Baldas, L. On the modeling of the switching mechanisms of a Coanda fluidic oscillator. Sens. *Actuator A Phys.* **2019**, *299*, 111618. [CrossRef]
39. Zhou, L.; Wang, S.; Wen, X. Working characteristics of a fluidic oscillator with high frequency and high speed. *J. Aerosp. Power* **2022**, *37*, 877–885.
40. Zhou, L.; Wang, S.; Son, J.; Li, Z.; Liu, X.; Peng, D.; Wen, X.; Liu, Y. Study of internal time-resolved flow dynamics of a subsonic fluidic oscillator using fast pressure sensitive paint. *Exp. Fluids* **2022**, *63*, 17. [CrossRef]
41. Dong, X.; Sun, D.; Li, F.; Sun, X. Stall margin enhancement of a novel casing treatment subjected to circumferential pressure distortion. *Aerosp. Sci. Technol.* **2018**, *73*, 239–255. [CrossRef]
42. Simões, E.W.; Furlan, R.; Bruzetti, R.E.; Gongora-Rubio, M.; Pereira, M.; Morimoto, N.; Jorge, J.; Avilés, S. Microfluidic oscillator for gas flow control and measurement. *Flow Meas. Instrum.* **2005**, *16*, 7–12. [CrossRef]
43. Cerretelli, C.; Gharaibah, E. An experimental and numerical investigation on fluidic oscillators for flow control. In Proceedings of the 37th AIAA Fluid Dynamics Conference and Exhibit, Miami, FL, USA, 25–28 June 2007.
44. Wang, W.; Chu, W.; Zhang, H.; Wu, Y. Numerical Investigation on the Effects of Circumferential Coverage of Injection in a Transonic Compressor With Discrete Tip Injection. In Proceedings of the ASME Turbo Expo 2014: Turbine Technical Conference and Exposition, Düsseldorf, Germany, 16–20 June 2014; Volume 2A: Turbomachinery.
45. Alone, D.B.; Kumar, S.S.; Thimmaiah, S.; Mudipalli, J.R.R.; Pradeep, A.M.; Ramamurthy, S.; Iyengar, V.S. Experimental Investigation on the Effect of Porosity of Bend Skewed Casing Treatment on a Single Stage Transonic Axial Flow Compressor. In Proceedings of the ASME Turbo Expo 2014: Turbine Technical Conference and Exposition, Düsseldorf, Germany, 16–20 June 2014; Volume 2A: Turbomachinery.

MDPI
St. Alban-Anlage 66
4052 Basel
Switzerland
www.mdpi.com

Actuators Editorial Office
E-mail: actuators@mdpi.com
www.mdpi.com/journal/actuators

www.ingramcontent.com/pod-product-compliance
Lightning Source LLC
Chambersburg PA
CBHW061520210926
51850CB00008B/1979
9783036589008

developments, such as the DIS:CO (design communication)-based application. This innovative application offers personalized, real-time feedback on tourists' health conditions and helps to build trust between consumers and service providers (Contribution 9). Unconventional tourism in the Metaverse is a phenomenon yet to be explored and understood. Contribution 10 provides an overview of the main tendencies in the domains of tourism and food and the consequences for research applications. According to the authors, the Metaverse can become an important tool for promoting and advancing tourism research and to foster virtual and digital collaboration among service providers, academia, and tourists.

Innovative product developments have played a key role in supporting unconventional tourist mobility. In making cycling tourism more experiential, a symbiosis of body (optimal physical exertion) and soul (effective personal development) was created on Hainan Island, China (Contribution 11). The increasingly intensive development of knowledge societies has also generated new products in the global tourism industry. Thus, the case study of the Chinese university city of Guangzhou can be of considerable help to destinations wishing to engage in scientific tourism in order to create their own evaluation matrix (Contribution 12).

The aim of this Special Issue is to raise awareness of unconventional, invisible, and uncounted forms of tourism—concepts not readily found in mainstream tourism research but which are extremely important, especially in certain destinations, such as borderlands, peripheral regions, and urban areas. The contributions in this theme issue illustrate that not all tourism is measurable, or even definable, in every context. Thus, human mobilities, more broadly as unconventional tourism, should become more prominent in tourism studies and in other disciplines that are concerned with human mobility and the use of space for leisure or utilitarian purposes.

List of Contributions

1 Timothy, D.; Irimiás, A.; Michalkó, G. Unconventional Tourist Mobility—A Geography-Oriented Theoretical Framework
2 Kovalcsik, T.; Elekes, Á.; Boros, L.; Könnyid, L.; Kovács, Z. Capturing Unobserved Tourists: Challenges and Opportunities of Processing Mobile Positioning Data in Tourism Research
3 Tóth, G.; Kincses Á. (In)Visible Tourism According to Online Cash Registers in Hungary, 2018–2020
4 Kádár, B.; Gede, M. The Measurable Predominance of Weekend Trips in Established Tourism Regions–The Case of Visitors from Budapest at Waterside Destinations
5 Garrido, B.; Sebrek, Sz.; Semenova, V.; Bal, D.; Michalkó, G. Addressing the Phenomenon of Overtourism in Budapest from Multiple Angles Using Unconventional Methodologies and Data
6 Sallay, Á.; Mikházi, Zs.; Gecséné Tar, I.; Takács, K. Cemeteries as a Part of Green Infrastructure and Tourism
7 Ásványi, K.; Miskolczi, M.; Jászberényi, M.; Kenesei, Zs.; Kökény, L. The Emergence of Unconventional Tourism Services Based on Autonomous Vehicles (AVs)–Attitude Analysis of Tourism Experts Using the Q Methodology
8 Könnyid, L.; Váradi, Zs.; Nagy, Zs.; Ilyés, N.; Horváth, H.O. The Changes in the Demographic Characteristics and Spatial Structure of Tourism Demand in the West Balaton Region's Spa Cities
9 Megyeri, G.; Boros, K.; Fekete, B. A Theoretical Concept of an Innovative and Sustainable Product Based on an Unconventional Approach to Design Development
10 Monaco, S.; Sacchi, G. Travelling the Metaverse: Potential Benefits and Main Challenges for Tourism Sectors and Research Applications
11 Lin, M.; Xu, H. Subjective Bodily Experiences of Island Cyclists in Different Contexts: The Case of Hainan Island, China

12 Guo, W.; Wu, D-F.; Li, Y.; Wang, F-X.; Ye, Y-Q.; Lin, H-W.; Zhang, Ch-F. Suitability Evaluation of Popular Science Tourism Sites in University Towns: Case Study of Guangzhou University Town

Author Contributions: Conceptualization—G.M., A.I., D.J.T.; Writing—original draft preparation, A.I., G.M., D.J.T.; Writing—review and editing, D.J.T., G.M., A.I. All authors have read and agreed to the published version of the manuscript.

Acknowledgments: This editorial was done in the framework of the OTKA K134877 research project supported by National Research, Development and Innovation Office (NKFIH). The special issue was supported by The Hungarian Tourism Association Foundation (MTSZ).

Conflicts of Interest: The authors declare no conflict of interest.

References

1. Buhalis, D. The Tourism Phenomenon: The New Tourist and Consumer. In *Tourism in the Age of Globalization*; Wahab, S., Cooper, C., Eds.; Routledge: London, UK, 2001; pp. 69–96.
2. MacCannell, D. The Making of "The Tourist". *Via—Tour. Rev.* **2018**, *13*. [CrossRef]
3. Mustonen, P. *Postmodern Tourism—Alternative Approaches*; Turku School of Economics: Tampere, Finland, 2006.
4. Dujmovic, M.; Vitasovic, A. Postmodern Society and Tourism. *Tour. South. East. Eur.* **2015**, *3*, 49–62.
5. Timothy, D.J. (Ed.) *Handbook of Globalisation and Tourism*; Edward Elgar Publishing: Cheltenham, UK, 2019.
6. Irimiás, A.; Michalkó, G. Hosting while being Hosted: A Perspective of Hungarian Migrant Hospitality Workers in London, UK. *Tour. Hosp. Res.* **2016**, *16*, 172–183. [CrossRef]
7. Zátori, A.; Michalkó, G.; Nagy, J.; Kulcsár, N.; Balizs, D. The Tourist Experience of Domestic VFR Travellers: The Case of Hungary. *Curr. Issues Tour.* **2019**, *22*, 1437–1459. [CrossRef]
8. Michalkó, G.; Timothy, D.J. Cross-border Shopping in Hungary: Causes and Effects. *Vis. Leis. Bus.* **2001**, *20*, 4–22.
9. Mitev, A.; Irimiás, A. Travel Craving. *Ann. Tour. Res.* **2021**, *90*, 103111. [CrossRef] [PubMed]

sustainability

MDPI

Hypothesis

Unconventional Tourist Mobility: A Geography-Oriented Theoretical Framework

Dallen J. Timothy [1,2,3], Gábor Michalkó [4,5,*] and Anna Irimiás [4]

1 School of Community Resources and Development, Arizona State University, Phoenix, AZ 85004, USA; dtimothy@asu.edu
2 School of Tourism and Hospitality, University of Johannesburg, Johannesburg 2092, South Africa
3 College of Tourism, Hunan Normal University, Changsha 410081, China
4 Marketing Institute, Tourism Department, Corvinus University of Budapest, 1093 Budapest, Hungary; anna.irimias@uni-corvinus.hu
5 Geographical Institute, CSFK (MTA Centre of Excellence), 1112 Budapest, Hungary
* Correspondence: michalko.gabor@csfk.org

Abstract: Prior to the COVID-19 pandemic, tourism had permeated all spaces of experience, reaching nearly every country, region, community, and corner of the globe. In recent decades, the meanings, implications, and roles of tourism have also expanded significantly. This article focuses on unconventional tourism mobilities, including same-day visits, which are an important but often neglected part of the tourism system, constantly challenging both scholars and tourism industry stakeholders. Unconventional tourism is an umbrella term that covers most kinds of unregistered or unaccounted tourist mobilities, some of which might not appear to be 'tourism' but should be in certain localities and under certain conditions. Given the growth of unregistered tourist flows and unaccounted leisure (or utilitarian) mobilities, there is a need in tourism studies to apply innovative research methods and to reconceptualize the meanings of tourism in different geographical and social contexts. It is expected that people's desire to travel in the post-pandemic era will educe new spatial and temporal travel experiences and behaviours in which unconventional tourisms will play an important role. To better understand this phenomenon and to evaluate the development of new approaches to travel and behavioural spatialities, new ways of thinking, new theoretical constructs, and new methodologies are needed. This article seeks to explore certain hidden or invisible tourism mobilities, focusing on the geographical patterns, processes, and hidden aspects of unconventional tourism.

Keywords: unregistered tourism; unconventional tourism; unaccounted tourism; short-haul travel; same-day travel; cross-border mobility; VFR tourism; sharing economy; shopping tourism

Citation: Timothy, D.J.; Michalkó, G.; Irimiás, A. Unconventional Tourist Mobility: A Geography-Oriented Theoretical Framework. *Sustainability* 2022, *14*, 6494. https://doi.org/10.3390/su14116494

Academic Editor: Anna Mazzi

Received: 19 April 2022
Accepted: 24 May 2022
Published: 26 May 2022

1. Introduction

The COVID-19 pandemic led to a drastic decline in world tourism [1]. According to World Tourism Organization (UNWTO) data [2], there was an 87% drop in international tourism demand in 2020 and a 56% drop in 2021 compared to 2019, which was a record high year. The crisis in the global tourism industry provides an opportunity to assess the development of tourism in the 21st century and to carefully plan for its revitalisation [3]. The spread of the pandemic closely correlates with the lifestyle of our time and with travelling becoming an ordinary practice [4]. Intensive tourist mobility contributed to the spread of the coronavirus to almost all countries on Earth in the first three months of 2020 [5]. In 2019, the UNWTO [2] recorded 1.5 billion international tourist arrivals, double the number of arrivals in 2000. Multiple changes underlie the dynamic growth of tourism over the last two decades; understanding these mechanisms will help construct post-pandemic scenarios and realise the vision of sustainable tourism [6]. The COVID-19 pandemic has brought unprecedented silence to popular tourist destinations around the world. Deserted streets, empty terraces, and closed restaurants, cafes and bars—a doomsday atmosphere was realised by globetrotters in previously crowded neighbourhoods. The waves of the COVID-19 pandemic have raised hopes for greater sustainable uses of places but have also shown the downside of the temporal collapse of the experience economy and experience society [7]. The extent of lockdowns, quarantines, and travel restrictions differed by country and region, but the reduction of environmental pressures was greatest where both long-term and same-day tourist mobility had previously been present [8]. In the resurrected mobilities of the post-pandemic period, innovative solutions will likely emerge to help promote and support sustainable destination management [9].

The impressive statistics produced by the UNWTO are primarily representative of measured and recognized tourist flows, not of invisible mobilities that also fit under the auspices of tourism [10]. For example, domestic same-day travel, cross-border and transit same-day trips, and overnight stays outside registered accommodations, including the use of second homes, courtesy lodging, and informal accommodations, are for the most part outside the scope of official tourism statistics [11]. 'Invisible tourism' estimates and periodic surveys are done in some countries, but there are no databases for international comparisons [12], yet this is a salient part of the tourism system worldwide.

The European Parliament and the Council of Europe adopted regulations concerning European tourism statistics in 2011 [13], clearly stating that (5) "The changing nature of tourism behaviour since the entry into force of Council Directive 95/57/EC of 23 November 1995 on the collection of statistical information in the field of tourism, with the growing importance of short trips and same-day visits contributing substantially in many regions or countries to the income from tourism, the increasing importance of non-rented accommodation or accommodation in smaller establishments, and the growing impact of the Internet on the booking behaviour of tourists and on the tourism industry, means that the production of tourism statistics should be adapted."

The European Commission has required Eurostat, the statistical office of the European Union, to draw up a methodological manual that must be regularly updated. However, even the political will to make invisible tourism visible has failed to produce tangible results. The UNWTO and the World Travel and Tourism Council (WTTC) estimate that domestic tourism is several times larger than international demand but, due to data collection gaps, few numbers are available for scientific analysis of this segment [14].

Unconventional (or invisible, or hidden) tourism mobility refers to trips involving the use of accommodation outside of regular statistical data collection and trips of less than 24 h in duration (Figure 1). Unconventional tourism mobility as conceived here differs from conventional tourism in essentially two ways: the nature of the accommodation used and the short length of stay. Unconventional tourism is, therefore, statistically invisible, but the characteristics of travel consumer behaviour are relatively well captured and essentially the same as conventionally measured tourism. In a given destination, the same tourist activity can be carried out in both unconventional and conventional modalities (e.g., a day visit to a ski resort or a beach holiday in a second home property provided free of charge by friends) [15]. Unconventional tourism mobility thus overlaps with conventional tourism mobility, in both space and time, and managing the two phenomena together poses significant challenges for planners, developers and managers.

The phenomena associated with unconventional tourism mobility have been known to tourism researchers for decades, yet they remain under-researched due to the estimated volume, assumed impacts, and the difficulty in drawing parameters for analysis. The lack of a conceptual framework to explore unconventional tourism mobilities has been an additional obstacle to the discussion of the phenomenon. However, profound changes in the nature of tourism (in particular overtourism) have drawn attention to the economic, social and scientific potential of unconventional tourism mobility [16]. There is also a growing recognition that hidden tourism is symbiotic with the trends that are driving its broader uptake so the need for invisible tourism to operate smoothly has implications for the formation of trends, and in some cases the trends themselves can be interpreted as unconventional tourism (e.g., CouchSurfing) [17].

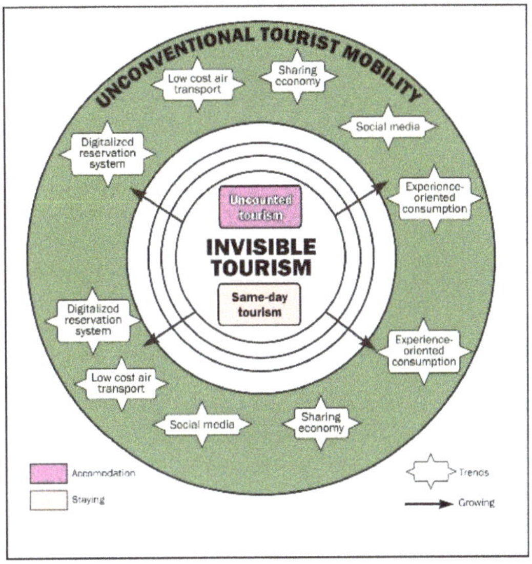

Figure 1. Model of unconventional tourist mobility. Source: authors' conceptualization (design: Livia Kaiser).

It is difficult to draw boundaries that mark the changing nature of tourism and to determine the period which has now generated and increased the need for a complex exploration of unconventional tourism mobility. The emergence and market penetration of low-cost airlines [18]; the widespread use of the internet and online retail spaces that facilitate the purchase of most tourism services [19]; the ability to share travel experiences instantaneously on social media [20]; the diffusion of the sharing economy [21]; the fact that consumption is not only a means of satisfying needs but also a means of accumulating experiences and avoiding boredom [22] all have played a significant role in this process.

If we look at the above processes in their historical context, the turn of the 20th and 21st centuries is when the very nature of tourism changed on a global scale, affecting almost all segments of society and generating measurable results in national economies. Conventional market processes in tourism have been joined by unconventional ones driven by several business and technological innovations. First, the alternative to expensive, convenient, and privileged air travel has become cheap and democratised air travel [23]. Second, online booking systems have made it possible to bypass travel agents and ticket offices when purchasing travel products, even with immediate validity [24]. Third, experience narratives have become real-time, spreading across the web through visuals, posts, hashtags, comments and 'likes' on Facebook and Instagram platforms, and transforming marketing communication [25]. Fourth, because of Airbnb, private homes have become commercial places to stay and attract tourists in droves with the promise of 'living like a local' [26]. Finally, historically popular tourist destinations have become victims of overtourism, and simultaneously, many previously unknown geographical locations have entered the world tourism marketplace, offering an almost inexhaustible pool of experiences [27].

In Europe, unconventional tourist mobility has also been catalysed by policy changes, such as the expansion of the European Union, facilitating the free movement of people, and the expansion of the Schengen Zone, which abolished most border controls, thereby contributing to the reduction of the unpleasantness often associated with border crossings [28]. In a market dominated by planned trips, mostly related to the possibility of taking some time off work, the share of spontaneous impulse trips generated by internet-based marketing communications has also increased [29]. Discount airfares that facilitate even one-day return trips have in many cases led to hyper-fast consumption of a destination/product/service without necessitating overnight accommodations [30]. Changes in the world of work and learning, especially due to the advance in digital technology, have created opportunities for blurring the boundaries between the private and the official spheres and between leisure and work. The use of second homes as home offices and 'workations' (work during vacation) both broaden the conceptual framework of unconventional tourism mobility [31].

As the issue of spatiality is a primary concern in understanding the characteristics of unconventional tourism mobility, geography and related disciplines have a major responsibility to address our gaps in knowledge. The challenge is two-fold: to strengthen the theoretical background on the one hand and the methodological arsenal on the other. The theoretical framework can be built up through a critical literature review on different issues linked to unconventional tourism mobility, while the methodological framework needs to be improved through an evaluation of the procedures used in previous studies. The primary objective of this paper is to outline the theoretical concept of unconventional tourism mobility by analysing and evaluating the literature on the common set of trends leading to invisible tourism and overtourism. Emphasis is placed on the geographic aspects of mobility and on the methodology, which can support the recognition of invisible tourism.

2. Invisible Tourism

Tourism statistics and databases facilitate the systematic monitoring of tourism phenomena and processes [32]. In principle, official statistical data have been collected at border crossings and from lodging establishment records [33,34]. Data from these sources support the UNWTO's assertation that tourists are people who spend at least one night away from home. In certain countries, border and/or accommodation statistics are complemented by surveys using different methodologies [35,36]. The global tourism statistics published annually by the UNWTO [2], which are representative of international tourism flows and comparable over space and time, are a combination of the above data collection methods. Given the statistical definition of a tourist and typical data source points, tourism statistics are not suitable for monitoring tourism mobility of less than 24 h in duration and travel that does not entail the use of registered accommodation—together termed here 'invisible tourism', 'unconventional tourism', or 'hidden tourism'. Statistically, invisible tourism generates tourist flows that affect the socioeconomic, sociocultural and natural environments in the destination [37], even if it is not counted directly. Although no official statistical data are collected on invisible tourism, service providers' statistics, onsite surveys and estimates (e.g., number of tickets sold, one-day city card sales, and parking receipts) are collected for different purposes [38]. Because the methodologies in data collection differ significantly, it is almost impossible to compare data.

2.1. Unobserved Tourism

De Cantis et al. [39] define unobserved tourism as visits where tourists stay in a destination for more than 24 h and spend the night in unofficial establishments or in official establishments but are deliberately not registered by the host (mainly due to tax evasion) (Figure 2). The former is also called unmeasured tourism, the latter underground tourism. Based on their studies in Sicily, De Cantis and Ferrante [11] found that the number of tourists staying on the island is almost the same in official and in unofficial establishments, but they did not attempt to estimate the number of visitors concealed by the managers of official establishments. Unobserved tourism stays of at least one night generate a noticeable turnover in the destination [40]. On the one hand, this translates into increased revenues from the use of non-accommodation tourism services, and, on the other hand, it undermines the principle of sustainability [41]. Parocco et al. [42] point out that both unobserved and short-haul tourism pose serious challenges to data-driven planning.

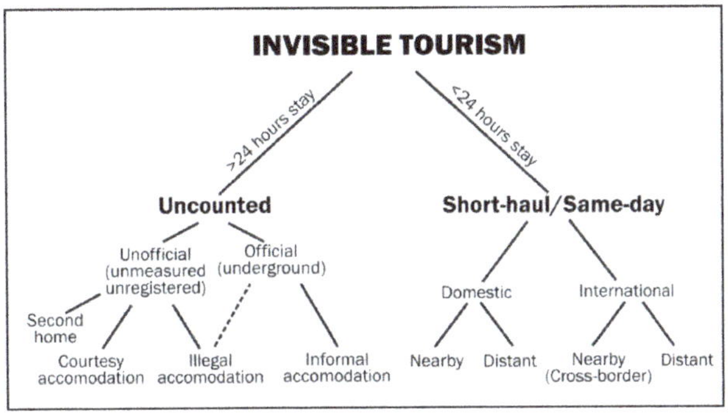

Figure 2. Model of invisible tourism. Source: authors' conceptualization (design: Livia Kaiser).

Within unobserved tourism, a distinction is made between unofficial and official tourism. Unofficial accommodations include second homes for owners' use or for family members, the total or partial letting of owned or rented property to distant relatives or friends on a courtesy basis, and illegal accommodations operated specifically for business purposes but completely hidden from the authorities. One of the characteristics of second homes is that they are usually specifically intended for leisure purposes, whether they are located in prominent or secluded tourist destinations, and thus increase the demand for local or regional tourist services despite their stay not being recorded in any database [43]. The full or partial letting of owned or rented property on a courtesy basis is most often carried out in the context of VFR tourism [44]. In these cases, the owner or tenant offers guests the temporary use of a dedicated space (e.g., guest room) without payment or offers overnight accommodation in other parts of the property without registration (e.g., in the living room, garage, garden tent). A key feature of VFR tourism is that the guest and the host usually spend as much time together as possible, and so a significant proportion of meals are also served in the host's household. Although VFR tourism is not necessarily linked to tourist destinations, the guests often use tourist services as leisure activities, in which hosts also are involved [45]. Some prominent tourist destinations specifically control and sanction the use of properties (including second homes) disguised as VFR tourism. The CouchSurfing movement is an organised form of accommodation by courtesy, a reduced form of home-swapping, where members of an online community stay overnight in each other's homes during their travels, with no statistical reporting obligations on the part of the host [46]. Illegal accommodation—accommodation that is operated without a licence or a declaration of the existence of the accommodation—can occur anywhere in the world, but is more likely to be found in regions where income concealment is an integral part of everyday life [47]. Spending nights outdoors 'under the sky', such as sleeping in tents, sleeping bags, caravans, or cars, should also be considered under the umbrella of non-official accommodation [48].

Official accommodations become part of unobserved tourism when managers conceal from authorities some of their income, mainly for tax evasion purposes [49]. By doing so, they operate informally, either seasonally or for a certain proportion of the actual turnover. For example, they offer their services legally, with official permits (subject to registration), but the recorded bed-nights do not match reality.

A very specific type of informal accommodation is represented by sea- or river cruise ships with several hundred or thousands of passengers on board, which stay parked at least one night in a destination's harbour. These stays are not included in official destination statistics, and cruise passengers are considered day-trippers [50].

Likewise, hotels which once operated legally but have had their licences withdrawn or surrendered, have retained their marketing and/or sales channels, and still receive guests at certain intervals, operate on the conceptual borderline between formal and informal accommodations. This also includes operators who run several different types of accommodation (hotel-apartment or camping-bungalow/caravan) but conceal the turnover generated in one or the other.

2.2. Same-Day Tourism

Travel within 24 h without an overnight stay is part of the broader phenomenon of short-haul tourism [51,52]. Official statistical counts of same-day visitors, or day-trippers, can only be made at border posts, where government controls are enacted and travel documents presented (e.g., passports, visas, vaccination documents) as a condition for entry into the country. Same-day tourism takes place both internationally and domestically, with the length of the journeys (in time and distance) typically being rather short, both subject to a return to the place of origin within 24 h, or more precisely without an overnight stay (Figure 2). The length of the trip and the return journey essentially depend on the means of transport used [53]. Despite the fact that day trips are popularly perceived as trips to areas near one's residence, there are also trips of less than 24 h to destinations hundreds or even thousands of kilometres from the place of origin, thanks to efficient air travel and high-speed rail. Although climatic conditions can influence short-haul tourism, their seasonal effect is somewhat lower, as water-based holidays, which attract large crowds and require long hot summer periods, or ski tourism, which is linked to cold, wet winter weather, are by definition peripheral to day trips.

The archetypes of same-day tourism in international contexts are cross-border shopping tourism, health tourism, VFR tourism and business tourism. Shopping tourists travel to purchase goods and certain repair and maintenance services available on the other side of a national border, motivated by lower prices, better quality, a greater product selection, a more attractive sales environment, the purchase of certain goods not available at the point of origin, and faster, more efficient service [54]. Day-trip international health tourism involves the use of services that are either linked to health insurance or self-financed [55]. Health care providers in Mexican municipalities bordering the US are visited by Americans seeking less-expensive care and Mexican citizens living in the US without adequate health insurance [56]. In Hungary's border region with Austria, day visits by Austrian health tourists are driven by Hungarian dental services, which are both low-cost and high-quality by international standards [57]. In border regions where history has divided families and communities that had lived together for generations, international day visits to relatives are a common travel activity [58]. While business tourism usually assumes a stay of at least one night with premium services, brief visits of less than 24 h to offices or worksites relatively close but in another country are common, especially where good transport links make day-tripping easy [59].

In the case of short-haul, domestic tourism, it would be difficult to highlight archetypal tourism products similar to international ones. However, it can be noted that the most common activities during domestic day trips are visits to natural environments offering a wide variety of recreational activities, visits to iconic sites of cultural and heritage tourism, gastronomic strongholds, sport events, amusement parks, and visits to relatives and friends [60,61]. The distance scale of domestic trips shorter than 24 h depends to a large extent on the size of the country's territory, because the length of a trip within a country's borders is incomparably different in a microstate than it is in a very large country [62,63]. It is also related to the available transport infrastructure. Where domestic flights are available, it is obviously possible to cover a return trip over much greater distances within 24 h than in regions where only land connections are available (although high-speed rail may be an alternative to air transport) [64].

3. Geographical Factors and Methodological Challenges

Time and distance are closely connected and are among the most important attributes of tourism mobility; thus, understanding the geographical factors of unconventional tourism is important [65–67]. Tourism entails changing one's environment, seeking experiences, and using services. This supposes, according to traditional understandings, that tourism-motivated consumption is tied to at least one night away from the place of residence, at a registered lodging service [68]. Although unconventional tourism mobility is not a new phenomenon, there is a need for its statistical recording, scientific examination and more conscious management [69]. The tourism trends of the 21st century contribute, on one hand, to the increased demand for shorter journeys, providing more intensive experiences and, on the other hand, to the horizontal and vertical extension of tourism space [70]. The fact that early-morning and late-night flights have become fashionable means that tourism destinations are accessible within a one-day round trip, evoking the need to re-define tourism in certain situations [23]. If destinations that have previously been almost unknown among international tourists now appear on airline route maps and in marketing campaigns, and within a destination, parts of a city located away from the traditional tourist zones are garnering more tourist attention, then the appearance of new tourism geographies defined by unconventional mobility becomes more evident [71,72]. The increasing desire (generated largely by social media) for life satisfaction through travel (a need once outlined by Maslow), the rapid expansion of transportation in both techniques and technologies, and the rational use of shared accommodation, including second homes and courtesy accommodation, have led to such high numbers of invisible tourists that characteristics of change in consumer behaviour in time and space are almost impossible to examine [73].

The wide range of methodological tools at researchers' disposal can potentially help develop databases that extend over and above simple border-crossing and accommodations statistics. The exploration of, and better acquaintance with, the macro- and micro-changes of unconventional tourism can contribute both to a deeper understanding of certain phenomena, as well as to their effects and management. Representative surveys undertaken from time to time among foreign tourists in a country and among domestic tourists in their own homelands can help refine and triangulate statistical data collected at border crossings and lodging establishments, as they also explore the volume, characteristics and geographical extent of unconventional tourism [74,75]. These surveys are extremely costly and their databases are incomplete, even from the point of view of macro-changes. Surveys dealing with travel of less than 24 h are typically done at a very local or regional level. Thus, they are rarely representative of broader populations and are less comparable, but the results may be adequate to evaluate unknown or hidden factors of tourism consumption in a given destination [76,77]. Research undertaken by a local authority or destination management organisation can provide opportunities to assess the micro-mobility of tourists within the community by having visitors list the sites visited or show them on a map [78]. Even before the spread of digital cameras and smartphones, photo elicitation studies and photograph content analyses enabled researchers to understand tourists' consumer behaviours, especially the extent of their mobility in the destination. For the most part, there were two consistent methods. One was the use of tourists' personal photographs; the other was asking tourists to take pictures using disposable cameras provided by the researcher [79]. Digital photography provided a breakthrough in two main ways. First, film cameras were no longer necessary and the digital images were easier to store and analyse. Secondly, the exact location and time of the photograph could be digitally stamped with the help of GPS coordinates [80]. Digital photographs energized the use of social media and online photo-sharing platforms. Because of this technology, using 'geotags' (certain cartographic algorithms), it became possible to assemble databases to better understand the spatialities of tourism regions [81,82].

Exploring unconventional tourism mobility is possible through regular, countrywide quota sampling surveys among international and domestic tourists, paralleled with the tourism demand of the relevant communities and/or regions at the same time. The costs of such a massive endeavour would likely be returned with a greater understanding of tourism satellite accounts and in the results of efforts to stimulate the economy. However, such surveys would not produce an exact picture of the spatial and temporal characteristics of macro- and micro-sized changes in tourism mobility [83], but they might produce some level of actionable data. Using mobile cellular data from smartphones might be part of the solution [84]. Cellular transmission towers are located relatively near one another and record mobile phone use within their radius, constantly producing large databases daily [85]. Using big data technology and current analytical tools with programmed algorithms, tourists' mobility, including the locations visited and lengths of stay can be determined. Although the use of mobile phone data to explore unconventional tourism mobilities is still in a nascent phase, it has a great deal of potential for making invisible tourism more visible.

4. Conclusions

A common characteristic of economic and social development is the movement from unconventional to conventional. What was unconventional at one time becomes conventional later through processes of change and normalization. This cycle is also apparent in tourism, although the period of change to conventional from unconventional appears to be getting shorter [86]. In the history of tourism, technical and technological innovations are the most notable milestones that have effected the most change in travel patterns and tourist attitudes. For example, in 1841, Europe's first tour operator, Thomas Cook, started his global enterprise, which continues to operate today, built upon Stephenson's Rocket, an early steam locomotive. In 1958, the first Boeing jet airliner, the 707 model, initiated a breakthrough in intercontinental and transoceanic flights. At the turn of the 21st century, massive advances in information and communications technology pervaded everyday life in the Global North and Global South and ushered in the latest chapter of tourism history. The underlying ethos of tourism—seeking experiences, visiting other environments, and utilizing services, were common in all three periods of history and have remained the essence of tourism.

The unconventional features of tourism discussed in this paper are not new or previously unheard of; overnighting in non-registered accommodations or same-day journeys have long been acknowledged as part of the tourism system, but their volume and diversity have increased as a result of trends endemic to the first decades of the 21st century. As acknowledged earlier, unconventional tourism is defined largely by the non-use of registered lodging and a short length of stay, which does not necessarily fit the UNWTO's official definition of a tourist or tourism. Thus, it may be relegated to the margins of tourism by national or supranational authorities because of its 'invisibility' and position outside the normative scope of statistically measurable tourism.

Our attempt to introduce the concept of unconventional or invisible tourism mobility aims to advance the following: (1) to produce a succinct and simple conceptual umbrella to discuss the unobservable (and unmeasurable) mobilities and same-day tourism; (2) to extend and differentiate the scope of these phenomena; (3) to explore the potential for more systematically analysing these hidden manifestations of tourism in the face of a 'data desert'. Geography plays an important role in the conceptualization of unconventional tourism mobility, particularly as regards place, distance, and temporality, yet the spatial and chronological features of unconventional tourism have so far received little academic attention. Geographers have a special responsibility to explore patterns of horizontal and vertical mobility, as well as the length of stay associated with them. Scholars should not be content with the limited information and data available. Rather, new and creative methods need to be developed to create databases that can help researchers better understand otherwise 'unmeasurable' manifestations of tourism. Digital photographs, including those posted by internet users, are a potentially lucrative source of data, especially because they are digitally stamped with GPS coordinates. Likewise, mobile and cellular data generated by smartphones might be alternative sources of tourism statistics, especially in light of the scarcity of data otherwise. Studying unconventional and invisible tourism mobilities serves to create a deeper and more sophisticated understanding of new turns in 21st-century tourism and contributes to the development of data-driven tourism management. It also creates a more holistic view of tourism in its broadest sense, rather than what is normally measured and accounted for, and has the potential to support the concept of sustainable development as communities and industry leaders begin to recognize the potential socioeconomic and sociocultural contributions of alternative tourisms that traditionally have been unaccounted and overlooked.

The perspectives on forgotten tourism or unconventional tourism at the centre of this article have inspired previous research. Representatives from many disciplines have recognized the research potential on the topic, including management, urban studies, landscape architecture, statistics, and environmental science, yet it remains tangential to the thrust of mainstream tourism research. Our hope is that the concept of unconventional tourism will become more accepted among researchers, with the potential to generate additional international research. The most important task for the future is to develop, in cooperation with statistics offices and tourism management representatives (governmental decision makers), a comprehensive monitoring system that will be more inclusive in providing credible, comparable and regular data on the real and more holistic processes of tourism.

Author Contributions: Conceptualization, G.M., A.I. and D.J.T.; writing—original draft preparation, G.M.; writing—review and editing, A.I. and D.J.T.; visualization, G.M.; supervision, D.J.T.; project administration, A.I.; funding acquisition, G.M. All authors have read and agreed to the published version of the manuscript.

Funding: This research was funded by National Research, Development and Innovation Office (NKFIH), grant number OTKA K134877. The dissemination of research outcomes was supported by The Hungarian Tourism Association Foundation (MTSZ).

Institutional Review Board Statement: Not applicable.

Informed Consent Statement: Not applicable.

Acknowledgments: The authors would like to express gratitude to Márta Bakucz and Chadwick Clifford Sheperd furthermore to Livia Kaiser for their assistance.

Conflicts of Interest: The authors declare no conflict of interest.

References

1. DeMicco, F.; Cetron, M.; Davies, O.; Guzman, J. COVID-19 impact on the future of hospitality and travel. *J. Hosp. Tour. Res.* **2021**, *45*, 911–914. [CrossRef]
2. UNWTO. *Tourism Highlights 2020*; UNWTO: Madrid, Spain, 2021.
3. Kuščer, K.; Eichelberger, S.; Peters, M. Tourism organizations' responses to the COVID-19 pandemic: An investigation of the lockdown period. *Curr. Issues Tour.* **2022**, *25*, 247–260. [CrossRef]
4. Abu-Rayash, A.; Dincer, I. Analysis of mobility trends during the COVID-19 coronavirus pandemic: Exploring the impacts on global aviation and travel in selected cities. *Energy Res. Soc. Sci.* **2020**, *68*, 101693. [CrossRef] [PubMed]
5. Zhu, H.; Wei, L.; Niu, P. The novel coronavirus outbreak in Wuhan, China. *Glob. Health Res. Policy* **2020**, *5*, 6. [CrossRef] [PubMed]
6. Fotiadis, A.; Polyzos, S.; Huan, T.C. The good, the bad and the ugly on COVID-19 tourism recovery. *Ann. Tour. Res.* **2021**, *87*, 103–117. [CrossRef] [PubMed]
7. Abbas, J.; Mubeen, R.; Lorember, P.; Rasa, S.; Mamirkulova, G. Exploring the impact of COVID-19 on tourism: Transformational potential and implications for a sustainable recovery of the travel and leisure industry. *Curr. Res. Behav. Sci.* **2021**, *2*, 100033. [CrossRef]
8. Rosin, U.; Gombault, A. Venice in Crisis: The Brutal Marker of Covid-19. *Int. J. Arts Manag.* **2021**, *23*, 75–88.
9. Turoń, K.; Kubik, A. Business innovations in the new mobility market during the COVID-19 with the possibility of open business model innovation. *J. Open Innov. Technol. Mark. Complex.* **2021**, *7*, 195. [CrossRef]
10. Back, A.; Marjavaara, R. Mapping an invisible population: The uneven geography of second-home tourism. *Tour. Geogr.* **2017**, *19*, 595–611. [CrossRef]
11. De Cantis, S.; Ferrante, M. Tourism Statistics for Destination Management: The Trips/Arrivals Model. In *Statistical Methods and Applications from a Historical Perspective*; Studies in Theoretical and Applied, Statistics; Crescenzi, F., Mignani, S., Eds.; Springer: Cham, Switzerland, 2014. [CrossRef]
12. Rátz, T.; Kundi, V.; Michalkó, G. The hidden dimensions of cultural consumption within the framework of tourism mobility. In *Touring Consumption*; Sonnenburg, S., Wee, D., Eds.; Springer: Wiesbaden, Germany, 2015; pp. 285–302.

13. EU. Regulation (EU) No 692/2011 of the European Parliament and of the Council of 6 July 2011 Concerning European Statistics on Tourism and Repealing Council Directive 95/57/EC. 2011. Available online: https://eur-lex.europa.eu/legal-content/EN/TXT/HTML/?uri=CELEX:32011R0692&from=EN (accessed on 25 May 2022).
14. Nyaupane, G.; Paris, C.M.; Li, X. Introduction: Special Issue on Domestic Tourism in Asia. *Tour. Rev. Int.* **2020**, *24*, 1–4. [CrossRef]
15. McKercher, B.; Chan, A.; Lam, C. The Impact of Distance on International Tourist Movements. *J. Travel Res.* **2008**, *47*, 208–224. [CrossRef]
16. Dodds, R.; Butler, R. The phenomena of overtourism: A review. *Int. J. Tour. Cities* **2019**, *5*, 90. [CrossRef]
17. Sardak, S.; Dzhyndzhoian, V.; Samoilenko, A. Global innovations in tourism. *Innov. Mark.* **2016**, *12*, 45–50. [CrossRef]
18. Williams, A.; Baláž, V. Low-Cost Carriers, Economies of Flows and Regional Externalities. *Reg. Stud.* **2009**, *43*, 677–691. [CrossRef]
19. Goecke, R. The Evolution of Online Booking Systems. In *Handbook of e-Tourism*; Xiang, Z., Fuchs, M., Gretzel, U., Höpken, W., Eds.; Springer: Cham, Switzerland, 2020. [CrossRef]
20. Hays, S.; Page, S.; Buhalis, D. Social media as a destination marketing tool: Its use by national tourism organisations. *Curr. Issues Tour.* **2013**, *16*, 211–239. [CrossRef]
21. Altinay, L.; Taheri, B. Emerging themes and theories in the sharing economy: A critical note for hospitality and tourism. *Int. J. Contemp. Hosp. Manag.* **2019**, *31*, 180–193. [CrossRef]
22. Sharpley, R.; Stone, P. *Tourist Experience: Contemporary Perspectives*; Routledge: London, UK, 2010.
23. Haddad, R. Exploring Service Quality of Low Cost Airlines. *Serv. Mark. Q.* **2019**, *40*, 301–315. [CrossRef]
24. Jeng, C. The Role of Trust in Explaining Tourists' Behavioural Intention to Use E-booking Services in Taiwan. *J. China Tour. Res.* **2019**, *15*, 478–489. [CrossRef]
25. Volo, S.; Irimiás, A. Instagram: Visual methods in tourism research. *Ann. Tour. Res.* **2021**, *91*, 103098. [CrossRef]
26. Paulauskaite, D.; Powell, R.; Coca-Stefaniak, A.; Morrison, A. Living like a local: Authentic tourism experiences and the sharing economy. *Int. J. Tour. Res.* **2017**, *19*, 619–628. [CrossRef]
27. Capocchi, A.; Vallone, C.; Pierotti, M.; Amaduzzi, A. Overtourism: A Literature Review to Assess Implications and Future Perspectives. *Sustainability* **2019**, *11*, 3303. [CrossRef]
28. Gelbman, A.; Timothy, D.J. From hostile boundaries to tourist attractions. *Curr. Issues Tour.* **2010**, *13*, 239–259. [CrossRef]
29. Karl, M.; Chien, M.; Ong, F. Impulse buying behaviour in tourism: A new perspective. *Ann. Tour. Res.* **2021**, *90*, 103136. [CrossRef]
30. Minoia, P. Venice Reshaped? Tourist Gentrification and Sense of Place. In *Tourism in the City*; Bellini, N., Pasquinelli, C., Eds.; Springer: Cham, Switzerland, 2017. [CrossRef]
31. Matsushita, K. Workation and the Doubling of Time and Place. In *The Second Offline. Advances in Information and Communication Research*; Tomita, H., Ed.; Springer: Singapore, 2021. [CrossRef]
32. Bigano, A.; Hamilton, M.; Lau, M.; Tol, S.; Zhou, Y. A global database of domestic and international tourist numbers at national and subnational level. *Int. J. Tour. Res.* **2007**, *9*, 147–174. [CrossRef]
33. Frenţ, C. Measuring tourism at the border: A critical analysis of the Icelandic context. *Scand. J. Hosp. Tour.* **2016**, *16*, 87–97. [CrossRef]
34. Truong, N.; Shimizu, T.; Kurihara, T.; Choi, S. Accommodation statistics: The current issues and an innovation. *Curr. Issues Tour.* **2021**, *22*, 1731–1747. [CrossRef]
35. Škafar, A. Overview of tourism statistics at the Statistical office of the Republic Slovenia. *Tour. Hosp. Manag.* **1998**, *4*, 135–144. [CrossRef]
36. Michalkó, G.; Rátz, T.; Hinek, M.; Tömöri, M. Shopping tourism in Hungary during the period of the economic crisis. *Tour. Econ.* **2014**, *20*, 1319–1336. [CrossRef]
37. Bertocchi, D.; Camatti, N.; Giove, S.; Borg, J. Venice and overtourism: Simulating sustainable development scenarios through a tourism carrying capacity model. *Sustainability* **2020**, *12*, 512. [CrossRef]
38. Vystoupil, J.; Šauer, M.; Bobková, M. Spa, spa tourism and wellness tourism in the Czech Republic. *Czech J. Tour.* **2017**, *6*, 5–26. [CrossRef]
39. De Cantis, S.; Parroco, M.; Ferrante, M.; Vaccina, F. Unobserved Tourism. *Ann. Tour. Res.* **2015**, *50*, 1–18. [CrossRef]
40. Pavlić, I. Cruise tourism demand forecasting-the case of Dubrovnik. *Tour. Hosp. Manag.* **2013**, *19*, 125–142. [CrossRef]
41. Murillo, J.; Vayà, E.; Romaní, J.; Suriñach, J. How important to a city are tourists and day-trippers? The economic impact of tourism on the city of Barcelona. *Tour. Econ.* **2013**, *19*, 897–917. [CrossRef]
42. Parroco, M.; Vaccina, F.; De Cantis, S.; Ferrante, M. Multi-destination trips and tourism statistics: Empirical evidences in Sicily. *Economics* **2012**, *6*, 1–27. [CrossRef]
43. Hall, C.M.; Müller, K. *The Routledge Handbook of Second Home Tourism and Mobilities*; Routledge: London, UK, 2018.
44. Backer, E.; King, B. *VFR Travel Research: International Perspectives*; Channel View Publications: Clevedon, UK, 2015.
45. Backer, E. VFR travel: An examination of the expenditures of VFR travellers and their hosts. *Curr. Issues Tour.* **2007**, *10*, 366–377. [CrossRef]
46. Molz, J. Social networking technologies and the moral economy of alternative tourism: The case of couchsurfing.org. *Ann. Tour. Res.* **2013**, *43*, 210–230. [CrossRef]
47. Rădan-Gorska, M. Destinations without regulations: Informal practices in Romanian rural tourism. *J. Comp. Res. Anthropol. Sociol.* **2013**, *4*, 195–225.

48. Rantala, O.; Varley, P. Wild camping and the weight of tourism. *Tour. Stud.* **2019**, *19*, 295–312. [CrossRef]
49. Wahnschafft, R. Formal and informal tourism sectors: A case study in Pattaya, Thailand. *Ann. Tour. Res.* **1982**, *9*, 429–451. [CrossRef]
50. Papathanassis, A. The growth and development of the cruise sector: A perspective article. *Tour. Rev.* **2020**, *75*, 130–135. [CrossRef]
51. Crouch, I. Demand elasticities for short-haul versus long-haul tourism. *J. Travel Res.* **1994**, *33*, 2–7. [CrossRef]
52. Bao, Y.; Mckercher, B. The Effect of Distance on Tourism in Hong Kong: A Comparison of Short Haul and Long Haul Visitors. *Asia Pac. J. Tour. Res.* **2008**, *13*, 101–111. [CrossRef]
53. McKercher, B. The impact of distance on tourism: A tourism geography law. *Tour. Geogr.* **2018**, *20*, 905–909. [CrossRef]
54. Timothy, D.J. *Shopping Tourism, Retailing and Leisure*; Channel View Publications: Clevedon, UK, 2005.
55. Konu, H.; Smith, M. Cross-border health tourism collaborations. In *The Routledge Handbook of Health Tourism*; Puczkó, L., Smith, M., Eds.; Routledge: London, UK, 2016. [CrossRef]
56. Bustamante, A.; Ojeda, G.; Castañeda, X. Willingness to pay for cross-border health insurance between the United States and Mexico. *Health Aff.* **2008**, *27*, 169–178. [CrossRef]
57. Michalkó, G.; Rátz, T.; Hinek, M. Spatial differences in Hungarian medical tourism supply based on service providers' online presence. *Hung. Geogr. Bull.* **2012**, *61*, 31–47.
58. Butler, R.; Mao, B. Conceptual and theoretical implications of tourism between partitioned states. *Asia Pac. J. Tour. Res.* **1996**, *1*, 25–34. [CrossRef]
59. Moyano, A.; Rivas, A.; Coronado, J. Business and tourism high-speed rail same-day trips: Factors influencing the efficiency of high-speed rail links for Spanish cities. *Eur. Plan. Stud.* **2019**, *27*, 533–554. [CrossRef]
60. Božić, S.; Jovanović, T.; Tomić, N.; Vasiljević, D. An analytical scale for domestic tourism motivation and constraints at multi-attraction destinations: The case study of Serbia's Lower and Middle Danube region. *Tour. Manag. Perspect.* **2017**, *23*, 97–111. [CrossRef]
61. Alzboun, N. Domestic tourism in Jordan: Patterns, challenges and opportunities. *J. Environ. Manag. Tour.* **2019**, *10*, 281–291. [CrossRef]
62. Shiomi, E. Do faster trains challenge air carriers? *Jpn. Railw. Transp. Rev.* **1999**, *19*, 4–7.
63. Timothy, D.J. Domestic tourism: Challenging the notion through a geopolitical lens. *Tour. Rev. Int.* **2020**, *24*, 67–73. [CrossRef]
64. Timothy, D.J. *Tourism in European Microstates and Dependencies: Geopolitics, Scale and Resource Limitations*; CABI: London, UK, 2021.
65. Hall, C.M. Reconsidering the geography of tourism and contemporary mobility. *Geogr. Res.* **2005**, *43*, 125–139. [CrossRef]
66. Peeters, P.; Landré, M. The emerging global tourism geography—An environmental sustainability perspective. *Sustainability* **2012**, *4*, 42–71. [CrossRef]
67. Che, D. Tourism geography and its central role in a globalized world. *Tour. Geogr.* **2018**, *20*, 164–165. [CrossRef]
68. Leiper, N. The framework of tourism: Towards a definition of tourism, tourist, and the tourist industry. *Ann. Tour. Res.* **1979**, *6*, 390–407. [CrossRef]
69. Yu, X.; Kim, N.; Chen, C.; Schwartz, Z. Are you a tourist? Tourism definition from the tourist perspective. *Tour. Anal.* **2012**, *17*, 445–457. [CrossRef]
70. Buhalis, D. The tourism phenomenon: The new tourist and consumer. In *Tourism in the Age of Globalisation*; Wahab, S., Cooper, C., Eds.; Routledge: Clevedon, UK, 2005; pp. 83–110.
71. Richards, G.; Russo, A. Synthesis and Conclusions: Towards a new geography of tourism. In *Reinventing the Local in Tourism: Producing, Consuming and Negotiating Place*; Richards, G., Russo, A., Eds.; Channel View Publications: Bristol, UK, 2016; pp. 251–266.
72. Poon, A. The 'new tourism'revolution. *Tour. Manag.* **1994**, *15*, 91–92. [CrossRef]
73. Dwyer, L.; Gill, A.; Seetaram, N. *Handbook of Research Methods in Tourism: Quantitative and Qualitative Approaches*; Edward Elgar Publishing: Cheltenham, UK, 2012.
74. Leiper, N. Main destination ratios: Analyses of tourist flows. *Ann. Tour. Res.* **1989**, *16*, 530–541. [CrossRef]
75. Jansen-Verbeke, M.; Spee, R. A regional analysis of tourist flows within Europe. *Tour. Manag.* **1995**, *16*, 73–80. [CrossRef]
76. Tömöri, M. The Role of the "DebOra" Cross-Border Eurometropolis in the Hungarian-Romanian CBC Relations A Case Study of Shopping Tourism in Debrecen and Oradea. *Eurolimes* **2011**, *11*, 170–179.
77. Podhorodecka, K. Impact of tourist and one-day visitor arrivals on economic growth. Case study of the Cayman Islands. *Misc. Geogr.* **2014**, *18*, 16–25. [CrossRef]
78. Son, A. The measurement of tourist destination image: Applying a sketch map technique. *Int. J. Tour. Res.* **2005**, *7*, 279–294. [CrossRef]
79. Bandyopadhyay, R. A photo ethnography of tourism as neo-colonialism. *Ann. Tour. Res.* **2011**, *38*, 714–718. [CrossRef]
80. Barros, C.; Moya-Gómez, B.; Gutiérrez, J. Using geotagged photographs and GPS tracks from social networks to analyse visitor behaviour in national parks. *Curr. Issues Tour.* **2020**, *23*, 1291–1310. [CrossRef]
81. Kádár, B. Measuring tourist activities in cities using geotagged photography. *Tour. Geogr.* **2014**, *16*, 88–104. [CrossRef]
82. Kádár, B.; Gede, M. Tourism flows in large-scale destination systems. *Ann. Tour. Res.* **2021**, *87*, 103–113. [CrossRef]
83. Jones, C.; Munday, M. Tourism satellite accounts and impact assessments: Some considerations. *Tour. Anal.* **2008**, *13*, 53–69. [CrossRef]

84. Raun, J.; Ahas, R.; Tiru, M. Measuring tourism destinations using mobile tracking data. *Tour. Manag.* **2016**, *57*, 202–212. [CrossRef]
85. Saluveer, E.; Raun, J.; Tiru, M.; Altin, L.; Kroon, J.; Snitsarenko, T.; Aasa, A.; Silm, S. Methodological framework for producing national tourism statistics from mobile positioning data. *Ann. Tour. Res.* **2020**, *81*, 102895. [CrossRef]
86. Zuelow, E. *A History of Modern Tourism*; Macmillan International Higher Education: New York, NY, USA, 2015.

sustainability

MDPI

Article

The Measurable Predominance of Weekend Trips in Established Tourism Regions—The Case of Visitors from Budapest at Waterside Destinations

Bálint Kádár [1,*] and Mátyás Gede [2]

1 Department of Urban Planning and Design, Budapest University of Technology and Economics, 1111 Budapest, Hungary
2 Institute of Cartography and Geoinformatics, ELTE Eötvös Loránd University, 1117 Budapest, Hungary; saman@map.elte.hu
* Correspondence: kadarb@urb.bme.hu; Tel.: +36-204-102-858

Abstract: Short trips to weekend destinations are less researched than more conventional forms of tourism involving longer trips and overnight stays, because quantitative data are hard to procure on the behavior of such weekend tourists. As a result, the effects of these day trips on secondary destinations cannot be measured, yet weekend tourism does contribute to the economic sustainability of many tourism regions. In this study, we analyzed geotagged photography uploaded to Flickr.com in the Budapest metropolitan area, the Danube Bend north of the Hungarian capital, and the Northern Balaton Region. Analysis of the spatio-temporal activity of photographers revealed Flickr users who live in the analyzed regions or in foreign countries, identifying the locals, weekend visitors from Budapest, those from other Hungarian regions, and foreign tourists. The predominance of visitors from Budapest was measured in both of the water-side destinations, and the spatial patterns of such visitors were more dispersed than the more concentrated spatial patterns of foreign tourists. These results show how day-trippers spread out the economic effects of tourism to much wider geographic areas than conventional tourists. Therefore, more focus should be directed toward these previously invisible forms of visits among the scientific community, policy makers, and the tourism industry.

Keywords: visitor behavior; tracking tourists; Flickr; day trips; weekend destination

Citation: Kádár, B.; Gede, M. The Measurable Predominance of Weekend Trips in Established Tourism Regions—The Case of Visitors from Budapest at Waterside Destinations. *Sustainability* **2022**, *14*, 3293. https://doi.org/10.3390/su14063293

Academic Editors: Anna Rita Irimiás, Dallen J. Timothy and Gábor Michalkó

Received: 9 February 2022
Accepted: 9 March 2022
Published: 11 March 2022

1. Introduction

Since the Grand Tours of the 18th century, discourse on tourism has focused mainly on faraway trips and long vacations, where the differences between ordinary and outstanding experiences were evident [1]. This focus was needed in scientific research to formulate clear theories, but at the same time, it was also essential for the tourism industry in order to develop its offers and infrastructures. However, parallel with the development of conventional tourism, society also developed less visible forms of leisure travel: excursions and short weekend trips to nearby natural areas, waterfronts, or cultural destinations. Most of these trips do not use any form of paid accommodation. Some visitors make only day trips, others have their own holiday houses or family properties to use for a weekend. Still, these nonconventional tourists do participate in the visitor economy, eating out, visiting cultural destinations and national parks, and using local recreational services and transport. Their contribution is especially important in places out of the reach of mass tourism, where restaurants and services make a visible turnover without being accessible for international tourists [2,3]. Many businesses offering tourism-related services or any services building on the experience economy outside the large urban and tourism centers are fundamentally sustained by such short-term visitors, not having a large local consumer base. These businesses must build their services on local authenticity [4] and offer experiences competitive with those in mainstream destinations using innovative methods [5,6]. Successful strategies are mostly based on the transformation of local values into experiences, including local food and specific rural landscapes [7,8]. However, the sustainability of this segment of the experience economy in rural tourism regions relies on consumers whose activity cannot be measured with traditional methods in tourism statistics.

New methods must be developed in order to quantify the contribution of day-trippers and weekend visitors to rural tourism regions, as these groups do not use transport types (air travel and organized tours) and accommodation from where the main statistical data for tourism come from [9]. In the past few decades, tourism researchers and the operators and developers of the tourism industry started to realize how nonconventional tourism with short visits is an important factor, and therefore, destination marketing is focusing more and more on such visitors nowadays. New trends in tourism—especially the changes imposed by the COVID-19 crisis [10–12]—direct much more attention to local travel and short weekend visits, and new data sources are needed in order to establish solid research methods in this field [13–15].

The biggest social changes in the past decade have been linked to the spread of social media. Tourism systems are also deeply affected since the main platforms of destination branding became those where users share their experiences and influence other users in their travel choices. Millions of selfies, perpetually re-composed landscapes and cityscapes, food photos, and detail photos of attractive subjects have flooded the popular photo sharing services. Using photo sharing services has become an integral part of tourism, having a major impact on shaping the destination systems. While in the twentieth century, "Kodak Moments" were the highlights of once-in-a-lifetime vacation trips [16], today, an "Instagram moment" can be found anywhere [17], so proximity tourists do not even have to travel far to post about any leisure activity that can be popular on social media. The motivation for sharing anything on social media is very much the same as the motivation for travel photography since the beginning: tourism consumption (as social media presence) can only be fulfilled if evidence is produced of the experience and shared with others [1]. Since Kodak introduced its interchangeable film camera in 1888, photography has become the main accompaniment to travel [18]—a real holiday produced pictural memories in family albums [19,20]. The development of technology has expanded the activities of photography to even more areas. Digital photography, GPS positioning, internet access, and through it, social content generated by individuals (and companies) have become the mediums that have completely transformed the consumption and processing of tourist experiences in a few years. In addition, unlike family photo albums, these mediums can be accessed and researched from anywhere; there is currently no other accessible method for mapping the statistically "invisible" tourism of weekend trips.

Urry formulated the role of photography in tourism [1], also building on the works of Hall [21] and MacCannell [22]. The circle of representation for tourist destination images described by Jenkins [23] is more relevant than ever: according to this model, a desired image of a destination is formed by the tourist before the visit from pictures conveyed to him/her by marketing or other means, and when finding this image, the tourist also takes a photo of it during the trip and shares it on the (social) media used, thus strengthening the already-formed image of the destination for others. However, through the digital and social media revolution, imaging and sharing have become so widespread that there is a competition to publish new, unique, and attractive images, which have brought previously undiscovered sites into the public consciousness, but because of the functioning of the above circle, they have also become known and desired destinations [17,24]. Today, significant research focuses on how social media affects the perception and development of tourist destinations [25–28]. Such research confirms that UGC (user-generated content) on social media fundamentally influences the perception of a destination, and has a great marketing value especially for smaller, specialized businesses, mainly serving short visits [29]. Thus, these media and the short messages and visual content circulating on them promote the fragmentation of large tourism infrastructures and the viability of smaller-scale services at the local level.

Some social media platforms partially manage their service on the geographical position of photography uploaded. The "geotags" of images uploaded to sites such as Flickr.com (https://flickr.com/, accessed on 8 February 2022) allow for a new kind of research on the spatial behavior of tourists [30–32]. On Flickr, there are more than 1,000,000 photographs uploaded with a geotag only from the area of Hungary. This quantity of specific spatio-temporal data enabled the research of unconventional tourism in the two most popular rural tourism regions in Hungary, where the majority of visits comprise weekend and one-day trips, and also the activities of holiday home owners.

In this study, the spatial patterns of different tourist types are mapped for the rural tourism regions of the Hungarian Danube Bend and the Northern Balaton Region, revealing the activities of the previously invisible day-trippers and holiday home owners. Special attention was paid to the residents of the Budapest Metropolitan Area, as according to our hypothesis, most of the visitors to the two analyzed regions come from Budapest. Budapest is the main tourist hub of Hungary for international tourism; most international visitors to the two analyzed areas arrive to Budapest first, and they possibly spend time in the capital, too. Budapest had 4,578,000 tourist arrivals in 2019 [33], the last full year prior to the COVID-19 crisis, and 86.1% of these were international arrivals, with only 13.8% being Hungarians. Such detailed statistics do not exist for the two analyzed regions, because international tourists usually stay in hotels in the capital city of Budapest, taking day trips to the Danube Bend, or go for a lakeside holiday to the shores of Lake Balaton. Residents of Budapest also take day trips to the Danube Bend, and most Hungarians have had vacations on the shores of Balaton Lake, from where day trips to the villages and destinations in the Balaton Uplands National Park are popular. The proportion of weekend houses and holiday homes in these areas is extremely high; therefore, many Hungarians make short visits from their own or rented vacation house.

In the following section, we present a methodology to collect data on the tourism activities of day-trippers and weekend visitors from Flickr.com, comparing the different data analysis methods to separate the space usage patterns of different user groups. In the results section, we present previously unmeasurable findings on the origin and space usage of visitors to the two regions. We demonstrate the predominance of visitors from the capital city, the role of weekend visits, and the seasonality of tourism in such waterside rural destinations, summarizing the contribution of such findings to the field in the discussion and conclusion section.

2. Materials and Methods

The patterns of the tourist space use of the Danube Bend and the Balaton Uplands in the Northern Balaton Region have been mapped using the database of Flickr, which is still operating with an open API. Among the available social media databases, the one of Flickr.com is most commonly used by researchers [34–36]. Flickr research is beneficial not only because of its quantification, but this social media has users from the most diverse age and social groups, and this portal is most aimed at sharing images taken during tourist visits [37,38]. While only a quarter of Flickr images are geotagged with precise geographical coordinates, it was still possible to analyze more than half a million photos, using almost 100,000 for the study of the two regions. In fact, all photos with geotags in the Budapest agglomeration (including the Danube Bend) and around Lake Balaton were downloaded. This large dataset can combine the quantitative advantages of the huge number of social media users [37] with the geographical accuracy of GNSS technology [39,40]. Based on the work of Kádár and Gede [30,41], the method used analyzes spatial and temporal data obtained from the API of Flickr.com, grouping users who create the images into various categories based on spatio-temporal behavior patterns and the user data provided. The subject, quantity, and frequency of geotagged photos uploaded by users on Flickr reveals whether the person is a tourist or a local resident in the given area, retrieving the frequency of visits to determine whether someone lives locally or has only arrived for a short time. Based on the categorization of users, it is also possible to map which ones are the main tourist attractions in a given city and which places are mainly visited by local residents [30,42].

First, users who could be residents of the Budapest area were identified. For each user, the time tags for their geotagged images were examined in the Budapest agglomeration area. The Budapest agglomeration data (together with the data of the Danube Bend) were downloaded on 31 March 2020, resulting in 682,312 geotagged photos of 19,284 Flickr users. The photos of each user were grouped into time intervals in a way that the longest time gaps between photos in the same interval are less than 60 days. Users were called 'local' if they have an interval at least 30 days long or have at least 4 intervals. Locals in this case were residents from the Budapest Metropolitan region, and such users kept this attribute also in the analysis of the Northern Balaton Region (where another group of 'locals' was identified). It is possible to accurately identify the visiting patterns of the above user groups at various points of a given tourist region even with an accuracy of up to 10 m, due to the accuracy of the geotags provided by GNSS. After analyzing the Budapest agglomeration database, the samples of the Danube Bend were also analyzed based on data from 37,720 photos. The 'local' group was kept intact for this area, as determining who is 'local' from the Danube Bend area and who is from Budapest proved to be inaccurate as most of the residents of the towns of the Danube Bend commute for work to Budapest, and therefore, the photographing patterns of Budapest visitors and locals visiting Budapest showed very similar patterns. Keeping 'locals' as every user from the larger Budapest area, 'tourists' were further broken down into sub-groups. If all their photos were in one interval, the tourist was on his/her first visit to Budapest and the Danube Bend. If this interval was less than 5 days, then the user was placed in the group of tourists who made a short visit, and if it was longer, in the group of long-visit tourists. Some tourists' photos fall into two or three intervals; they were placed in the group of returning visitors. The 7 most popular clusters, which are the most touristy communities in the Danube Bend area, were highlighted, quantifying visitor numbers in each of the following sub-groups:

- Users living in the Budapest area.
- First-time tourists on a short trip.
- First-time tourists on a longer trip.
- Returning visitors to the area.

Further analysis of Flickr data could identify additional categories of visitors. In addition to visit space–time patterns, user data and profile analysis gave more information about the origin of visitors, but due to the manual inspection of all user profiles, this was a much slower process. In this study, we developed a methodology to check each user profile for affiliation information on a smaller sample in the Danube Bend; therefore, we selected the data from Szentendre and the Skanzen (open air ethnographic museum) nearby. Flickr has a tag for the nationality and the hometown of the user, but not every user fills out this information. Therefore, a manual check of users had to be performed, discarding those from the dataset whose affiliation to a country could not be verified. Out of the 809 users found in this area, 396 users had user profiles with enough data about their town of origin. A total of 8343 photos were analyzed after discarding those where the origin of the user was not possible to determinate. Four groups could be separated for the users of such photos:

- Locals who live in Szentendre.
- Foreign tourists with a foreign profile.
- Domestic first-time visitors.
- Returning domestic visitors who returned to Szentendre in different time intervals.

For patterns of the Balaton Uplands, the two previous methodologies were combined. Flickr metadata of the entire Balaton region were downloaded on 31 March 2020, resulting in a total of 60,535 photos. After separating 'locals' who most probably live in the area from 'tourists' using the methodology described above, a relevant 'tourist' sub-group was instantly formed by comparing the dataset of the Budapest agglomeration area with the Balaton dataset. This resulted in the tag 'tourist from Budapest', but for the next sub-groups of other Hungarian visitors and foreign tourists, a second step of analysis was needed with the verification of the nationality. After taking out the images of users without a secure affiliation, as well as all users who uploaded fewer than 5 images in the Balaton region, 57,974 geo-positioned photos were examined from 1432 users. In this way, the following groups could be isolated in the Balaton region:

- Visitors from Budapest (and its agglomeration).
- International tourists.
- Hungarian tourists (outside the capital).
- Locals from the Balaton region.

The focus of the study was the Balaton Uplands area, so the geotags were separated into 20 clusters which are the main communities in this region, measuring the visitor numbers in these clusters.

In the Balaton region, we also identified Flickr users staying in non-paying accommodation, typically holiday homes. This was achieved by separately analyzing the activity patterns of users who uploaded photos taken in at least two different time periods within a 50-m radius in a recreational area around Balaton—these are built-up areas where there are no residential buildings, only holiday homes. It was therefore necessary to qualitatively analyze all repeating photographs in such recreational areas: the content of these images typically depict a home environment, i.e., family, interior, private gardens, and related subjects. Only if such attributes were true users were assigned the 'holiday home owner' tag, and their activities outside their accommodation could be separately analyzed in both the northern and southern shore of Lake Balaton for comparison.

3. Results

3.1. Weekend Tourism in the Danube Bend

Analyzing all photos from the Budapest Metropolitan Area and using this database to narrow down the geographical territory to the towns of the Danube Bend made it possible to determinate the role of this region inside the tourism system of Budapest and also in relation to the residents of the larger region itself. Only 13.6% of the 19,284 users uploading geotagged photography in the area of the Budapest Metropolitan Area were locals; the rest of the users were in one of the groups of tourists (Table 1). The largest group was the first-time short trip visitors, with 57.9% of all users being in this category, and most of these are international tourists, according to official statistics [33].

A total of 1648 users (8.5% of all in the Budapest Metropolitan Area) left digital traces in the Danube Bend region. The majority of these users are not international tourists, but visitors from the Budapest region itself (55.5%). The residents of the seven analyzed areas of the Danube Bend are also represented among the 914 users from the greater Budapest area, but it was not methodologically possible to separate them, as their movement patterns are also intensive in the capital city, where many of these residents work. However, the results of the next section in this paper further analyzing the Flickr users in Szentendre clearly show how locals are outnumbered by visitors from Budapest.

Table 1. Flickr users taking photos in the Budapest Metropolitan Area and in the communities of the Danube Bend sorted into different user groups of locals and tourists.

	Locals from the Budapest Metro. Area			Tourists												Sum	
				All			First Time on Short Visit			Returning More Times			First Time on Long Visit				
	photos	*users*	photos /user	photos	*users*	photos /user	photos	*users*	photos /user	photos	*users*	photos /user	photos	*users*	photos /user	photos	*users*
Budapest Metropolitan	326,356	2628	124	355,963	16,656	21	194,410	11,170	17	45,512	2871	16	116,041	2615	44	682,319	19,284
%in Budapest		13.6%			86.4%			57.9%			14.9%			13.6%			100.0%
Danube Bend count	12,150	914	13	5952	734	8	2467	422	6	975	180	5	2510	132	19	18,102	1648
%in Danube Bend	67.1%	55.5%		32.9%	44.5%		13.6%	25.6%		5.4%	10.9%		13.9%	8.0%		100.0%	100.0%
%in Budapest	3.7%	34.8%		1.7%	4.4%		1.3%	3.8%		2.1%	6.3%		2.2%	5.0%		2.7%	8.5%
Visegrád count	4572	262	17	1391	213	7	668	114	6	334	63	5	389	36	11	5963	475
%in Visegrád	76.7%	55.2%		23.3%	44.8%		11.2%	24.0%		5.6%	13.3%		6.5%	7.6%		100.0%	100.0%
%in Budapest		10.0%			1.3%			1.0%			2.2%			1.4%		0.9%	2.5%
Szentendre count	4048	341	12	3582	383	9	1413	234	6	443	76	6	1726	73	24	7630	724
%in Szentendre	53.1%	47.1%		46.9%	52.9%		18.5%	32.3%		5.8%	10.5%		22.6%	10.1%		100.0%	100.0%
%in Budapest		13.0%			2.3%			2.1%			2.6%			2.8%		1.1%	3.8%
Skanzen count	1319	55	24	466	30	16	111	16	7	31	7	4	324	7	46	1785	85
%in Skanzen	73.9%	64.7%		26.1%	35.3%		6.2%	18.8%		1.7%	8.2%		18.2%	8.2%		100.0%	100.0%
%in Budapest		2.1%			0.2%			0.1%			0.2%			0.3%		0.3%	0.4%
Nagymaros count	922	90	10	197	41	5	131	23	6	44	12	4	22	6	4	1119	131
%in Nagymaros	82.4%	68.7%		17.6%	31.3%		11.7%	17.6%		3.9%	9.2%		2.0%	4.6%		100.0%	100.0%
%in Budapest		3.4%			0.2%			0.2%			0.4%			0.2%		0.2%	0.7%
Dunabogdány-Kisoroszi	656	88	7	156	34	5	60	18	3	87	12	7	9	4	2	812	122
%in Dunabogdány-Kiso.	80.8%	72.1%		19.2%	27.9%		7.4%	14.8%		10.7%	9.8%		1.1%	3.3%		100.0%	100.0%
%in Budapest		3.3%			0.2%			0.2%			0.4%			0.2%		0.1%	0.6%
Leányfalu count	347	35	10	86	22	4	47	8	6	29	9	3	10	5	2	433	57
%in Leányfalu	80.1%	61.4%		19.9%	38.6%		10.9%	14.0%		6.7%	15.8%		2.3%	8.8%		100.0%	100.0%
%in Budapest		1.3%			0.1%			0.1%			0.3%			0.2%		0.1%	0.3%
Tahitótfalu count	286	43	7	74	11	7	37	9	4	7	1	7	30	1	30	360	54
%in Tahitótfalu	79.4%	79.6%		20.6%	20.4%		10.3%	16.7%		1.9%	1.9%		8.3%	1.9%		100.0%	100.0%
%in Budapest		1.6%			0.1%			0.1%			0.0%			0.0%		0.1%	0.3%

Seven separate areas grouped around the settlements of the Danube Bend were visualized on diagrams juxtaposed on the map of the exact location of photos uploaded by first-time visitors and by locals from Budapest Metropolitan Area (Figure 1). It is evident from the spatial patterns of these user groups that first-time visitors (mainly foreign tourists) visit only the most important destinations and never explore the region for minor attractions. Visitors from the Budapest areas (including locals) take a lot more photographs in this tourism region than any tourist groups (double, on average), except long-term tourists staying for more than 5 days. The most tourists outside Budapest visit Szentendre, where 52.9% of users are tourists, but only 2.3% of all tourists visiting Budapest area go to Szentendre. Accepting the correlation between the number of Flickr users in the 'tourist' category group and the actual annual arrivals of tourists, Szentendre should have to 105,000 visitors from outside Budapest in 2019. The second most visited town is Visegrád, and 44.8% of users are tourists from outside Budapest, accounting for 1.3% of all tourists to Budapest (approx. 58,500 arrivals). The smaller destinations along the Danube have many more local users from Budapest and the region; on average, only 30% are 'tourists' from elsewhere. Moreover, 79.6% of photographers in Tahitótfalu are either locals or from Budapest.

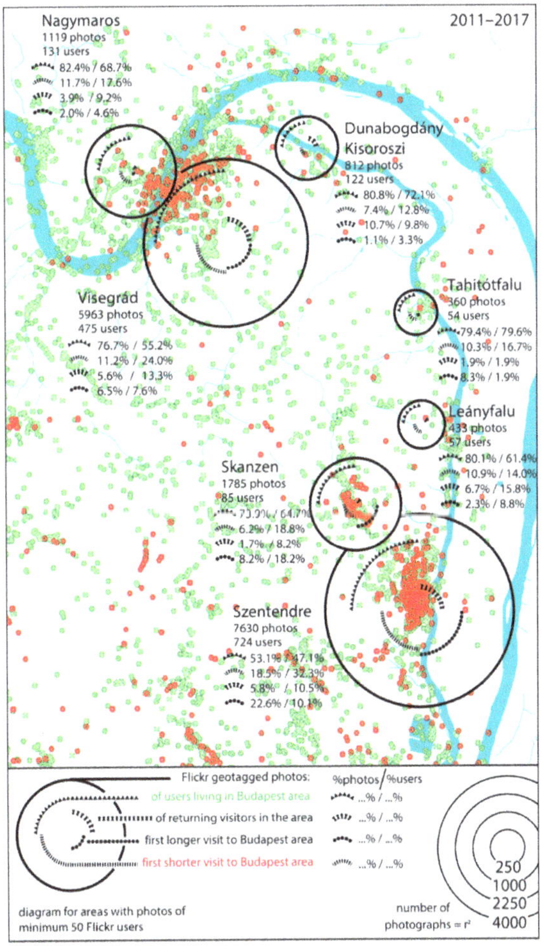

Figure 1. Visitors to the Danube Bend and Szentendre based on geotagged photos uploaded to Flickr (red dots represent photos from mostly foreign tourists who come to the Budapest region only once for a short time, green dots represent the photos from users living in Budapest or its agglomeration).

Some 34.8% of Flickr users living in the Budapest Metropolitan Area did visit the Danube Bend; 13% went to Szentendre and 10% to Visegrád. Those visiting the Budapest area for at least the second time went more frequently to the Danube Bend than those on a first visit to this metropolitan area. Approximately 3.8% of first-time visitors went to the Danube Bend area, 5.0% of longer-staying tourists and 6.3% of returning tourists.

It is possible to examine the temporal distribution of these groups of photographers in the Danube Bend area, both for the different months of the year (Figure 2) and the different days of the week (Figure 3). The figures display the percentage of total users of the groups taking photos in the given time periods. Although the general trend is that fewer users took photos during the winter months, locals and visitors from Budapest also visited the region in this season. For most of the groups, the months of May, August, and October were the most popular. The exception is the group of returning tourists, whose number constantly increases until September and then decreases until the February minimum. First-time visitors (also those on longer trips) follow similar patterns, and their sum of percentages through the 12 months is exactly 100%, showing how every first-time tourist visited the area just once. This is 108% for returning tourists, meaning that very few of these returning tourists also return to the Danube Bend. The sum is 239% for locals and visitors from Budapest, showing how users from this group took photos in two or three different seasons during their activity. This number is probably much higher for locals and stays around two for those from Budapest, but it is still safe to state that an average visitor from Budapest returned to the region at least twice during the past decade.

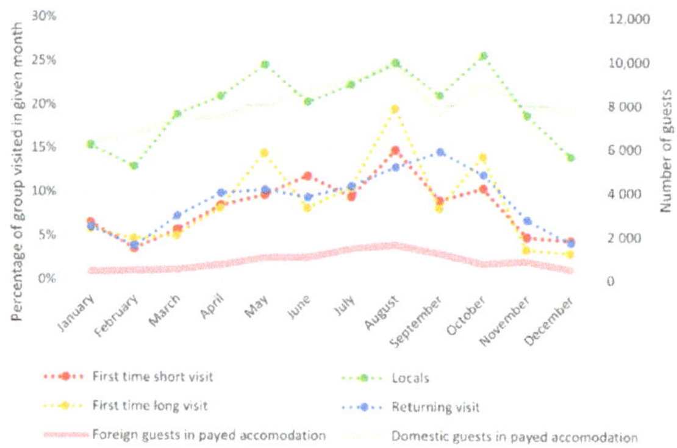

Figure 2. Temporal distribution of photographers in the Danube Bend region through the months of the year compared with official statistics of yearly average tourists in paid accommodation (Visegrád and Szentendre) between 2008 and 2019 [43].

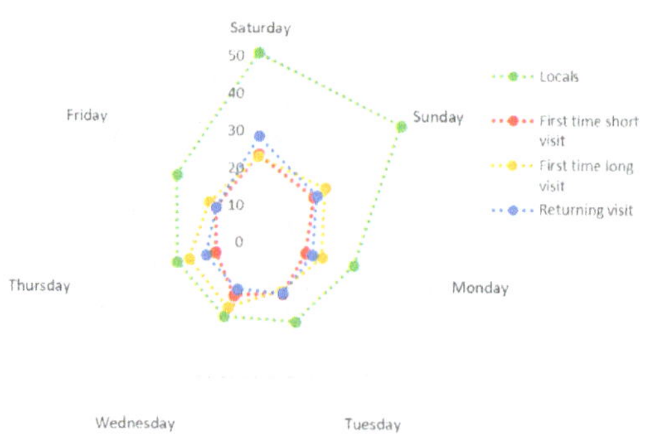

Figure 3. Temporal distribution of photographers in the Danube Bend region through the days of the week (% of users from the user groups who visited the area on the days of the week).

It is possible to compare the number of users with the official statistics of tourists in paid forms of accommodation in this region. The distribution along seasons correlates well in the case of domestic tourists. However, foreigners stay much less often in hotels here than domestic visitors, as most of these foreign visitors come to Budapest, stay there, and visit the Danube Bend only for a day trip.

Examining the number of users over the days of the week, the peaks are on Saturdays and Sundays. These peaks are especially outstanding in the case of visitors from Budapest, while tourists outside the capital area tend to visit in a more balanced distribution. Those visiting the capital city area on a longer trip were present in larger numbers also on Monday, Wednesday, and Thursday apart from the weekend. The average number of visits is much higher for locals and visitors from Budapest—a user from this group visited the Danube Bend for an average of 4.92 days, either living there or returning more times from Budapest during the past decade. Returning tourists spent 1.28 days, while first-time tourists on trips longer than 5 days spent an average of 1.41 days; therefore, more users from this group stayed overnight. In contrast, first-time visitors on short trips spent only 1.14 days here, meaning that very few of these visitors spent a night in the area. As stated before, most of the international tourists arriving to the region are in this group, taking only day trips to the region.

3.2. More Detailed Territorial Analysis around the Town of Szentendre

In this analysis, the verification of the behavior of foreign tourists and locals living in Szentendre was the focus, and the following users were identified:

- 10 locals who live in Szentendre;
- 175 international tourists with a foreign profile;
- 113 domestic first-time visitors; and
- 98 returning domestic visitors who returned to Szentendre in different time intervals.

In Szentendre, five spatial clusters were identified, all of which had more than 200 photos taken (Figure 4). The most visited cluster was the historical city and its riverbank, where 5290 photos were taken, of which 8.1% were taken by locals, the majority of the photos of this user group (54.7%). Most of the photos in the center were taken by foreigners (58.4%), and 83.3% of their photographing activities were concentrated here. Moreover, 53.1% of the photos by first-time Hungarian visitors and 38.9% of the photos by returning Hungarian visitors were taken in the center, making this area the tourism center of Szentendre. The second most visited cluster was the area of Szentendre Skanzen with 1182 photos, 13.6% of which were taken by foreigners, 28.3% by Hungarian tourists, and only 5.9% by returning visitors. Szentendre has a holiday area on the Danube, the Papp Island campsite, where 90.5% of the photos were taken by returning Hungarian visitors. Returning visitors were also the majority taking part in Pilis Park Forest excursions. In the Sztaravoda creek valley, the proportion of their photos was 91%, and in the Bükkös creek valley, it was 60.4%. The few local users were understandably significant in numbers (31.1%) at the mostly residential and holiday home areas around the center of Szentendre, where 39% of photos belonged to returning Hungarian visitors who were on holiday in Szentendre's holiday area or visiting the more out-of-town parts of the Danube (e.g., Postás beach). This detailed visitor analysis of Szentendre shows that foreign tourists mostly visited the historical city center only, with few of them reaching the Skanzen. Hungarian visitors had similar visit patterns, except for returning tourists, who visited Szentendre more regularly. They are hikers who followed the hiking trails of the Pilis Park Forest, holiday home owners who spent time in their weekend house in Szentendre, and returning visitors who were regularly on vacation on Papp Island or other sections of the Danube. All of the user groups spent at least a third of their time in the city center, so restaurants and services here were also used by weekend tourists not registered in the official statistics.

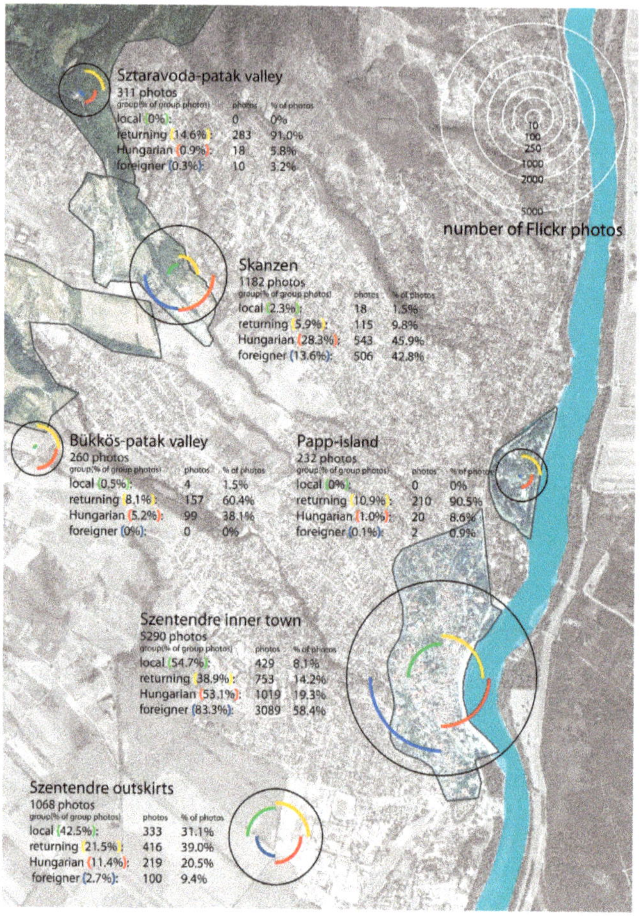

Figure 4. Number of photographs taken by different Flickr user groups in Szentendre, indicating the space usage of users living in Szentendre in green, regularly returning domestic tourists in yellow, first-time visiting domestic tourists in red, and international tourists in blue.

3.3. *Visitors of the Northern Balaton Region and Their Movements Based on Flickr Images*

The Balaton region is a much larger area than the Danube Bend, but due to the rigorous methodology discarding all users without a defined nationality, almost the same number of users was analyzed in the two regions. For the Balaton region, we found:

- 385 users (with 11,692 photos) clearly identified as foreigners who do not live in the country, as opposed to the 1046 Hungarian users;
- 106 users who could clearly be identified by their profile as inhabitants of the wider Balaton area, taking 5288 photos;
- 646 users active in this area classed as locals in the Budapest Metropolitan Area study, who took 32,297 photos; and
- 294 Hungarian users who were not from around Balaton or from Budapest, who took 8651 photos.

In addition, among the users from Budapest, 30 holiday home owners were identified who took more than five photos and stayed in non-commercial holiday homes around Lake Balaton, taking many more photos than other user groups, a total of 4055.

At first, it was surprising that only 7.4% of Flickr users who take photos around the Balaton Uplands are locals, while 45% are from Budapest. According to the data of the Hungarian Central Statistical Office [43], 262,293 people lived in the settlements of the Balaton recreation area in 2013. The Budapest Metropolitan Area had 2,541,835 inhabitants, while the rest of Hungary had 7,104,670 inhabitants. Therefore, correlated to the local inhabitants (106 users, 1/2474 inhabitants have Flickr activities = 100%), all users of Budapest have an activity level of 62.9% around Balaton (646), while other Hungarians (294) have an activity level of only 10.2%. In fact, if all domestic user groups had the same activity levels as locals, 1027 users from Budapest and 2871 Hungarian users outside Budapest should have been present in this area. The number of visitors from Budapest is very high, suggesting that more than half of all inhabitants from Budapest visited the Balaton region in a decade (given the timeframe of the Flickr dataset), while in the meantime, only 10% of other Hungarians did so. The data proved that Lake Balaton is primarily a recreation area of the capital city, even though it is located 100 to 200 km away from Budapest. International tourism was present in the region, but only 26.9% of users were foreigners. In addition, while the average foreign visitor took 30 photos, almost the same as the average Hungarian visitor (29 photos), guests from Budapest took almost 50 pictures, just as many per person as locals. This suggests that while a Hungarian tourist visiting Lake Balaton only visited once on average during the investigated period, visitors from Budapest spent much more time there, or reached Lake Balaton several times. The spatio-temporal patterns of visitors from Budapest are almost identical to those of locals in this region.

The temporal distribution of visitors in the Balaton region is observable in Figures 5 and 6. Instead of the absolute numbers, the percentage of the total number of users belonging to the different groups is shown because it reveals more clearly the different behavior of these groups. The highest numbers are during the two months of the beach season (July and August). The exception was the group of locals, whose numbers had two maximums in May and in October and were much more distributed around the year. The changes during the days of the week were similar to the ones in the Danube Bend region. Locals and foreigners were more balanced throughout the week: locals were in place all year around, while international tourists visited the lake once a year for a complete holiday week, as opposed to visitors from Budapest coming more times during the weekends. It is interesting again to compare the yearly distribution with official statistics. Flickr data of foreigners correlate well with international tourists staying in paid accommodation around Balaton Lake, and the same is true for domestic tourism. One interesting fact is that the statistical numbers correlate well in the summer season, while in all other months of the year, Flickr data show a great surplus of Hungarian visitors compared to domestic tourists in paid forms of accommodation. In fact, day trips and weekend visits to the area are frequent in autumn and spring, while in the summer, domestic tourists come for more time, using more hotels and campsites.

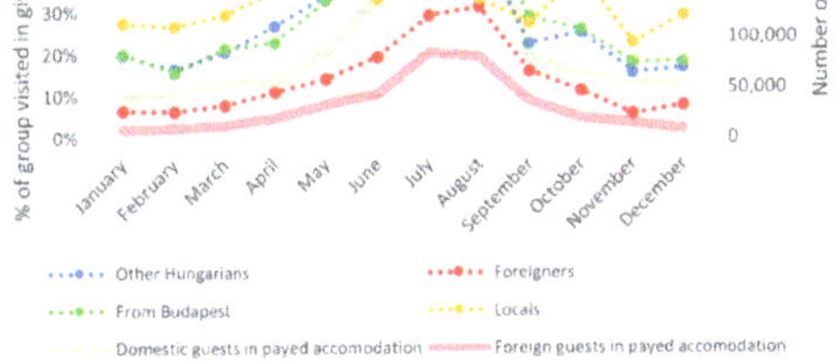

Figure 5. Monthly change in the number of Flickr users in the Balaton region compared with official statistics of yearly average tourists in paid accommodation between 2008 and 2019 [43].

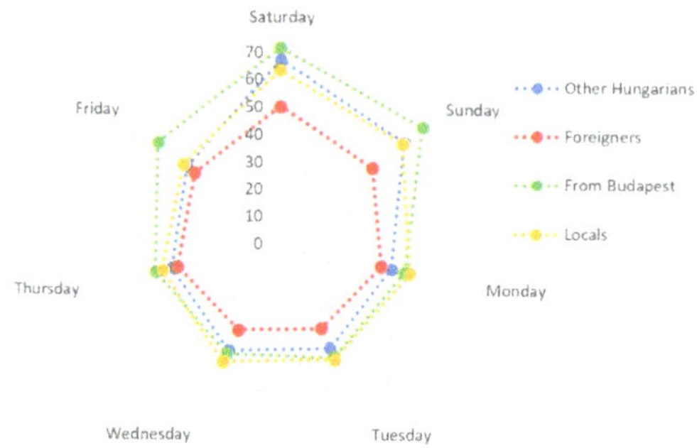

Figure 6. Change in the number of Flickr users through the days of the week in the Balaton region.

Part of the dataset in the Northern Balaton Region was aggregated into 20 microregions, as shown in Figure 7, highlighting the places photographed by Budapest users.

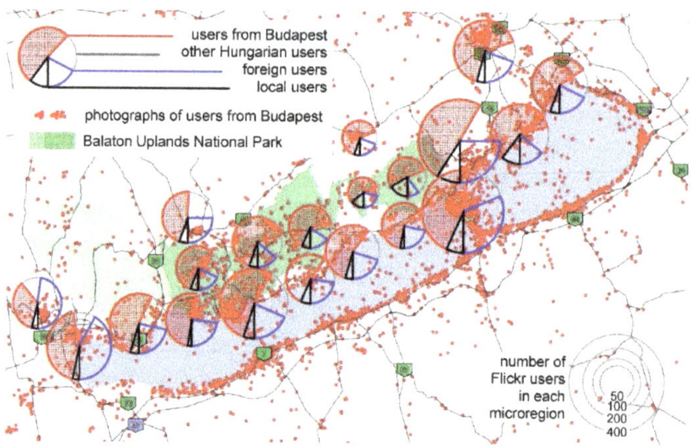

Figure 7. Map of the distribution of the photos by users from Budapest in the Northern Balaton Region, with diagrams showing the number and proportion of photographers in all the user groups identified.

Based on their Flickr photos, foreigners usually visit the settlements of the shores of Lake Balaton, where the beaches and ports can be found. In the Balaton Uplands, they visit the famous tourist destinations Veszprém (19.32% of all users), Balatonfüred (25.03%), Tihany (31.27%), Badacsony (18.78%), Szigliget (23.28%), and especially Keszthely (41.12%) and Hévíz (49.18%). Hévíz is the only settlement in the Balaton region where there are more international tourists than domestic tourists (46.7%). The place most neglected by foreign users is the most 'local' one: in the Pécsely Basin, 16.7% of Flickr users are locals, 51.7% are from Budapest, and only 11.67% are international tourists. Budapest users took photos in almost every settlement in the Balaton Uplands, with their share being between 50% and 60%, except where there are more foreigners, such as Keszthely (43.15%) and around Hévíz (39.34%), Tapolca (46.15%), Tihany (47.17%), and Balatonfüred (49.70%). Visitors from Budapest are highest in proportion in the Káli Basin: 66.27% of the users in Köveskál and Kővágóörs (international: 15.7%), and 65.09% in Szentbékkálla and Káptalantóti (international: 14.2%). The results show how the Northern Balaton region is prevalently the holiday area of the residents of the Budapest Metropolitan Area.

3.4. Tourist Consumption of Holiday Home Owners in Lake Balaton Based on Their Photos

The methodology identifying returning photographers in holiday home areas made it possible to analyze the tourist space usage of 20 residents of Budapest staying in private holiday homes on the northern shores of the lake, and 10 such users on the southern shores. The 4055 photos of these 30 users (135 on average) show that holiday home owners as a group use the region for tourism purposes most intensively (Figure 8). Every Flickr user staying in a holiday home in the Balaton area took photos in the shore strip of Lake Balaton: on beaches, piers, ferries, or promenades close to the water (53.93% of all photos). At the same time, 30.28% of the users also took photos in parts of the Balaton Uplands farther from the shore (at least 500 m from the water). Among the pictures of those staying in holiday homes on the north shore, 35% were taken in the Balaton Uplands, while half of these users did not take any photos on the south coast. In contrast, 90% of those staying on the south shore took photos in the north: 43.31% of their pictures were taken there (compared to 63.07% of those from the north) and 22% of the photos were taken in parts of the Balaton Uplands farther from the lake. Most of the holiday home owners travel all around the lake and the Balaton Uplands, visiting many secondary destinations where international tourists rarely arrive. Based on this dataset, it can be stated that a great number of holiday home owners around Lake Balaton travel around the region and the Uplands frequently instead of just staying in their holiday homes close to the beach. Even though this group is not included in the statistics, they have a much stronger presence and, presumably, consumption than many registered tourists.

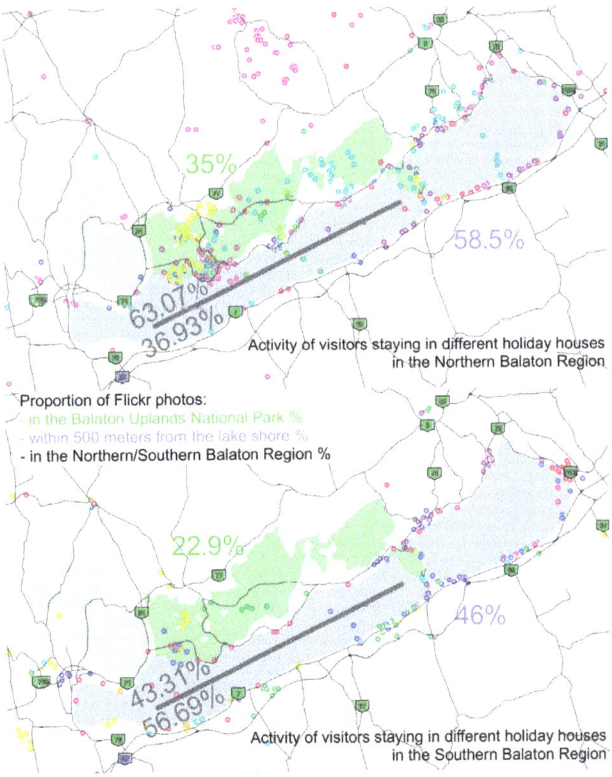

Figure 8. The movement of visitors (holidaymakers) staying in holiday homes around Lake Balaton, separately breaking down those who use holiday homes on the north shore and on the south coast.

4. Discussion and Conclusions

Geotagged photography retrieved from a social media site provided new data enabling us to have a more precise measurement of unconventional forms of tourism: day trips and weekend visits not using paid forms of accommodation. No other datasets equally available in multiple regions can describe the pattern of day trips and domestic tourism with such accuracy, but the selection of social media available for research is becoming more difficult [44]. Due to the protection of personal data, only data that are anonymized or that have been made freely available to users can be used for research, closing out large portions of social media usage. Moreover, even anonymized data are of value in the digital economy, so unlike the first services, having an open API (Application Programming Interface), today's social media data are kept in a closed system by the large service providers (Google, Facebook...) and cannot be researched [45]. For this reason, Twitter, which is not widespread in Hungary, and Flickr, which is used in Hungary, are used by most researchers for geo-positioned datasets [34,38,46]. Using Flickr has its limitations, not delivering statistically accurate data for all age and social groups. Kádár and Gede showed how Flickr is the most inclusive social media site for all age groups, meaning that older age groups are still under-represented but more active than on other photo sharing sites [30]. The higher photo sharing activity of tourists compared to locals in their domestic environments must also be taken into consideration. The experience of visiting a new environment is the main motivation for photo sharing [1], and therefore, visitors are always over-represented in such datasets. This study used Flickr data from before the outbreak of the COVID-19 pandemic, and the same datasets likely could not be used to measure the changing patterns in tourism consumption after the current crisis, as the declining userbase of this data source will provide less representative data.

However, we could deliver evidence on the predominance of domestic visitors in secondary tourism regions from before the COVID-19 crisis, when international tourism was still at its peak in Hungary and worldwide. The results partially explain why these rural destinations did not experience almost any decline in popularity during COVID-19 [47], unlike in Budapest, where most of the tourism facilities had serious difficulties [48,49].

The study bought evidence from tourism regions near watersides (Danube and Balaton), where beach tourism, nature tourism, and cultural tourism are all available, and some small- and medium-sized cities are the main tourist hubs. In the analyzed period, the main attractions and most branded destinations in these regions had a high proportion of international tourists (58.4% of all photography in the city center of Szentendre and 49.2% in Hévíz). However, at a regional level, only less than a quarter of all photography in both the Danube Bend and the Northern Balaton Region were taken by foreigners, and more than 80% only visited the primary attractions. The secondary attractions—sites farther away from important attractions, but with valuable landscapes, cultural attractions, and hospitality services—were mostly visited only by photographers from Hungary. Official statistics of paid accommodation also show the predominance of domestic tourists, and data from Flickr clearly showed correlation with such statistics (Figures 2 and 5). There is a great difference between the Danube Bend in the agglomeration area of Budapest and the Balaton region. Around the Balaton region, international tourists use paid forms of accommodation, and their numbers around the year correlate well with statistical data from these accommodation facilities. In fact, these statistics show that domestic tourists are underrepresented in the autumn, winter, and spring months, the times of year when many domestic tourists take day trips to the Balaton Uplands without using accommodation, or stay only for a night on the weekend. The Danube Bend shows reversed tendencies. Foreign tourists are almost completely missing from official statistics, as they book hotels in Budapest and only make day trips to this part of the Danube. Domestic tourists prefer not to go to Budapest, but stay in hotels in Visegrád or another town of the Danube Bend, but many visitors from Budapest also enjoy the thermal hotels of Visegrád. While in the Balaton region, the day trips and weekend visits invisible for statistics were taken mostly by residents of Budapest, in the Danube Bend, foreigners were also active day-trippers.

The most important finding from this dataset is that even if only one-fourth of the Hungarian population lives in the metropolitan area of the capital, well above 50% of all Flickr users active in both areas came from Budapest (in Nagymaros 68.7%, in the Kali Basin 65.5%). The visitors from Budapest—unlike international tourists—took photos of all primary and secondary destinations, and the least accessible ones in the two regions. The most active tourists in the Balaton region were the holiday home owners. The analysis of the spatial behavior of 30 such Flickr users reinforced previous findings about the role of this form of tourism [14]. Holiday home owners took 135 photos on average compared to the 50 of normal visitors from Budapest and 30 of other tourists; therefore, it can be assumed that this group extensively uses the paid services of the Balaton region such as restaurants, wine bars, and leisure services, and even more intensively in the Balaton Uplands, where even those based south of the lake prefer to take day trips. The exact proportion of holiday home owners compared to those staying in paying accommodation cannot be estimated from this analysis, but it can be stated with certainty that those staying in their own holiday homes contribute to the tourism economy much more than was measurable. The same is true for all day-trippers and weekend visitors—the present study reinforced similar studies with alternative methods on their tourism consumption [2,15].

The analysis of the photo sharing activities of Flickr users did not only reveal which domestic visitor groups visited the two waterside tourism regions the most, but also highlighted the role of social media and online image sharing in the function and sustainability of the economies and tourism systems in such destinations. Image taking and sharing became a main driving force in the experience economy, and local destinations can be competitive with global attractions if they offer authentic experiences users want to capture and share on social media. Studies of this phenomenon are important to understand the basis of economic sustainability of local tourism-related businesses outside of large destinations. The success of these local tourism regions contributes to the environmental sustainability of tourism as well; if local destinations can give memorable tourism experiences to domestic visitors, such visitors will have less motivation to seek out long-distance travel destinations, thereby reducing the environmental footprint of travel.

Author Contributions: Conceptualization, validation, and writing, B.K.; methodology, software, and data curation, M.G. All authors have read and agreed to the published version of the manuscript.

Funding: The dissemination of research outcomes was supported by the Hungarian Tourism Association Foundation.

Institutional Review Board Statement: Not applicable.

Conflicts of Interest: The authors declare no conflict of interest.

References

1. Urry, J. *The Tourist Gaze*, 2nd ed.; SAGE Publications Ltd.: London, UK, 2002.
2. Wynen, J. An Estimation of the Determinants of Same-Day Visit Expenditures in Belgium. *Tour. Econ.* **2013**, *19*, 161–172. [CrossRef]
3. Murillo, J.; Vayà, E.; Romaní, J.; Suriñach, J. How important to a city are tourists and day-trippers? The economic impact of tourism on the city of Barcelona. *Tour. Econ.* **2013**, *19*, 897–917. [CrossRef]
4. Pine, B.J.; Gilmore, J.H. *Authenticity: What Consumers Really Want*; Harvard Business Review Press: Brighton, MA, USA, 2007.
5. Pine, B.J.; Gilmore, J.H. A leader's guide to innovation in the experience economy. *Strategy Leadersh.* **2014**, *42*, 24–29. [CrossRef]
6. Tung, V.W.S.; Ritchie, J.R.B. Exploring the essence of memorable tourism experiences. *Ann. Tour. Res.* **2011**, *38*, 1367–1386. [CrossRef]
7. Basile, G.; Cavallo, A. Rural identity, authenticity, and sustainability in Italian inner areas. *Sustainability* **2020**, *12*, 1272. [CrossRef]
8. Sidali, K.L.; Kastenholz, E.; Bianchi, R. Food tourism, niche markets and products in rural tourism: Combining the intimacy model and the experience economy as a rural development strategy. *J. Sustain. Tour.* **2015**, *23*, 1179–1197. [CrossRef]
9. Nyns, S.; Schmitz, S. Using mobile data to evaluate unobserved tourist overnight stays. *Tour. Manag.* **2022**, *89*, 104453. [CrossRef]
10. Tomassini, L.; Cavagnaro, E. The novel spaces and power-geometries in tourism and hospitality after 2020 will belong to the 'local'. *Tour. Geogr.* **2020**, *22*, 713–719. [CrossRef]
11. Kumar, V.; Alshazly, H.; Idris, S.A.; Bourouis, S. Evaluating the Impact of COVID-19 on Society, Environment, Economy, and Education. *Sustainability* **2021**, *13*, 13642. [CrossRef]

12. Jeon, C.Y.; Yang, H.W. The structural changes of a local tourism network: Comparison of before and after COVID-19. *Curr. Issues Tour.* **2021**, *24*, 3324–3338. [CrossRef]
13. Lohmann, M. Evolution of shortbreak holidays. *Tour. Rev.* **1991**, *46*, 14–23. [CrossRef]
14. Back, A.; Marjavaara, R. Mapping an invisible population: The uneven geography of second-home tourism. *Tour. Geogr.* **2017**, *19*, 595–611. [CrossRef]
15. Wynen, J. Explaining travel distance during same-day visits. *Tour. Manag.* **2013**, *36*, 133–140. [CrossRef]
16. Munir, K.A.; Phillips, N. The Birth of the 'Kodak Moment': Institutional Entrepreneurship and the Adoption of New Technologies. *Organ. Stud.* **2005**, *26*, 1665–1687. [CrossRef]
17. Fatanti, M.N.; Suyadnya, I.W. Beyond User Gaze: How Instagram Creates Tourism Destination Brand? *Procedia—Soc. Behav. Sci.* **2015**, *211*, 1089–1095. [CrossRef]
18. Sontag, S. *On Photography*; Penguin Books: London, UK, 1977.
19. Chalfen, R.M. Photography's Role in Tourism Some Unexplored Relationships. *Ann. Tour. Res.* **1979**, *6*, 435–447. [CrossRef]
20. Urry, J.; Larsen, J. *The Tourist Gaze 3.0*; SAGE Publications Ltd.: London, UK, 2011.
21. Stuart, H. *Representation: Cultural Representations and Signifying Practices*; SAGE Publications Ltd.: London, UK, 1997.
22. MacCannell, D. *The Tourist: A New Theory of the Leisure Class*, 3rd ed.; University of California Press: Berkeley, CA, USA; Los Angeles, CA, USA; London, UK, 1999.
23. Jenkins, O.H. Photography and travel brochures: The circle of representation. *Tour. Geogr.* **2003**, *5*, 305–328. [CrossRef]
24. Hanan, H.; Putit, N. Express marketing of tourism destinations using Instagram in social media networking. In *Hospitality and Tourism*; Taylor and Francis Group: London, UK, 2014; pp. 471–474. ISBN 978-1-138-00151-0.
25. Narangajavana, Y.; Fiol, L.J.C.; Tena, M.Á.M.; Artola, R.M.R.; García, J.S. The influence of social media in creating expectations. An empirical study for a tourist destination. *Ann. Tour. Res.* **2017**, *65*, 60–70. [CrossRef]
26. Chung, N.; Koo, C. The use of social media in travel information search. *Telemat. Inform.* **2015**, *32*, 215–229. [CrossRef]
27. Sotiriadis, M.D. Sharing tourism experiences in social media. *Int. J. Contemp. Hosp. Manag.* **2017**, *29*, 179–225. [CrossRef]
28. Jabreel, M.; Moreno, A.; Huertas, A. Semantic comparison of the emotional values communicated by destinations and tourists on social media. *J. Destin. Mark. Manag.* **2017**, *6*, 170–183. [CrossRef]
29. Jones, N.; Borgman, R.; Ulusoy, E. Impact of social media on small businesses. *J. Small Bus. Enterp. Dev.* **2015**, *22*, 611–632. [CrossRef]
30. Kádár, B.; Gede, M. Where Do Tourists Go? Visualizing and Analysing the Spatial Distribution of Geotagged Photography. *Cartogr. Int. J. Geogr. Inf. Geovis.* **2013**, *48*, 78–88. [CrossRef]
31. Marine-Roig, E.; Clavé, S.A. Tourism analytics with massive user-generated content: A case study of Barcelona. *J. Destin. Mark. Manag.* **2015**, *4*, 162–172. [CrossRef]
32. Donaire, J.A.; Camprubí, R.; Galí, N. Tourist clusters from Flickr travel photography. *Tour. Manag. Perspect.* **2014**, *11*, 26–33. [CrossRef]
33. TourMIS. City Tourism in Europe. 2021. Available online: http://www.tourmis.info (accessed on 8 February 2022).
34. Girardin, F.; Fiore, F.D.; Blat, J.; Ratti, C.; Dal Fiore, F. Understanding of Tourist Dynamics from Explicitly Disclosed Location Information. *J. Locat. Based Serv.* **2008**, 2. [CrossRef]
35. Wood, S.A.; Guerry, A.D.; Silver, J.M.; Lacayo, M. Using social media to quantify nature-based tourism and recreation. *Sci. Rep.* **2013**, *3*, 2976. [CrossRef] [PubMed]
36. Straumann, R.K.; Çöltekin, A.; Andrienko, G. Towards (Re)Constructing Narratives from Georeferenced Photographs through Visual Analytics. *Cartogr. J.* **2014**, *51*, 152–165. [CrossRef]
37. Lo, I.S.; McKercher, B.; Lo, A.; Cheung, C.; Law, R. Tourism and online photography. *Tour. Manag.* **2011**, *32*, 725–731. [CrossRef]
38. Önder, I.; Koerbitz, W.; Hubmann-Haidvogel, A. Tracing Tourists by Their Digital Footprints: The Case of Austria. *J. Travel Res.* **2014**, *55*, 566–573. [CrossRef]
39. Shoval, N.; Isaacson, M. Tracking tourists in the digital age. *Ann. Tour. Res.* **2007**, *34*, 141–159. [CrossRef]
40. Shoval, N.; Isaacson, M. *Tourist Mobility and Advanced Tracking Technologies*; Routledge: New York, NY, USA, 2010.
41. Kádár, B.; Gede, M. Tourism flows in large-scale destination systems. *Ann. Tour. Res.* **2021**, *87*, 103113. [CrossRef]
42. Kádár, B. Measuring tourist activities in cities using geotagged photography. *Tour. Geogr.* **2014**, *16*, 88–104. [CrossRef]
43. Central Statistical Office. Hungarian Central Statistical Office Dissemination Database. 2020. Available online: https://statinfo.ksh.hu/Statinfo/themeSelector.jsp?&lang=en (accessed on 8 February 2022).
44. Perriam, J.; Birkbak, A.; Freeman, A.; Birkbak, A.; Freeman, A. Digital methods in a post-API environment. *Int. J. Soc. Res. Methodol.* **2019**, *23*, 277–290. [CrossRef]
45. Smith, M.; Szongott, C.; Henne, B.; von Voigt, G. Big Data Privacy Issues in Public Social Media. In Proceedings of the 2012 6th IEEE International Conference on Digital Ecosystems and Technologies (DEST), Campione d'Italia, Italy, 18–20 June 2012; pp. 1–6.
46. Zhang, W.; Tan, G. Detecting tourist attractions using geo-tagged photo clustering. *Chin. Sociol. Dialogue* **2018**, *3*, 3–16. [CrossRef]
47. Kupi, M.; Szemer, E. Impact of the COVID-19 on the Destination Choices of Hungarian Tourists: A Comparative Analysis. *Sustainability* **2021**, *13*, 13785. [CrossRef]
48. Boros, L.; Kovalcsik, T. A COVID-19-járvány hatása a budapesti Airbnb-piacra. *Terul. Stat.* **2021**, *61*, 380–402. [CrossRef]
49. Simonovits, B.; Kondorosy, C. Participation, trust, and risks associated with peer-to-peer accommodation platforms: How did the COVID-19 crisis affect Airbnb Budapest in 2020? *Intersect. East Eur. J. Soc. Polit.* **2021**, *7*, 178–200. [CrossRef]

sustainability

MDPI

Article

(In)Visible Tourism According to Online Cash Registers in Hungary, 2018–2020

Géza Tóth [1,2] and Áron Kincses [1,2,*]

[1] Institute of World and Reginal Economics, University of Miskolc, 3515 Miskolc, Hungary; geza.toth@ksh.hu
[2] Hungarian Central Statistical Office, Statistics Directorate, 1024 Budapest, Hungary
* Correspondence: aron.kincses@ksh.hu

Abstract: The aim of this paper is to share innovations and some key lessons learned in the use of non-traditional data sources to improve data quality and enable more accurate and efficient data use in the field of tourism. Research on visitor traffic is based on classical statistical measures, but it may be worth expanding it with alternative data sources, such as databases based on online cash register (OCR) data. These data can be particularly useful for analysing tourism-related consumption habits in a given area. The study introduces the "invisible", tourism-related, non-accommodation spending characteristics of transit traffic in Hungary, the possibilities of its analysis and spatial aspects, using online cash register data (includes all retail sales in Hungary, except for motorcycle purchases), and additionally, we identify the most affected municipalities which are invisible for traditional data sources. The results show that invisible tourism, linked to transit traffic, has significant economic potential. The analysis of this new type of database will provide a more accurate and faster picture of consumption associated with hidden tourism, which can be an important input for economic and marketing development.

Keywords: invisible tourism; transiting; transit traffic; geospatial information; Hungary

Citation: Tóth, G.; Kincses, Á. (In)Visible Tourism According to Online Cash Registers in Hungary, 2018–2020. *Sustainability* **2022**, *14*, 3038. https://doi.org/10.3390/su14053038

Academic Editors: Anna Rita Irimiás, Dallen J. Timothy and Gábor Michalkó

Received: 4 February 2022
Accepted: 1 March 2022
Published: 4 March 2022

1. Introduction

Hungary is a transit country, not only for freight but also for passengers. About one third of foreigners entering Hungary each year only pass through the country. This is explained by Hungary's central location in East-Central Europe, its proximity to seven countries and its access to the European road network. Transit does not necessarily occur in a hermetically sealed corridor—such as motorways across national borders—almost isolated of the socio-economic environment, but often becomes part of it. The nature of the transit implies stops, longer or shorter breaks for the driver and passengers, meals and refuelling of the vehicle, which take place both in motorway car parks with appropriate infrastructures and in the living areas of the local population within residential areas.

Classic survey-based data sources are available for the travel activity of foreigners not involved in freight transport, so we have a relatively reliable picture of the macro-level characteristics of transit flows. According to data from the Hungarian Central Statistical Office, the share of transit passengers arriving in Hungary increased from 33% in 2017 to 34% in 2020. Their number exceeded 20 million in 2019, before the COVID-19 pandemic, and decreased to 11 million in 2020. Spending by transit passengers accounted for 6% of total spending in 2017, increasing to 9% by 2020. Spending was HUF 116 billion in 2017, rising to HUF 174 billion in 2019, and then decreased to just over HUF 104 billion in 2020. The indicators speak for themselves: traffic of this magnitude is capable of generating changes at the micro-level, with social, economic and physical impacts [1] on the life of the municipalities concerned. However, in order to examine these effects, it is essential to define the municipalities involved in transit traffic, i.e., to identify the towns and villages where the demand of transit passengers is realised.

2. Materials and Methods

2.1. Online Cash Register Data in Hungary

The challenges faced by statistics in the 21st century are manifold. We are surrounded by systems that are becoming substantially more and more complex. With the emergence of new phenomena (e.g., globalisation, global demographic trends or sustainable development) and complex realities that need to be meaningfully and timely captured by statistics, new types of data have also been emerging, offering opportunities to improve the relevance of statistics.

Fortunately, there is currently an abundance of data sources: questionnaire-based statistical data, census data, big data, smart data, machine data, administrative data, privately held data, etc. In many cases, statistical domains are based on traditional data sources that are reaching their limits with respect to timeliness, relevance and compliance.

For the purpose of reducing the abuses committed during the retail trade turnover, the government of Hungary decided to introduce an online connection of cash registers with the National Tax Authority in October 2014. According to the OECD [2], electronic cash registers are used (among others) in Argentina, Austria, Belgium, the Czech Republic, Hungary, Italy, the Netherlands, Portugal, Russia and Sweden. As such, cash machines involved in the online cash register system send online retail chain sales information to the National Tax Authority. This means that data concerning the sales of retail stores can be accessed in real time from the NTCA (National Tax and Customs Administration) database. Immediately after the purchase of goods or services, invoice details are automatically sent online to the National Tax and Customs Administration. The Hungarian Central Statistical Office receives this in full. The database contains the OCR number, the date of commissioning, the period of validity, the tax number of the company, the name and address of the business, the economic activity of the company, the number and turnover value of invoices, cancellation and returns for the given period.

In this paper, we use OCR data from 2018, 2019 and 2020 to illustrate methodological approaches that map out the crystallisation points of transit traffic in Hungary. While the study primarily aims to capture the phenomenon of transit at the municipal level, we draw attention to the fact that the impacts are economy-wide in their multiplier effect.

The OCR data are available up-to-date, with the added advantage of being disaggregated by economic activity and geographical location. The advantage of online cash registers is that tax authorities can determine the amount of tax payable more accurately and quickly based on the data they receive, and if they see suspicious traffic data, they will be able to filter out fraudsters in a matter of seconds. The system developed from a tax point of view also provides much more accurate data for analysis, so we used this data source. According to the OECD [2], the use of online cash registers has the following advantages: better tax compliance, protection of fair competition, reduction of compliance burdens, and enhanced consumer protection. The other advantage of OCR is that we have quasi-real-time information about retail processes, the time to market is much faster than the classic questionnaire survey, and it reduces the statistical reporting burden on businesses.

The disadvantage of OCR data is that due to possible failures of the cash machines, NTCA will not receive data for another 11 months, so the first data transmission cannot be considered as final information. However, the differences between data sets are now negligible.

For the purpose of this study, we consider tourism-related expenditures based on the international recommendations (https://unstats.un.org/unsd/publication/Seriesm/SeriesM_83rev1e.pdf) (accessed on 5 January 2022) presented in Appendix A Table A1.

Transiting is a relatively rarely analysed part of tourism, despite the unquestionable importance of connectivity in tourism systems [3]. One of the reasons for this, according to certain opinions, is that transit is partly seen as a low revenue generator and partly as a necessary inconvenience for tourists [4].

2.2. Theoretical Background

Although transit is a phenomenon that affects many countries in Europe, the study of this kind of mobility is not part of mainstream national and international research on transport, migration and tourism. Some of the studies on transit issues focus on freight transport infrastructure and its development [5,6] and logistics [7], while others focus on cost implications [8–11], environmental issues [12] and geopolitical aspects [13,14]. Researchers are much more interested in exploring the specificities of traffic flowing through large cities [15–17] than that between national borders.

Transiting is a specific form of mobility, as its main purpose is to cover the distance between the origin and the destination in the shortest possible time, while travellers usually do not consider the tourist experiences they may have on the way. However, transit does not exclude the use of the tourist services of the areas concerned, since transit passengers may interrupt their journey to buy fuel or to visit retail shops, but we should not forget the possibility of visiting certain tourist attractions and, in some cases, of using accommodation facilities [18–21]. Same-day movements are distinguished more by their duration and less by their radius. With minimal investment, same-day trips will meet tourist demand, and thus contribute to the development of tourism in general [22].

Tourism researchers have a surprisingly lenient approach to the issue of transit, which is certainly related to the statistical interpretation of the concept [23]. There are hardly any studies that have recognised the tourism implications of the behaviour of transit passengers. This topic is mainly addressed in studies on tourism in the former socialist countries of Central and Eastern Europe and South-Eastern Europe, where the flow of guest workers to the West is also highlighted [24–26].

Transiting is a key phenomenon not only in land transport but also in water [27] and air transport, with retail stores available from airport transit lounges, serving both to spend time and to encourage spending by transit passengers [28–31]. Researchers are also interested in the measurement of transit traffic, which, while not a potential substitute for classical traffic counting, can provide a useful complementary source of information from the spatial- and time-based recording of call volumes over a wide range of mobile communication devices [32]. Last, but not least, the specific tourist behaviour of caravans and motorhomes is unexplored in the relationship between tourism and transit [33].

Although our study tries to delimit the range of settlements that primarily benefit from transit tourism, we also consider it important to mention that transit has its drawbacks, the management of which is definitely a challenge for decision-makers. Such a challenge could be increased traffic, the resulting congestion, high levels of pollution, etc. This, in turn, leads us to the issue of tourism penetration [34], which may be the subject of further research.

2.3. Transiting as a Crypto Mobility Activity

The tourism aspects of this phenomenon have also been recognised by relevant EU experts, who have suggested, among other things, the need to measure transit travel. Recital 5 of the EU regulation (Regulation (EU) No 692/2011 of the European Parliament and of the Council of 6 July 2011 concerning European statistics on tourism and repealing Council Directive 95/57/EC) to modernise the methodology of tourism statistics states that:

"The changing nature of tourism behaviour ... with the growing importance of short trips and same-day visits contributing substantially in many regions or countries to the income from tourism [...] means that the production of tourism statistics should be adapted."

The previous definition of tourism, which assumed a minimum one-night stay, therefore needs to be amended to take into account not only so-called "excursion" traffic (shopping, visiting relatives), which is mainly concentrated in border areas and lasts less than 24 h, but also transit traffic. Since the paradigm of thinking about tourism [35] excludes the discussion of the demand arising from the needs of freight transport actors and the supply created to satisfy them under the umbrella of tourism, it is still appropriate to ignore freight transport.

However, it should be noted that road corridors for freight transport include a number of infrastructure and suprastructure objects that are also used by lorry drivers. Thus, petrol stations offering complex services, roadside accommodation and catering outlets meet the demand not only of passengers but also of freight transport. Therefore, in a certain context, it is appropriate to carry out an analysis in this respect, even if the demand generated by freight transport is not considered as tourism expenditure.

If passenger transit is discussed as part of tourism, it is assumed that similar to conventional tourist mobility; it has its clear manifestations. However, while the fact of staying in a registered accommodation is included in tourism statistics, services used in unconventional tourism mobility rarely become part of a database that can be compared over space and time. The exploration of the territorial aspects of transiting, therefore, requires the creation of a methodological approach which, on the one hand, takes into consideration the specificities of the phenomenon under study and, on the other hand, is based on some statistically measurable fact. The challenge is not a minor one, as we need to make visible a crypto mobility, a quasi-invisible travel phenomenon. In our approach, we consider areas that are covered by official statistical data collection as visible tourism areas, and those that are not covered by official data collection as invisible tourism areas (for example: trips shorter than 24 h, border trips, the use of friendly accommodation). Within these, the present study deals mainly with the issue of transit.

For this, it is worth starting with the characteristic feature of transit that captures the moment of stopping for whatever reason. Buying fuel or shopping in retail stores is definitely one of them, but we should also bear in mind visiting certain tourist attractions and, in some cases, staying in accommodation. However, there is no statistical data on the number of visitors to each municipality, so in these cases only the OCR can provide a useful indication.

2.4. *Delineation of the Municipalities Involved in Transiting Using GIS Methods*

The first step of the analysis was to identify the Hungarian settlements that could be affected by transit traffic. The delineation was carried out using geographic information methods and taking into account the following criteria:

(a) When a transit traveller enters Hungary at a road border crossing point, he or she does not leave Hungary at the same border crossing point from which he or she arrived.
(b) Due to the nature of transit, it was also assumed that a person transiting through Hungary would use the shortest route in time between the entry and exit points.

Based on these criteria, the shortest travel time to all border crossing points in all neighbouring countries was determined. The municipalities affected by the intersection of each route were selected and summarised. The possibility was considered that transit traffic might benefit not only the settlements directly affected by transport routes, but also neighbouring settlements. However, we considered this to be an isolated and ad hoc phenomenon and the original delineation was therefore considered sufficient. Based on this delineation, 956 of Hungary's 3152 settlements were included in the scope of the study, i.e., this was the number of municipalities considered as potential crystallisation points for transit traffic, and further investigations were carried out on this population of municipalities.

3. Outcomes, Results

3.1. *Exploring the Role of Tourism by Expert Assessment*

The 956 municipalities included in the study were classified into three groups based on expert assessment of the role of the tourist attractions located in their administrative territory. Given that presenting the methodology of the expert assessment of tourism attractiveness would go far beyond the scope of this study, we will only outline the main points of the assessment carried out. Our work was based on the recommendations of the methodological handbook (compiled by Gábor Michalkó and Tamara Rátz), which was prepared in 2006 under the co-ordination of Hungarian Tourism Ltd. In this case, for the classification of individual attractions, legislation (e.g., Act CXII of 2000 on the adoption of the Spatial Planning Plan of the Balaton Special Tourist Region and on the establishment of the Balaton Spatial Planning Code), official databases (e.g., the National Public Health and Medical Officer's Service, the Registry of the National Directorate General for Spas and Thermal Baths), surveys and registers of the HCSO (e.g., the number of overnight stays and offer of spa and wellness hotels), ratings published by professional organisations (e.g., the list of certified festivals of the Hungarian Festival Association) and the websites of the municipalities were used. For example, international and national ratings were awarded to those municipalities whose accommodation demand was dominated by foreign demand, which had the highest ranking in the various databases (e.g., settlement by Lake Balaton; spa; wellness hotel; festival with excellent rating). Municipalities with regional attractiveness include those towns and villages that were rated lower in the sources reviewed (e.g., settlement further from Lake Balaton, no spa, but there was thermal water for bathing; well-rated festival). The municipalities that were not included in the databases we processed and whose attractiveness could not be identified from the information on their websites were grouped into municipalities without significant tourist attractions. The first group included international and national attractiveness, the second group was made up of regional attractiveness and the third group contained municipalities without significant tourist attractions (see Figure 1). The data used in the following analysis were derived from the above settlement classification of the role of tourism attraction and the settlement aggregation of the OCR data related to tourism (see Tables 1–3).

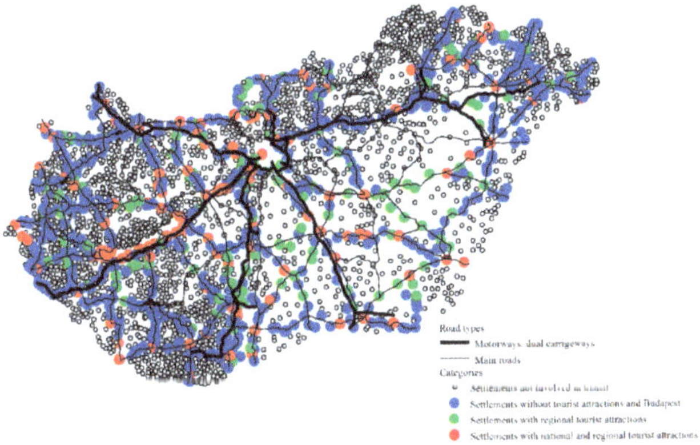

Figure 1. The Hungarian municipalities according to their tourism attractiveness.

Table 1. Number of surveyed municipalities as a function of tourism-related revenue and the tourism attractiveness category in 2018.

Category/Revenue	HUF 0–50 Million	HUF 50–250 Million	HUF 250–500 Million	HUF 500–1000 Million	HUF 1000–5000 Million	HUF over 5000 Million	Total
1 (municipalities without significant attractiveness)	481	184	31	12	6	0	714
2 (municipalities with regional attractiveness)	27	39	17	18	21	3	125
3 (municipalities with national, international attractiveness)	11	20	12	21	34	19	117
Total	519	243	60	51	61	22	956

Sources: Own calculations, based on OCR register.

Table 2. Number of surveyed municipalities as a function of tourism-related revenue and the tourism attractiveness category in 2019.

Category/Revenue	HUF 0–50 Million	HUF 50–250 Million	HUF 250–500 Million	HUF 500–1000 Million	HUF 1000–5000 Million	HUF over 5000 Million	Total
1 (municipalities without significant attractiveness)	465	196	30	15	7	1	714
2 (municipalities with regional attractiveness)	27	34	17	20	22	5	125
3 (municipalities with national, international attractiveness)	11	19	9	20	37	21	117
Total	503	249	56	55	66	27	956

Sources: Own calculations, based on OCR register.

Table 3. Number of surveyed municipalities as a function of tourism-related revenue and the tourism attractiveness category in 2020.

Category/Revenue	HUF 0–50 Million	HUF 50–250 Million	HUF 250–500 Million	HUF 500–1000 Million	HUF 1000–5000 Million	HUF over 5000 Million	Total
1 (municipalities without significant attractiveness)	484	187	25	13	5		714
2 (municipalities with regional attractiveness)	27	38	16	19	22	3	125
3 (municipalities with national, international attractiveness)	11	20	12	22	35	17	117
Total	522	245	53	54	62	20	956

Sources: Own calculations, based on OCR register.

The strength of stochastic relationships between qualitative and quantitative criteria can be measured by association measures [35,36]. The independence of the two variables (X and Y) is then

$$\Phi^2 = \sum_{i=1}^{I}\sum_{j=1}^{J} \frac{(p_{ij} - p_{i.} \cdot p_{j.})^2}{p_{i.} \cdot p_{j.}} \tag{1}$$

measured by statistics, where $P_{ij} = P(X = x_i, Y = y_j)$, $i = 1,2, \ldots .I; j = 1,2, \ldots ,J$.

If independent, the value is 0.

Noteworthy statistics are obtained by different normalizations of the quantity Φ^2 [37]:

Pearson's formula:

$$P = \sqrt{\frac{\Phi^2}{\Phi^2 + 1}} \tag{2}$$

Cramer's formula:

$$T = \sqrt{\frac{\Phi^2}{\sqrt{(I-1)\cdot(J-1)}}} \tag{3}$$

(P, $T = 0$ the two variables are independent of each other; P, $T = 1$ means a functional relationship, while $0 < P, T < 1$ indicates a stochastic relationship.)

Based on our calculations (Table 4) using the above theory, we found only a moderately strong stochastic relationship between tourism attractiveness and revenue. Hence, it has been demonstrated that there is no functional relationship between the importance of tourism attractiveness and tourism-related revenues. In fact, there is only a weak stochastic relationship.

Table 4. Association results between tourism attractiveness and corresponding OCR revenues in the surveyed municipalities.

	2018	2019	2020
P	0.566	0.564	0.567
T	0.386	0.385	0.387

Sources: Own calculations.

3.2. *Identification of the Municipalities Involved in Transiting*

In most cases, OCR and attractiveness were proportional at the level of municipalities, but there was a significant anomaly (Tables 1–3). The differences between the above criteria could be distinguished in two directions. In the first case, tourism attractiveness was much higher than the revenue, while in the second case, the municipalities where the revenue significantly exceeded the tourism attractiveness are included. In this study, we will focus on the latter group in more detail.

From Tables 1–3, we consider the municipalities as the crystallisation points of invisible tourism when:

- there was no significant tourist attraction and the corresponding OCR-revenue was above HUF 250 million;
- there was a regional tourist attraction and the above revenue reached or exceeded HUF 1000 million.

These municipalities clearly generate more revenue than would be expected from the role of tourist attractiveness in the area, according to the paradigms of the sector [38], and thus we have identified a group of municipalities that are likely to show a different attitude from conventional tourism.

In 2018, 2019 and 2020, we counted 73, 80 and 68 such municipalities, respectively. The number of municipalities varied slightly over the period. The frequencies of occurrence are shown in Figure 2.

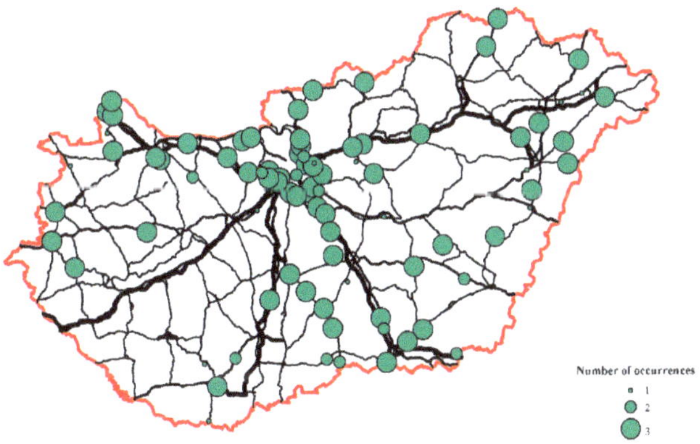

Figure 2. Geographical locations of the surveyed municipalities.

In Hungary, the total value added by the activities included in tourism satellite accounts was 3.4 times higher than the gross value added by tourism in the narrow sense, i.e., accommodation and food service. Based on OCR data, the national ratio was 4.2 for revenues. In 2020, the share of revenues was already 4.7 times for the 956 municipalities analysed. Moreover, in the 68 municipalities identified in 2020, it was more than 10 times higher. This underlines the relevance of the identified municipalities in invisible tourism.

3.3. *Characteristics of the Municipalities Affected by Transit Tourism*

In this section we will further examine the municipalities defined in the previous chapter as the focal points of invisible tourism, in order to identify the characteristics, features and geographical locations of the municipalities where transit tourism may become relevant, and the characteristics that increase the chances of transit tourism in a given municipality.

The municipalities affected by transit tourism were highly heterogeneous in terms of population and socio-economic characteristics. We found that the economic and accessibility characteristics of the municipalities did not have a significant impact on the focal points of transit tourism. Therefore, it was assumed that the geographical location of the municipality may play a greater role in transit tourism. In fact, the spatial distribution of the municipalities theoretically affected by transit tourism (956) showed a strong link with road network nodes.

The table below shows: Category 1—there is a motorway slip road in the municipality, i.e., a main road intersects the motorway; Category 2—intersection of main roads; and Category 3—all other municipalities located at a point not important for road traffic (Figure 3).

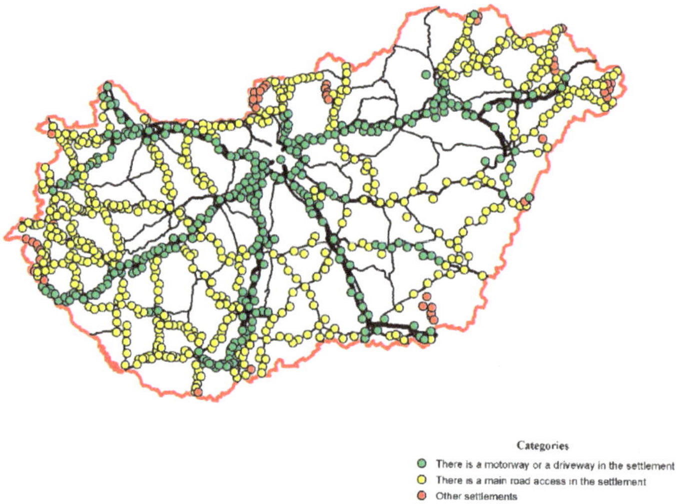

Figure 3. Categorisation of Hungarian settlements by road network.

Tables 5–7 provides a breakdown of the surveyed municipalities by OCR revenue and tourism attractiveness, broken down by the above node categories for the year 2020, highlighting in bold, italic number the municipalities most affected by invisible tourism on the basis of previous analyses. The results for 2018 and 2019 are presented in Appendix A Tables A2–A7.

Table 5. Number of municipalities with motorways and main roads as a function of tourist attractiveness and OCR tourism-related income in 2020.

Category/Revenue	HUF 0–50 Million	HUF 50–250 Million	HUF 250–500 Million	HUF 500–1000 Million	HUF 1000–5000 Million	HUF over 5000 Million	Total
1 (municipalities without significant attractiveness)	115	64	*14*	*10*	5		208
2 (municipalities with regional attractiveness)	7	8	7	5	*11*	2	40
3 (municipalities with national, international attractiveness)	5	6	3	5	20	13	52
Total	127	78	24	20	36	15	300

Table 6. Number of municipalities with main roads as a function of tourist attractiveness and OCR tourism-related income in 2020.

Category/Revenue	HUF 0–50 Million	HUF 50–250 Million	HUF 250–500 Million	HUF 500–1000 Million	HUF 1000–5000 Million	HUF over 5000 Million	Total
1 (municipalities without significant attractiveness)	338	118	11	3			470
2 (municipalities with regional attractiveness)	18	28	8	14	11	1	80
3 (municipalities with national, international attractiveness)	4	13	8	17	15	4	61
Total	360	159	27	34	26	5	611

Table 7. Number of municipalities located at a point not important for road traffic as a function of tourist attractiveness and OCR tourism-related income in 2020.

Category/Revenue	HUF 0–50 Million	HUF 50–250 Million	HUF 250–500 Million	HUF 500–1000 Million	HUF 1000–5000 Million	HUF over 5000 Million	Total
1 (municipalities without significant attractiveness)	31	5					36
2 (municipalities with regional attractiveness)	2	2	1				5
3 (municipalities with national, international attractiveness)	2	1	1				4
Total	35	8	2				45

Sources: Own calculations, based on OCR register.

This means (Table 8) that while invisible tourism was detected in 0% of the municipalities belonging to Category 3 (located at a point not important for road traffic), in Category 2 (existence of main roads), it was already 4–5%, while in Category 1 (motorway slip roads) it was 14–16% of the 956 municipalities originally delimited (results for 2018 and 2019 are presented in Appendix A Table A8). In other words, in these cases, the correlation between tourism attractiveness and OCR data cannot be explained.

Table 8. Number of municipalities affected by invisible tourism as a function of road network characteristics in 2020.

Road Network Characteristics	Number of Municipalities Affected by Invisible Tourism	Total Number of Municipalities Surveyed	%
1	42	300	14.00
2	26	611	4.26
3	0	45	0.00
Total	68	956	7.11

Sources: Own calculations, based on OCR register.

The results demonstrate that transit traffic generates significant invisible tourism revenue, primarily in municipalities with motorway slip roads and secondarily in municipalities bordering main roads (in municipalities with a large tourist attractiveness, these revenues are "visible", but for smaller ones, only the OCR data provide a guide). The lack of adequate connectivity practically excludes the emergence of invisible tourism.

4. Discussion

Hungary is one of Europe's transit countries. It has a significant role in both East–West and North–South passenger traffic on the continent. Transit passengers, most of who arrive by car, use motorways and/or trunk roads to cover the distance between two border sections, stopping off for a variety of needs. Transit is often disrupted in settlements where service providers have built their businesses specifically on transit. To ensure that these towns and villages are excluded from traditional tourism destinations, it is necessary to identify the characteristics that reliably indicate the likelihood of transit travel. This required the inclusion of a new data source (OCR) with sufficiently detailed spatial, temporal and activity level information to identify the tourism-related expenditure of transit travel. In the study, a geospatial delineation was carried out, based on which the groups were analysed according to their statistical characteristics It was assumed that a foreign transit passenger would also stop in a municipality that has no significant tourist attraction, but that its favourable location would contribute to its choice as a place for a break. Therefore, if a municipality located at the intersection of the shortest routes between border sections has above-average OCR tourism-related revenue, even though there is no significant tourism attraction, then the presence of transit traffic is probable. We were able to identify 86 municipalities that met the above criteria, which will be the focus of further research in the future.

We concluded that being located at the junction of a main road and a motorway or two main roads already has a significant impact on spending.

5. Conclusions

Overall, it can be concluded that the settlements that benefit most from the transit traffic of foreigners arriving in Hungary are those that:

- are located at the intersection of routes that allow the shortest time between each border section
- are located at the intersections of access from main roads to motorways, or at the junction of two main roads.

These results are certainly due to the psychological component of the nature of transit, in which intersections and nodes are seen as stages of the journey, the fulfilment of the desire to get somewhere, the physical completion of a stage (a well-deserved rest), and a place of preparation for the next stage.

We consider it important that the needs of the local residents and tourist service providers of the demarcated settlements, as well as transit passengers, be taken into account in spatial and transport planning. A complex approach is needed in order to ensure that the benefits of transit tourism do not only accrue to the service providers concerned, but also to minimize the damage or inconvenience it causes. Thus, instead of destroying tourist destinations, transit tourism becomes sustainable, even in the longer term. This, in turn, requires a holistic approach and long-term planning, the beginning of which we have tried to contribute to with our work.

Author Contributions: Conceptualization, G.T. and Á.K.; methodology, G.T. and Á.K.; validation, G.T. and Á.K.; formal analysis, G.T. and Á.K.; writing—original draft preparation, Á.K.; writing—review and editing, G.T. and Á.K.; visualization, G.T.; supervision, G.T. and Á.K. All authors have read and agreed to the published version of the manuscript.

Funding: The dissemination of research outcomes was supported by The Hungarian Tourism Association Foundation (MTSZ).

Institutional Review Board Statement: Not applicable.

Informed Consent Statement: Not applicable.

Data Availability Statement: Data available on request due to restrictions privacy. The data presented in this study are available on request from the corresponding author. The data are not publicly available due to data protection aspects.

Conflicts of Interest: The authors declare no conflict of interest.

Appendix A

Table A1. Internationally recommended groupings of economic sectors related to tourism.

IRTS/TSA: RMF Tourism Industries [a]	ISIC Rev. 4	NACE Rev. 2 (TEÁOR'08) [b]	Description (TEÁOR'08)
Accommodation	5510	5510	Hotels and similar accommodation
	5520	5520	Holiday and other short-stay accommodation
		5530	Camping grounds, recreational vehicle parks and trailer parks
	5590	5590	Other accommodation
Food and beverage serving activities	5610	5610	Restaurants and mobile food service activities
	5629	5629	Other food service activities
	5630	5630	Beverage serving activities
Railway passenger transport	4911	4910	Passenger rail transport, interurban
Road passenger transport	4922	4932	Taxi operation
		4939	Other passenger land transport
Water passenger transport	5011	5010	Sea and coastal passenger water transport
	5021	5030	Inland passenger water transport
Air passenger transport	5110	5110	Passenger air transport
Renting and leasing of motor vehicles	7710	7711	Renting and leasing of cars and light motor vehicles
		7712	Renting and leasing of trucks (over 3.5 t)
Travel agency and tour operator activities	7911	7911	Travel agency activities
	7912	7912	Tour operator activities
	7990	7990	Other reservation services and related activities
Cultural activities	9000	9001	Performing arts
		9002	Support activities to performing arts
		9003	Artistic creation
		9004	Operation of arts facilities
	9102	9102	Museum activities
		9103	Operation of historical sites and buildings
	9103	9104	Botanical and zoological gardens and nature reserve activities
Sports and recreational activities	7721	7721	Renting and leasing of recreational and sports goods
	9200	9200	Gambling and betting activities
	9311	9311	Operation of sports facilities
		9313	Fitness facilities
	9321	9321	Activities of amusement parks and theme parks
	9329	9329	Other amusement and recreation activities
	Country-specific tourism characteristic activities		
Other personal service activities (spa)		9604	Physical well-being activities
		8690	Other human health activities
Support activities for transportation		5221	Service activities incidental to land transportation
		5222	Service activities incidental to water transportation
		5223	Service activities incidental to air transportation

[a] Source: International Recommendations for Tourism Statistics 2008—Annex 3. [b] TEÁOR'08 is the Hungarian activity classification identical to the European one, NACE Rev.2. Statistical Classification of Economic Activities in the European Community, 2008.

Table A2. Number of municipalities with motorways and main roads as a function of tourist attractiveness and OCR tourism-related income in 2018.

Category/Revenue	HUF 0–50 Million	HUF 50–250 Million	HUF 250–500 Million	HUF 500–1000 Million	HUF 1000–5000 Million	HUF over 5000 Million	Total
1 (municipalities without significant attractiveness)	111	66	*15*	*10*	*6*		208
2 (municipalities with regional attractiveness)	7	9	8	4	*10*	*2*	40
3 (municipalities with national, international attractiveness)	5	5	4	5	19	14	52
Total	123	80	27	19	35	16	300

Table A3. Number of municipalities with main roads as a function of tourist attractiveness and OCR tourism-related income in 2018.

Category/Revenue	HUF 0–50 Million	HUF 50–250 Million	HUF 250–500 Million	HUF 500–1000 Million	HUF 1000–5000 Million	HUF over 5000 Million	Total
1 (municipalities without significant attractiveness)	338	114	*16*	*2*			470
2 (municipalities with regional attractiveness)	18	28	8	14	*11*	*1*	80
3 (municipalities with national, international attractiveness)	4	13	8	16	15	5	61
Total	360	155	32	32	26	6	611

Table A4. Number of municipalities located at a point not important for road traffic as a function of tourist attractiveness and OCR tourism-related income in 2018.

Category/Revenue	HUF 0–50 Million	HUF 50–250 Million	HUF 250–500 Million	HUF 500–1000 Million	HUF 1000–5000 Million	HUF over 5000 Million	Total
1 (municipalities without significant attractiveness)	32	4					36
2 (municipalities with regional attractiveness)	2	2	1				5
3 (municipalities with national, international attractiveness)	2	2					4
Total	36	8	1				45

Table A5. Number of municipalities with motorways and main roads as a function of tourist attractiveness and OCR tourism-related income in 2019.

Category/Revenue	HUF 0–50 Million	HUF 50–250 Million	HUF 250–500 Million	HUF 500–1000 Million	HUF 1000–5000 Million	HUF over 5000 Million	Total
1 (municipalities without significant attractiveness)	102	72	14	12	7	1	208
2 (municipalities with regional attractiveness)	7	8	7	4	10	4	40
3 (municipalities with national, international attractiveness)	5	5	3	6	18	15	52
Total	114	85	24	22	35	20	300

Table A6. Number of municipalities with main roads as a function of tourist attractiveness and OCR tourism-related income in 2019.

Category/Revenue	HUF 0–50 Million	HUF 50–250 Million	HUF 250–500 Million	HUF 500–1000 Million	HUF 1000–5000 Million	HUF over 5000 Million	Total
1 (municipalities without significant attractiveness)	333	118	16	3			470
2 (municipalities with regional attractiveness)	18	24	9	16	12	1	80
3 (municipalities with national, international attractiveness)	4	13	5	14	19	6	61
Total	355	155	30	33	31	7	611

Table A7. Number of municipalities located at a point not important for road traffic as a function of tourist attractiveness and OCR tourism-related income in 2019.

Category/Revenue	HUF 0–50 Million	HUF 50–250 Million	HUF 250–500 Million	HUF 500–1000 Million	HUF 1000–5000 Million	HUF over 5000 Million	Total
1 (municipalities without significant attractiveness)	30	6					36
2 (municipalities with regional attractiveness)	2	2	1				5
3 (municipalities with national, international attractiveness)	2	1	1				4
Total	34	9	2				45

Table A8. Number of municipalities affected by invisible tourism as a function of road network characteristics, 2018–2019.

2018			
Road Network Characteristics	**Number of Municipalities Affected by Invisible Tourism**	**Total Number of Municipalities Surveyed**	**%**
1	43	300	14.33
2	30	611	4.91
3	0	45	0.00
Total	73	956	7,64
2019			
Road Network Characteristics	**Number of Municipalities Affected by Invisible Tourism**	**Total Number of Municipalities Surveyed**	**%**
1	48	300	16.00
2	32	611	5.24
3	0	45	0.00
Total	80	956	8.37

References

1. Puczkó, L.; Rátz, T. *A Turizmus Hatásai*; Aula-Kodolányi: Budapest, Hungary, 1998; ISBN 963-9345-11-3.
2. OECD. Implementing Online Cash Registers: Benefits, Considerations and Guidance OECD, Paris. 2019. Available online: www.oecd.org/tax/forum-on-tax-administration/publications-and-products/implementing-online-cash-registers-benefits-considerations-and-guidance.htm (accessed on 5 January 2022).
3. Leiper, N. The framework of tourism: Towards a definition of tourism, tourist, and the tourist industry. *Ann. Tour. Res.* **1979**, *6*, 390–407. [CrossRef]
4. Weaver, D.B.; Lawton, L.J. *Tourism Management*, 5th ed.; John Wiley & Sons: Milton, QLD, Australia, 2014; ISBN 978-1-118-64481-2.
5. Hall, P.; Landis, J.; Cervero, R. Transit joint development in the USA: An inventory and policy assessment. *Environ. Plan. C Govern. Policy* **1991**, *9*, 431–452. [CrossRef]
6. Erdősi, F. Oroszország nemzetközi közlekedési infrastruktúráinak kényszerei, globális és regionális szerepei. *Tér és Társadalom* **2005**, *19*, 1–20. [CrossRef]
7. Kerschner, H.; Petrovitsch, H. Alpentransit auf der Schiene—das Beispiel Brennerachse. *Geogr. Rundsch.* **1998**, *50*, 580–586.
8. Kraus, M. Discomfort externalities and marginal cost transit fares. *J. Urban. Econ.* **1991**, *29*, 249–259. [CrossRef]
9. Fielding, G. Congestion pricing and the future of transit. *J. Transp. Geogr.* **1995**, *3*, 239–246. [CrossRef]
10. Bulis, A.; Skapars, R. Development of international freight transit in Latvia. *Proced. Soc. Behav. Sci.* **2013**, *99*, 57–64. [CrossRef]
11. Liu, Y.; Yang, L.; Chau, K.W. Impacts of Tourism Demand on Retail Property Prices in a Shopping Destination. *Sustainability* **2020**, *12*, 1361. [CrossRef]
12. Kalas, G. A déli tranzitautópálya csak alibi. Javaslat a déli tranzitautópálya hatásvizsgálataihoz, megvalósíthatósági tanulmány. *ÖKO Ökológia Környezetgazdálkodás Társadalom* **1992**, *3*, 30–38.
13. Fleischer, T. Transzeurópai folyosók: A meglévők hosszabbítgatása vagy egy összeurópai hálózat kialakítása? In *Balkán és Magyarország: Váltás a Külpolitikai Gondolkodásban*; Glatz, F., Ed.; MTA Társadalomkutató Központ Európa Intézet: Budapest, Hungary, 2007; pp. 365–379.
14. Ruppert, L. Az EU délkeleti irányú közlekedésének Magyarországot érintő hatásai. In *Balkán és Magyarország: Váltás a Külpolitikai Gondolkodásban*; Glatz, F., Ed.; MTA Társadalomkutató Központ Európa Intézet: Budapest, Hungary, 2007; pp. 357–363.
15. Viton, P. Consolidations of scale and scope of urban transit. *Reg. Sci. Urban. Econ.* **1992**, *22*, 25–50. [CrossRef]
16. Knowles, R. Transit oriented development in Copenhagen, Denmark: From the Finger Plan to Ørestad. *J. Transp. Geogr.* **2012**, *22*, 251–261. [CrossRef]
17. Wang, Y.; Chen, H.; Wu, X. Spatial Structure Characteristics of Tourist Attraction Cooperation Networks in the Yangtze River Delta Based on Tourism Flow. *Sustainability* **2021**, *13*, 12036. [CrossRef]
18. Michalkó, G.; Rátz, T. Rejtett dimenziók a Kárpát-medence turizmusában. In *Kárpát-Medence*; Frisnyák, S., Gál, A., Eds.; Természet, Társadalom, Gazdaság Bocskai István Gimnázium, Nyíregyházi Főiskola Turizmus és Föld-rajztudományi Intézete: Szerencs, Hungary, 2013; pp. 463–476.
19. Kincses, Á.; Tóth, G.; Tömöri, M.; Michalkó, G. Characteristics of transit tourism in Hungary with a focus on expenditure. *Reg. Stat.* **2016**, *6*, 129–148. [CrossRef]

20. Kincses, Á.; Tóth, G.; Tömöri, M.; Michalkó, G. Identifying settlements involved in Hungary's transit traffic. *Reg. Stat.* **2016**, *6*, 193–216. [CrossRef]
21. Kincses, Á.; Tóth, G.; Tömöri, M.; Michalkó, G. Az átutazó turizmus magyarországi sajátosságai, különös tekintettel a költésre. *Területi Stat.* **2016**, *56*, 455–476. [CrossRef]
22. Stetic, S.; Simicevic, D.; Stanic, S. Same-day trips: A chance of urban destination development. *UTMS J. Econ.* **2011**, *2*, 113–124. Available online: https://ideas.repec.org/a/ris/utmsje/0033.html (accessed on 10 October 2021).
23. Probáld, Á. (Ed.) *Proceedings of the Sixth International Forum on Tourism Statistics;* Hungarian Central Statistical Office, Services Statistics Department: Budapest, Hungary, 2002.
24. Bakic, O. Yugoslavian tourism: Strategic planning of destination product. *Tour. Manag.* **1988**, *9*, 145–151. [CrossRef]
25. Johnson, M. Czech and Slovak tourism: Patterns, problems and prospects. *Tour. Manag.* **1995**, *16*, 21–28. [CrossRef]
26. Baláz, V.; Mitsutake, M. Japanese tourists in transition countries of Central Europe: Present behaviour and future trends. *Tour. Manag.* **1998**, *19*, 433–443. [CrossRef]
27. Santos, M.; Radicchi, E.; Zagnoli, P. Port's Role as a Determinant of Cruise Destination Socio-Economic Sustainability. *Sustainability* **2019**, *11*, 4542. [CrossRef]
28. Achen, M.; Klein, K. Retail trade in transit areas: Introduction to a new field of research. *Die Erde* **2002**, *133*, 19–36.
29. McKercher, B.; Tang, E. The challenges of developing transit tourism. *Asia Pac. J.Tour. Res.* **2017**, *9*, 151–160. [CrossRef]
30. Tang, C.; Weaver, D.; Lawton, L. Can stopovers be induced to revisit transit hubs as stayovers? A new perspective on the relationship between air transportation and tourism. *J. Air Transp. Manag.* **2017**, *62*, 54–64. [CrossRef]
31. Poon, P.C.M.; Ho, G.K.S. Opportunities in Transit Tourism: A Case Study of Hong Kong as a Transit Destination. *J. Travel Tour. Market.* **2021**, *38*, 31–43. [CrossRef]
32. Ma, X.; Wu, Y.; Wang, Y.; Chen, F.; Liu, J. Mining smart card data for transit riders' travel patterns. *Transp. Res. Part. C* **2013**, *36*, 1–12. [CrossRef]
33. Green, F.B. Recreation vehicles: A perspective. *Ann. Tour. Res.* **1978**, *5*, 429–439. [CrossRef]
34. McElroy, J.L.; Albuquerque, K. Tourism penetration index in small Caribbean islands. *Ann. Tour. Res.* **1998**, *25*, 145–168. [CrossRef]
35. Michalkó, G. *Turizmológia;* Akadémiai Kiadó: Budapest, Hungary, 2012; ISBN 978-963-05-9717-3.
36. Cramer, H. *Mathematical Methods of Statistics;* Princeton University Press: Princeton, NJ, USA, 1946; ISBN 978-0-691-00547-8.
37. Everitt, B.S. *The Cambridge Dictionary of Statistics;* Cambridge University Press: Cambridge, UK, 2002; ISBN 978-0-521-76699-9.
38. Davenport, E.; El-Sanhury, N. Phi/Phimax: Review and Synthesis. *Educ. Psychol. Meas.* **1991**, *51*, 821–828. [CrossRef]

sustainability MDPI

Article

Capturing Unobserved Tourists: Challenges and Opportunities of Processing Mobile Positioning Data in Tourism Research

Tamás Kovalcsik [1,*], Ábel Elekes [2], Lajos Boros [1], László Könnyid [2] and Zoltán Kovács [1,3]

1 Department of Economic and Social Geography, University of Szeged, 6722 Szeged, Hungary
2 Hungarian Tourism Agency, 1027 Budapest, Hungary
3 Geographical Institute, Research Centre for Astronomy and Earth Sciences, Eötvös Loránd Research Network, 1112 Budapest, Hungary
* Correspondence: kovalcsik.tamas@geo.u-szeged.hu

Abstract: Even though tourism (both domestic and international) is one of the main triggering factors of human mobility worldwide, some of its forms are unexplored. This can be partly linked with the lack of reliable data and obstacles related to data processing and interpretation. Mobile Positioning Data (MPD) allows us to identify various forms of tourism that are undetectable through traditional data sources such as accommodation statistics. Using MPD, not only same-day tourists but also the real time mobility patterns of tourists among various destinations can be revealed, and even hidden (i.e., unobserved) forms of tourism can be detected. However, despite the obvious benefits of such data, very few comprehensive studies exist to date on the processing, and interpretation of MPD in tourism research. In this paper, a case study is presented on the challenges and opportunities of processing MPD from raw to good quality researchable data offering a baseline tool for MPD-based research in the field of tourism. With the methodology introduced in this paper, it is possible to provide a more accurate picture of tourist flows regarding unobserved tourists, including same-day visitors.

Keywords: mobile positioning data (MPD); tourism mobility; unobserved tourism; same-day tourists; international tourism; domestic tourism

Citation: Kovalcsik, T.; Elekes, Á.; Boros, L.; Könnyid, L.; Kovács, Z. Capturing Unobserved Tourists: Challenges and Opportunities of Processing Mobile Positioning Data in Tourism Research. *Sustainability* **2022**, *14*, 13826. https://doi.org/10.3390/su142113826

Academic Editors: Dallen J. Timothy, Gábor Michalkó and Anna Rita Irimiás

Received: 29 July 2022
Accepted: 21 October 2022
Published: 25 October 2022

1. Introduction

With skyrocketing numbers of tourists and intensifying mobility flows, the tourism industry has experienced an extraordinary growth over recent decades [1] and, despite the significant fracture caused by the recent COVID-19 pandemic [2], the trend is likely to increase in the future. Therefore, unfolding and better understanding the spatial behavior of tourists is becoming increasingly important. On the one hand, for national and local governments it is crucial to have up-to-date information on the spatio-temporal dynamics and real time characteristics of international and domestic tourists at different destinations. Such knowledge can contribute to a more efficient planning and management of tourism-related infrastructure and transport facilities and can help elaborate national and local policies to increase tourism revenues [3–6]. On the other hand, alongside the obvious economic benefits, the rapid growth of tourism also brings challenges [7–9]. The negative impacts of tourism include the increasing carbon emission or damage to the natural environment [10–12], over-tourism and the discontent of local communities with mass tourism [13,14]. To make tourism development more sustainable, it is worth identifying tourist destinations with rapidly expanding ecological footprint, congestion and social conflict in order to mitigate the negative environmental and social impacts of tourism [15,16]. In addition, private actors, such as representatives of the hospitality industry, event organizers etc., can also benefit from a more comprehensive knowledge regarding the spatial and temporal behavior of tourists [17,18].

While traditional data sources (accommodation and border crossing statistics) are useful for macro-level analyses (countries, regions, larger administrative units) of the spatio-temporal aspects of tourism [19–21], such data have several limitations when it comes to a lower geographical scale such as the analysis of intra-urban tourist flows, intra-regional moves of same-day visitors or cross-border day trippers [22,23]. To capture such statistically unobserved forms of tourism, GIS-based data from a variety of sources can be used [24–30], such as GPS-based tracking systems [31,32], wireless, Bluetooth [33] and geotagged social media photos and posts [34], or mobile positioning data (MPD) [35–37]. With the spread of new Information and Communication Technologies (ICT), a more accurate tracking of human movements has become possible, especially with data obtained from mobile positioning data (MPD).

Previous studies using mobile phone datasets have focused predominantly on the possible methodological approaches of determining anchor points of users (most typically home and work) in order to assess the spatio-temporal mobility of individuals [38]. For instance, based on the analysis of communication data of 100,000 anonymized and randomly chosen individuals in Portugal and with the help of clustering methods, Csáji et al. [39] identified home and office locations of users and compared the results with official census data. In the same vein, many studies using mobile positioning data have been concerned, with either home detection [40–43] or flow detection [44–48] as a prerequisite of locating users in real time.

Despite the growing body of literature on the possible use of MPD in tourism research, there are still several methodological challenges and obstacles in the field [49]. According to Li et al. [33], this can be linked partly with the fact that most of the studies focus only on international visitors [37,50–53], which is also confirmed by Grassini et al., who highlight the overrepresentation of the international context in MPD tourism research [49]. One of the reasons behind the scarcity of MPD analysis of domestic tourist flows is the limitation regarding definitions, benchmarks and methodological considerations regarding how to process, filter and analyze such data. To date, we know only one comprehensive work focusing on methodological considerations related to MPD by Saluveer et al. [54]; however, domestic tourism is not considered by these authors either. This is the point of departure for this paper, which aims to introduce a data processing method regarding various forms of domestic and international tourism. This is a challenging task, because the overwhelming majority of MPD are generated by domestic users, resulting in log files that are Terabytes in size, which in turn makes data processing only possible if using big data compatible algorithms. This methodological advancement is important also because the role of domestic tourism has been on the rise recently due to the COVID-19 pandemic [55].

In addition to the analysis of domestic tourist flows, this paper also aims to provide new methodological considerations regarding international visitors. In this study, each event is attached to a municipality (i.e., settlement), instead of cell towers, resulting in more interpretable and geographically precise information. Via this method, we are also able to identify same-day visitors, day tripper shopping tourists from neighboring countries, extending the method used by Saluveer et al. [54]. Our aim here is to provide a data processing guideline through which not only domestic tourist flows can be understood better, separating touristic and non-touristic daily movements of local people [56–58], but also a more comprehensive picture on international tourist flows can be achieved, separating 'tourism' 'from transit traffic' or 'migrant workers'[54].

Before introducing the methodology and describing the steps in data processing, it is important to conceptualize and operationalize the terms used in this paper. We use the UNWTO recommendations [59]—along with Mamei and Colonna [58]—to define tourism-related terms, based on which a visitor is a person who travels for business, leisure or other personal purposes for less than a year. The domestic tourist is the visitor if his/her trip includes at least one overnight stay, while the same-day tourist (day trippers or excursionist) is the visitor who does not spend a night, so the day of arrival and departure is the same. With this interpretation, we follow the approach described by Saluveer et al. [54]. However, using MPD we can identify significantly more tourists than appear in official statistics, also described by De Cantis et al. [60] as unobserved tourism, the extent of which has already been quantified by Nyns and Schmitz [61] based on AirDNA and CDR data. During the research, we apply the theoretical framework developed by De Cantis et al. for approaching unobserved tourism.

In the remaining part of the paper, first the data acquisition procedure is presented, with a general overview of the raw data. This is followed by the description of data processing methodology. In the results and discussion section, we assess the final database and shed light on the limitations of the proposed method. We use Hungary as a study area, but the purpose is not a case-study unto itself. Instead, we intend to enrich the literature with a detailed insight into every step of the data life cycle, focusing more on different aspects of the data processing procedure, such as the identification of unobserved forms of tourism, [50] rather than providing case-specific research findings.

2. Data Description

This research is based on a dataset obtained by the Hungarian Tourism Agency (HTA) from one of Hungary's mobile network operators (provider in the following), which had a market share of 26% of domestic subscribers and 28% of international users in 2018. Through a contract, HTA was able to get access to the full MPD of the provider stored between 1 June and 31 December 2018. Thus, the dataset contains the full log data of each SIM card that connected to the provider's network in the 6-month period as well as anonymized user meta information for the SIM cards. The MPD in this case is passive call detail records (CDR), i.e., log events that are generated each time a user makes a phone call, sends a text message or uses cellular data. In such events, the mobile phone connects to a nearby mobile cell tower, usually to the closest one, which logs the event. While connecting to the tower, the network is able to locate the position of the cell phone, with an accuracy that depends on various attributes, most importantly on the distance between the mobile phone and the tower.

The SIM cards present in the dataset are of two kinds: domestic and international. Domestic cards belong to users living in Hungary who have an active subscription with the provider. The international SIM cards correspond to foreigners who travelled to (or through) Hungary and their phones connected to the provider's network. These users are usually subscribers of the providers' partner networks in their home country. However, tourists also have the option to manually choose among operators. Furthermore, there is a random factor, determining which network a cell phone connects to when staying in a foreign country.

In the raw data we have two comma separated value (CSV) files without headers for each day in the 6-month period. One of them contains the log events for a particular day, while the other file contains the anonymized user information for each user that had at least one log event on that day. The log events file contains columns in the following order separated by semicolons:

- The date of the event with minute precision in datetime format;
- The id of the user, which is a 38-digit number;
- The type of the event, i.e., whether it is a voice call, text message (SMS) or data usage;
- The latitude coordinates;
- The longitude coordinates;
- The precision of the location.

The exact location of an event was fixed inside a circle with the center defined by the latitude and longitude coordinates and the radius defined by the precision variable (in meters). Figure 1 shows a sample of the raw log data. Because of this, the use of Voronoi tessellation algorithm (which was used through other CDR data processing [62–64]) is not necessary, since the CDR data contained the precision value, so we were able to correctly estimate the errors of the coordinates.

datetime	user_id	type	latitude	longitude	precision
2018.12.01 15:19	81222529376966600796154844505105996882	VOICE	17.569418204684844	47.54763045972998	2959.508
2018.12.01 10:20	214813588485372106808716444206441703666	VOICE	17.569418204684844	47.54763045972998	2959.508
2018.12.01 07:11	96042799109531519307714776903629345085	DATA	19.17403106762232	47.427257417511576	320.59
2018.12.01 13:20	160609052934802405245811113475113718662	VOICE	19.03216743781843	47.54260828890025	250
2018.12.01 12:27	248831390322046716224108500176753789029	VOICE	16.86894928734427	47.23863248277103	4884.144
2018.12.01 11:46	29615459468153898810790806871866412358	VOICE	19.815512008766433	48.09907214171443	771.213
2018.12.01 09:25	216606290945961548852797768200616391361	VOICE	18.892408078569424	47.412078273665635	670.82
2018.12.01 11:45	45446894552881422340015926547817832374	VOICE	21.63564063890104	47.44998755328685	2771.018
2018.12.01 08:19	166080050413456223770593309883051886028	SMS	19.073341218094413	47.52956448708487	353.553

Figure 1. Sample of the raw log data.

The user information files contain the following columns. Day; user id; whether the user has a pre- or postpaid subscription plan; business type; business type code; the country code of the SIM card; the name of the country where the SIM card was bought; whether the SIM is domestic or international; age; gender; country of residence; postal code; brand of the cell phone, if the user has a subscription, whether it is a smartphone or not; the amount of last bill in Hungarian Forints; whether mobile parking, motorway ticket, or other services have been bought through the subscription. However, not all information is available for every user. In addition, the data were anonymized, so it was not possible to identify a concrete person based on the attributes, and only the user id was used during the data processing, along with the data management regulations set by the European Union's General Data Protection Regulation (GDPR).

One log event file is around 2.5–4 GB in size when compressed and around 6–9 GB when uncompressed. For the user information files, these numbers are 110–140 and 500–600 MB, respectively. On average, a log event file contains around 100 million events, while a user information file contains around 3–4 million users. This means, the uncompressed raw log dataset was around 1.8 Terabytes, comprising around 18 billion events, which is the final raw dataset that we rely on in this paper (Figure 2).

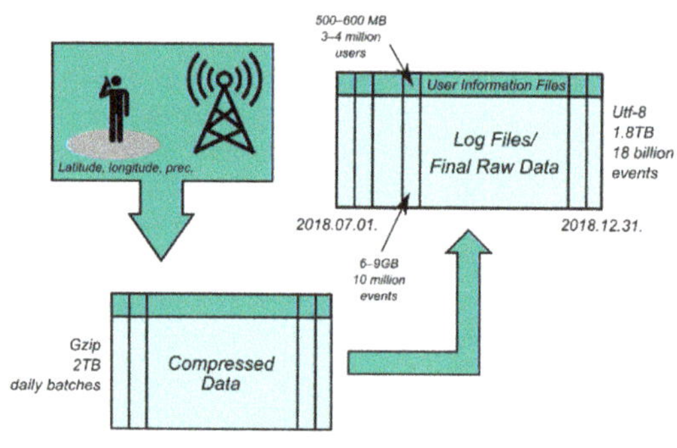

Figure 2. The uncompression of the raw data.

3. Data Processing and Cleaning

Prior to the assessment of mobile positioning data, it is important to describe the logic of data processing resulting in a reliable dataset. First, the steps of processing raw data are summarized which can be useful for similar research in the future. Then the steps of extending the database with user information and splitting it into international and domestic parts are described. For research purposes it is necessary to separate the domestic tourism-related movements from usual activities, for which the concept of network of usual places (NUP) is applied, based on the existing literature. Finally, a detailed description is provided regarding the filtering and cleaning of the two (domestic vs. international) datasets, removing possible errors (Figure 3).

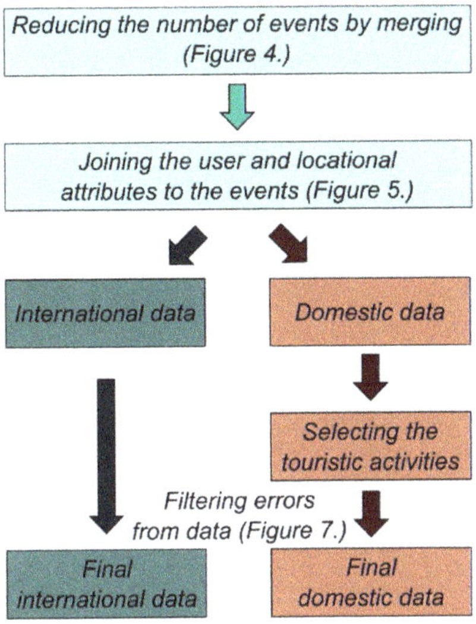

Figure 3. The steps of the data processing and cleaning.

3.1. Loading the Data into the SQL Database

As the first step of data processing, data were loaded into the sql database. To do this, a data collector script was used that downloads and uncompresses the daily batch files in chronological order, while splitting the data into rows and columns based on the separator characters. Then, the script inputs were applied in the Kafka stream processing application. The Kafka application uses a persistent storage, which means that it does not delete the raw data while processing it.

During data processing, the log and user information data were handled separately. Since the size of the user information files was relatively small, we were able to load them simply into a postgresql database using the raw data columns without any further modification. Next, the daily user information tables were merged into one table since the daily tables contain many repetitions. The resulting table contained all the 7.3 million users that appeared in the log events during the studied period.

To be able to load the huge log data into the database, the raw daily batch files had to be processed. Using the Kafka application, data were split into 12 partitions based on the user ids. This means that every event of a user was loaded into the same partition. A random sampling method of the user ids was used to partition them, based on the assumption that the ids are coming from a normal distribution.

Since the daily data batches are not grouped by user id, nor are they ordered by time, during processing a grouping and ordering step for each batch had to be included. This was done after partitioning the daily events based on the user ids, i.e., the events were grouped and ordered for each user chronologically in each partition. This made the processing significantly faster later, since the merge and heap ordering methods that were used have quadratic time complexity in terms of the number of events. Since the daily packages were processed in chronological order, in the final raw data tables the events of every user appeared in daily batches and were ordered by time.

When events were merged during the grouping and sorting process, the number of distinct events were also reduced for scalability reasons. Our aim was to merge several events into one robust event, which in itself contains all the information that the merged events contained, but in a denser format. Since the whole time series of a user could not be processed in one step due to the enormous size of the data, the merging was done right after processing each daily batch log file. In this way we were able to work with the grouped and sorted daily events while they were still in the Kafka stream and loaded only the merged data into the sql database. The idea was that successive events of a user that happen at the same place are merged. A counter was kept that tracked how much time the user spent at a specific location in order not to lose any information from the raw data.

3.2. Merging Events

The input data for our proposed algorithm is a time series of circles on the map defined by their center and their radius. The output is a time series of points on the map with an extra attribute that refers to the time spent at the corresponding location. Events were iterated through each user and each day when the user had at least one event.

Our algorithm used four variables. The first one is the 'current place circle', where the user is assumed to be currently. The second is a 'time spent' variable that measures how much time the user spent at the current place. Third, a 'temporary circle' was used that was compared to the current place variable. Finally, a 'temporary set of circles' was used to calculate the exact place of the user in the merged event (Figure 4).

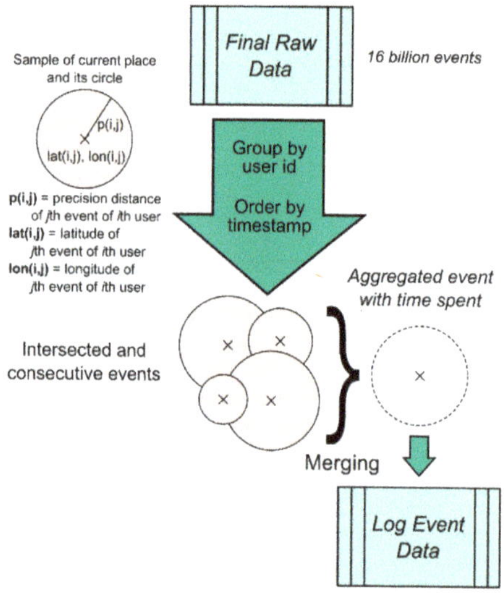

Figure 4. Visualization of merging algorithm.

For each user, the first circle was started with the first occurrence in the daily log data. The current place was set to the first circle and the time spent attribute to zero. This circle was also added to the temporary set of circles. Then, whether the subsequent circle of the user in the data intersected with the current place circle was checked (In practice, we used the maximum norm instead of the Euclidean norm, since it scales much better). If they intersected, this circle was merged into the current place; this circle was inserted into the temporary set of circles and updated the time spent attribute to the time difference between this and the current place event. The merging process was continued until a circle did not intersect with the current place circle or the user did not have any circles left. When this happened, first a new data point was created for the user, based on the current set of temporary circles and the time spent variable. The latitude and longitude coordinates for this new point were set to be the averages of the latitude and longitude coordinates of the temporary set of circles, respectively. The time spent attribute for this data point was set to be the current value of the time spent variable. Second, the current place variable was set to the temporary circle that did not intersect with the previous current place. The circles were also deleted from the temporary set of circles while adding this circle to the set, and the time spent variable was set to zero. The merging process was iterated for all the circles in the input for each user (Algorithm 1).

Algorithm 1. The pseudo code of the merging algorithm.

```
forall USER do:
  Input: USER_circles[]
  Output: USER_points_with_time_spent[]

  current_place_circle <- USER_circles [0]
  time_spent <- 0
  temporary_circle <- USER_circles[0]
  temporary_set_of_circles.append(USER_circles[0])

  for circle_i in USER_circles[1:] do:
    if circle_i INTERSECT temporary_circle do:
      temporary_set_of_circles.append(circle_i)
      time_spent += circle_i.time - temporary_circle.time
      temporary_circle <- circle_i
    else:
      current_location <- AVG(temporary_set_of_circles.location)
      USER_points_with_time_spent.append(current_location, time_spent)
      current_place_circle <- circle_i
      time_spent <- 0
      temporary_circle <- circle_i
      temporary_set_of_circles.clear()
      temporary_set_of_circles.append(circle_i)
    end if;
  end for;
  return USER_points_with_time_spent;
end forall;
```

Those events that were close to each other were merged, while keeping track of how much time the user spent at the place of the merged events. In summary, our algorithm compressed the data from 16 billion events to around 3.5 billion events, with almost a one to five compression rate. After the merging process the output data was loaded into the postgresql database and split into partitions.

3.3. Adding User Information and Interpreting Coordinates

Although the dataset became denser after the first stage of processing, further steps were needed to make it more compact and researchable. For example, the data types of the user id and the coordinate variables had to be changed, since they were not efficient, which made querying slow. Thus, a new user id was created that is a 32bit integer, instead of the large integer used previously; in addition, a 16 bit floating point number was used for the coordinates moving forward. The information contained in the log events and the user information tables were joined, in order to make querying possible based on, for example, the users' country of origin. At this point, although it was possible, no additional information on the users was added to the tables in order to keep the size relatively small.

The next step was to connect the coordinates of each event to a local municipality. This enabled later querying and filtering of events based on their location. This was done by using an additional data source that contained the polygon coordinates of all municipalities (3155) in Hungary and each event was assigned to a polygon of a settlement (i.e., municipality). Finally, based on the municipality where the event took place, the code of the county and the region of the event also added. These attributes helped us later to easily find all the tourists in a certain region or to identify tourist flows between counties or municipalities.

Consequently, there were 11 attributes for the final baseline sql tables; a new user id that is an 8-digit integer; the time of the event; the day of the event; the country of origin of the user; the time spent attribute; the latitude and longitude coordinates; the municipality, the postal code, the county and the region.

3.4. *Identifying Domestic and International Users*

After processing the raw data and loading the results into the sql database using the Kafka stream processing framework, the data processing was continued in the postgresql environment. First of all, the data were split into two groups according to the nationality of the users: domestic and international. This was important because the overwhelming majority of the domestic data was not tourism-related, while most of the foreign user data belonged to this category. Hence, these two groups had to be handled separately and processed differently.

Figure 5 shows that the datasets of domestic and foreign users differ significantly. First of all, the number of events generated by domestic users (3.5 billion) is almost a hundred-fold that of foreign users (37 million). However, the number of distinct users is almost the same in the two groups (3.8 and 3.5 million users, respectively). The average number of events per user is 947 with a median of 387 for the domestic users and 10.7 with a median of 4 for the foreign users. This difference is, however, quite intuitive, since domestic users were present in the country for the whole period, while international users were captured only for a shorter period of time. Domestic users had events in 68 different days on average with a median of 62 days, while the same value for international users was 4 and 2, respectively. The event density (i.e., the number of events per day) was 14 for the domestic and 3 for the international users. This means, that not only the number of events, but their density is also significantly higher among domestic users. More than half of the international users had only less than 5 events (Figure 6). For domestic users, the numbers were much more balanced, but in this case also more than one-third of the users had less than 180 events in total, which is less than one event per day.

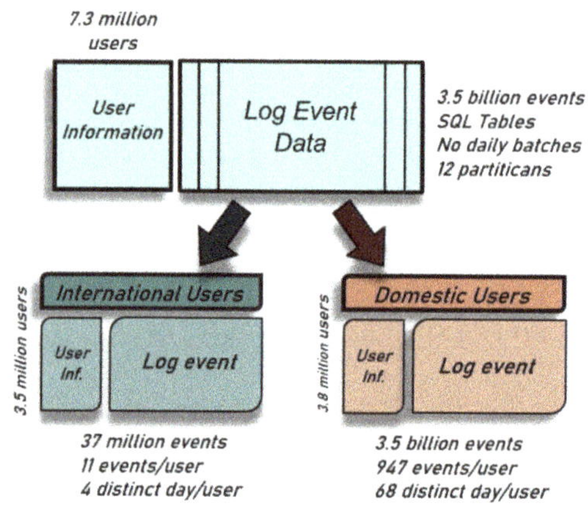

Figure 5. Sorting process of final raw dataset.

3.5. *Defining the Touristic Activities of Domestic Users*

Since our aim was to create a database that would provide meaningful insights on tourism-related movements, not only for international but also for domestic users, we had to clean the dataset further to get rid of the bad quality and noisy data. For domestic users, those intervals had to be defined that could be related to tourism activities and they had to be separated from the usual environment, following the UNWTO approach [59]. A conceptual framework has already been developed for the detection of the usual environment [65,66], which we also used to select tourism events, but we had to adapt it to our own database and purposes. A table was created to support the labelling as follows: first, the number of events was counted for each user as well as the number of distinct days at every location the user visited. Based on these data, we could define for each user the Usual Place ('UP'), Near Usual Place ('NUP') or Touristic Place ('T') and label the municipalities. The motivation of the labelling was to identify those places where a user usually spent longer time and more frequently, such as the users' home and workplaces or places they regularly visited. In defining these places, existing conceptual frameworks and the principles of the concept of network of usual places (NUP) [65,66] were considered, although the thresholds were determined based on the natural breaks shown in our dataset. These events should not be considered as touristic movements. At the same time, all other events that were not at or near usually visited places were considered as tourism-related activity.

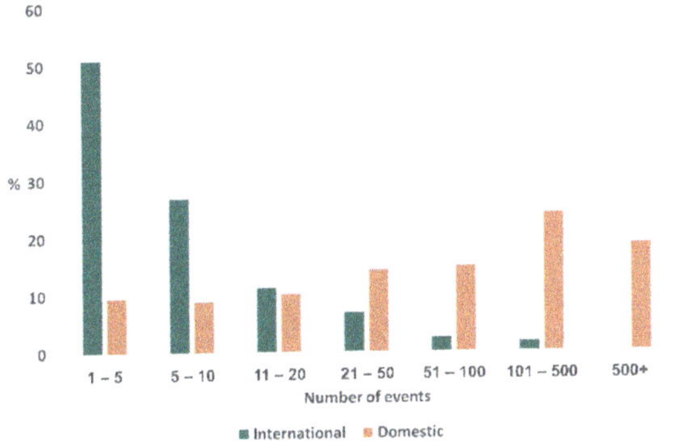

Figure 6. The distribution of the average number of events by international and domestic tourists.

For a user, we labelled a location UP if at least one of the following held:

- It was (one of) the users most visited place(s), based on the number of days the user spent there;
- The user had an event on at least 25 distinct days;
- The user had an event on at least 20% of his/her total number of days.

A location was labelled NUP, if it was less than 15 km away from the place labelled as UP. The intuition with the NUP label was that certain movements in the narrower home/work environment could not be considered as tourism-related events, such as, for example, if the user went home on a different route or visited a nearby shopping mall or sport facility. Every other place was labelled as T.

As a next step, in order to separate tourism-related activities of a user, events that took place in a location labelled either UP or NUP were deleted. Doing so, the users' timeline was split into touristic intervals, i.e., such intervals that were consecutive events with T labels. As a result, every interval could be considered as a possible touristic visit for each user. Each such visit could be considered a separate entity, similar to an international visit, and a special ID was assigned for each of them. This means that a distinct touristic visit could be examined without considering the users' other visits.

In order to detect the usual places properly, every user was filtered out with less than 100 distinct days. This was necessary to be able to work with only those users that had events throughout most of the examined period. Users with less distinct days also tended to have proportionally more touristic visits, because of the definition of the usual place.

Obviously, this labelling is not perfect, for example, rare occasions such as family visits or work trips are not distinguishable from touristic visits, since they have very similar patterns. Although the original aim of the research was precisely to explore unobserved tourism (which contains these activities), and in practice our methodology works well for identifying touristic visits, and because of the size of the domestic data, more sophisticated filtering methods would not be scalable.

As a final step of the labelling of domestic data, the resulting dataset was divided into two tables: same-day visitors and tourists, based on whether the user had at least one sleeping event (stay overnight) during the visit or not. For some research questions, these two tables were merged; however, for certain questions it was more meaningful to work with two separate tables.

3.6. Filtering the Data

The next step was to clean the datasets, extracting the touristic movements of domestic users and international visitors from roaming activities. For this, the following heuristic was used to filter bad quality users and events. First, the variables of 'time difference', 'distance' and 'velocity' were created. The time difference variable refers to the time difference between the current and the previous event. The distance variable is the distance between the locations of the current and the previous event. The velocity variable equals the distance divided by the time difference. These parameters were created to filter such errors in the dataset where the users travelled very long distances during a very short period of time. These events usually came in bunches for certain users, with the user jumping back-and-forth between places far from each other.

Next, different variables were created with the help of experts from the Hungarian Tourism Agency (HTA) that helped identify meaningful events at certain places, such as: '1 h spent', '4 h spent' and 'sleeping'. These variables were experimented and calculated beforehand based on the time spent variable, which indicated for a particular event how much time the user stayed at a certain place. If the time spent variable for an event was greater than 1 h or 4 h, then we assigned the value 'True' to these variables. The value 'False' was assigned otherwise. As for the sleeping event, we assigned 'True' to either the very first event after 4 am in the morning or the last event before 4 am, based on which place is more common for the users, based in turn on his/her visit. If these two events were at the same place then the user obviously slept there. However, when the place of these two events differed, it was not obvious which place should be chosen as a sleeping place. In these cases, the heuristic above proved to work the best in finding the sleeping place.

These variables could be used in further steps of the research to answer the questions: 'Where did those people stay overnight who spent at least 4 h in a specific location but did not sleep there?' or 'Have the tourists staying at a particular location taken same-day trips in neighboring places?'.

As a first step, those visits were filtered for which at least one of the following criteria was fulfilled:

- It has less than five events,
- It does not contain at least one event where the 1 h spent variable is true,
- The difference between the latest and earliest event is less than 12 h,
- The speed variable in one of the events exceeds 500 km/h,
- The speed variable exceeds 300 km/h and the time difference variable is at most 3 in multiple events.

The last two conditions aimed to filter visits that contained errors. On the one hand, where the speed exceeded 500 km/h, the user travelled way too fast for the event not to be an error. On the other hand, the last condition filtered out those back-and-forth movements that usually are the result of two users having the same user id.

All the thresholds above are purely based on heuristics and can be changed when using other datasets. At the same time, these numbers were those that worked the best for us on this particular dataset in terms of the resulting data quantity and quality trade-off. To answer specific research questions, obviously other threshold settings can be used. For example, to examine a place with only few tourists, the settings can be loosened, allowing consideration of more users in the filtered dataset. However, analyzing such problems where the number of users was big enough, stricter filtering settings could be applied for the sake of better data quality.

For the international users, our assumption was that, except for some special cases (e.g., truck drivers), every visit can be considered as tourism-related activity. However, similarly to the domestic users, we had to filter the data to keep only those that reflect tourist visits. Every international user was disregarded for whom at least one of the following was fulfilled:

- The user had less than five events in total,
- The user had an event per day ratio less than 1.5,
- The user had events on more than 25 distinct days,
- The user did not have a sleeping event.

The first two conditions are straightforward data quality requirements. The third one makes sure that users who stay in the country for a longer period of time (e.g., Erasmus students) are not considered, since they are not tourists by strict definition. Finally, the last criterion filters out the same-day visitors. Separate tables were created for the users with at least 25 distinct days and one day tourists in order to be able to gain insight also into their spatial behavior [6]. Figure 7 shows the overall event distribution for the filtered international and domestic events and users within Hungary.

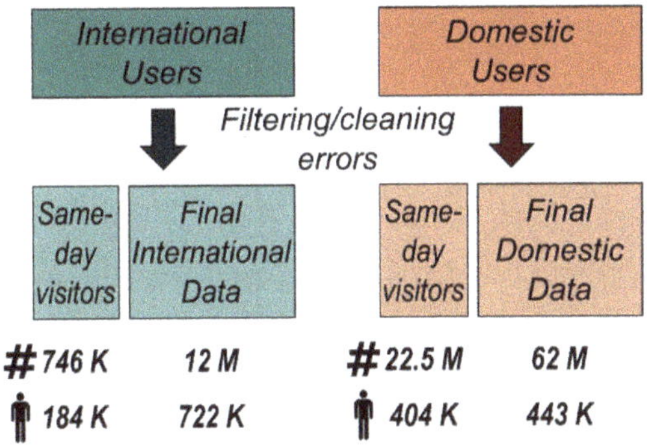

Figure 7. The cleaned event and user numbers of the datasets (# represents the number of events).

3.7. Validation of the Final Database

In order to assess the reliability and representativeness of the filtered database, official tourism-related statistics of the Hungarian Central Statistical Office (HCSO) were used. During the assessment, the official data of guests and spent guest nights of a given period at the municipal level was compared to the dataset based on MPD. While the former contains only data of tourists who were officially registered by commercial accommodation services, the latter includes the data of those who used the services of the provider (between 27 and 45% of all domestic mobile users [67]). However, we must keep in mind that in the case of international tourists it is not straightforward which network their mobile phone connects to when they arrive to a given country. Therefore, the shares of the total visitors that connected to our network provider for each country had to be calculated. In addition to representativeness at the macro-scale, spatial and temporal validation of domestic and international tourists is also important. For this purpose, correlation and regression coefficients were calculated regarding the spatial (settlement level) and temporal (monthly) distribution of guests and guest nights derived from the HCSO and the MPD datasets, in order to determine to what extent the filtered database differs from official statistics. However, certain biases do not necessarily mean that the database is of bad quality, as several tourism-related activities can be detected with the MPD, which are not included in commercial statistics (e.g., unobserved tourism which can be linked with family relations, or private accommodation services, or same-day visits).

4. Results and Discussion: Assessment of the Final Database

In this section we aim to assess the reliability and representativeness of the final database. For the validation, we use official statistics collected by the Hungarian Central Statistical Office at the settlement level. Data on international and domestic tourists staying at commercial accommodation establishments (hotels, pensions, camp-sites, hostels, bed-and-breakfasts, etc.) cover the period 1 July–31 December 2018. The share of officially recorded international tourists captured by MPD differs very much by country (Table 1). According to our dataset, more than two thirds of tourists arriving from France, Cyprus and Bulgaria were connected to the network of our provider, whereas only 6% from the USA, and even fewer from Japan and Norway. This means that MPD of a single service provider can be highly representative for some countries, whereas it has limitations for others. Our estimation is that, below 20% representativeness, the data of foreign visitors should be handled with caution. For a full coverage of international tourism, obviously, data of all mobile network operators (in Hungary there are three) should be collected and processed, which means not only a methodological, but also a financial and organizational challenge.

Table 1. Share of tourists by countries tracked by MPD.

Country	Ratio (%)	Country	Ratio (%)	Country	Ratio (%)
France	77.38	Estonia	36.41	Turkey	23.7
Cyprus	74.04	Croatia	35.52	Israel	20.29
Bulgaria	73.36	Switzerland	35.16	Serbia	16.66
Poland	58.28	Ukraine	35.07	United Kingdom	14.94
Belgium	53.13	Latvia	34.06	Sweden	12.73
Slovenia	50.3	Italy	33.68	Canada	11.81
Portugal	47.39	India	33.18	Brazil	9.87
South Africa	46.75	Russia	32.2	South Korea	9.67
Romania	41.55	Ireland	29.59	Australia	8.13
Iceland	39.49	Slovakia	28.99	Finland	8.03
Malta	38.89	Denmark	28.94	China	7.28
Austria	38.8	Czech Republic	27.66	USA	6.34
Greece	38.05	Spain	26.38	Japan	2.15
Netherlands	37.38	Germany	25.12	Norway	1.16

As a next step, the filtered datasets (both domestic and international tourists) were evaluated with correlation coefficients at the settlement level (Table 2). Regarding the number of domestic tourists, the correlation coefficient is 0.733 and the slope of the regression line is 0.24. The latter means that, on average, 24% of tourists were detected using MPD, while the former shows how accurately the tourist pattern was captured by the filtered database when compared to the official statistics. The degree of correlation can be considered good, especially since the figure shows that the MPD tended to overestimate the number of tourists at those settlements where the number of visitors was low, and hence they are probably more exposed to unobserved tourism (i.e., tourists not appearing in accommodation statistics) [68]. The values were similar in the case of the number of guest nights, but this was estimated with a slightly higher efficiency ($R = 0.766$), but to a lesser degree (19%). With the exclusion of those municipalities where the number of guests was zero in commercial accommodation, but the MPD detected tourist movements, the efficiency of the estimation increases ($R = 0.771$ and $R = 0.779$).

The spatial concentration of international tourists is very high in Hungary, but Budapest, the capital city, attracts a disproportionately large share of foreign visitors, which thus, distorts the correlation and regression calculations which are highly sensitive to outliers. Therefore, the efficiency of the estimation was also examined without Budapest. Nonetheless, the calculated correlation coefficients and the slope of the regression lines also provided useful information, as according to both official statistics and MPD more than half of the international tourists concentrated in the capital city and 21% of these guests and 17% of guest-nights were captured by MPD (Table 2). These values significantly decrease with the exclusion of Budapest (16% and 7%, respectively).

Table 2. Correlation coefficients.

	R	Constant	Slope
Domestic tourists	0.733 **	708.793	0.24
Domestic tourists (without zero values)	0.771 **	1500.525	0.23
Domestic nights	0.766 **	1417.592	0.19
Domestic nights (without zero values)	0.779 **	2117.036	0.19
International tourists	0.998 **	109.760	0.21
International tourists without Budapest	0.764 **	137.980	0.16
International nights	0.996 **	9.723	0.17
International nights without Budapest	0.711 **	194.776	0.07

** Correlation is significant at the 0.01 level (2-tailed).

We were also interested in the geographical pattern of unobserved tourism. Based on Figure 8, our preliminary assumption was that unobserved tourists concentrate predominantly outside major tourist hotspots. To test this hypothesis, we separated the data of Budapest and the 100 biggest destinations based on the number of officially registered tourists (both international and domestic) within the country. As Table 3 shows, the 101 most important tourist destinations concentrated 93.5% of the international and 76.6% of the domestic tourists officially registered in the country in the second half of 2018. Our MPD dataset captured 28.2% of the officially recorded international visitors in the period, which is very similar to the market share of the provider. However, the spatial distribution of international SIM cards was significantly different from national statistics on tourism, since, as opposed to 93.5%, only 70.9% of the international tourists could be linked with major tourist destinations. In our opinion, a substantial part of international visitors travelling to regions/settlement less frequented by tourism normally stay with friends, relatives, or at private accommodation facilities. In this way most of them remain hidden for tourism statistics. According to our data, the share of unobserved visitors is even higher in domestic tourism. There were 3.7 million domestic tourists in Hungary in the studied period, out of which MPD recorded 3.1 million, again predominantly outside the most frequented tourist destinations. These results shed light on the limitations of accommodation statistics and confirm the usefulness of MPD as a proxy for the actual number of tourists in a given location/region.

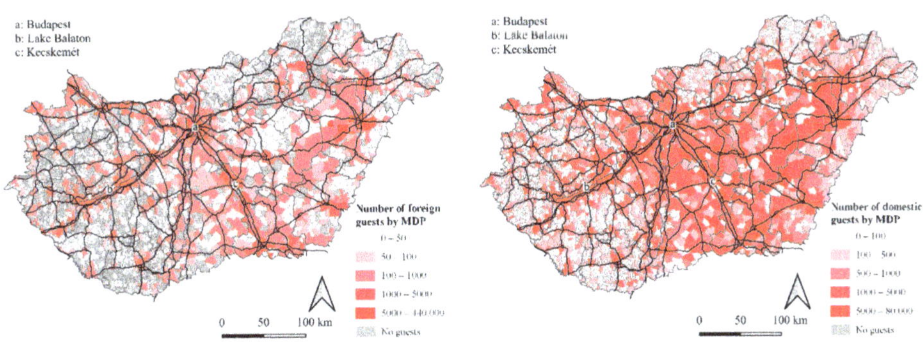

(**a**) (**b**)

Figure 8. Spatial distribution of international (**a**) and domestic (**b**) tourists.

Table 3. Official and unobserved tourism in Hungary.

	Major Tourist Destinations		Rest of the Country		Total
	Number	Ratio (%)	Number	Ratio (%)	
Number of municipalities	101	3.2	3054	96.8	3155
Number of domestic tourists by official statistics	2,848,648	76.6	871,708	23.4	3,720,356
Number of domestic tourists by MPD	942,569	30.1	2,189,202	69.9	3,131,771
Number of international tourists by official statistics	3,097,826	93.5	215,854	6.5	3,313,680
Number of international tourists by MPD	664,806	70.9	272,533	29.1	937,339

The correlation coefficients calculated on a monthly basis reflect the spatial accuracy of the estimation in time. Based on this calculation, we can say that there was no significant change in the case of domestic tourism, i.e., the MPD database estimated the official statistics with similar efficiency in every month (Table 4). However, the accuracy of international tourism estimates is steadily declining over time in terms of both the number of guests and the guest nights. This may be the result of the high seasonality of international tourism in Hungary [69], i.e., it is easier to create a geographically accurate database from MPD for a larger number of tourists in the summer period (Figure 9). Thus, for the late fall and winter months, our dataset is less accurate.

Table 4. Correlations between the spatial distribution of tourism data by official statistics and MPD.

Time Period	Domestic Tourists	Domestic Nights	International Tourists *	International Nights *
All	0.733 **	0.766 **	0.764 **	0.711 **
July	0.738 **	0.735 **	0.790 **	0.785 **
August	0.768 **	0.771 **	0.786 **	0.746 **
September	0.751 **	0.748 **	0.769 **	0.676 **
October	0.761 **	0.767 **	0.717 **	0.638 **
November	0.766 **	0.757 **	0.693 **	0.629 **
December	0.803 **	0.777 **	0.613 **	0.540 **

* Without Budapest. ** Correlation is significant at the 0.01 level (2-tailed).

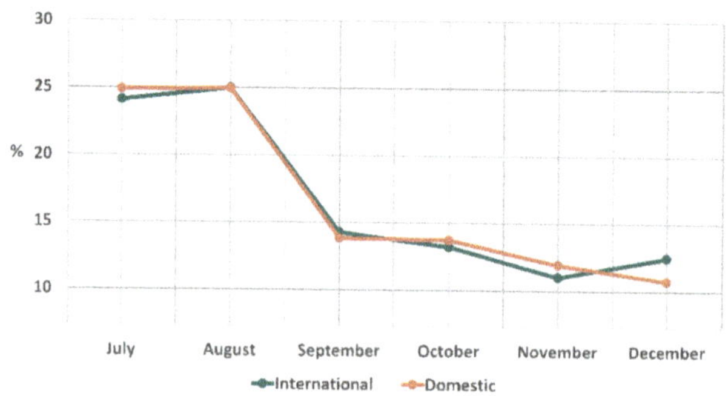

Figure 9. Temporal distribution of users by MDP.

The difference between the number of guests predicted by the regression line and detected by the MPD has distinct spatial variations for both domestic and international tourists (Figure 10a,c). In the case of domestic tourists, the inaccuracy of the estimate is minimal for major tourist destinations (i.e., regional centres), while in the case of medium-size towns and smaller settlements, where the phenomenon of unobserved tourism is more likely, the number of guests was overestimated (Figure 10a). This can also be observed in the case of Lake Balaton (for the estimation of international tourists too), where a significant part of guest nights is not recorded in commercial accommodations [70,71]. The spatial configuration of international tourists shows a similar but much more concentrated pattern, where flows related to the concentrations of foreign direct investments (i.e., business and logistic centres) may play a role in the overestimation (for example, the Mercedes plant in Kecskemét) (Figure 10c). In addition, there are several settlements in Hungary without commercial accommodation services (where no guests can officially be registered), but we could detect tourists on the basis of MPD (Figure 10b,d). Out of the 3155 Hungarian settlements, international tourists were captured in 1531, and domestic tourists in 1404 locations, i.e., 48.5% and 44,5% of the settlements, respectively. According to our understanding, they can be defined as unobserved tourists. This phenomenon is more prevalent in the central part of the country, in the wider urban agglomeration of Budapest, and in the eastern part of the country [72]. In our opinion, most of these visitors are either people staying with relatives or friends, or use some alternative (non-registered) accommodation, or they are transit travelers (especially near the border crossings) (Figure 10b).

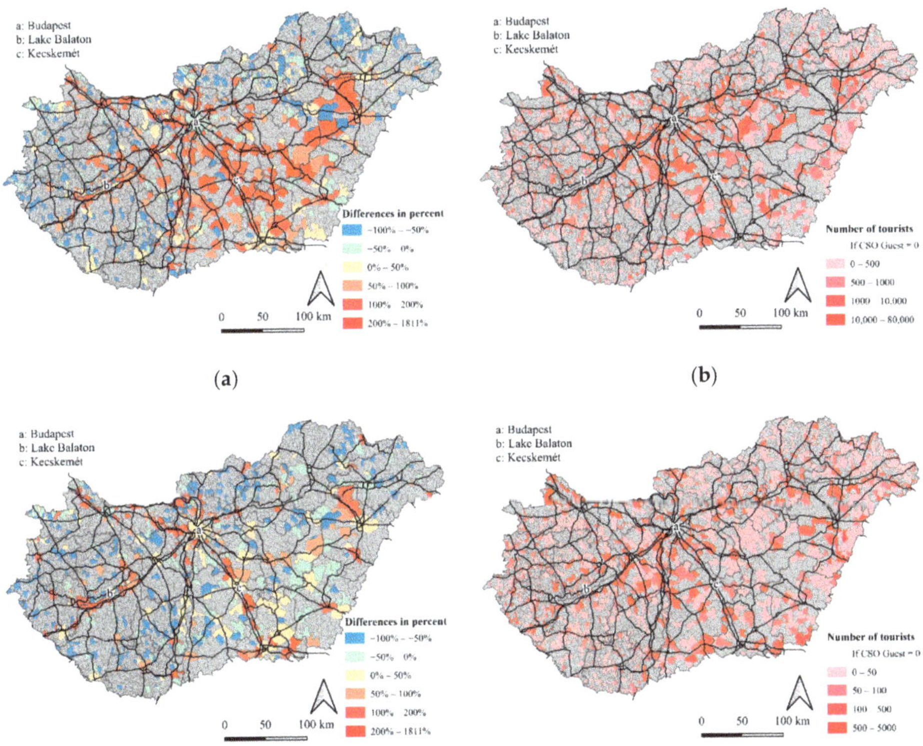

(**a**) (**b**)

(**c**) (**d**)

Figure 10. The spatiality of under- and overestimation in the case of domestic (**a**) and international (**c**) tourists and unobserved tourists in the case of domestic (**b**) and international (**d**).

5. Conclusions

Even though this research had clear methodological considerations, the presented results also have some conceptual and policy implications. As was demonstrated, official statistics on tourism have many shortcomings and ambiguities that can be eliminated by new research technologies, such as mobile positioning data. Knowledge of the real-time activities of tourists at a certain location and their spatio-temporal behavior can provide much added value for sustainable tourism policy and destination management [73]. As was highlighted in the introduction of this paper, despite the growing importance of Big Data sources in tourism research, a comprehensive overview of the processing and filtering of mobile positioning data to capture various unobserved forms of tourist movements is by and large missing from the literature. Therefore, the main aim of this paper was to present a case-study on the processing and cleaning mechanisms of mobile positioning data to gain reliable information about the actual number of visitors (both domestic and international) at a certain location. Such data sources can extend the knowledge provided by accommodation statistics, improving the timeliness and spatiotemporal resolution of the data.

While the majority of previous studies in the field used only data for a shorter period of time (e.g., two weeks or the time of specific events) [74] or focused exclusively on inbound international tourism [75], this research was based on the full dataset of one of Hungary's mobile network operators for six months in 2018. Our main aim was to shed light on the challenges and opportunities of MPD processing, especially regarding statistically unobserved forms of tourism (e.g., day trippers, family visitors), and to provide possible benchmarks and solutions. On the one hand, we presented solutions to challenges caused by the size and structure of the dataset (e.g., data management, filtering, cleaning, etc.), and on the other hand, we demonstrated how a more accurate picture of the spatio-temporal dimensions of tourist movements can be achieved. The novelty of the research lays in (1) the definition of thresholds for data filtering, (2) a more sophisticated and accurate definition of possible errors in the dataset, (3) the introduction of new variables that facilitate data reduction in tourism-related analyses, (4) an accurate identification of unobserved (same day) tourism. The definition of the thresholds is well-grounded due to the size and the time-span of the dataset. After defining the key variables, it was possible to identify and analyze the flows of various user groups (i.e., domestic and international tourists). As was shown with the inclusion of the 1-h and 4-h spent variables (and the merging of these events), a significant data reduction could be achieved, while the new dataset offered good opportunities for measuring hidden forms of touristic activities (e.g., cross border trips, one day visitors) [76]. Thus, we can say that through the application of our approach, the study of various forms of unobserved tourism at the intra-urban and intra-regional level becomes possible.

Furthermore, the delimitation of secondary destinations without officially registered tourist visitors became possible and the spatio-temporal characteristics of the related tourist flows could also be measured [6]. Such locations can gain useful information about the actual number of tourists (both international and domestic) visiting them, their seasonality and spatio-temporal behavior. This knowledge can help less frequented tourist destinations elaborate their own tourism policies and pursue sustainable tourism planning and destination development. Altogether, the methodological procedure presented in this study can help increase the efficiency and reliability of the use of MPD and contribute to the standardization of the methods used in Big Data tourism research. The methodology can support the analysis of local tourism flows; thus, it may lead to better management and more optimal policy formulations at the local level. At the macro-geographical (national, regional) scale, the analysis of MPD with the suggested methodology can help track tourism flows more accurately, identifying secondary tourism hot spots that remain very often hidden in statistics [16]. Challenges related to the recent COVID-19 pandemic also increased the importance of such analyses [15,77]. Tracking the tourism-related spread of infections or monitoring the effects of changing travel regulations on tourism activities has great potential in future studies [2]. Future research could also refine the data cleaning process, for example, filtering transit traffic (e.g., truck drivers) that cross the country without any tourism-related activity.

As with other research in the field, this study has also certain limitations. First of all, we used data only from one network provider and the clienteles of the other two operators have not been covered. As was shown by our assessment, using the data of one single operator may lead to bias. This is important to note because the structure of the data and the problems arising during data processing may be different in the case of other providers. In addition, the sending countries are not equally represented either, because this depends on the provider's international collaborations, so this distortion should be considered in the interpretation of results. Furthermore, the simultaneous use of data from various providers may positively affect the reliability and comparability of such data. Last, but not least, while MPD offers useful insights to tourism flows, their underlying causes and motivations remain unexplored, thus the presented Big Data method should be extended by conventional methodologies (e.g., surveys) to obtain a comprehensive picture of the driving forces of tourism related mobility.

Author Contributions: Conceptualization, Á.E., T.K. and Z.K.; methodology, Á.E. and T.K.; software, Á.E.; validation, Á.E., L.B. and T.K.; formal analysis, T.K.; investigation, Á.E. and T.K.; resources, Z.K. and L.K.; data curation, L.K.; writing—original draft preparation, Á.E., L.B. and T.K.; writing—review and editing, Z.K. and T.K.; visualization, Á.E. and T.K.; supervision, Z.K.; project administration, Z.K. and L.K.; funding acquisition, Z.K. and L.K. All authors have read and agreed to the published version of the manuscript.

Funding: This research was funded by the National Research, Development and Innovation Fund of the Ministry of Innovation and Technology, Hungary grant number NKFIH-1279-2/2020 and TKP2021-NVA-09; and the Hungarian Scientific Research Fund (OTKA), grant number K135546.

Institutional Review Board Statement: Not applicable.

Informed Consent Statement: Not applicable.

Data Availability Statement: Due to confidentiality agreements, supporting data can only be made available to collaborative researchers subject to a non-disclosure agreement. Details of the data and how to request access are available from László Könnyid at Hungarian Tourism Agency.

Acknowledgments: The authors are grateful to the Hungarian Tourism Agency for providing the mobile-cell data. This research was conducted in a collaboration between the University of Szeged and the Hungarian Tourism Agency. Project no. TKP2021-NVA-09 has been implemented with the support provided by the Ministry of Innovation and Technology of Hungary from the National Research, Development and Innovation Fund, financed under the TKP2021-NVA funding scheme.

Conflicts of Interest: The authors declare no conflict of interest. The funders had no role in the design of the study; in the collection, analyses, or interpretation of data; in the writing of the manuscript, or in the decision to publish the results.

References

1. UNWTO. International Tourism Growth Continues to Outpace the Global Economy. Available online: https://www.unwto.org/%0Ainternational-tourism-growth-continues-to-outpace-the-economy (accessed on 16 April 2021).
2. Boros, L.; Dudás, G.; Kovalcsik, T. The effects of COVID-19 on Airbnb. *Hung. Geogr. Bull.* **2020**, *69*, 363–381. [CrossRef]
3. Buhalis, D.; Law, R. Progress in information technology and tourism management: 20 years on and 10 years after the Internet—The state of eTourism research. *Tour. Manag.* **2008**, *29*, 609–623. [CrossRef]
4. Roberts, J.H.; Kayande, U.; Stremersch, S. From academic research to marketing practice: Some further thoughts. *Int. J. Res. Mark.* **2014**, *31*, 144–146. [CrossRef]
5. e Silva, F.B.; Marín Herrera, M.A.; Rosina, K.; Barranco, R.R.; Freire, S.; Schiavina, M. Analysing spatiotemporal patterns of tourism in Europe at high-resolution with conventional and big data sources. *Tour. Manag.* **2018**, *68*, 101–115. [CrossRef]
6. Kovács, Z.; Vida, G.; Elekes, Á.; Kovalcsik, T. Combining Social Media and Mobile Positioning Data in the Analysis of Tourist Flows: A Case Study from Szeged, Hungary. *Sustainability* **2021**, *13*, 2926. [CrossRef]
7. Garau-Vadell, J.B.; Gutierrez-Taño, D.; Diaz-Armas, R. Economic crisis and residents' perception of the impacts of tourism in mass tourism destinations. *J. Destin. Mark. Manag.* **2018**, *7*, 68–75. [CrossRef]
8. Lee, J.W.; Syah, A.M. Economic and Environmental Impacts of Mass Tourism on Regional Tourism Destinations in Indonesia. *J. Asian Financ. Econ. Bus.* **2018**, *5*, 31–41. [CrossRef]
9. Marzouki, M.; Froger, G.; Ballet, J. Ecotourism versus Mass Tourism. A Comparison of Environmental Impacts Based on Ecological Footprint Analysis. *Sustainability* **2012**, *4*, 123–140. [CrossRef]
10. Lee, J.W.; Brahmasrene, T. Investigating the influence of tourism on economic growth and carbon emissions: Evidence from panel analysis of the European Union. *Tour. Manag.* **2013**, *38*, 69–76. [CrossRef]

11. Scott, D.; Peeters, P.; Gössling, S. Can tourism deliver its "aspirational" greenhouse gas emission reduction targets? *J. Sustain. Tour.* **2010**, *18*, 393–408. [CrossRef]
12. Lenzen, M.; Sun, Y.-Y.; Faturay, F.; Ting, Y.-P.; Geschke, A.; Malik, A. The carbon footprint of global tourism. *Nat. Clim. Chang.* **2018**, *8*, 522–528. [CrossRef]
13. Gursoy, D.; Chi, C.G.; Dyer, P. Locals' Attitudes toward Mass and Alternative Tourism: The Case of Sunshine Coast, Australia. *J. Travel Res.* **2010**, *49*, 381–394. [CrossRef]
14. Martínez-Garcia, E.; Raya, J.M.; Majó, J. Differences in residents' attitudes towards tourism among mass tourism destinations. *Int. J. Tour. Res.* **2017**, *19*, 535–545. [CrossRef]
15. Więckowski, M. Will the Consequences of COVID-19 Trigger a Redefining of the Role of Transport in the Development of Sustainable Tourism? *Sustainability* **2021**, *13*, 1887. [CrossRef]
16. Więckowski, M.; Saarinen, J. Tourism transitions, changes, and the creation of new spaces and places in central-eastern Europe. *Geogr. Pol.* **2019**, *92*, 369–377. [CrossRef]
17. Lane, B. Sustainable rural tourism strategies: A tool for development and conservation. *J. Sustain. Tour.* **1994**, *2*, 102–111. [CrossRef]
18. Law, R.; Buhalis, D.; Cobanoglu, C. Progress on information and communication technologies in hospitality and tourism. *Int. J. Contemp. Hosp. Manag.* **2014**, *26*, 727–750. [CrossRef]
19. Eugenio-Martin, J.L.; Cazorla-Artiles, J.M.; González-Martel, C. On the determinants of Airbnb location and its spatial distribution. *Tour. Econ.* **2019**, *25*, 1224–1244. [CrossRef]
20. Sarrión-Gavilán, M.D.; Benítez-Márquez, M.D.; Mora-Rangel, E.O. Spatial distribution of tourism supply in Andalusia. *Tour. Manag. Perspect.* **2015**, *15*, 29–45. [CrossRef]
21. Roman, M.; Roman, M.; Niedziółka, A. Spatial Diversity of Tourism in the Countries of the European Union. *Sustainability* **2020**, *12*, 2713. [CrossRef]
22. Rátz, T.; Michalkó, G.; Keszeg, R. Educational tourism and nation building: Cross-border school trips in the carpathian basin. *Hung. Geogr. Bull.* **2020**, *69*, 57–71. [CrossRef]
23. Tóth, B.; Vida, G.; Lados, G.; Kovács, Z. The Potentials of Cross-Border Tourism Development in the Lower-Tisa Valley. *Geoj. Tour. Geosites* **2020**, *28*, 360–375. [CrossRef]
24. Buhalis, D.; Licata, M.C. The future eTourism intermediaries. *Tour. Manag.* **2002**, *23*, 207–220. [CrossRef]
25. Frihida, A.; Marceau, D.J.; Thériault, M. Development of a Temporal Extension to Query Travel Behavior Time Paths Using an Object-Oriented GIS. *Geoinformatica* **2004**, *8*, 211–235. [CrossRef]
26. Lew, A.; McKercher, B. Modeling Tourist Movements. *Ann. Tour. Res.* **2006**, *33*, 403–423. [CrossRef]
27. Noguera, J.M.; Barranco, M.J.; Segura, R.J.; Martínez, L. A mobile 3D-GIS hybrid recommender system for tourism. *Inf. Sci.* **2012**, *215*, 37–52. [CrossRef]
28. García-Palomares, J.C.; Gutiérrez, J.; Mínguez, C. Identification of tourist hot spots based on social networks: A comparative analysis of European metropolises using photo-sharing services and GIS. *Appl. Geogr.* **2015**, *63*, 408–417. [CrossRef]
29. Cetin, M.; Zeren, I.; Sevik, H.; Cakir, C.; Akpinar, H. A study on the determination of the natural park's sustainable tourism potential. *Environ. Monit. Assess.* **2018**, *190*, 167. [CrossRef]
30. Timothy, D.J.; Michalkó, G.; Irimiás, A. Unconventional Tourist Mobility: A Geography-Oriented Theoretical Framework. *Sustainability* **2022**, *14*, 6494. [CrossRef]
31. Hardy, A.; Hyslop, S.; Booth, K.; Robards, B.; Aryal, J.; Gretzel, U.; Eccleston, R. Tracking tourists' travel with smartphone-based GPS technology: A methodological discussion. *Inf. Technol. Tour.* **2017**, *17*, 255–274. [CrossRef]
32. McKercher, B.; Shoval, N.; Ng, E.; Birenboim, A. First and Repeat Visitor Behaviour: GPS Tracking and GIS Analysis in Hong Kong. *Tour. Geogr.* **2012**, *14*, 147–161. [CrossRef]
33. Li, J.; Xu, L.; Tang, L.; Wang, S.; Li, L. Big data in tourism research: A literature review. *Tour. Manag.* **2018**, *68*, 301–323. [CrossRef]
34. Gombos, K.; Herczeg, R.; Erőss, B.; Kovács, S.Z.; Uzzoli, A.; Nagy, T.; Kiss, S.; Szakács, Z.; Imrei, M.; Szentesi, A.; et al. Translating Scientific Knowledge to Government Decision Makers Has Crucial Importance in the Management of the COVID-19 Pandemic. *Popul. Health Manag.* **2021**, *24*, 35–45. [CrossRef] [PubMed]
35. Ahas, R.; Mark, Ü. Location based services—New challenges for planning and public administration? *Futures* **2005**, *37*, 547–561. [CrossRef]
36. Ratti, C.; Frenchman, D.; Pulselli, R.M.; Williams, S. Mobile Landscapes: Using Location Data from Cell Phones for Urban Analysis. *Environ. Plan. B Plan. Des.* **2006**, *33*, 727–748. [CrossRef]
37. Ahas, R.; Aasa, A.; Mark, Ü.; Pae, T.; Kull, A. Seasonal tourism spaces in Estonia: Case study with mobile positioning data. *Tour. Manag.* **2007**, *28*, 898–910. [CrossRef]
38. Ahas, R.; Silm, S.; Järv, O.; Saluveer, E.; Tiru, M. Using Mobile Positioning Data to Model Locations Meaningful to Users of Mobile Phones. *J. Urban Technol.* **2010**, *17*, 3–27. [CrossRef]
39. Csáji, B.C.; Browet, A.; Traag, V.A.; Delvenne, J.-C.; Huens, E.; Van Dooren, P.; Smoreda, Z.; Blondel, V.D. Exploring the mobility of mobile phone users. *Phys. A Stat. Mech. Appl.* **2013**, *392*, 1459–1473. [CrossRef]
40. Bojic, I.; Massaro, E.; Belyi, A.; Sobolevsky, S.; Ratti, C. Choosing the Right Home Location Definition Method for the Given Dataset. In *Social Informatics, 7th International Conference, SocInfo 2015, Beijing, China, 9–12 December 2015*; Springer: Cham, Switzerland, 2015; pp. 194–208.

41. Xu, Y.; Shaw, S.-L.; Zhao, Z.; Yin, L.; Fang, Z.; Li, Q. Understanding aggregate human mobility patterns using passive mobile phone location data: A home-based approach. *Transportation* **2015**, *42*, 625–646. [CrossRef]
42. Vanhoof, M.; Reis, F.; Ploetz, T.; Smoreda, Z. Assessing the Quality of Home Detection from Mobile Phone Data for Official Statistics. *J. Off. Stat.* **2018**, *34*, 935–960. [CrossRef]
43. Pappalardo, L.; Ferres, L.; Sacasa, M.; Cattuto, C.; Bravo, L. Evaluation of home detection algorithms on mobile phone data using individual-level ground truth. *EPJ Data Sci.* **2021**, *10*, 29. [CrossRef]
44. Pappalardo, L.; Simini, F.; Rinzivillo, S.; Pedreschi, D.; Giannotti, F.; Barabási, A.-L. Returners and explorers dichotomy in human mobility. *Nat. Commun.* **2015**, *6*, 8166. [CrossRef] [PubMed]
45. Diao, M.; Zhu, Y.; Ferreira, J.; Ratti, C. Inferring individual daily activities from mobile phone traces: A Boston example. *Environ. Plan. B Plan. Des.* **2016**, *43*, 920–940. [CrossRef]
46. Jiang, S.; Ferreira, J.; Gonzalez, M.C. Activity-Based Human Mobility Patterns Inferred from Mobile Phone Data: A Case Study of Singapore. *IEEE Trans. Big Data* **2017**, *3*, 208–219. [CrossRef]
47. Mamei, M.; Bicocchi, N.; Lippi, M.; Mariani, S.; Zambonelli, F. Evaluating Origin–Destination Matrices Obtained from CDR Data. *Sensors* **2019**, *19*, 4470. [CrossRef] [PubMed]
48. Yin, L.; Lin, N.; Zhao, Z. Mining Daily Activity Chains from Large-Scale Mobile Phone Location Data. *Cities* **2021**, *109*, 103013. [CrossRef] [PubMed]
49. Grassini, L.; Dugheri, G. Mobile phone data and tourism statistics: A broken promise? *Natl. Account. Rev.* **2021**, *3*, 50–68. [CrossRef]
50. Raun, J.; Ahas, R.; Tiru, M. Measuring tourism destinations using mobile tracking data. *Tour. Manag.* **2016**, *57*, 202–212. [CrossRef]
51. Nilbe, K.; Ahas, R.; Silm, S. Evaluating the Travel Distances of Events Visitors and Regular Visitors Using Mobile Positioning Data: The Case of Estonia. *J. Urban Technol.* **2014**, *21*, 91–107. [CrossRef]
52. Kuusik, A.; Tiru, M.; Ahas, R.; Varblane, U. Innovation in destination marketing: The use of passive mobile positioning for the segmentation of repeat visitors in Estonia. *Balt. J. Manag.* **2011**, *6*, 378–399. [CrossRef]
53. Tiru, M.; Kuusik, A.; Lamp, M.L.; Ahas, R. LBS in marketing and tourism management: Measuring destination loyalty with mobile positioning data. *J. Locat. Based Serv.* **2010**, *4*, 120–140. [CrossRef]
54. Saluveer, E.; Raun, J.; Tiru, M.; Altin, L.; Kroon, J.; Snitsarenko, T.; Aasa, A.; Silm, S. Methodological framework for producing national tourism statistics from mobile positioning data. *Ann. Tour. Res.* **2020**, *81*, 102895. [CrossRef]
55. Arbulú, I.; Razumova, M.; Rey-Maquieira, J.; Sastre, F. Can domestic tourism relieve the COVID-19 tourist industry crisis? The case of Spain. *J. Destin. Mark. Manag.* **2021**, *20*, 100568. [CrossRef]
56. Lambiotte, R.; Blondel, V.D.; de Kerchove, C.; Huens, E.; Prieur, C.; Smoreda, Z.; Van Dooren, P. Geographical dispersal of mobile communication networks. *Phys. A Stat. Mech. Appl.* **2008**, *387*, 5317–5325. [CrossRef]
57. Šveda, M.; Barlík, P. Daily commuting in the Bratislava metropolitan area: Case study with mobile positioning data. *Pap. Appl. Geogr.* **2018**, *4*, 409–423. [CrossRef]
58. Mamei, M.; Colonna, M. Analysis of tourist classification from cellular network data. *J. Locat. Based Serv.* **2018**, *12*, 19–39. [CrossRef]
59. UNWTO. *International Recommendations for Tourism Statistics 2008*; United Nations Publication: New York, NY, USA, 2008; ISBN 9789211615210.
60. Cantis, S.D.; Parroco, A.M.; Ferrante, M.; Vaccina, F. Unobserved tourism. *Ann. Tour. Res.* **2015**, *50*, 1–18. [CrossRef]
61. Nyns, S.; Schmitz, S. Using mobile data to evaluate unobserved tourist overnight stays. *Tour. Manag.* **2022**, *89*, 104453. [CrossRef]
62. Song, C.; Qu, Z.; Blumm, N.; Barabási, A.-L. Limits of Predictability in Human Mobility. *Science* **2010**, *327*, 1018–1021. [CrossRef]
63. Trasarti, R.; Olteanu-Raimond, A.-M.; Nanni, M.; Couronné, T.; Furletti, B.; Giannotti, F.; Smoreda, Z.; Ziemlicki, C. Discovering urban and country dynamics from mobile phone data with spatial correlation patterns. *Telecommun. Policy* **2015**, *39*, 347–362. [CrossRef]
64. Fiadino, P.; Ponce-Lopez, V.; Antonio, J.; Torrent-Moreno, M.; D'Alconzo, A. Call Detail Records for Human Mobility Studies. In Proceedings of the Workshop on Big Data Analytics and Machine Learning for Data Communication Networks, Los Angeles, CA, USA, 21 August 2017; ACM: New York, NY, USA, 2017; pp. 43–48.
65. Schönfelder, S.; Axhausen, K.W. Activity spaces: Measures of social exclusion? *Transp. Policy* **2003**, *10*, 273–286. [CrossRef]
66. Flamn, M.F.; Kaufmann, V. The concept of personal network of usual places as a tool for analysing human activity spaces: A quantitative exploration. In Proceedings of the 6th Swiss Transport Research Conference, Monte Veritá, Ascona, Switzerland, 15–17 March 2006.
67. NMIA. Mobile Market Report of the National Media and Infocommunications Authority. Available online: https://nmhh.hu/dokumentum/211976/NMHH_mobilpiaci_jelentes_2015Q42019Q4.pdf (accessed on 16 April 2021).
68. Pénzes, J.; Demeter, G. Peripheral areas and their distinctive characteristics: The case of Hungary. *Morav. Geogr. Rep.* **2021**, *29*, 217–230. [CrossRef]
69. Marton, G.; Hinek, M.; Kiss, R.; Csapó, J. Measuring seasonality at the major spa towns of Hungary. *Hung. Geogr. Bull.* **2019**, *68*, 391–403. [CrossRef]
70. Dudás, G.; Kovalcsik, T.; Vida, G.; Boros, L.; Nagy, G. Price determinants of airbnb listing prices in lake balaton touristic region, Hungary. *Eur. J. Tour. Res.* **2020**, *24*, 2410. [CrossRef]

71. Medarić, Z.; Sulyok, J.; Kardos, S.; Gabruč, J. Lake Balaton as an accessible tourism destination—The stakeholders' perspectives. *Hung. Geogr. Bull.* **2021**, *70*, 233–247. [CrossRef]
72. Drotár, N.; Kozma, G. A new element of tourism in north-eastern part of Hungary—Steps to attract jewish pilgrims to Tokaj-Hegyalja Region. *Folia Geogr.* **2021**, *63*, 19–39.
73. Schmücker, D.; Reif, J. Measuring tourism with big data? Empirical insights from comparing passive GPS data and passive mobile data. *Ann. Tour. Res. Empir. Insights* **2022**, *3*, 100061. [CrossRef]
74. Lamp, M.-L.; Ahas, R.; Tiru, M.; Saluveer, E.; Aasa, A. Mobile Positioning Data in Emergency Management: Measuring the Impact of Street Riots and Political Confrontation on Incoming Tourism. In *Principle and Application Progress in Location-Based Services*; Liu, C., Ed.; Springer: New York, NY, USA, 2014; pp. 295–314.
75. Raun, J.; Shoval, N.; Tiru, M. Gateways for intra-national tourism flows: Measured using two types of tracking technologies. *Int. J. Tour. Cities* **2020**, *6*, 261–278. [CrossRef]
76. Jászberényi, M.; Miskolczi, M. Danube Cruise Tourism as a Niche Product—An Overview of the Current Supply and Potential. *Sustainability* **2020**, *12*, 4598. [CrossRef]
77. Szocska, M.; Pollner, P.; Schiszler, I.; Joo, T.; Palicz, T.; McKee, M.; Asztalos, A.; Bencze, L.; Kapronczay, M.; Petrecz, P.; et al. Countrywide population movement monitoring using mobile devices generated (big) data during the COVID-19 crisis. *Sci. Rep.* **2021**, *11*, 5943. [CrossRef]

sustainability

MDPI

Article

The Emergence of Unconventional Tourism Services Based on Autonomous Vehicles (AVs)—Attitude Analysis of Tourism Experts Using the Q Methodology

Katalin Ásványi *, Márk Miskolczi, Melinda Jászberényi, Zsófia Kenesei and László Kökény

Institute of Marketing, Corvinus University of Budapest, 1093 Budapest, Hungary; mark.miskolczi@uni-corvinus.hu (M.M.); jaszberenyi@uni-corvinus.hu (M.J.); zsofia.kenesei@uni-corvinus.hu (Z.K.); laszlo.kokeny2@uni-corvinus.hu (L.K.)
* Correspondence: katalin.asvanyi@uni-corvinus.hu

Abstract: The spread of autonomous vehicles (AVs) could fundamentally change tourism-related mobility in the near future. However, the empirical research on tourism impacts is still very limited. This research aims to systematize the expected tourism impacts of AVs and to explore how experts in different fields of tourism view the technological innovation ahead. The opinions of tourism experts (n = 21) involved in the research were analyzed using the Q methodology. Statements (n = 40) were formulated in topics derived from the literature. Based on the analysis, we distinguished four groups of opinions. An optimistic, technology-oriented group of experts suggested that AV-based sightseeing may emerge as a prominent unconventional service. An accessibility-focused group of experts predicting slow progress stressed that the spread of AVs could improve access to infrastructurally advanced destinations. A mobility-service-oriented group considered that the use of AVs is becoming conventional in terms of mobility, but its appearance remains unconventional for other tourism services. Due to the potential negative effects, a skeptical group of experts believes that AVs will not become conventional in tourism. The value of the research is the creation of groups based on tourism experts' attitudes, which can help prepare strategic tourism decisions in the future.

Keywords: autonomous vehicles (AVs); unconventional tourism services; tourism experts' perception; Q methodology; future of tourism

Citation: Ásványi, K.; Miskolczi, M.; Jászberényi, M.; Kenesei, Z.; Kökény, L. The Emergence of Unconventional Tourism Services Based on Autonomous Vehicles (AVs)—Attitude Analysis of Tourism Experts Using the Q Methodology. *Sustainability* **2022**, *14*, 3691. https://doi.org/10.3390/su14063691

Academic Editor: Fabio Carlucci

Received: 14 February 2022
Accepted: 15 March 2022
Published: 21 March 2022

1. Introduction

Innovative technologies and new mobility services of the fourth industrial revolution (Industry 4.0) are rapidly transforming the supply of tourism services. Although fully autonomous vehicles (AVs) have not yet entered the market, partial automation has already had an impact on tourism. It is, therefore, essential to be prepared for the spread and potential impacts of AVs [1]. In contrast, the literature on tourism does not focus much on the potential risks and benefits of AVs. This is supported by previous research [2] that addressed the impact of AVs on commuting travel. The exploration of this topic area was typically in the Global North, which highlights future research opportunities.

The relevance of this research is also supported by the fact that AVs are expected to be first adapted for long-distance trips, hence the need to explore the tourism side of the topic [2]. Furthermore, since COVID-19, the use of personal vehicles to get to and from a destination has also become even more valued [3].

To the best of our knowledge, except for one study [4], no research examines the attitudes of tourism service providers to the emergence of AVs, even though this technology may lead to the creation of new, unconventional tourism services in the future.

There are several approaches to unconventional (or non-conventional) tourism in the literature. It can refer to tourism solutions that primarily benefit local communities and not visitors to the destination. It can be called unconventional if it represents an alternative route to capitalist tourism. It can also be unconventional if it goes against and criticizes mass tourism or traditional tourism. However, the approaches all represent something unique in tourism services [1].

With all these in mind, our research seeks to clarify the expected role of AVs in tourism and, in contrast to previous research, draws together the potential changes, including threats and opportunities for conventional tourism services. For this, the attitude of tourism experts to the impact of AVs on the tourism industry has been analyzed by applying the Q methodology. We completed our research with 21 Hungarian tourism experts from the subsectors of accommodation, hospitality, attraction management, and mobility fields. As global research assumes that we still need ten more years to reach a mainstream adoption of fully automated vehicles [4,5], our research is an exploratory study of the impact of AVs on the tourism sector.

Our results have both theoretical and practical contributions. On the one hand, the findings of our research (four groups of experts identified: mobility-service oriented, accessibility focused, skeptic, optimists) provide a new perspective on the study of AVs in the social sciences and thus provide important input for academics. In addition, our findings might help tourism stakeholders to prepare for the impact of the technological revolution ahead (e.g., transforming conventional tourism services with AVs to increase competitiveness).

2. Expected Impacts of AVs on Tourism and Mobility

Since automation is a quite complex and incremental innovation, the basic definition of the technology needs to be clarified. First, we explain the different levels of AVs and then introduce some current initiatives and the expected impacts of AVs on tourism.

2.1. The Incremental Nature of Automation (SAE Levels)

Automation is an innovation of the fourth industrial revolution (Industry 4.0), which is rooted in the acceleration of information and communication technologies (ICTs) in the second half of the 20th century. The questions around AVs are numerous and variable; therefore, the knowledge of industrial and consumer behavior is limited. Regarding AVs, we discuss the incremental technology that is not only free of human intervention but is able to transport by itself. The current technology described as self-driven still depends on decisions made by a human driver. The real self-driving experience is getting closer, however, with the help of intense development.

Based on international standards, we can distinguish five levels of automation as suggested by the Society of Automotive Engineers (SAE) [6]:

- Level 0: The only controller of the vehicle is the human driver; the phase is described as a complete lack of automation.
- Level 1: The only controller of the vehicle is still the human driver, but there are supporting functions (e.g., change in the direction or speed, occasional automated steering of the wheel).
- Level 2: The only controller of the vehicle is still the human driver, but the supporting functions can be applied simultaneously. Cars currently available for purchase belong to this category (e.g., Tesla Model 3).
- Level 3: The role of the human controller is necessary, but the continuous observation of the surroundings is not required; the car is able to handle the driving operations. Yet when the car notifies the driver, the driver must take control of the vehicle.
- Level 4: The car can control every task for the duration of the journey. The presence of the human driver is optional in this phase as the system does not require them to take control of the vehicle. Based on the predictions of top companies in the automotive industry (Tesla, BMW, Google Waymo), cars with a high level of automation can be expected in the mid-2020s.
- Level 5: All aspects of driving can be owned and sustained by the car. Whether steering wheels and pedals will be needed for manual control remains an open

question at this stage of development. How future road users will react to the complete removal of the driving experience is uncertain.

Based on SAE levels, we can see that lower automation (SAE Levels 2–3) does not change mobility patterns much, while higher automation (SAE Levels 4–5) might completely change tourists' travel preferences and conventional tourism services. Tourists are typically the first users of this new technology, while for commuters, higher driving frequency and experience result in lower AV level preferences [7].

There are already destinations where AVs have been introduced and are undergoing public trials. The Lake District National Park (UK) is planning to introduce AVs on SAE Level 3 as a sustainable transport solution [8]. Gatwick Airport (London, UK) offers the autonomous solution as a shuttle service for passengers [9]. In Beijing's Haidan Park (China), the Apollo minibus is operated on SAE Level 4 automation, where tourists can experience 700 m of autonomous driving, free of charge. Seven passengers are allowed to use the vehicle at a time, on a predetermined route, at a limited speed (15 km/h), and there is a supervisor on board [10]. In the city of Sion (Switzerland), AVs are being tested on public roads, carrying 11 passengers at 20 km/h. As these are SAE Level 4 vehicles, a safety supervisor is also on board [4]. The Dubai Roads and Transport Authority (RTA), the local roads and transport authority, is set to become the first taxi operator in the emirate to switch to pure electric and autonomous vehicles. The first vehicles are planned to be on the road in 2023, and the company "Cruise" will have exclusive service rights for autonomous taxis until 2029. Dubai will thus become the first non-US city in the world to operate electric autonomous vehicles. The fleet should reach the target of 4000 vehicles by 2030 [11].

2.2. *Potential Benefits of AVs from the Tourist Perspective*

Previous research has shown that AVs will primarily be used for leisure [2] rather than non-leisure activities [10]. Research on the topic [12–14] revealed that respondents would use autonomous taxis more as tourists than as residents, which also demonstrates the significant impact of AVs on tourism-related mobility.

While traveling, tourists can also engage in other activities [15–19]: watching TV, relaxing, sleeping, reading, taking photos, eating, playing games, and gathering information. On SAE Level 5, tourists are even able to travel alone [20,21], which will be a major opportunity for people with disabilities [22]. Importantly, it is also expected that tourists will travel more often and greater distances by AVs [21]. The proliferation of AVs might help overcome barriers in terms of unfamiliar traffic rules and environments [23], making car rental easier for international tourists solving the problem of jet lag or the anxiety that comes from unusual traffic rules or conditions [9]. It also provides freedom of travel for people without a driver's license [24], is a more convenient solution for people under stress [25], and reduces isolation by providing easier access to services [15].

2.3. *Changes in Tourism-Related Mobility*

In terms of tourist mobility, there are two main categories of changes expected from AVs: accessing a destination and intra-destination mobility.

2.3.1. Access to a Destination for Tourism Purposes

In infrastructurally developed destinations, AVs can compete primarily with short-distance mobility services (e.g., rail, bus, public transportation) [2] and become alternatives to taxis [16,26] as well. As AVs can be a viable alternative for tourists on short-haul trips, it is expected that low-cost airlines will switch to long-haul trips [9] to preserve their competition. Due to the rapid and early market growth of AVs [27], traditional taxis will have to fulfill a new function, i.e., they will have to provide additional services (e.g., tour guiding) to compete with AVs.

2.3.2. Intra-Destination Tourism Mobility

AVs could have both positive and negative impacts on urban traffic in the future. It is important to note that AVs are primarily suited to urban transport and will therefore spread first to urban areas and then to rural areas. Initially, they can only travel along a defined route. Adverse weather conditions can cause problems [4], which will make the service less accessible to areas with low infrastructure [26].

In urban areas, however, there are several transport benefits that can be gained from the spread of AVs. Through efficient route selection and more efficient use of traffic lanes [28], congestion can be reduced [29], leading to improved urban traffic flows [30]. However, scholars [9] predict this is for shared vehicle use. Individual vehicle use tends to increase traffic, and as the number of AVs increases, this may cause congestion around the must-see tourist attractions.

In response to this, there might be some car-free destinations where tourists can only park outside the center and access the city by shared AVs [4]. As AVs spread, the design of cities may also change. Research suggests that in urban spaces, fewer parking spaces will be needed [2,31]. By eliminating parking spaces, green spaces can be created [32], which can increase the livability of cities. Thus, the spread of AVs can also increase the sustainability of tourism, thus having a positive impact on the environment, which is a priority for the sector these days [33]. The freed-up areas can be used for pedestrian and cycling facilities [28] or urban parks. The development of check-in points will also transform shopping districts, as they drop shoppers off at the store and pick them up later [9].

2.4. Expected Changes in Conventional Tourism Services Due to AVs

2.4.1. The New Dimension of Car Use

With the spread of full automation (SAE Level 5), driving as an activity will disappear, transforming it into a unique experience [26]. Destinations, where AVs appear for the first time, can enhance their image and strengthen their attractiveness [4]. At the same time, the spread of AVs can also promote cooperation between operators and destinations. Due to the seasonality of tourist destinations, it may make sense to share AVs between destinations with different seasons, e.g., ski areas in winter and lakes in summer. This will require flexibility and that these areas are geographically close to each other [4].

2.4.2. Sightseeing and City Tours

AVs can transform many areas of conventional tourism services. One of the most likely changes relates to sightseeing tours. Several studies have discussed pre-planned, AV-based sightseeing tours [9,26] and the resultant expected decline in walking and hop-on-hop-off tours [26]. However, with the proliferation of AVs currently, conventional bikes, Segways, and walking tours may become niche products in the longer term [26]. The new types of sightseeing tours will be algorithm-based, favoring those operators who pay more to be included on the itinerary, i.e., multinational operators will be favored over local businesses [9]. However, the automation of hiking trails can also raise problems such as the possibility that some of the less attractive, but the real face of destinations may remain hidden from tourists.

2.4.3. Hotel Industry, Restaurants, and MICE Tourism

The use of AVs will not only affect tourism during the day. As AVs are also suitable for overnight travel [34] and can be used for sleeping, the significance of accommodation services might be decreased in the long run. In addition to hotels selling rooms for a couple of hours, AVs can provide a more convenient alternative to the airport's sleeping pods [26] and can be applied as mobile motels by both business and leisure travelers [9].

AVs will also be suitable for holding meetings [25,26], which might result in the displacement of smaller meeting and conference rooms from the market. AVs could function as mobile restaurants in the future, combining sightseeing and restaurant services like dinner cruises [9]. The creation of mobile restaurants could also boost wine tourism, as the consumer does not have to drive the vehicle.

There are diverging views on what pricing can be expected for AVs. In the airport environment, they have emerged as a free service in the areas tested, but in the near future, it may also represent a price premium for users due to their benefits and novelty.

2.5. Research Gap

Based on the literature, we can see that the spread of AVs, especially at higher levels of automation (SAE Levels 4–5), could change the mobility and experience of tourists, passenger transport to and within the destination, and the content of traditional tourism services. The literature suggests three main categories of expected changes (Figure 1, [2–40]).

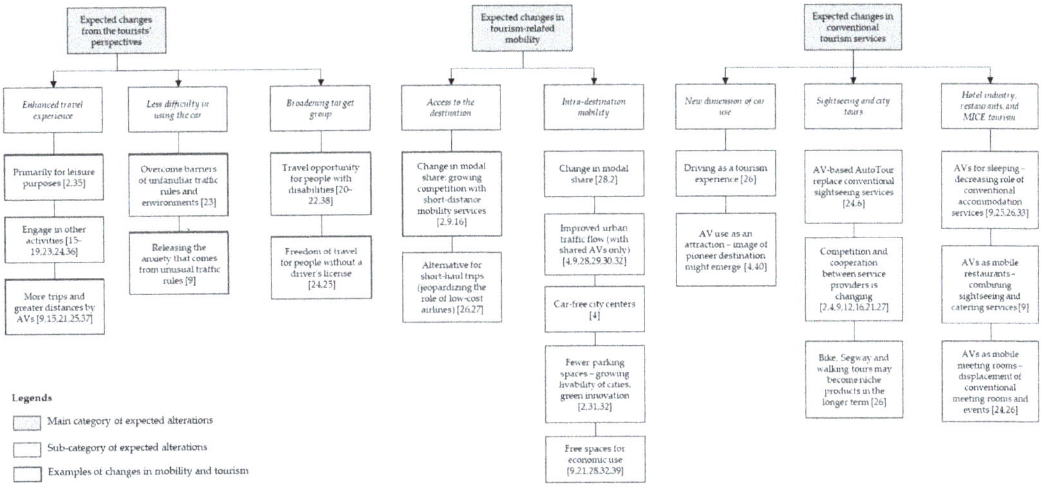

Figure 1. Expected alterations in tourism with the spread of autonomous vehicles (AVs). Source: Authors' own editing, based on the literature review [2–40].

Our analysis revealed that in previous research, attitudes toward expected changes have usually been examined from the perspective of potential consumers. In contrast, the views of experts working in specific sub-sectors of tourism are an under-explored topic. With this in mind, we have developed our primary research directions.

3. Methodology

3.1. Overview of Q Methodology

Since the main objective of our research is to identify the attitudes of tourism professionals, we applied the Q methodology. It is a useful method for identifying technological opportunities [41] and is also applicable to the field of transport [42]. The Q methodology was developed in the 1930s by William Stephenson. A distinctive feature of the methodology is its positioning between qualitative and quantitative methodologies. It is appropriate for exploring patterns along with the different preferences of respondents and reducing the complexity of opinions while also providing numerical results. Its advantage over the Likert scale is that it does not rate statements separately but ranks the respondent's preferences in relation to each other, which results in more sophisticated opinions [43]. While the methodology has been previously applied to exploring the field of AVs, it has been used to investigate user attitudes and adoption primarily [44], rather than in a tourism context [45]. Our research, therefore, is methodologically novel in the field of studying the impact of AVs on tourism from the experts' perspective.

3.2. Steps in Q Methodology

We implemented the Q methodology in six steps (Figure 2).

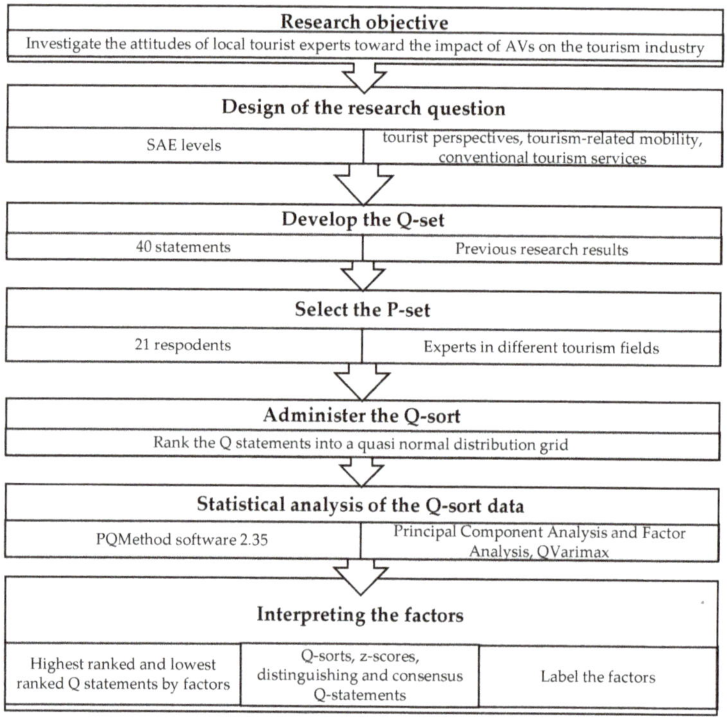

Figure 2. Steps of Q methodology. Source: Authors' own editing.

As a first step, we defined our research objective, which was formulated along the gap identified in the literature review, namely, to explore the impact of AVs on tourism not from the tourists' point of view but to analyze the views of tourism professionals. In the second step, we analyzed the peer-reviewed sources on the subject [45–47] and interpreted them along with a novel structure. However, previous articles on the topic conducted their research and drew their conclusions assuming SAE Levels 4–5, and therefore there were fewer results for SAE Levels 2–3. The effects identified in the various sources are therefore grouped according to Figure 1. For the three themes (changes from the tourists' perspectives, in tourism-related mobility, and in conventional tourism services), we formulated a total of 98 claims. We stopped at this number of statements because the content of the remaining statements had become repetitive. To make the layout of the method meaningful to respondents, it was necessary to reduce the number of statements [48], which previous research has shown to be satisfactory at between 40 and 50 [42] and 30 and 60 [49]. The number of statements was eventually reduced to 40 (see in attachments: 8.1); those with similar content were grouped together. To check comprehensibility, statements were pre-tested and, where appropriate, shortened or supplemented with illustrative examples (Appendix A).

To learn more about the tourism experts' attitudes to the potential impacts of AVs, the Q method statements were completed by key figures from the Hungarian tourism sector, from different areas of tourism (accommodation, hospitality, attraction management, and mobility fields), including market, non-profit, and public actors.

The tourism experts were limited to the domestic destination (Hungary, which is a growing car market in Europe [7]), representing a wide range of opinions and all major tourism service and expert groups. No knowledge of AVs was required. A particular feature of the Q methodology is that the participants themselves are the variables; they are not selected randomly but according to their theoretical knowledge [46]. Typically, 10–40 participants are asked to arrange the set of statements [50] since, with this number of participants, the factors are comparable. In our case, a total of 21 people participated in the study. When completing the questionnaire, respondents were asked to place the statements in a quasi-normally distributed table (Table 1), which forced them to rank them in relative order of preference. Quantitative analysis and qualitative interpretation of the responses are presented in the next section.

Table 1. Sorting distribution.

	Most Disagree			Neutral			Most Agree
Value	−3	−2	−1	0	+1	+2	+3
Frequency	4	5	7	8	7	5	4

Source: Own construction.

4. Results

The ratings by respondents are called Q-scores; we analyzed them using PQMethod version 2.35 software [51]. We performed principal component analysis, followed by varimax rotation, and then designed four factors along with the following criteria: an eigenvalue above 1 [47], minimum of two opinions per factor [49], variance level above 50%, each factor should account for at least 10% of the total variance [52], and low correlation between factors (below 0.5).

The distribution of the number of opinions varied between the factors, with five in Factor 1 and Factor 3, three in Factor 2, and eight in Factor 4 (Table 2). The factors were characterized according to the highest- and lowest-rated statements, considering separately those with significantly different opinions from the other factors.

Table 2. Rotated factor score matrix and explained variance.

Respondents	Factor 1	Factor 2	Factor 3	Factor 4
SZCE	0.0567	0.7638X	−0.0982	0.0917
FZS	0.6259X	−0.0139	0.2900	0.1151
NK	0.6469X	0.0741	0.0002	0.2116
GP	0.2218	−0.3163	0.1423	0.7275X
HG	0.9237X	−0.0523	0.1112	0.0658
HD	0.9237X	−0.0523	0.1112	0.0658
BA	−0.2495	0.7940X	0.1610	−0.1005
MAJ	0.2007	0.0230	−0.0639	0.7002X
MZ	0.1718	0.3105	0.3071	0.5223X
LP	0.1749	0.5869X	0.2865	0.1498
NJS	0.4381	0.2867	0.5907X	0.0727
RB	−0.0474	0.0093	0.1781	0.4678X
KN	0.1169	0.1120	0.1201	0.5504X
FD	−0.0741	0.2574	0.1209	0.5618X
HH	0.1785	0.0286	0.6882X	0.2507
BE	0.2267	0.4655	−0.4729	0.5440X
HD	0.1132	0.1744	0.4850X	0.1076
VE	0.2465	−0.1234	0.4883X	0.3670
OA	−0.0113	−0.0031	0.8016X	0.1107
AA	0.2939	−0.0718	0.2117	0.5930X
FM	0.4237X	0.1396	0.1180	0.3742
Number of respondents	5	3	5	8
% Explained Variance	16	11	13	15
Name of factor	Mobility-service oriented	Accessibility focused	Skeptics	Optimist, technology-oriented

Source: Own edition.

4.1. Mobility-Service Oriented Group

The first factor sees the impacts of AVs on tourism primarily in terms of mobility. They see the need for door-to-door mobility coming to the fore as opposed to public transport, but they also see SAE Levels 4–5 vehicles as a substitute for airport transfers and taxis. These experts predict a strong preference of tourists for individual travel; they think that lower demand for shared AVs can be expected. As they see the rapid spread of AVs, they no longer consider driving in a foreign environment as an obstacle. They also agree that the potential for the development of wine tourism as a designated driver will no longer be needed. They agree the most that conventional driving will become a tourist experience. However, it is interesting that they only see the changes in terms of mobility but do not agree that AVs could replace hotels, restaurants, or event venues. In sum, AVs can appear as an unconventional mobility service for tourists. Therefore, we named this factor mobility-service oriented.

4.2. Accessibility Focused

Factor 2 sees the main impact of AVs on tourism as the change in accessibility. They agree that geographically the opportunities are widening; more distant and less-known destinations can be included in tourism, which can be understood on two levels: tourists prefer to visit a destination farther from their home when traveling to a tourist attraction, and they prefer shorter distances within the destination, giving greater visibility and accessibility to attractions. At the same time, there is a risk that areas that are difficult to reach and less developed in terms of infrastructure could be at a disadvantage because they are less accessible to AVs. Factor 2 believes that closer cooperation between destinations sharing AVs is feasible. Based on the responses, AVs will not undermine the role of public transport and will rather be used for longer-distance (inter-destination) transport. Factor 2 alone disagrees with the emergence of car-free destinations and does not believe that traffic would be less or cities more livable because of the advent of AVs. These experts think that wine tourism and barrier-free tourism will develop, which could result in increased over-tourism. In terms of tourism services, this factor is not in favor of the development of AutoTour services. Their order of preference reflects the fact that traveling to more remote places by AVs would become conventional, but they see it less so within cities, where the emergence of AVs would remain unconventional. Therefore, we call this factor accessibility-focused.

4.3. Skeptics

The opinion of Factor 3 reflects the negative impact of AVs as a safety issue from a transport, privacy, or data point of view. Tourists are reluctant to use shared AVs because they do not want to travel with strangers. As it is not safer for pedestrians to use AVs, they do not agree with the increase in the number of walking tours. Factor 3 considers AVs to be suitable for both shorter and longer distance trips, and these experts also believe that they will be used more for leisure and visiting friends and relatives (VFR) by tourists. They agree that new AutoTour services could be created but also emphasize the risk that small local businesses will be at a disadvantage compared to multinationals. Factor 3 also rejects the idea of destinations cooperating and sharing AVs. They also see an opportunity for taxis to survive if they develop additional services (e.g., tour guide service). This factor is therefore labeled skeptics, as they are less likely to believe that the uptake of AVs can become conventional in tourism due to the threats listed.

4.4. Optimist, Technology-Oriented Group

Factor 4 is optimistic about the emergence of AVs and positive about their impact on tourism. They think that AVs will be a viable alternative to public transport, taxis, and airport transfers. This factor is the most likely to see AVs as meeting rooms or mobile motels but does not feel that this threatens conventional hotel services. They see a potential for the development of wine tourism because of the elimination of driving, which could increase demand, but they do not fear the emergence of over-tourism. Factor 4 agrees that AVs will also have a positive impact on passenger transport and that the transformation of urban spaces will make cities more livable, making them even more attractive to tourists. Experts in this group believe that AVs can be suitable in both the short and long term and can therefore influence destination choices. However, they agree that, for the time being, AVs can offer an unconventional tourist experience, which can also translate into a price premium. Based on these opinions, this factor has been labeled an optimist, technology-oriented group.

5. Discussion

To obtain a complete picture of respondents' views, we have highlighted the distinguishing, consensus, and neutral statements for each factor, which helps us draw conclusions and make recommendations.

5.1. Distinguishing Statements

As we have seen in the literature review, there was no common understanding and approach in all areas of previous research on the impact of AVs on tourism. In the present research, we also found areas where tourism experts have quite different views. There is no uniform picture as to whether shared AVs can be as successful with tourists as individual AVs, i.e., whether both can become conventional. Factors 1 and 3, supporting the previous results [2,11] that tourists prefer individual travel for leisure trips, while contrary to other findings [16], Factors 2 and 4 do not see sharing with strangers as a problem. There are experts who say that the issue of data security will be a problem (Factor 3), while others do not consider it a matter of concern (Factors 1 and 4), so this will not prevent the use of AVs from becoming conventional. The transformation of urban spaces can have several consequences, as it depends on whether the freed-up parking areas are used for civic or business purposes. According to Factors 3 and 4, creating green spaces and parks would not only make a city more attractive but also more livable, a view also expressed in previous research [9]. Although previous research predicts AV use for longer distances [21], respondents are also divided on this issue. While some believe that removing barriers, such as traveling in unfamiliar environments [23], will increase demand, fueled by expanding accessibility [2,21], there is no consensus on whether this will lead to over-tourism. However, accessibility could be a key factor in making AVs increasingly part of conventional tourism in terms of mobility.

5.2. Consensus Statements

On certain issues, tourism experts think very much alike. They consider AVs a less flexible mobility service since they can only travel along a predefined path, which takes away access to destinations with poorer infrastructure, as pointed out previously [24]. It would therefore be important to prepare these destinations for the roll-out of AVs so they can meet the required conditions as soon as possible and take advantage of the opportunities offered by AVs (e.g., 5G network extension). At the same time, AVs will broaden the range of destinations available [2], bringing tourists to parts of the destination that may have been less frequented or even considered undesirable due to inappropriate surroundings [4]. All four factors agree that, since there is no need for a driver in SAE Level 4–5 AVs, this represents an opportunity for the development of wine tourism. While there was not complete agreement on the use of parking spaces, as mentioned in the distinguishing statements, there was consensus on the idea that these spaces are less used for economic purposes; previous research [9,21,28] highlighted many utilization options that would be worthwhile for tourism experts to rethink. By becoming a place for rest and sleep [9], AVs could replace the function of some hotels, which is questioned by our domestic experts. Factors 2 and 4 are more concerned with the expansion of available destinations, and all four factors reject the conventionalization of AVs as mobile motels.

Thus, not only are central hotels part of conventional tourism, but the use and booking of the peripheral areas also become conventional.

5.3. Neutral Statements

While several previous studies confirm that tourists will benefit more from the emergence of AVs [2], domestic tourism practitioners have paid less attention to the issue of inequality between locals and tourists [8]. All factors put these claims in the neutral category. However, the involvement of locals plays a key role in the acceptance of AVs; AVs may even provide an opportunity to connect locals and tourists [8], a demand that is becoming increasingly conventional nowadays. From the perspective of locals, technology will have an impact not only at the residential level but also at the entrepreneurial level [8]. Except for Factor 3, our experts consider competition between small local services and multinational services in tour planning to be less of an issue. The issue of growth in demand for overnight experiences is also neutral across all factors, implying that overnight experiences will become less conventional, not affecting changes in demand for tourism services to the same extent, contradicting previous findings [26]. In terms of service change, factors were neutral regarding the replacement of the restaurant, suggesting that they do not see the mobile restaurant function of AVs as unimaginable but that it will not become conventional in the near future (2030).

6. Conclusions

The research has shown how experts in different areas of tourism perceive the changes that we are facing as a result of the spread of AVs. An important result of our research is that we have explored the role of conventional tourism services from an expert perspective, thus improving the accuracy of previous predictions. Based on the four groups of opinions generated, we can see that the role of AV-based mobility and AV-based sightseeing is likely to increase in the near future, which could greatly improve the tourism experience offered by destinations with well-developed infrastructure.

An important question is whether tourism experts want to play a leader or a follower role in the emergence of AVs. The academic sphere of tourism has a responsibility to help tourism service providers understand the possible changes that may affect them. Policymakers and investors should also place greater emphasis on assessing the impacts on tourism, as leisure travel can be the first area using AVs.

The opinion groups generated by the Q methodology can be useful to quickly identify different opinions and to develop different communication strategies based on them to increase the acceptance of AVs among tourism experts. For the skeptics (Factor 3), it is worth emphasizing the safety of the technology, providing evidence that the lack of a human factor does not reduce confidence or even offer them the opportunity to try the technology.

A limitation of our research along the Q methodology is that we must limit the number of statements (30–60) to be transparent to respondents, so some topics must be combined or even omitted. The sample is also a limitation, as we only asked tourism experts in Hungary; cultural differences may influence the formation of opinion groups. Another limitation of the research is that new developments may emerge in the coming years, leading to new insights and opinions.

For further research, there is a need to explore opportunities for cooperation between tour companies (e.g., Hop-on Hop-off) and automotive companies to develop the details of an AutoTour service based on self-driving vehicles. It should also be a great contribution to the current findings to conduct a consumer attitude analysis based on real experiences (e.g., participation in living lab surveys) to verify the validity of the variables I have identified that influence the technology acceptance of autonomous vehicles.

Author Contributions: Conceptualization, K.Á., M.M. and M.J.; methodology, K.Á.; validation, K.Á. and M.M.; formal analysis, K.Á., M.M. and M.J.; writing—original draft preparation, K.Á., M.M. and M.J.; writing—review and editing, Z.K. and L.K.; supervision, M.J.; funding acquisition, M.J. All authors have read and agreed to the published version of the manuscript.

Funding: Project no. NKFIH-869-10/2019 has been implemented with support provided by the National Research, Development, and Innovation Fund of Hungary, financed under the Tématerületi Kiválósági Programme Funding Scheme.

Institutional Review Board Statement: Not applicable.

Informed Consent Statement: Not applicable.

Conflicts of Interest: The authors declare no conflict of interest.

Appendix A

Table A1. Q statements.

N	Q Statements
1.	Tourists prefer fully autonomous vehicles to public transport.
2.	The spread of autonomous vehicles could lead to over-tourism.
3.	Autonomous vehicles will be used more for recreation and VFR (visiting friends and relatives) tourism.
4.	Tourists will be less open to the use of autonomous vehicles because they are concerned about their personal data.
5.	Autonomous vehicles will improve urban transport and reduce congestion.
6.	It will be more difficult for autonomous vehicles to travel on unimproved roads, making it harder to reach destinations with less developed infrastructure.
7.	It is predicted that autonomous vehicles could replace the role of traditional shuttle buses and taxi services, completely redefining urban passenger transport. The role of airport shuttles and taxis will also be reduced.
8.	Autonomous vehicles will be used as mobile motels, replacing hotel rooms rented for only a few hours.
9.	Autonomous vehicles will make door-to-door mobility more in demand than public transport (e.g., between rail and accommodation).
10.	Shared and autonomous vehicles will be less used by tourists than individually owned vehicles, as they are more sensitive to the presence of strangers on a leisure trip.
11.	With the spread of autonomous vehicles, there will be more interest in leisure and recreational services than in shopping and administration.
12.	The availability of autonomous vehicles will not be a determining factor in the choice of destination.
13.	Autonomous vehicles will be free of charge during the test period.
14.	Shared and autonomous vehicles will reduce costs, so only those will be allowed into cities, and other vehicles will have to park outside the city. Car-free tourist destinations could be created.
15.	Shared and autonomous vehicles can create a new type of sightseeing, called AutoTour services, which are more flexible and fully customizable, thus replacing conventional walking and bus tours.
16.	Local businesses are marginalized by the funding of multinational companies (disadvantage of smaller attractions).
17.	Tourists will travel to more distant destinations as the partially/fully autonomous system makes longer distance travel more comfortable.
18.	Autonomous vehicles offer the possibility to reach new destinations and attractions, allowing tourists to use hotel or restaurant services in more and more places.
19.	Urban spaces transformed by autonomous vehicles will be less attractive to tourists.
20.	There is no sense of security towards autonomous vehicles; it is like a tourist traveling in a driverless „box," the possibility of crime is more likely to arise.
21.	Testing fully autonomous vehicles could increase travel motivation and be a stand-alone tourist experience in places where the technology is not yet widespread.
22.	Solving parking problems will create more liveable urban centers with more green spaces.
23.	Autonomous vehicles can be used as mobile offices/meeting rooms.
24.	Sightseeing tours with autonomous vehicles will easily hide the negative side of the urban environment, masking the reality.

Table A1. *Cont.*

N	Q Statements
25.	The demand for accessible tourism will increase, as more people will be able to travel alone (due to the rise of autonomous vehicles, tourists without a driving license and with health problems can now travel alone).
26.	The number of evening sightseeing/night-time tourist experiences is increasing.
27.	Autonomous vehicles will be used by tourists for longer distances rather than short commutes.
28.	Autonomous vehicles will make travel safer.
29.	For tourists, driving a conventional vehicle will become a tourist experience as autonomous vehicles become more common.
30.	Less parking places will be needed, their space will be used for economic purposes (e.g., hotel to expand the number of rooms, event venue, bicycle paths).
31.	Restaurants will be in competition with mobile restaurants or vehicles combining sightseeing with dining, such as dinner boats.
32.	The spread of autonomous vehicles will benefit wine tourism as there is no problem with driving after drinking alcohol.
33.	The unfamiliar surroundings will no longer be a limiting factor; autonomous vehicles help by eliminating the barriers that international tourists face when traveling in unfamiliar surroundings.
34.	The number of walking tours may increase because there are fewer pedestrian accidents.
35.	Autonomous vehicles can create inequalities in urban transport if the interests of tourists take precedence over residents. Self-driving taxis will be used more by tourists than by residents.
36.	Tour management will become easier as the route becomes more flexible.
37.	Shared and autonomous vehicles operate only on pre-defined routes and under strictly controlled conditions (e.g., as airport shuttles, connecting airports or train stations to city centers, on urban sightseeing routes, or in off-road locations such as zoos), and therefore the system is not flexible.
38.	Traditional taxis and city bus tours can be maintained if they provide an additional service (guided tours), as a different type of staff will be needed.
39.	Hotels on the motorway will disappear because passengers will be able to sleep in the autonomous vehicle and will not have to stop for a rest during a long journey.
40.	Depending on seasonality, destinations can share their fleet of autonomous vehicles, reducing environmental pollution and congestion (e.g., making autonomous vehicles available in ski resorts in winter and in beach destinations in summer).

Source: Own edition.

References

1. Chatterton, P.; Pickerill, J. Everyday activism and transitions towards post-capitalist worlds. *Trans. Inst. Br. Geogr.* **2010**, *35*, 475–490.
2. Kimber, S.; Siegel, L.; Cohen, S.; Thomopoulos, N. The wider use of autonomous vehicles in non-commuting journeys. *Adv. Transp. Policy Plan.* **2020**, *5*, 125–148.
3. Ribeiro, M.A.; Gursoy, D.; Chi, O.H. Customer Acceptance of Autonomous Vehicles in Travel and Tourism. *J. Travel Res.* **2022**, *61*, 620–636.
4. Grèzes-Bürcher, S.; Grèzes, V.; Fux, M.; Ramseyer, R.; Wilk, R. The Potential of Public Autonomous Vehicles in Alpine Tourism Destinations. *Téoros* **2021**, *40*, 2.
5. Miskolczi, M.; Földes, D.; Munkácsy, A.; Jászberényi, M. Urban mobility scenarios until the 2030s. *Sustain. Cities Soc.* **2021**, *72*, 103029. [CrossRef]
6. SAE. SAE International Releases Updated Visual Chart for Its "Levels of Driving Automation" Standard for Self-Driving Vehicles. 2021. Available online: https://www.sae.org/news/press-room/2018/12/sae-international-releases-updated-visual-chart-for-its-%E2%80%9Clevels-of-driving-automation%E2%80%9D-standard-for-self-drivingvehicles (accessed on 21 January 2022).
7. Syahrivar, J.; Gyulavári, T.; Jászberényi, M.; Ásványi, K.; Kökény, L.; Chairy, C. Surrendering personal control to automation: Appalling or appealing? *Transp. Res. Part F Traffic Psychol. Behav.* **2021**, *80*, 90–103.
8. Mogg, T. Driverless Pods Could Be Used to Ferry Tourists around a U.K. National Park. Digital Trends: Energing Technologies. 2018. Available online: https://www.digitaltrends.com/cool-tech/driverless-pods-lake-district/ (accessed on 21 January 2022).
9. Cohen, S.A.; Hopkins, D. Autonomous vehicles and the future of urban tourism. *Ann. Tour. Res.* **2019**, *74*, 33–42.

10. Dai, J.; Li, R.; Liu, Z. Does Initial Experience Affect Consumers' Intention to Use Autonomous Vehicles? Evidence from a Field Experiment in Beijing. *Accid. Anal. Prev.* **2021**, *149*, 105778. [CrossRef]
11. Roads and Transport Authority (RTA). About SDT. 2022. Available online: https://www.rta.ae/links/sdt/en/index.html (accessed on 21 January 2022).
12. Tussyadiah, I.P.; Zach, F.K.; Wang, J. Attitudes Toward Autonomous on Demand Mobility System: The Case of Self-Driving Taxi. In *Information and Communication Technologies in Tourism*; Springer: Rome, Italy, 2017; pp. 755–766.
13. Laidlaw, K.; Sweet, M.N. Estimating consumer demand for autonomous vehicles in the greater Toronto-Hamilton Area: 2016 survey and model results. In Proceedings of the Transportation Research Board 97th Annual Meeting, Washington, DC, USA, 7–11 January 2018.
14. Miskolczi, M.; Kökény, L.; Ásványi, K.; Jászberényi, M.; Gyulavári, T.; Syahrivar, J. Impacts and potential of autonomous vehicles in tourism. *Deturope* **2021**, *13*, 34–51.
15. IFMO Autonomous Driving—The Impact of Vehicle Automation on Mobility Behaviour. 2016. Available online: https://elib.dlr.de/110337/ (accessed on 21 January 2022).
16. Krueger, R.; Rashidi, T.H.; Rose, J.M. Preferences for Shared Autonomous Vehicles. *Transp. Res. Part C Emerg. Technol.* **2016**, *69*, 343–355.
17. Kyriakidis, M.; Happee, R.; de Winter, J.C.F. Public opinion on automated driving: Results of an international questionnaire among 5000 respondents. *Transp. Res. Part F* **2015**, *32*, 127–140. [CrossRef]
18. Platt, M. *Drivers Cautious but Curious over Automated Cars: First Canadian Study Shows*; University of Calgary: Calgary, AB, Canada, 2017.
19. Pudane, B.; Molin, E.J.E.; Arentze, T.A.; Maknoon, Y.; Chorus, C.G. A Time-Use Model for the Automated Vehicle Era. *Transp. Res. Part C Emerg. Technol.* **2018**, *93*, 102–114. [CrossRef]
20. Haboucha, C.J.; Ishaq, R.; Shiftan, Y. User preferences regarding autonomous vehicles. *Transp. Res. Part C Emerg. Technol.* **2017**, *78*, 37–49. [CrossRef]
21. Kellett, J.; Barreto, R.; Van Den Hengel, A.; Vogiatzis, N. How Might Autonomous Vehicles Impact the City? The Case of Commuting to Central Adelaide. *Urban Policy Res.* **2019**, *37*, 442–457. [CrossRef]
22. Carr, N.K. As the Role of the Driver Changes with Autonomous Vehicle Technology, so, Too, Must the Law Change. *St. Mary's Law J.* **2020**, *51*, 817–843.
23. Prideaux, B.; Ping, Y. The disruptive potential of autonomous vehicles (AVs) on future low-carbon tourism mobility. *Asia Pac. J. Tour. Res.* **2019**, *24*, 459–467. [CrossRef]
24. Anderson, J.M.; Kalra, N.; Stanley, K.D.; Sorensen, P.; Samaras, C.; Oluwatola, O. *Autonomous Vehicle Technology—A Guide for Policymakers*; Rand Corporation: Santa Monica, CA, USA, 2016.
25. Maleki, M.; Chan, Y.; Arani, M. Impact of Autonomous Vehicle Technology on Long Distance Travel. *arXiv*. 2021. Available online: http://arxiv.org/abs/2101.06097 (accessed on 21 January 2022).
26. Bainbridge, A. Autonomous Vehicles & Auto-Tours: What Is an Auto-Tour and How Will Autonomous Vehicles Impact Tours, Attractions & Cities? 2018. Available online: https://www.destinationcto.com/docs/AutoTour.pdf (accessed on 21 January 2022).
27. Greenblatt, J.B.; Saxena, S. Autonomous taxis could greatly reduce greenhouse-gas emissions of US light-duty vehicles. *Nat. Clim. Change* **2015**, *5*, 860–863. [CrossRef]
28. Alessandrini, A.; Campagna, A.; Delle Site, P.; Filippi, F.; Persia, L. Automated vehicles and the rethinking of mobility and cities *Transp. Res. Procedia* **2015**, *5*, 145–160. [CrossRef]
29. Alam, M.J.; Habib, M.A. Investigation of the impacts of shared autonomous vehicle operation in halifax, canada using a dynamic traffic microsimulation model. *Procedia Comput. Sci.* **2018**, *130*, 496–503. [CrossRef]
30. Freudendal-Pedersen, M.; Kesselring, S.; Servou, E. What is smart for the future city? Mobilities and automation. *Sustainability* **2019**, *11*, 221. [CrossRef]
31. Faisal, A.; Yigitcanlar, T.; Kamruzzaman, M.; Currie, G. Understanding autonomous vehicles: A systematic literature review on capability, impact, planning and policy. *J. Transp. Land Use* **2019**, *12*, 45–72. [CrossRef]
32. Cavoli, C.; Phillips, B.; Cohen, T.; Jones, P. *Social and Behavioural Questions Associated with Automated Vehicles. A Literature Review*; Department for Transport: London, UK, 2017.
33. Zhakanova, I.A.; Isiksal, H.; Shynar Zhakanovna, R.; Armanay Sagatbayevna, S. Ecological development impact on tourism in Pavlodar region. *Econ.—Innov. Econ. Res.* **2018**, *6*, 113–124.
34. Pudāne, B.; Rataj, M.; Molin, J.E.; Mouter, N.; van Cranenburgh, S.; Chorus, C.G. How will automated vehicles shape users' daily activities? Insights from focus groups with commuters in the Netherlands. *Transp. Res. Part D Transp. Environ.* **2019**, *71*, 222–235. [CrossRef]
35. Ashkrof, P.; Homem de Almeida Correia, G.; Cats, O.; van Arem, B. Impact of Automated Vehicles on Travel Mode Preference for Different Trip Purposes and Distances. *Transp. Res. Rec.* **2019**, *2673*, 607–616. [CrossRef]
36. Wadud, Z.; Huda, F.Y. Fully automated vehicles: The use of travel time and its association with intention to use. *Proc. Inst. Civ. Eng. Transp.* **2021**. [CrossRef]
37. Dannemiller, K.A.; Mondal, A.; Asmussen, K.E.; Bhat, C.R. Investigating autonomous vehicle impacts on individual activity-travel behavior. *Transp. Res. Part A* **2021**, *148*, 402–422. [CrossRef]

38. Fagnant, D.J.; Kockelman, K.M. Preparing a nation for autonomous vehicles: Opportunities, barriers and policy recommendations for capitalizing on Self-Driven vehicles. *Transp. Res. Part A Policy Pract.* **2015**, *77*, 167–181. [CrossRef]
39. Henderson, J.; Spencer, J. Autonomous vehicles and commercial real estate. *Cornell Real Estate Rev.* **2016**, *14*, 44–54.
40. Rong, H.H.; Tu, W.; Duarte, F.; Ratti, C. Employing waterborne autonomous vehicles for museum visits: A case study in Amsterdam. *Eur. Transp. Res. Rev.* **2020**, *12*, 1–13. [CrossRef]
41. Irawan, N.; Geisler, S.; Pawlowski, J.M. Why Should the QMethod Be Integrated into the Design Science Research? A Systematic Mapping Study. 2019. Available online: https://aisel.aisnet.org/scis2019/9/ (accessed on 21 January 2022).
42. Van Exel, J.; de Graaf, G. Q Methodology: A Sneak Preview. 2005. Available online: http://qmethod.org/articles/vanExel.pdf (accessed on 21 January 2022).
43. Zabala, A.; Sandbrook, C.; Mukherjee, N. When and how to use q methodology to understand perspectives in conservation research. *Conserv. Biol.* **2018**, *32*, 1185–1194. [CrossRef]
44. Detjen, H.; Irawan, N.; Geisler, S. Attitudes towards autonomous public transportation. In Proceedings of the 13th International Conference on Automotive User Interfaces and Interactive Vehicular Applications (AutomotiveUI '21 Adjunct), Leeds, UK, 9–14 September 2021; ACM: New York, NY, USA, 2021.
45. Rodak, A.; Jamson, S.; Kruszewski, M.; Pedzierska, M. User requirements for autonomous vehicles—A comparative analysis of expert and non-expert-based approach. In Proceedings of the AEIT International Conference of Electrical and Electronic Technologies for Automotive (AEIT AUTOMOTIVE), Turin, Italy, 18–20 November 2020.
46. Brown, S.R. *Political Subjectivity: Applications of Q Methodology in Political Science*; Yale University Press: London, UK, 1980.
47. McKeown, B.; Thomas, D. *Q-Methodology*; Sage Publisher: London, UK, 2013.
48. Webler, T.; Danielson, S.; Tuler, S. *Using Q Method to Reveal Social Perspectives in Environmental Research*; Social and Environmental Research Institute: Greenfield, MA, USA, 2009.
49. Watts, S.; Stenner, P. Doing Q methodology: Theory, method and interpretation. *Qual. Res. Psychol.* **2005**, *2*, 67–91. [CrossRef]
50. Dryzek, J.S. Handle with care: The deadly hermeneutics of deliberative instrumentation. *Acta. Polit.* **2005**, *40*, 197–211. [CrossRef]
51. Schmolck, P.; Atkinson, J. PQMethod Software, Version 2.35. 2014. Available online: http://schmolck.userweb.mwn.de/qmethod (accessed on 21 January 2022).
52. Sneegas, G.; Beckner, S.; Brannstrom, C.; Jepson, W.; Lee, K.; Seghezzo, L. Using Q-methodology in environmental sustainability research: A bibliometric analysis and systematic review. *Ecol. Econ.* **2021**, *180*, 106864. [CrossRef]

sustainability

MDPI

Article

Addressing the Phenomenon of Overtourism in Budapest from Multiple Angles Using Unconventional Methodologies and Data

Betsabé Pérez Garrido [1,*], Szabolcs Szilárd Sebrek [2], Viktoriia Semenova [3], Damla Bal [3] and Gábor Michalkó[4,5]

1 Department of Computer Science, Institute of Information Technology, Corvinus University of Budapest, 1093 Budapest, Hungary
2 Corvinus Institute for Advanced Studies, Corvinus University of Budapest, 1093 Budapest, Hungary; sebrek@uni-corvinus.hu
3 Doctoral School of Business and Management, Corvinus University of Budapest, 1093 Budapest, Hungary; viktoriia.semenova@stud.uni-corvinus.hu (V.S.); damla.bal@stud.uni-corvinus.hu (D.B.)
4 Institute of Marketing, Corvinus University of Budapest, 1093 Budapest, Hungary; gabor.michalko@uni-corvinus.hu
5 CSFK Geographical Institute, 1112 Budapest, Hungary
* Correspondence: perez.betsabe@uni-corvinus.hu; Tel.: +36-1482-7478

Abstract: This paper addresses the phenomenon of overtourism in Budapest from multiple perspectives, starting with an overview that uses information collected from news, media, and academic tourism literature. Further, the phenomenon of overtourism is addressed quantitatively using different indicators, including tourism density and intensity. According to these indicators, the center of Budapest (formed by districts I, V, VI, VII, VIII, and IX) has been strongly affected by the presence of tourists, while districts physically far from the center have been less affected. This fact suggests the heterogeneity of the city in terms of overtourism. The number one catalyst of the negative impacts of foreign visitors' behavior is party tourism ('ruin pub' tourism), which involves an unconventional use of the Hungarian capital. Finally, using an unconventional optimization method called fuzzy linear programming, we attempt to explore the challenging problem of identifying the optimal number of tourists for the city. The results of the study have important theoretical, methodological, and practical implications. On the theoretical side, we offer a comprehensive understanding of the phenomenon of overtourism in Budapest. Methodologically, the integrated approach in terms of data gathering and unconventional analytical methodologies (comprised of a case study analysis, the assessment of effective indicators for measuring the discussed phenomenon, and the demonstration of the sustainable number of visitors) represents a novel perspective about the extent of overtourism in Budapest. On the practical side, our findings provide valuable guidance for policymakers to help mitigate the problem of overtourism in the city. With regard to future research, we suggest extending and updating the results presented in this study to develop more sustainable tourism strategies.

Keywords: overtourism; Budapest; tourism carrying capacity; unconventional data gathering; unconventional analytical methodology

Citation: Pérez Garrido, B.; Sebrek, S.S.; Semenova, V.; Bal, D.; Michalkó, G. Addressing the Phenomenon of Overtourism in Budapest from Multiple Angles Using Unconventional Methodologies and Data. *Sustainability* **2022**, *14*, 2268. https://doi.org/10.3390/su14042268

Academic Editor: Chia-Lin Chang

Received: 29 January 2022
Accepted: 15 February 2022
Published: 16 February 2022

1. Introduction

In 2019, Budapest was named the best destination in Europe, outranking classic urban destinations such as Paris, London, and Barcelona [1]. Within the same calendar year, the capital of Hungary ranked second on the "Best in Travel 2020" list, being awarded the title of the world's most affordable large city [2]. Indeed, the year 2019 witnessed unprecedented tourist traffic in Budapest and the Hungarian tourism sector enjoyed "a golden age" [3] before the COVID-19 pandemic. The contribution of tourism to GDP reached 13.2% in 2019 and the growth rate of the tourism sector exceeded both the EU and worldwide average [4]. Being among the most popular tourist destinations, however, has its downsides. Earlier,

Budapest was ranked fifth among the European cities most affected by overtourism in 2017 [5]. Followed by Barcelona and Amsterdam, Budapest is third on the list of cities most impacted by overtourism at night [6].

Budapest is divided into 23 districts that have their own political, economic, social, and cultural structures. The fragmentation of the city in terms of district management, and uncoordinated urban planning in the post-socialist years have led to few attempts at planning for tourism [7]. Previous research [8] concluded that tourism planning and management in Budapest is fragmented. Moreover, one city agency responsible for tourism was closed, and the remaining state-owned Hungarian Tourism Agency is focused on marketing and communications rather than planning or development [7]. As Smith et al. [9] noted, unplanned urban processes may result in many adverse effects from tourism-related activity. For instance, the unstructured and somewhat unregulated urban planning system in Budapest was one of the reasons for the expansion of the ruin bar market and the formation of the whole party area, as well as the growth in Airbnb accommodation [7]. Additionally, the city is extremely congested due to the existence of sections of national roads and a Budapest-centric transport infrastructure [10]. Consequently, the 'Pearl of the Danube' is considered one of the most affected urban areas.

The phenomenon of overtourism in the context of Budapest has recently been investigated in the tourism literature. For instance, Pinke-Sziva et al. [11] analyzed the phenomenon of overtourism with specific reference to the night-time economy in District VII—the so-called "party quarter of Budapest." With a focus on the same District VII, Smith et al. [7] explored the two closely connected phenomena—ruin bars, and the high density of Airbnb apartments—which are local and global examples of tourism consumption. Smith et al. [9] explored the role of tourism in growing resident discontent and resistance, while Remenyik et al. [6] assessed the extent of overtourism in Budapest. Party tourism is an unconventional tourism product in an urban environment which was first developed for cultural, health, and conference tourists [12]. Due to the divergent interests of inhabitants and the management of flows on the territory, scholars have highlighted the difficulty of measuring overtourism in such urban environments [13].

The excessive use of resources, infrastructure, and the facilities of a destination is among the causes of overtourism, implying that cities have a carrying capacity that tourism can exceed [14]. As per Camatti et al. [15], the tourism carrying capacity of a destination is an essential reference point for developing responses to the complex phenomenon of overtourism. Hence, we refer to the concept of tourism carrying capacity (TCC), which reflects the dynamics of the destination environment [16,17]. TCC is a management tool [18] designed for tourism planning that enables the development of initiatives for managing tourist flows via the balanced redistribution and segmentation of demand, introducing regulations or reservation systems, and creating alternative itineraries or novel tourist attractions [15,17].

Prior research has shed some light on the problems of overtourism in Budapest. However, we still lack an integrated view that links its emergence and comprehensive measurement in a city setting. In our study, we aimed to identify factors that made overtourism possible in Budapest, to apply recognized indicators of overtourism to better understand the city's carrying capacity and flows of different types of visitors, and to determine the sustainable number of tourists in the Hungarian capital. To address the research objectives, three research questions were raised: (1) What are the factors that led to the emergence of overtourism in Budapest? (2) How can one quantify the phenomenon based on verified overtourism-tailored indicators? (3) How can one approximate the optimal number of visitors in the city?

As suggested by Remenyik et al. [6], the overtourism phenomenon and the capability of an area may be assessed via the use of complex methodologies. Therefore, we adopted an integrated view built upon a case study of urban destination indicators to assess the degree of overtourism in Budapest, and calculated the sustainable number of visitors with the use of unconventional optimization methods such as Fuzzy Linear Programming (FLP).

The application of different methods for data collection and data analysis opens up fresh lines of inquiry into the phenomenon of overtourism in Budapest and its districts. The following section presents the theoretical discussion of the overtourism phenomenon and its measurement, with a focus on TCC.

2. Literature Review

2.1. Definition of Overtourism

Over the last two decades, the number of global outbound tourists has more than doubled. International tourist arrivals increased from 673 million in 2000 to 1460 million in 2019 [19], harming many popular destinations [20]. Even though the term 'overtourism' has attracted increased interest among researchers in recent times, the issue of the crowding of destinations has been addressed in the literature since the mid-1960s [21–26]. Along with rapid urbanization, tourism mobility, low-cost flights, increases in income, the growing popularity of Airbnb, and the proliferation of social media and information communication technologies have increased the demand for city tourism immoderately [27,28]. To define the unfavorable consequences of this excessive demand, Skift created the term 'overtourism' in 2016 [27], which has been increasingly used in the literature [14]. Nevertheless, there is no academically accepted definition of the term; thus, it remains open to multiple interpretations [29]. According to the UNWTO definition, overtourism is "the impact of tourism on a destination, or parts thereof, that excessively influences perceived quality of life of citizens and/or quality of visitors experiences in a negative way" [27] (p. 4).

One of the main economic problems caused by overtourism is the increase in the price of housing, goods, and services in destination areas [30] which fuels gentrification. Gentrification is the process of the displacement of low-income households from a transforming destination that often entails great expense [31]. The other element that drives gentrification is short-term rentals, which decrease the housing stock for long-term rentals [32]. The social and cultural issues associated with overtourism include depopulation, conflict between residents and tourists, insecurity, decreasing quality of life, and the destruction of local culture. With the increase in the touristification of urban centers, living conditions change and become less suitable for residents, which leads to depopulation [33]. An excessive number of tourists per resident decreases the quality of life of residents, resulting in tourismphobia [34]. Young travelers may be driven to seek out new experiences, causing a severe deterioration in perceived security because of their poor behavior, alcohol/drug use, and engagement in prostitution and gambling that diminish moral values and cultural traditions [35].

In terms of negative environmental impacts, overtourism drastically increases water, land, air, noise, and aesthetic pollution, and is associated with solid waste, littering and sewage treatment issues, infrastructure degradation, natural resource depletion, biodiversity loss, and climate change [36,37]. Additionally, although tourists may treat cultural assets with respect, an excessive number of visitors can threaten the physical condition of historical and archaeological sites through wear and tear, together with insufficient regulation and poor management [38].

2.2. The Measurement of Overtourism

The literature has defined several indicators for measuring the phenomenon of overtourism, which affects the wellbeing of both tourists and residents. For instance, Simancas Cruz and Peñarrubia Zaragoza [39] estimated tourist accommodation density, an indicator for determining a state which is undesirable for tourists and, thus, defines the optimal limit for accommodation saturation. Table 1 presents six indicators of overtourism as defined by Peeters et al. [40].

Table 1. Indicators of overtourism. Source: [40].

Indicator	Definition	Description
Tourism density	Bed-nights/km^2	Annual number of bed-nights per km^2
Tourism intensity	Bed-nights/resident	Annual number of bed-nights per resident in the destination
Sharing economy: Airbnb	Number of Airbnb offers	Number of Airbnb offers in a destination
Share of tourism contribution to GDP		
Air transport intensity	Air passengers/Bed-nights	Ratio of the number of air passengers to the number of bed-nights
Closeness to airports Closeness to cruise ports World Heritage Sites closeness		Arrivals within 50 km Number within 10 km Number within 30 km

The indicators above are designed to estimate whether a destination may be at risk of overtourism from an economic or social perspective. However, the authors [40] point to the inability of assigning a general value to an indicator that infers when a state of overtourism is likely to develop.

Additionally, to avoid negative consequences (e.g., overcrowding, congestion, or environmental degradation) arising from tourist activities and flows, how tourism develops in relation to the principles of sustainability should be considered [41]. As Coccossis et al. [41] note, planning and management for tourism growth are of particular importance in the context of sustainable development. Hence, it is especially important for tourism destinations to apply a minimum number of consistent indicators for assessing sustainable tourism [42] and the phenomenon of overtourism.

2.3. Tourism Carriying Capacity

In measuring overtourism, the concept of TCC is considered a valuable operational tool for identifying the limits to sustainable tourism activity [18,43]. On the one hand, a growing number of visitors positively affects the income and employment levels of some of the population in the tourist destination. On the other hand, growth in visitor numbers can generate negative effects. Understanding the number of tourists a destination can tolerate has become one of the major challenges which policymakers and scholars have been trying to address.

The topic of TCC prevails in the overtourism debate. The concept was defined by the World Tourism Organization as "the maximum number of tourists that a space can absorb without a lowering of the quality of the visitor's experience and without serious consequences for its ecology and its socio-economic structures" [44] (p. 5). The literature outlines three main dimensions of carrying capacity, including the physical–ecological, socio–demographic, and political–economic dimensions which are derived from problems such as the threat to the physical environment of a destination, the loss of a local community's character, and the dependence of the local economy on tourism [15,18].

An early description of TCC referred to the maximum number of visitors that could be supported without an unacceptable decline in the quality of visitor experience and the quality of the environment. Subsequently, the definition of TCC was complemented with reference to other types of capacity, including the capacity of supporting facilities, among them the maximum number of accommodation units, sitting places associated with the catering sector, parking spaces, the capacity for solid waste disposal, and other services which cater to the needs of visitors. These facilities can constrain the tourist capacity of the destination and impose additional costs on residents [45]. Previous research has widely applied the TCC model to determine the optimal number of visitors, especially in the context of cities, which have been exposed to negative externalities because of overtourism over recent years. Several scholars have estimated the carrying capacity of different destinations—including for the historical center of Venice [16,45], Rome [46], a coastal destination on the Costa del Sol, Spain [43], and Dubrovnik in Croatia [15].

In terms of the practical calculation of TCC, as Bertocchi et al. [16] highlighted, one of the prominent research streams is dedicated to the use of fuzzy linear programming. Costa and Van der Borg [47] and Canestrelli and Costa [45] were among the first to determine the social–economic carrying capacity of Venice's historical center by applying fuzzy linear programming. Fuzzy linear programming permits the estimation of the optimal number of tourists by considering several constraints and using approximate data [45,46]. The advantage of this unconventional method of optimization is its capacity to incorporate imprecise data.

However, there is no single and universal method of calculating TCC [16], since TCC can be studied in relation to its specific components and through a holistic approach involving the application of different methodologies and approaches. Some indicators cannot be operationalized because of the lack of data. The concept of TCC has also been criticized in the literature, as this type of measurement simplifies the problem of overtourism [14,43,48,49]. Some scholars assume that the concentration on calculating carrying capacity may be misleading as the estimated number of tourists may be too large and suboptimal [50]. Hence, the focus should be on visitors' experiences and behavior, rather than on the number of people [9,14,51].

Notwithstanding the difficulty of establishing the carrying capacity of destinations [9] and the critiques of the approach, McCool and Lime [49] confirm that in some cases the carrying capacities for facilities (e.g., parking lots or cultural sites) may be identified. In general, proponents believe that TCC can help stimulate sustainable development scenarios and the better management of tourist flows [16]. In this study, we adopt the commonly accepted indicators for measuring the carrying capacity of a given city by applying FLP to identify the optimal number of visitors to Budapest. Thus, we attempt to contribute to an under-researched area of urban sustainable tourism [52] that may help create the highest possible utility for tourism actors and the local community and economy.

This section has attempted to provide a summary of the literature related to the phenomenon of overtourism and its measurement. Much of the research on overtourism has focused on identifying and evaluating the reasons for the adverse effects of tourist activities and streams. To date, various methods have been developed and introduced to measure overtourism. However, the previous studies indicate the difficulty of measuring this phenomenon, especially in urban environments, outlining the need for comprehensive methodology to analyze the extent of overtourism in a city setting. Collectively, those studies highlight the critical role of determining the tourism carrying capacity of a destination and adopting an integrated approach in terms of data gathering and unconventional analytical methods. To address the issues related to overtourism in the context of Budapest, we, therefore, implemented the complex methodologies which are carefully described in the next section.

3. Materials and Methods

3.1. Methodological Triangulation

The phenomenon of overtourism in Budapest has been investigated by applying a combination of multiple methods of data collection and data analysis. The use of different data collection methodologies, called triangulation [53,54], has been theorized and adopted in tourism literature [55–57], including research about tourism-related issues in Budapest [7,9,11]. A triangulation strategy increases the credibility and validity of research [56,58], and provides better conclusions through the convergence and collaboration of findings [59]. However, as Koc and Boz reported, tourism literature lacks studies for which more than one method of data collection was used [56].

Following the recommendation of Koc and Boz [56] to increase the use of data triangulation, we triangulated different data sources from several independent sources to explore the studied phenomenon. First, we analyzed the academic literature and online media news on the appearance of overtourism in Budapest to understand the problems the city experienced before the outbreak of the coronavirus pandemic and subsequent border closures. Second, one of the authors collected photographs depicting the negative impacts of tourism activities in Budapest during 2019. Previous studies have confirmed the practicality of using images of tourist destinations as a research tool [60,61], as such images encapsulate the visual look of a place, its atmosphere, and the emotions it evokes [62].

The adoption of qualitative techniques preceded the development of further quantitative research [53,55]. Hence, third, we collected longitudinal tourism data for Budapest and its 23 districts. The data was complemented with the results of a survey distributed among Hungarian tourism industry experts. Similar to the studies of Bertocchi et al. and Camatti et al. [15,16], we adopted the criteria of TCC to study pressure on the urban destination (i.e., Budapest) to define the sustainable number of tourists staying in three types of accommodation. Unlike in Venice and Dubrovnik [15,16], Budapest has geographical characteristics that make it more difficult to capture the phenomenon of overtourism. Subsequently, eight tourist-supporting facilities (see their description in Section 4.3) were selected as relevant as they cater to the needs of tourists, and were analyzed with the application of FLP. Based on the integrated approach, the unconventional analytical methods applied in the current research offer new lenses and perspectives with which to understand the extent of overtourism in Budapest.

3.2. Tourism Carrying Capacity Model: Fuzzy Linear Programming

This section presents the mathematical background behind the estimation of the ideal maximum number of tourists in Budapest. Considering previous research on TCC [16], we use the unconventional optimization method called FLP to estimate the optimal number of tourists in the city. As an optimization tool, FLP helps identify the optimal number of tourists in a destination by considering, on the one hand, the need to maximize the total revenue generated by tourists in the destination, and on the other, the inherent limitations of the destination in terms of physical–ecological, socio–demographic, and political–economic constraints.

Mathematically, the FLP problem can be formulated as:

$$\begin{aligned} &\text{maximize} \widetilde{\mathbf{z}} \approx \widetilde{\mathbf{c}}\mathbf{x} \\ &s.t. \widetilde{\boldsymbol{A}}\mathbf{x} \preccurlyeq \widetilde{\boldsymbol{b}} \\ &\mathbf{x} \geq \mathbf{0} \end{aligned} \tag{1}$$

where the symbols $\approx$ and $\preccurlyeq$ mean equality and inequality, respectively, with respect to a given linear ranking function, F; $\widetilde{\mathbf{c}}^T = (\widetilde{c}_1, \ldots, \widetilde{c}_n)^T \epsilon R^n$ is the cost vector; $\mathbf{x} = (x_1, \ldots, x_n)^T \epsilon R^n$ is the vector of decision variables; $\widetilde{\boldsymbol{b}} = \left(\widetilde{b}_1, \ldots, \widetilde{b}_m\right)^T \epsilon R^m$ is the right-hand side vector of the constraints, and $\widetilde{\boldsymbol{A}} = \left[\widetilde{a}_{ij}\right]_{mxn} \epsilon R^{mxn}$ is the constraints matrix, being $\widetilde{c}_j$, $\widetilde{b}_i$, and $\widetilde{a}_{ij}$ fuzzy numbers, for $i = 1, \ldots, m, j = 1, \ldots, n$.

In the context of this study, given the revenue generated by each type of tourist in Budapest (the cost vector), the goal is to maximize total revenue, denoted as $\widetilde{\mathbf{c}}\mathbf{x}$, subject to the inherent limitations of the city represented by $\widetilde{\boldsymbol{A}}\mathbf{x} \preccurlyeq \widetilde{\boldsymbol{b}}$. The use of fuzzy numbers indicates that information is only an approximation (rather than exact information).

The solution to (1) uses the concept of the ranking function, where the original problem is transformed into its equivalent crisp problem and then standard methods such as the Simplex Method are applied [63]. This study considers the following ranking functions:

$$\begin{aligned} \text{Yager's } F_1 \text{ function}: \ F_1(\widetilde{a}) &= \tfrac{1}{3} \cdot \frac{\left[\left(a^R\right)^2 - \left(a^L\right)^2\right] + \left(R^2 - L^2\right) + \left(Ra^R - La^L\right)}{\left(a^R - a^L\right) + (R - L)} \\ \text{Yager's } F_3 \text{ function}: \ F_3(\widetilde{a}) &= \frac{L + a^L + a^R + R}{4} \end{aligned}$$

where $\widetilde{a} = \left(a^L; a^R; L; R\right)$ is a trapezoidal fuzzy number, being $L < a^L < a^R < R$ real numbers.

4. Results

4.1. Appearance of Overtourism in Budapest

With EU accession in 2004, the tourism industry in Budapest experienced rapid growth in international visitor numbers [11]. Low-cost airlines and numerous Airbnb units leveraged the number of tourists in Budapest, particularly younger tourists whose main motives were having fun and partying in the so-called Jewish ghetto of Budapest, located in the city center [9]. Besides these factors, the government and Hungary's tourism marketing organization, which covers Budapest, share responsibility for this boom in party tourism due to the lack of a consistent and strong marketing campaign between the years of 1990 and 2010 that resulted in an ambiguous brand image for the city [64,65]. The figure presented below (Figure 1) demonstrates the marketing strategies of Budapest; note that the focal point in 2012 differs from the image creation strategy in 2018.

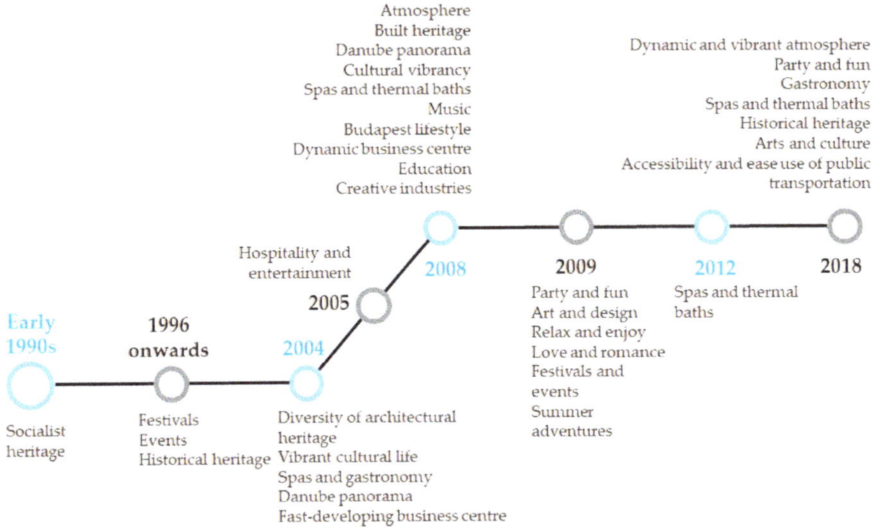

Figure 1. Main developments in the destination marketing strategy of Budapest. Source: Adapted from Ref. [66].

The other factor that boosted demand for the night-time economy was news about cheap alcohol and entertainment opportunities. In the mid-2000s, budget airlines promoted the low price of beer in Hungary in their marketing campaigns [9]. News on websites until 2016 similarly portrayed Budapest as the cheapest urban destination among European cities with a focus on alcoholic beverages. Problems caused by heavy alcohol consumption and party tourism created resistance among locals against tourism development in the city [9]. Since 2018, the news has centered upon overtourism and its negative consequences. In order to mitigate the phenomenon, the government introduced new regulation that authorizes municipalities to decide on or limit the maximum period of Airbnb rentals. Figure 2 exemplifies the evolution of online news about the capital city for 2013–2020.

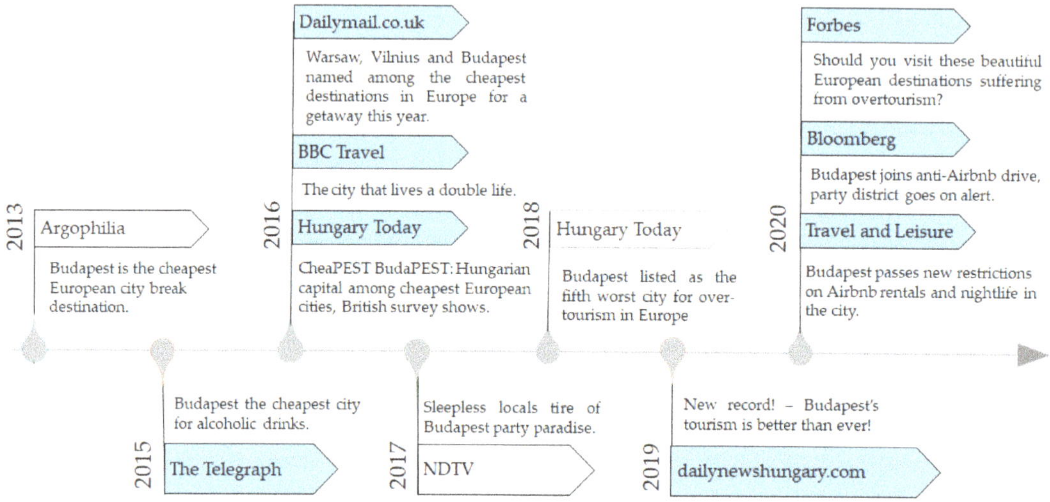

Figure 2. Online news about Budapest between 2013 and 2020. Sources: [67–77].

Pinke-Sziva et al. [11] conducted interviews with locals and tourists, and identified that most of the complaints arising from overtourism came from the city center of Budapest (District VII), which became a party and alcohol-centered quarter that provided an authentic visitor experience due to its ruin bars and the traces of Jewish cultural heritage. The concentration of short-term vacation rentals on Airbnb in this area is, likewise, influential in terms of leveraging the number of non-cultural tourists [78]. The touristification of the city center rapidly increased housing prices [9]. Based on data from the Hungarian Central Statistical Office (HCSO), housing prices increased 80% from 2007 to 2019 [6], which generated gentrification [9]. Other complaints about overtourism included public urination, street crime, litter, dirty streets, and the number of drunk people [11]. The photos below (Figure 3), taken in April–May 2019, demonstrate the negative environmental impacts of overtourism (cluttered streets caused by overloaded garbage containers, vomit-stained streets, and the social issue of overcrowding).

(a)

(b)

Figure 3. *Cont.*

(c) (d)

Figure 3. Impacts of overtourism in District VII, Budapest. (**a**) Overloaded garbage container; (**b**) dirty street; (**c**) overcrowding in Szimpla Garden (a ruin pub); and (**d**) overcrowding in Gozsdu Court. Source: Authors.

4.2. Indicators of Overtourism

This section explores the phenomenon of overtourism in Budapest considering four indicators: (1) tourism density, (2) tourism intensity, (3) the sharing economy, Airbnb, and (4) total occupation of hotels. The first three indicators are described in Table 1, while the last one describes the historical evolution of the total occupation rate in hotels.

4.2.1. Tourism Density and Intensity

According to Peeters et al. [40], the phenomenon of overtourism is primarily associated with tourism density (tourists per square kilometer) and tourism intensity (tourists per resident). Table 2 shows the calculation of both indicators considering Budapest (as a city) and five selected districts.

Table 2. Tourism density and intensity, Budapest and selected districts, 2019. Source: HCSO, calculations by authors.

Region	Bed-Nights (Annual Number)	Estimated Area (km^2)	Estimated Number of Residents *	Tourism Density	Tourism Intensity
				(Bed-Nights/km^2)	(Bed-Nights/Resident)
Budapest	1,405,548	525.14	1,751,251	26,765.20	8
District I	837,986	3.41	25,181	245,743.70	33.3
District V	2,688,192	2.59	25,975	1,037,912.00	103.5
District VI	2,050,451	2.38	38,670	861,534.00	53
District VII	2,758,598	2.09	51,896	1,319,903.30	53.2
District VIII	1,240,092	6.85	76,784	181,035.30	16.2

* Resident population in the middle of the year.

The results show that both indicators of overtourism vary significantly from district to district (see full table in the Supplementary Material). For instance, the minimum value for tourism density is associated with District XVII (value 65.3), while the maximum occurs in District VII (value 1,319,903.3). In the case of tourism intensity, the minimum also occurs in District XVII (value zero; after rounding 0.0408), while the maximum occurs in District V (value 103.5). Figure 4 shows both indicators graphically.

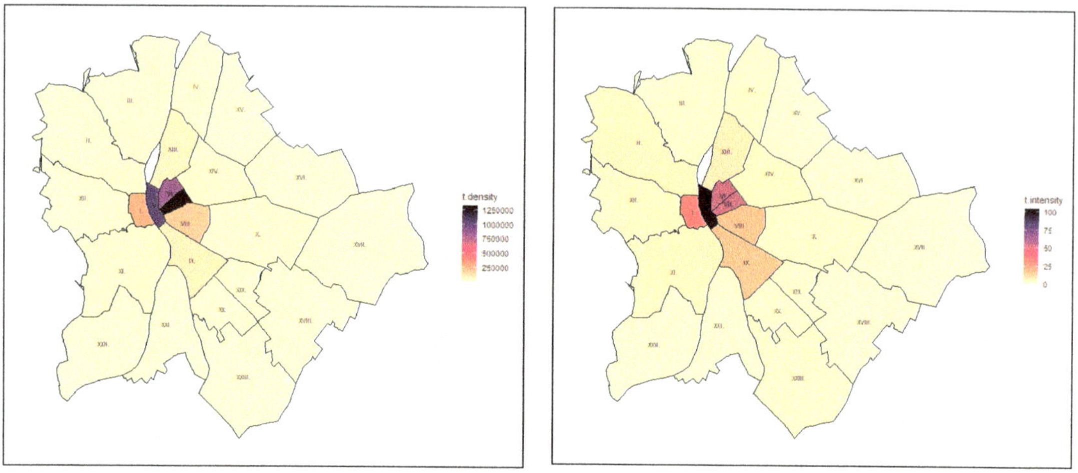

(**a**) (**b**)

Figure 4. Tourism density and intensity by district: (**a**) tourism density by district; and (**b**) tourism intensity by district. Source: HCSO, calculations by authors.

Based on Figure 4, the area most affected in terms of both measures is the center of Budapest (formed by districts I, V, VII, VII, and VIII). Moreover, tourism intensity also affects the nearest districts, such as districts II, IX, XI, XII, and XIII.

Table 3 presents the data per percentile group. Note that the last group (the 5th percentile) captures most of the range of values—approximately 90% in the case of tourism density and 85% in terms of tourism intensity (skewed distribution). It also shows that the districts within the 1st, 2nd, and 5th group are the same for both indicators.

Table 3. Percentiles for the 23 districts of Budapest, 2019. Source: HCSO, calculations by authors.

Percentile	Tourism Density (Bed-Nights/km^2)	Districts	Tourism Intensity (Bed Nights/Resident)	Districts
1st	65.3–709.4	XVI, XVII, XVIII, XXI, XXII	0–0.2	XVI, XVII, XVIII, XXI, XXII
2nd	709.4–6292.3	XV, XIX, XX, XXIII	0.2–2.3	XV, XIX, XX, XXIII
3rd	6292.3–13,855.3	II, III, IV, X, XII	2.3–4.0	III, IV, X, XII, XIV
4th	13,855.3–138,265.1	IX, XI, XIII, XIV	4.0–15.9	II, IX, XI, XIII
5th	138,265.1–1,319,903.3	I, V, VI, VII, VIII	15.9–103.5	I, V, VI, VII, VIII

4.2.2. Sharing Economy: Airbnb

Peeters et al. [40] also suggested monitoring the number of Airbnb offerings in the destination since this might be an indicator of overtourism (e.g., in the form of uncontrolled offerings in destinations, or their uncontrolled growth). Figure 5 presents the number of offerings in private accommodation (including Airbnb) by district.

Figure 5 shows that private accommodation is highly concentrated in the center of Budapest (formed by districts I, V, VI, VII, and VIII). Indeed, this zone concentrates 79.1% of the total offerings of the city. It also shows that nearby districts were also affected, such as districts II, IX, XI, and XIII.

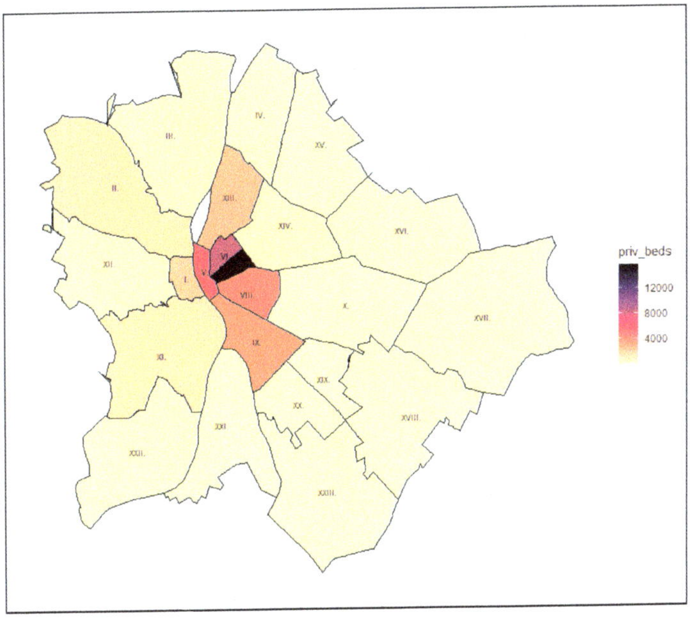

Figure 5. Number of units of private accommodation (including Airbnb). Budapest and its districts in 2019. Source: HCSO, calculations by authors.

4.2.3. Total Occupation in Hotels

The last indicator, as exemplified by Figure 6, illustrates the historical evolution of the total occupation of hotels (from 2010 to 2019).

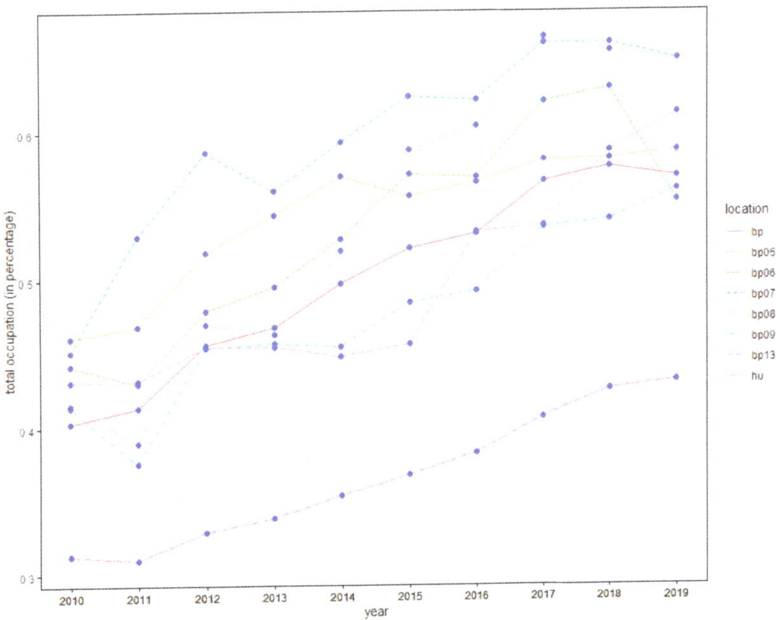

Figure 6. Occupation rate in hotels from 2010 to 2019. Source: HCSO, calculations by authors.

The figure displays the information at various geographical levels. At the national level, denoted as 'hu', the occupation rate in hotels increased by 37.7% (up from 31.3% in 2010 to 43.1% in 2019). In Budapest, denoted as 'bp,' the rate increased by 41.2% (from 40.3% in 2010 to 56.9% in 2019). However, at the district level the occupation rate increased unevenly. For instance, consider two districts located in the center of Budapest: in District VI, the occupation rate increased by 25.1% (from 44.2% in 2010 to 55.3% in 2019), while in District VIII the rate was 56.6% (from 41.5% in 2010 to 64.9% in 2019). Based on Figure 6, we conclude that the hotel occupation rate in Budapest displayed a robust increasing trend in the 2010s, and the rates for inner districts reached a high value by the end of the decade.

4.3. Estimation of Tourist Carrying Capacity

This section presents the first attempt to identify the optimal number of tourists in Budapest using an unconventional optimization method called FLP. The FLP problem can be formulated as

$$\begin{aligned} &\text{maximize} \widetilde{z} \approx \widetilde{c}\mathbf{x} \\ &s.t. \widetilde{\mathbf{A}}\mathbf{x} \preccurlyeq \widetilde{\boldsymbol{b}} \\ &\mathbf{x} \geq \mathbf{0} \end{aligned} \tag{2}$$

where $\mathbf{x} = (\text{TH}, \text{TO}, \text{TP})^T$ represents three types of tourists visiting Budapest: tourists staying in hotels (TH), tourists staying in other types of commercial accommodation excluding hotels (TO), and tourists staying in private accommodation (TP). The vector $\widetilde{c}^T = (\widetilde{c}_1, \widetilde{c}_2, \widetilde{c}_3)^T$ represents the economic benefit per day generated by each type of tourist. To capture the uncertainty associated with these values we use fuzzy numbers:

$$\widetilde{c}_1 = (25{,}200;\ 28{,}000;\ 30{,}800), \widetilde{c}_2 = (18{,}900;\ 21{,}000;\ 23{,}100), \widetilde{c}_3 = (22{,}050;\ 24{,}500;\ 26{,}950)$$

The limitations of the city, represented by $\widetilde{\mathbf{A}}\mathbf{x} \preccurlyeq \widetilde{\boldsymbol{b}}$, have been selected based on the available information. Table 4 summarizes the selected constraints.

Table 4. Selected constraints for Budapest.

	Tourism System	Description	Source	Utilization Rates			Maximum Daily Capacity
				TH	TO	TP	
1	Commercial accommodation: Hotels	Number of bed places in hotels (from 1 to 5 stars)	HCSO	1	0	0	$\widetilde{b}_1$
2	Commercial accommodation: Excluding hotels	Number of bed places excluding hotels	HCSO	0	1	0	$\widetilde{b}_2$
3	Private accommodation	Number of bed places in private accommodation (e.g., Airbnb platform, etc.)	HCSO	0	0	1	$\widetilde{b}_3$
4	Public transportation (Buses, Trams, Trolleybuses and Subway)	Daily maximum capacity for tourists (persons)	BKK Zrt. and authors' calculations	1	1	1	$\widetilde{b}_4$
5	Tourist attractions (The Hungarian Parliament Building)	Maximum capacity (persons)	Tourism Departments, Office of the Hungarian National Assembly	$\widetilde{a}_{51}$	$\widetilde{a}_{52}$	$\widetilde{a}_{53}$	$\widetilde{b}_5$
6	Tourist attractions: (Széchenyi thermal bath)	Maximum capacity (persons)	Budapest Spas cPlc.	$\widetilde{a}_{61}$	$\widetilde{a}_{62}$	$\widetilde{a}_{63}$	$\widetilde{a}_6$
7	Tourist attractions (Gellért thermal bath)	Maximum capacity (persons)	Budapest Spas cPlc.	$\widetilde{a}_{71}$	$\widetilde{a}_{72}$	$\widetilde{a}_{73}$	$\widetilde{b}_7$
8	Environment	Waste production per person (in kilograms)	HCSO and authors' calculations	$\widetilde{a}_{81}$	$\widetilde{a}_{82}$	$\widetilde{a}_{83}$	$\widetilde{b}_8$

Note: the acronym HCSO stands for Hungarian Central Statistical Office.

Table 4 presents some limitations of the city in terms of tourism. The first three constraints are associated with the capacity of the city in terms of accommodation. This study considers three types of accommodation: commercial accommodation in hotels, commercial accommodation excluding hotels, and private accommodation (including Airbnb). The fourth constraint shows the most frequently used forms of public transportation in the city: buses, trams, trolleybuses, and subway lines. The fifth constraint is related to the capacity of the Hungarian Parliament Building, which is considered one of the most scenic buildings in Budapest, and it symbolizes Hungary itself [79]. The following constraints are associated with two attractive places in the city: the Széchenyi and Gellért thermal baths. Typically, 90% of visitors who purchase daily tickets to these baths are foreign tourists [80]. Finally, the last constraint is related to an environmental aspect of the city in terms of waste production.

Table 4 also displays the utilization rates considering our three types of tourists: TH, TO, and TP. The utilization rates associated with tourist attractions (the Hungarian Parliament Building, and the Széchenyi and Gellért thermal baths) were estimated using a survey. We distributed a survey among Hungarian experts in the field of tourism [questionnaire in the Supplementary Material]. The survey contains four questions. For each question we asked for an approximation of the minimum, central, and maximum value. After the responses were collected, we calculated the means of the latter estimations. The resulting mean values were entered into the model through the following fuzzy numbers:

$$\widetilde{a}_{51} = (0.31; 0.41; 0.50), \widetilde{a}_{61} = (0.20; 0.30; 0.40), \widetilde{a}_{71} = (0.20; 0.30; 0.40), \widetilde{a}_{81} = (0.8; 1.1; 1.4)$$

$$\widetilde{a}_{52} = (0.28; 0.39; 0.50), \widetilde{a}_{62} = (0.16; 0.28; 0.39), \widetilde{a}_{72} = (0.16; 0.28; 0.39), \widetilde{a}_{82} = (0.7; 1.0; 1.4)$$

$$\widetilde{a}_{53} = (0.30; 0.43; 0.55), \widetilde{a}_{63} = (0.20; 0.31; 0.42), \widetilde{a}_{73} = (0.20; 0.31; 0.42), \widetilde{a}_{83} = (0.7; 1.0; 1.3)$$

The last column in Table 4 shows the elements of vector $\widetilde{\boldsymbol{b}} = \left(\widetilde{b}_1, \ldots, \widetilde{b}_8\right)^T$. The fuzzy numbers, $\widetilde{b}_{i's}$, were created in line with the following intuitive idea: consider the maximum daily capacity associated with the first constraint, $\widetilde{b}_1$. From HCSO, we know that the maximum number of bed places in hotels is 45,546. Based on this value, we created the fuzzy number (36,892; 40,991; 45,546), which indicates that the desirable number of occupied bed places in hotels is 90% of the maximum capacity (a value of 40,991), the minimum corresponds to 80% of maximum capacity (36,892), and the maximum represents 100% of maximum capacity (45,546). Using the same idea, we created the fuzzy numbers:

$$\widetilde{b}_1 = (36,892; 40,991; 45,546), \widetilde{b}_5 = (2734; 3038; 3375), \widetilde{b}_8 = (40,500; 45,000; 50,000)$$

$$\widetilde{b}_2 = (5311; 5901; 6557), \widetilde{b}_6 = (6184; 6872; 7635)$$

$$\widetilde{b}_3 = (36,357; 40,397; 44,885), \widetilde{b}_7 = (2809; 3121; 3468)$$

The fourth fuzzy number, $\widetilde{b}_4$, was created slightly differently. In this case, $\widetilde{b}_4$ represents the proportion of public transportation in Budapest (buses, trams, trolleybuses, and subway trips) associated with tourists. To our knowledge, there is no such information; thus, we asked experts to provide an approximation [questionnaire in the Supplementary Material]. According to the experts, the approximate proportion of public transportation accounted for by tourists in Budapest ranges from 7.3% to 18.8% of maximum daily capacity. Finally, based on the information provided by BKK Zrt., the estimated maximum daily capacity of public transportation in Budapest is 3,898,800 places, yielding the fuzzy number $\widetilde{b}_4$ = (304,106; 518,540; 732,974).

The solution of the FLP problem (2) was generated using the package FuzzyLP implemented in R [81,82]. Since the solution requires the use of ranking functions, we considered the following ones:

$$\text{Yager's } F_1 \text{ function}: \ R(\widetilde{a}) = \frac{1}{3} \cdot \frac{\left[\left(a^R\right)^2 - \left(a^L\right)^2\right] + \left(R^2 - L^2\right) + \left(Ra^R - La^L\right)}{\left(a^R - a^L\right) + (R - L)} \quad (3)$$

$$\text{Yager's } F_3 \text{ function}: \ R(\tilde{a}) = \frac{L + a^L + a^R + R}{4} \quad (4)$$

where $\tilde{a} = (a^L, a^R, L, R)$ is a trapezoidal fuzzy number, being $L < a^L < a^R < R$ real numbers.

Before introducing the results of our model, we reviewed the official information provided by HSCO. According to this institution, in total 5,694,027 tourists visited Budapest in 2019, staying at one of our three types of accommodation (TH, TO, or TP). This means that approximately 15,600 tourists visited the city per day (omitting the seasonality factor). Of these 15,600 tourists, approximately 11,269 (72.24%) stayed in hotels (TH); 1,378 (8.83%) stayed in other types of commercial accommodation excluding hotels (TO); and 2953 (18.93%) tourists stayed in private accommodation (TP).

Table 5 presents the estimation of the maximum number of tourists in Budapest using FLP. Considering our current information, the model suggests that the optimal number of visitors staying in hotels in Budapest is 7560 (right-hand side of Table 5) per day. Furthermore, we conducted a simulation study to explore under which circumstances it may be possible to accept more types of tourists into the city. Table 5 also displays the result of two selected simulations. In both cases, we modified the utilization rates associated with tourist attractions (the Hungarian Parliament Building, and the Széchenyi and Gellért baths), affecting three out of eight constraints of the model. The simulated utilization rates reported in Simulation 1 yield the optimal number of visitors: 9026 tourists staying in hotels, and 2643 tourists staying in other types of commercial accommodation. Similarly, the simulated utilization rates reported in Simulation 2 yield the optimal number of visitors per day: 8103 tourists staying in hotels, 5923 tourists staying in another type of commercial accommodation, and 991 tourists staying in private accommodation.

Table 5. Optimal number of tourists. Estimated result vs. Simulations.

Tourism System		Utilization Rates: Estimated and via Simulations			Optimal Number of Tourists		
		TH	TO	TP	TH	TO	TP
Tourist attractions (The Hungarian Parliament Building)	From experts	(0.31; 0.40; 0.50)	(0.28; 0.39; 0.50)	(0.30; 0.42; 0.55)	7560	0	0
	Simulation 1	(0.24; 0.28; 0.32)	(0.16; 0.20; 0.24)	(0.17; 0.24; 0.32)	9026	2643	0
	Simulation 2	(0.16; 0.23; 0.30)	(0.13; 0.17; 0.20)	(0.15; 0.20; 0.25)	8103	5923	991
Tourist attractions (Services of baths: Széchenyi and Gellért)	From experts	(0.20; 0.30; 0.40)	(0.16; 0.27; 0.39)	(0.20; 0.31; 0.42)			
	Simulation 1	(0.16; 0.23; 0.30)	(0.38; 0.40; 0.41)	(0.16; 0.23; 0.30)			
	Simulation 2	(0.13; 0.17; 0.20)	(0.18; 0.26; 0.35)	(0.16; 0.23; 0.30)			

The challenging problem of defining the optimal number of tourists in Budapest seems to be complex. In this section, we have made a first attempt to solve this problem using FLP.

5. Discussion and Conclusions

An initial objective of the study was to examine how the selected urban destination is affected by overtourism. First, the authors discussed the factors that caused overtourism in Budapest and presented the evolution of the marketing strategy in Budapest from the early 1990s to 2018 that resulted in unsustainable tourism in the city, associated with an ambiguous brand that focused on party tourism and cheap alcoholic beverages.

Second, the presence of tourists was estimated based on indicators such as tourism density and tourism intensity, the number of offerings in private accommodation, and total occupation in hotels. The calculation of tourism density and intensity at the district level revealed the uneven distribution of the former in the city. The center of Budapest (formed by districts I, V, VI, VII, VIII, and IX) was most affected by the presence of tourists in terms of the number of tourists per square kilometer and the number of tourists per resident. In contrast, districts XVI, XVII, XVIII, XXI, and XXII, which are physically far from the center, were less affected. This finding is consistent with that of Koens et al. [14], who stated that the impact of overtourism was not city-wide and could predominantly be observed in the more popular parts of the city. In the context of Budapest, Remenyik et al. [6] detected the rapid development of tourism in all 23 districts, with the downtown districts most affected. In this area, the mass of tourists and their behavior reached intolerable levels for locals.

Another important finding is related to accommodation. The number of offers of private accommodation, including Airbnb, is highly concentrated in the center of Budapest (districts I, V, VI, VII, and VIII). According to our calculations, this zone concentrated 79.1% of all offerings. This fact might partly explain the phenomenon of overtourism in these districts. This outcome was reported by Smith et al. [7], who highlighted the highest density of Airbnb accommodation in District VII and its neighboring districts.

In reviewing the literature, no data was found in relation to determining the sustainable number of visitors in Budapest. However, studies have noted the importance of defining the optimal number of tourists [15,16,43,45,46] in order to plan and manage their influx [41]. In the present study we have attempted to specify the optimal number of tourists in Budapest using an unconventional optimization method called FLP. The aim was to maximize the total revenue generated by tourists in Budapest considering a set of limitations associated with the city—specifically, in terms of accommodation, public transportation, tourist attractions (the Hungarian Parliament building, and the Széchenyi and Gellért thermal baths), and waste production. The results suggest that the optimal number of tourists per day is 7560 staying in hotels, compared to the actual, approximately 15 thousand in the peak year of 2019.

The major limitation of this approach is associated with the availability of data. Hence, we suggest incorporating additional information that was not available for the present study. For instance, this could include more types of tourists (e.g., one-day tourists or cruise tourists that transit along the Danube River), or more constraints (e.g., number of seating places available in restaurants, bars, cafes, pubs, number of available parking spaces, capacity in theaters or museums, and the like).

The complexity of the city environment demands that more attention be paid to estimating the optimal number of tourists in Budapest. One policy implication is that the implementation of a centralized system might contribute to the periodic collection of the information required for analysis. The use of technologies can be helpful in the data collection process. For example, Camatti et al. [15] proposed employing new sources of data, including sensors, telecommunication data, cameras, and data from connected objects, for constructing new indicators. Pásková et al. noted that the utilization of TCC could be facilitated by recent advances in tourism studies and information technologies—namely, in the fields of big data collection, advanced data analysis, and modeling [17]. Another policy recommendation is monitoring the places visited by tourists in Budapest using different technologies. This information could be useful for thoroughly evaluating the phenomenon of overtourism in the city.

The data-driven management of visitors may be appropriate for regulating the tourist flows, local traffic, and housing sector. The regulation of private accommodation offers is seen as a significant means of alleviating the situation of overtourism in the downtown area of Budapest. We encourage the local authorities to implement more sustainable tourism strategies and policies by restricting Airbnb apartments in Districts V, VI, and VII, but encouraging their operation in Districts VIII, IX, and XI, and other places, as well as opening new hotels and other types of commercial accommodation on the periphery of the city. Overall, visitors should be dispersed to the less crowded districts of the capital. Thereby, the heterogeneous character of overtourism in Budapest calls for a careful investigation of this phenomenon at the district level.

In agreement with Koens et al. [14], we conclude that overtourism cannot be alleviated by focusing on tourism alone, insofar as the wider usage of the city should be considered. With the help of the insight and knowledge obtained from the application of unconventional analytical methodology, local authorities would be able to better monitor their policy responses and improve the urban infrastructure—for example, by promoting alternative routes and attractions on the city outskirts. Increasing cooperation between multiple city departments and other stakeholders, including residents, is also advised.

Supplementary Materials: The following are available online at https://www.mdpi.com/article/10.3390/su14042268/s1: Table S1, Tourism density and intensity in Budapest and its districts, 2019. Questionnaire S1, Survey for experts.

Author Contributions: Conceptualization, S.S.S., B.P.G. and G.M.; methodology, V.S. and B.P.G.; code preparation and formal analysis, B.P.G.; investigation, V.S., S.S.S., B.P.G., G.M. and D.B.; resources, G.M., V.S., S.S.S., B.P.G. and D.B.; writing—original draft preparation, V.S., S.S.S., B.P.G. and D.B.; writing—review and editing, V.S., G.M., S.S.S., B.P.G. and D.B.; visualization, B.P.G.; supervision, S.S.S., B.P.G. and G.M.; project administration, S.S.S. All authors have read and agreed to the published version of the manuscript.

Funding: V.S. would like to thank the National Bank of Hungary for financial support under the Research Excellence Award. S.S.S. and B.P.G. are obliged to Project No. TKP2020-NKA-02 that was implemented with support from the National Research, Development, and Innovation Fund of Hungary, financed under the Thematic Excellence Programme funding scheme. The contribution of G.M. is supported by the OTKA K134877 project. S.S.S. and B.P.G. also wish to acknowledge research support from the Corvinus Institute for Advanced Studies (CIAS). The dissemination of research outcomes was supported by The Hungarian Tourism Association Foundation.

Institutional Review Board Statement: Ethical review and approval were waived for this study since the survey used only collects opinions from experts. The participation in the survey was voluntary. All participants were informed about the anonymity and confidentiality of their answers. The survey was performed in compliance with the Helsinki Declaration guidelines and supervised by the Corvinus University of Budapest.

Informed Consent Statement: Not applicable.

Data Availability Statement: Data was requested from the Hungarian Central Statistical Office, website: https://www.ksh.hu/?lang=en (accessed on 20 January 2022). BKK Centre for Budapest Transport, website: https://bkk.hu/en/. Budapest Spas cPlc., (accessed on 20 January 2022). Budapest Spas cPlc., website: https://www.spasbudapest.com/ (accessed on 20 January 2022). Office of the Hungarian National Assembly, website: https://www.parlament.hu/web/visitors/contact-us (accessed on 20 January 2022).

Conflicts of Interest: The authors declare no conflict of interest.

References

1. Best Places to Travel in 2019. Available online: http://www.europeanbestdestinations.com/european-best-destinations-2019/ (accessed on 27 January 2022).
2. Butler, A. These Are the Best-Value Destinations for an Affordable Adventure in 2020. Available online: https://www.lonelyplanet.com/articles/affordable-destinations-best-in-travel (accessed on 27 January 2022).
3. Kovács, Z. Year in Review: Hungary's Tourism Industry Had a Booming Year. Available online: https://abouthungary.hu//blog/2019-year-in-review-hungarys-tourism-industry-had-a-booming-year (accessed on 27 January 2022).
4. Hungarian Tourism Agency. Report on the Record Year. Tourism in Hungary. 2019. Available online: https://mtu.gov.hu/documents/prod/Report-on-the-record-year-2019..pdf (accessed on 27 January 2022).
5. Lock, S. European Cities Ranked Worst for Over-Tourism 2017. Available online: https://www.statista.com/statistics/778687/overtourism-worst-european-cities/ (accessed on 27 January 2022).
6. Remenyik, B.; Barcza, A.; Csapó, J.; Szabó, B.; Fodor, G.; Dávid, L.D. Overtourism in Budapest: Analysis of Spatial Process and Suggested Solutions. *Reg. Stat.* **2021**, *11*, 179–197. [CrossRef]
7. Smith, M.K.; Egedy, T.; Csizmady, A.; Jancsik, A.; Olt, G.; Michalkó, G. Non-Planning and Tourism Consumption in Budapest's Inner City. *Tour. Geogr.* **2018**, *20*, 524–548. [CrossRef]
8. Smith, M.; Puczkó, L.; Rátz, T. Twenty-Three districts in search of a city: Budapest—The capitaless capital? In *City Tourism: National Capital Perspectives*; Maitland, R., Ritchie, B.W., Eds.; CABI: Wallingford, UK, 2009; pp. 201–213. ISBN 9781845935467.
9. Smith, M.K.; Sziva, I.P.; Olt, G. Overtourism and Resident Resistance in Budapest. *Tour. Plan. Dev.* **2019**, *16*, 376–392. [CrossRef]
10. Miskolczi, M.; Jászberényi, M.; Munkácsy, A.; Nagy, D. Accessibility of Major Central and Eastern European Cities in Danube Cruise Tourism. *Deturope* **2020**, *12*, 133–150.
11. Pinke-Sziva, I.; Smith, M.; Olt, G.; Berezvai, Z. Overtourism and the Night-Time Economy: A Case Study of Budapest. *Int. J. Tour. Cities* **2019**, *5*, 1–16. [CrossRef]

12. Michalkó, G. Social and Geographical Aspects of Tourism in Budapest. *Eur. Spat. Res. Policy* **2001**, *8*, 105–118.
13. Khomsi, M.R.; Fernandez-Aubin, L.; Rabier, L. A Prospective Analysis of Overtourism in Montreal. *J. Travel Tour. Mark.* **2020**, *37*, 873–886. [CrossRef]
14. Koens, K.; Postma, A.; Papp, B. Is Overtourism Overused? Understanding the Impact of Tourism in a City Context. *Sustainability* **2018**, *10*, 4384. [CrossRef]
15. Camatti, N.; Bertocchi, D.; Carić, H.; van der Borg, J. A Digital Response System to Mitigate Overtourism. The Case of Dubrovnik. *J. Travel Tour. Mark.* **2020**, *37*, 887–901. [CrossRef]
16. Bertocchi, D.; Camatti, N.; Giove, S.; van der Borg, J. Venice and Overtourism: Simulating Sustainable Development Scenarios through a Tourism Carrying Capacity Model. *Sustainability* **2020**, *12*, 512. [CrossRef]
17. Pásková, M.; Wall, G.; Zejda, D.; Zelenka, J. Tourism Carrying Capacity Reconceptualization: Modelling and Management of Destinations. *J. Destin. Mark. Manag.* **2021**, *21*, 100638. [CrossRef]
18. Mexa, A.; Coccossis, H. Tourism carrying capacity: A theoretical overview. In *The Challenge of Tourism Carrying Capacity Assessment: Theory and Practice*; Coccossis, H., Mexa, A., Eds.; Ashgate Publishing: Aldershot, UK, 2004; pp. 37–54. ISBN 9780754635697.
19. World Tourism Organization (UNWTO). *International Tourism Highlights, 2020 ed.*; World Tourism Organization (UNWTO): Madrid, Spain, 2021; ISBN 9789284422456.
20. Yuval, F. To Compete or Cooperate? Intermunicipal Management of Overtourism. *J. Travel Res.* **2021**, 00472875211025088. [CrossRef]
21. Wagar, J.A. The Carrying Capacity of Wild Lands for Recreation. *For. Sci.* **1964**, *10*, a0001-24. [CrossRef]
22. Plog, S.C. Why Destination Areas Rise and Fall in Popularity. *Cornell Hotel Restaur. Adm. Q.* **1974**, *14*, 55–58. [CrossRef]
23. Emerson, R.M. Social Exchange Theory. *Annu. Rev. Sociol.* **1976**, *2*, 335–362. [CrossRef]
24. Boissevain, J. Tourism and Development in Malta. *Dev. Change* **1977**, *8*, 523–538. [CrossRef]
25. Williams, T.A. Impact of Domestic Tourism on Host Population. *Tour. Recreat. Res.* **1979**, *4*, 15–21. [CrossRef]
26. Butler, R.W. The Concept of a Tourist Area Cycle of Evolution: Implications for Management of Resources. *Can. Geogr. Géographe Can.* **1980**, *24*, 5–12. [CrossRef]
27. World Tourism Organization (UNWTO); Centre of Expertise Leisure, Tourism & Hospitality; NHTV Breda University of Applied Sciences; NHL Stenden University of Applied Sciences. *'Overtourism'?—Understanding and Managing Urban Tourism Growth beyond Perceptions, Executive Summary*; UNWTO: Madrid, Spain, 2018. [CrossRef]
28. Bock, K. The Changing Nature of City Tourism and Its Possible Implications for the Future of Cities. *Eur. J. Futur. Res.* **2015**, *3*, 20. [CrossRef]
29. Mihalic, T. Conceptualising Overtourism: A Sustainability Approach. *Ann. Tour. Res.* **2020**, *84*, 103025. [CrossRef]
30. Szromek, A.R.; Kruczek, Z.; Walas, B. The Attitude of Tourist Destination Residents towards the Effects of Overtourism—Kraków Case Study. *Sustainability* **2020**, *12*, 228. [CrossRef]
31. Wilhelmsson, M.; Ismail, M.; Warsame, A. Gentrification Effects on Housing Prices in Neighbouring Areas. *Int. J. Hous. Mark. Anal.* **2021**. *ahead-of-print*. [CrossRef]
32. Celata, F.; Romano, A. Overtourism and Online Short-Term Rental Platforms in Italian Cities. *J. Sustain. Tour.* **2020**, 1–20. [CrossRef]
33. Casagrande, M. Heritage, Tourism, and Demography in the Island City of Venice: Depopulation and Heritagisation. *Urban Isl. Stud.* **2016**, *2*, 121–141. [CrossRef]
34. Fontanari, M.; Traskevich, A. Consensus and Diversity Regarding Overtourism: The Delphi-Study and Derived Assumptions for the Post-COVID-19 Time. *Int. J. Tour. Policy* **2021**, *11*, 161–187. [CrossRef]
35. Szromek, A.R.; Hysa, B.; Karasek, A. The Perception of Overtourism from the Perspective of Different Generations. *Sustainability* **2019**, *11*, 7151. [CrossRef]
36. Sunlu, U. Environmental impacts of tourism. In *Local Resources and Global Trades: Environments and Agriculture in the Mediterranean Region*; Camarda, D., Grassini, L., Eds.; CIHEAM: Bari, Italy, 2003; pp. 263–270.
37. Atzori, R. Destination Stakeholders' Perceptions of Overtourism Impacts, Causes, and Responses: The Case of Big Sur, California. *J. Destin. Mark. Manag.* **2020**, *17*, 100440. [CrossRef]
38. Frey, B.S.; Briviba, A. Revived Originals—A Proposal to Deal with Cultural Overtourism. *Tour. Econ.* **2021**, *27*, 1221–1236. [CrossRef]
39. Simancas Cruz, M.; Peñarrubia Zaragoza, M.P. Analysis of the Accommodation Density in Coastal Tourism Areas of Insular Destinations from the Perspective of Overtourism. *Sustainability* **2019**, *11*, 3031. [CrossRef]
40. Peeters, P.; Gössling, S.; Klijs, J.; Milano, C.; Novelli, M.; Dijkmans, C.; Eijgelaar, E.; Hartman, S.; Heslinga, J.; Isaac, R.; et al. *Research for TRAN Committee—Overtourism: Impact and Possible Policy Responses*; European Parliament, Policy Department for Structural and Cohesion Policies: Brussels, Belgium, 2018.
41. Coccossis, H.; Mea, A.; Collovini, A.; Parpairi, A. *Defining, Measuring and Evaluating Carrying Capacity in European Tourism Destination*; University of the Aegean: Lesvos, Greece, 2000.
42. Tanguay, G.A.; Rajaonson, J.; Therrien, M.-C. Sustainable Tourism Indicators: Selection Criteria for Policy Implementation and Scientific Recognition. *J. Sustain. Tour.* **2013**, *21*, 862–879. [CrossRef]
43. Navarro Jurado, E.; Damian, I.M.; Fernández-Morales, A. Carriying Capacity Model Applied in Coastal Destinations. *Ann. Tour. Res.* **2013**, *43*, 1–19. [CrossRef]

44. World Tourism Organization (UNWTO). *Tourist Makerts, Promotion and Marketing, Saturation of Tourist Destinations*; World Tourism Organization (UNWTO): Rome, Italy, 1981.
45. Canestrelli, E.; Costa, P. Tourist Carrying Capacity. *Ann. Tour. Res.* **1991**, *18*, 295–311. [CrossRef]
46. Feliziani, V.; Miarelli, M. How Many Visitors Should There Be in the City? The Case of Rome. *Rev. Eur. Stud.* **2012**, *4*, 179–187. [CrossRef]
47. Costa, P.; van der Borg, J. Un Modello Lineare per La Programmazione Del Turismo. *CoSES Inf.* **1988**, *18*, 21–26.
48. Liberatore, G.; Biagioni, P.; Talia, V.; Francini, C. Overtourism in Cities of Art: A Framework for Measuring Tourism Carrying Capacity. *SSRN Electron. J.* **2019**, 1–28. [CrossRef]
49. McCool, S.F.; Lime, D.W. Tourism Carrying Capacity: Tempting Fantasy or Useful Reality? *J. Sustain. Tour.* **2001**, *9*, 372–388. [CrossRef]
50. Marsiglio, S. On the Carrying Capacity and the Optimal Number of Visitors in Tourism Destinations. *Tour. Econ.* **2017**, *23*, 632–646. [CrossRef]
51. Wang, J.; Huang, X.; Gong, Z.; Cao, K. Dynamic Assessment of Tourism Carrying Capacity and Its Impacts on Tourism Economic Growth in Urban Tourism Destinations in China. *J. Destin. Mark. Manag.* **2020**, *15*, 100383. [CrossRef]
52. Lu, J.; Nepal, S.K. Sustainable Tourism Research: An Analysis of Papers Published in the Journal of Sustainable Tourism. *J. Sustain. Tour.* **2009**, *17*, 5–16. [CrossRef]
53. Jick, T.D. Mixing Qualitative and Quantitative Methods: Triangulation in Action. *Adm. Sci. Q.* **1979**, *24*, 602. [CrossRef]
54. Scandura, T.A.; Williams, E.A. Research Methodology in Management: Current Practices, Trends, and Implications for Future Research. *Acad. Manage. J.* **2000**, *43*, 1248–1264. [CrossRef]
55. Decrop, A. Triangulation in Qualitative Tourism Research. *Tour. Manag.* **1999**, *20*, 157–161. [CrossRef]
56. Koc, E.; Boz, H. Triangulation in Tourism Research: A Bibliometric Study of Top Three Tourism Journals. *Tour. Manag. Perspect.* **2014**, *12*, 9–14. [CrossRef]
57. Oppermann, M. Triangulation? A Methodological Discussion. *Int. J. Tour. Res.* **2000**, *2*, 141–145. [CrossRef]
58. Flick, U. Triangulation in Data Collection. In *The SAGE Handbook of Qualitative Data Collection*; SAGE Publications Ltd.: London, UK, 2018; pp. 527–544.
59. Johnson, R.B.; Onwuegbuzie, A.J. Mixed-methods research: A research paradigm whose time has come. In *Mixed-Methods*; Bryman, A., Ed.; Sage Publications Ltd.: London, UK, 2006; Volume 2, pp. 3–24.
60. Garrod, B. Exploring Place Perception a Photo-Based Analysis. *Ann. Tour. Res.* **2008**, *35*, 381–401. [CrossRef]
61. Steen Jacobsen, J.K. Use of Landscape Perception Methods in Tourism Studies: A Review of Photo-Based Research Approaches. *Tour. Geogr.* **2007**, *9*, 234–253. [CrossRef]
62. Jenkins, O.H. Understanding and Measuring Tourist Destination Images. *Int. J. Tour. Res.* **1999**, *1*, 1–15. [CrossRef]
63. Nasseri, S.H.; Ebrahimnejad, A.; Cao, B.-Y. Fuzzy linear programming: Solution techniques and applications. In *Studies in Fuzziness and Soft Computing*; Springer: Cham, Switzerland, 2019; ISBN 9783030174217.
64. Puczko, L.; Ratz, T.; Smith, M. Old City, New Image: Perception, Positioning and Promotion of Budapest. *J. Travel Tour. Mark.* **2007**, *22*, 21–34. [CrossRef]
65. Smith, M.; Puczko, L. Out with the Old, in with the New? Twenty Years of Post-Socialist Marketing in Budapest. *J. Town City Manag.* **2010**, *1*, 288–299.
66. Smith, M.; Puczkó, L. Budapest: From Socialist Heritage to Cultural Capital? *Curr. Issues Tour.* **2012**, *15*, 107–119. [CrossRef]
67. Budapest Is the Cheapest European City Break Destination. Available online: https://www.argophilia.com/news/budapest-cheapest-destination/210019/ (accessed on 27 January 2022).
68. Budapest the Cheapest City for Alcoholic Drinks. Available online: https://www.telegraph.co.uk/travel/destinations/europe/hungary/budapest/articles/Budapest-the-cheapest-city-for-alcoholic-drinks/ (accessed on 27 January 2022).
69. Diebelius, G. Warsaw, Vilnius and Budapest Named Top Three Cheapest Destinations in Europe for a Getaway This Year. Mail Online. Available online: https://www.dailymail.co.uk/travel/travel_news/article-3500066/Warsaw-Vilnius-Budapest-named-cheapest-destinations-Europe-getaway-year.html (accessed on 27 January 2022).
70. The City That Lives a Double Life. Available online: https://www.bbc.com/travel/article/20160831-the-city-that-lives-a-double-life (accessed on 27 January 2022).
71. CheaPEST BudaPEST: Hungarian Capital among Cheapest European Cities, British Survey Shows. 2016. Available online: https://hungarytoday.hu/cheapest-budapest-hungarian-capital-cheapest-city-travel-barometer-90235/ (accessed on 27 January 2022).
72. Sleepless Locals Tire of Budapest Party Paradise. Available online: https://www.ndtv.com/world-news/sleepless-locals-tire-of-budapest-party-paradise-1789785 (accessed on 27 January 2022).
73. Budapest Listed as the Fifth Worst City for Over-Tourism in Europe. 2018. Available online: https://hungarytoday.hu/hungary-listed-as-the-fifth-worst-city-for-over-tourism-in-europe/ (accessed on 27 January 2022).
74. New Record!—Budapest's Tourism Is Better than Ever! Available online: https://dailynewshungary.com/new-record-budapests-tourism-is-better-than-ever/ (accessed on 27 January 2022).
75. Should You Visit These Beautiful European Destinations Suffering from Overtourism? Available online: https://www.forbes.com/sites/ceciliarodriguez/2020/01/30/to-go-or-not-to-go-to-these-beautiful-european-places/ (accessed on 27 January 2022).
76. Budapest Joins Anti-Airbnb Drive, Party District Goes on Alert. Available online: https://www.bloomberg.com/news/articles/2020-07-15/budapest-joins-anti-airbnb-drive-party-district-goes-on-alert (accessed on 27 January 2022).

77. Budapest Passes New Restrictions on Airbnb Rentals and Nightlife in the City. Available online: https://www.travelandleisure.com/travel-news/budapest-airbnb-nightlife-overtourism-restrictions (accessed on 27 January 2022).
78. Dudás, G.; Boros, L.; Kovalcsik, T.; Kovalcsik, B. The Visualization of the Spatiality of Airbnb in Budapest Using 3-Band Raster Representation. *Geogr. Tech.* **2017**, *12*, 23–30. [CrossRef]
79. Técsi, Z.; Orczy, Z.; Naszradi-Mészáros, O. *Csodálatos Budapest: Wonderful Budapest*; Pannon-Literatúra Kft: Kisújszállás, Hungary, 2014; ISBN 9789632514291.
80. Record Number of Visitors at Budapest Baths. Available online: https://www.budapestinfo.hu/record-number-of-visitors-at-budapest-baths (accessed on 27 January 2022).
81. R Core Team. R: A Language and Environment for Statistical Computing: R Foundation for Statistical Computing, Vienna, Austria. 2020. Available online: http://www.R-project.org/ (accessed on 27 January 2022).
82. Villacorta, P.J.; Rabelo, C.A.; Pelta, D.A.; Verdegay, J.L. FuzzyLP: An R package for solving fuzzy linear programming problems. In *Granular, Soft and Fuzzy Approaches for Intelligent Systems*; Kacprzyk, J., Filev, D., Beliakov, G., Eds.; Springer International Publishing: Cham, Switzerland, 2017; Volume 344, pp. 209–230. ISBN 9783319403120.

sustainability

Article

A Theoretical Concept of an Innovative and Sustainable Product Based on an Unconventional Approach to Design Development

Gábor Megyeri *, Kitti Boros and Balázs Fekete

Institute of Marketing, Corvinus University of Budapest, Fővám tér 8, H-1093 Budapest, Hungary; kitti.boros@uni-corvinus.hu (K.B.); education@balazsfekete.com (B.F.)
* Correspondence: gabor.megyeri@uni-corvinus.hu

Abstract: Health security measures have become increasingly important in tourism, as there is a heightened need to rebuild the trust of tourists in destinations and tourism services. Innovative product development might help respond to the emerging health-related needs of tourists. We address the lack of the application of Designcommunication (DIS:CO)-based approaches in the field of tourism, present a process for the theoretical development of a smart device, and explore its applications. Combining the results of the qualitative inquiry and applying DIS:CO in research and development, the concept of 4S—traveling Safe, Secure, Smart, and Sustainable—emerges in a socio-cultural and economic context, supporting the foundation of the material realisation of the product. As a result, the 4S concept presents how customised, continual feedback on the health condition of the user before and during a trip, and notifications about possible health risks in different tourist areas, might facilitate the process of trust-building. Although data on unconventional tourists is not included in tourism statistics, the designed product can assist in gathering information about them, too. With the permission of users, the built-in functions of the designed product can be used to detect their position and direction of movement. The product also helps monitor the health of tourists by providing up-to-date data on infection levels to tourism service providers. In its passive mode, as a souvenir, the device strengthens travel desire.

Keywords: tourism safety; design concept; Designcommunication; unconventional approach; cross-border mobility

Citation: Megyeri, G.; Boros, K.; Fekete, B. A Theoretical Concept of an Innovative and Sustainable Product Based on an Unconventional Approach to Design Development. *Sustainability* **2022**, *14*, 3022. https://doi.org/10.3390/su14053022

Academic Editors: Anna Rita Irimiás, Dallen J. Timothy and Gábor Michalkó

Received: 1 February 2022
Accepted: 28 February 2022
Published: 4 March 2022

1. Introduction

The novelty in our research is that we present a process of how Designcommunication (DIS:CO) and tourism can be connected in a way that results in the creation of the design concept of the product. The term "Designcommunication" is a patented expression (abbrev. DIS:CO), intentionally written in compound form referring to the definition of the term provided by its author, Attila Róbert Cosovan: Designcommunication = DIS:CO = communication integrated into development, according to the Hungarian Gazette for Patents and Trademarks (113. no. 12./I., 2008.12.15. Registration number: 196961). Interpreting the field of tourism as a context for design has become an increasingly common phenomenon. The field of service design seems to be a rapidly developing area. A strong customer-centred approach, combined with an academic understanding of both tourism and service design research, enhances both fields [1,2]. The branch of design theory that we work in can be identified as an example of human-centred design, applied in tourism in this case. The definition of Tussyadiah (2013) [3] (p. 548) of such approaches is that they can be characterised by an extensive awareness of the needs, wants, and expectations of humans, and the goal is to establish "a direct connection between the internal state of users, design characteristics, and the contexts of interactions between users and design". The DIS:CO theory and method fit this description, and thus can become excellent tools for development in the field of tourism. The application of DIS:CO facilitates concept formulation in a way that utilises the capacity for creativity in the form of interdisciplinary collaboration without the limitations of a rigid process.

The travel restrictions put in place by governments as a response to the COVID-19 pandemic have not only affected tourism revenues worldwide [4,5] but have also changed the behaviour of tourists [6]. According to Irimiás and Michalkó (2016) [7], attitude is shaped by enduring beliefs and tendencies (such as hate or longing) in relation to objects or persons. The researchers also found that mood and attitude can greatly differ with the levels of emotion [7]. Travel craving can be interpreted as a context-specific construct that appears when people are unable to travel, either for economic or health reasons or due to external constraints [8]. Travel craving is not a precursor of actual travel, but a cognitive-emotional event in time that focuses on travel when it is not possible [8]. Previous tourism practices will no longer meet the changing demands of tourists and, as a result, actors in the tourism supply chain need to be prepared for the challenges of the post-COVID-19 era. This is by no means a return to 'the normal' we experienced before, but an adaptation to a changing and developing world that will never be the same as it was before the pandemic. In the post-COVID-19 era, tourism practices need to be redesigned, and an unconventional approach based on product design and development will play an important role in the process. Based on the findings of Freire-Gibb and Lorentzen (2011) [9] (p. 165), we created the theoretical concept of an innovative product that is built on the idea that "taking unconventional approaches leads to the creation of something special. When we work together, across all kinds of boundaries, we can achieve things that we could never accomplish alone." The B-bridge product concept [10] has been expanded on this basis in this paper. The tool theoretically created as part of the B-bridge concept, a bracelet, may play an important role in the restoration of tourist traffic to Budapest, may contribute to the incentivisation of domestic tourists to visit the Hungarian capital, and may result in longer lengths of stay. The idea of the B-bridge concept—the name of which was inspired by the bridges of Budapest—has served as the basis for the 4S (Safe, Secure, Smart, and Sustainable) concept of our new product created in response to the pandemic. The new product (which is also a bracelet), while active, constantly monitors the state of health of the tourist and, when inactive, functions as an accessory and souvenir. The designed product has the capacity to influence both the mood and the attitude of the wearer. In its active mode, it shapes their mood as a smart device that monitors their state of health and offers various services, while in its passive mode, it shapes attitudes by evoking the longing for the pleasant memories it recalls, which is a reflection of the phenomenon explored in the literature.

The paper is structured as follows: a description of the reasons that justify the need for the research and of the current situation of international tourism is followed by a presentation of the trends of new tourism such as safety, smart technology, and sustainability, based on the literature (Section 2). In the Methodology section, the research design (the formulation of research questions and the explanation of the qualitative research method) is described in detail. With the interdisciplinary approach of the scientific fields that our research is rooted in, we have identified six dilemmas that helped us formulate our research question. Our research question is: "How can sustainable product design and development support the prosperity of different tourism products in the new normal?" (Section 3). The next section discusses the results, for which the groundwork is laid by the findings of the qualitative research, and the 4S product concept is created. The idea behind the product is presented both from a product designer's and a user's perspective, supported by theoretical examples (Section 4). The paper is then closed with the conclusions that can be drawn from our research, the description of the recommended managerial implications (Section 5), and a review of the limitations of the study and of the possible directions for future research (Section 6).

2. Literature Review

2.1. Designcommunication (DIS:CO) as a Bridge between Society, Economy, and Design

Design, in general, translates to design art, creative design, and creative behaviour—it is now an integral part of business planning as well, and the popularity of the design thinking method is no accident [11] (p. 233). In the case of Designcommunication, it is combined with communication integrated into development, as the definition of DIS:CO [12] suggests. Studies linking design and tourism prefer to employ the specific and rigid process of Design Thinking (e.g., [13–15]). In the present study, we attempted to go beyond this iterative linear development model and chose the DIS:CO method, with possible exponential development in mind. As Cosovan et al. (2018) [11] suggest in the anthropological analysis of Designcommunication: it is a process where value creation takes place during a transitional rite, a value-oriented process that forms the configuration of designers' roles and the outcome at the end of the process. In the case of design thinking, this could only take place through a fixed routine and within the boundaries of pre-set creator roles. Designcommunication refers to utilising the perspectives and creative abilities of designers in order to solve open-ended problems within a divergent and unstrained process. The person or group utilising the DIS:CO approach and methodology gains transdisciplinary and interprofessional methods that can be used to create real-time links between education, research, and business [16,17]. DIS:CO is a relationship-building concept that emerges like a bridge between the phenomena of the various disciplines of society and the economy [11] and [17] (p. 36).

2.2. Safe and Secure Traveling

It is essential to clarify context in order for value creation to be realised through Designcommunication; thus, the theoretical background of related travel trends such as tourism safety, smart technologies, and sustainability are described in detail below. Efficient customer communication, strategy creation, the training of staff, and cooperation to ensure the quickest possible containment of the pandemic have become essential in this new situation and, as a result, tourism has needed to find new pathways since 2020 [18]. The popularity of domestic tourism grew dynamically in 2020 and 2021 since visitors were much more willing to choose nearby destinations during the crisis than more distant locations that were deemed less safe. Following the elimination of lockdown measures, the primary motivation behind touristic travel was visiting friends and relatives (VFR), which, beyond the reinforcement of personal relationships, also contributed to an increase in tourist traffic [19]. The COVID-19 pandemic also influenced the behaviour of tourists, since certain travellers chose to refrain from travelling in order to avoid the risks of the pandemic [20]. Thus, residents need to be encouraged to travel locally and visit local attractions, as it is their own country's or region's pandemic status that they know best, and they and their family can travel safely within the destination that they are residents in [21]. International research agrees that tourism first needs to be boosted through the promotion of domestic travel, and that will be followed by the growth of international tourism markets [22–24].

The popularity of local travel is not a new phenomenon—the concept of the "staycation" had already been present during the world economic crisis. The phrase was coined from the words "stay" and "vacation", and its message is that people can recharge even while staying at home. In this touristic genre, visitors do not leave their homes during the vacation, and make one-day trips to local tourist attractions or participate in locally available recreational activities [25]. According to the 2021 report of the UNWTO, staycations and holidays spent near the home will regain their popularity in the future, as they provide a sense of safety for visitors. In the post-COVID-19 era, travel will continue to play an important role in stress reduction, socialisation processes, and the reinforcement of interpersonal relationships, but the exploration of local recreational options will be prioritised, as proven by the research of Steen Jacobsen et al. (2021) [26] through the territorial example of the metropolis of Oslo. The research of Weman-Josefsson (2021) [27] also confirms the market penetration of the staycation: 86% of the Swedish population reported that in 2020 they spent their summer vacation domestically, at a destination close to their home, instead of previously preferred destinations. With the gradual return of the demand for international travel, personal safety and the health safety status of touristic destinations (which have been increasingly important factors even before the pandemic) are becoming top factors in destination choices. In the future, the public health situation of locations will cease to be an independent tourism safety factor, and it will need to be harmonised with classic sustainability goals which will also need to be a reference point for regulation and educational strategies [28]. The pandemic has highlighted not only the difficulties around public health but housing inequalities and the related health effects as well. The restrictions put in place for the containment of the pandemic changed people's attitudes toward their homes, which became the sole physical location of their everyday lives, and as a result, became multifunctional spaces [29].

The restrictions that have regulated international travel during the pandemic will contribute to increased tourist traffic during the restart period, as people wish to remedy the frustration caused by long periods of confinement by travelling. According to the conclusions of Crompton (1979) [30], after the end of the pandemic, the tourism sector will continue to need to prioritise product development and marketing activities focused on unmet socio-psychological needs in order to increase tourist traffic. However, the longing for novelty may decrease the willingness to return, as a destination that has been once visited can no longer be considered a novelty. At the same time, even novelty-seeking tourists can be induced to visit again through efficient product development. The COVID-19 pandemic and its effects will not override the two dimensions of the process of choosing a destination, as internal psychological and external non-psychological considerations [31] continue to be important factors. However, the two dimensions will presumably be broadened by additional elements, and the relative ratio of their importance in the decision-making process will change. The internal factors that influence decision-making constitute the push factors during decision-making, which include (1) psychological considerations, (2) physical factors, (3) social interactions, (4) and searching/discovering. These include 11 additional subcriteria as well, such as escapism, prestige, the desire for recreation, and the maintenance and improvement of health. External factors draw visitors to destinations as pull factors, as tangible immaterial and intangible factors. The two-pull type motivational factors include an additional nine subfactors, such as transport options, personal safety, the attractiveness of the destination, and expected benefits [32].

2.3. Smart Technologies

Besides the negative effects of the pandemic, certain consequences can be interpreted as positive, such as technological innovation, the fast-paced development of the information technology skills of humanity, and the penetration of virtual reality in tourism. These have been proven to have contributed to and enhanced the wellbeing and satisfaction of people during lockdowns and travel restrictions [33]. Although virtual tourism has alleviated people's frustration during lockdowns, studies [34–36] have shown that it cannot replace in-person tourism in the long run; thus, new tourism provides destinations an opportunity for renewal. Currently, predominant trends include the increasing advancement of extended reality systems, condensed experience-seeking, and interactivity, since new and innovative forms of tourism have emerged as a result of the travel restrictions of 2020 [37]. Extended reality (XR) includes all immersive technologies. It covers all of those that already exist, such as augmented reality (AR), virtual reality (VR), and mixed reality (MR), and also includes those that are yet to be created [38]. Virtual reality transfers users into an environment that is entirely different from physical reality, namely, into a virtual environment created and displayed by computers. This, contrary to augmented reality, does not include elements of the physical environment [39]. The role of hyper-reality and virtual communities continues to grow and, as a result, traditional interpretative methods are obsolete for today's tourists, particularly for generation Z, who are familiar with the full spectrum of audio-visual technology. Tourists wish to participate in this active, entertaining, and informative process with all of their senses. One of the leading segments that spearheaded innovative solutions in response to the pandemic was the meetings and events industry; countless events were transferred into the online space fully or partially (hybrid events) or were postponed to later dates. The 2020 Tokyo Olympics were eventually held in the summer of 2021, and individuals entering the venues of the event and potential close contacts were identified through innovative biometric face recognition software [40].

The challenges posed by the COVID-19 pandemic have also created an opportunity for accepting certain rules as the norm. These can be adjusted as needed during a potential future crisis situation and, as a result, humanity can be more prepared to combat infectious diseases. Wearing masks to cover one's nose and mouth, regular hand sanitisation, physical distancing, and contactless technology all make it possible for business activities to continue, albeit under different conditions. In tourism, these contactless technologies can replace physical menu cards through scannable QR codes, and make contactless hotel check-ins possible through mobile room keys, contactless payment methods, and orders that can be submitted through mobile applications [41]. Contactless, easily adaptable, and customisable automated tools and devices facilitate and efficiently support both front-office (such as check-in/check-out) and back-office (such as guest follow-up, aftercare, and marketing) tasks. In the new normal, companies and organisations in the tourism sector need to have solutions that can help them manage uncertainty through the utilisation of automated and contactless services [42]. Even after the COVID-19 pandemic subsides, travellers will need to remain prepared, keep the appropriate distance from one another, sanitise their hands regularly, and pay extra attention to personal hygiene, while touristic service providers also need to be prepared for having these rules respected.

2.4. Aspects of Sustainable Development

The concept of sustainable development was defined in the 1980s, and its generally accepted and widely known definition was developed by the Brundtland Commission [43]. The goal of sustainable development is meeting the needs of modern-day society in a way that does not jeopardise the needs of future generations, creating equality between the generations [44]. The idea of sustainable development is comprised of three main aspects: environmental, economic, and sociocultural dimensions. However, there is no unified consensus yet among researchers regarding the synthetisation and measurement of sustainability [45–48]. Members of the scientific community agree that the time horizon of sustainability is long-term [49,50] since sustainable practices—although they are capable of generating minor development even in the short term—can only yield real results in the field of touristic products and services if applied in the long term. Adopting a sustainable approach as widely as possible has become important for the mitigation of the negative effects of tourism; thus, the question of sustainability is being considered increasingly consciously by academics, tourism experts, and the tourists themselves [48].

The pandemic, despite the damage it has caused to tourism, is creating the opportunity for conscious change so that the sector can ensure long-term environmental, economic, and sociocultural sustainability. In the literature of tourism, sustainability was first understood as environmental sustainability alone, and researchers dedicated special attention to studying the subject [46]. Studies exploring environmental sustainability were increasingly focused on the investigation of the effects of climate change, and the ecological footprint of the tourism sector has proven to be rather large [51,52]. However, beyond the environmental factor, the economic dimension of sustainability also needs to be explored, as these issues can only be tackled jointly. The economic dimension of sustainability is the protection of the economic needs and the standard of living of the population [48]; in the context of tourism, this means that tourism-derived revenues need to benefit local residents as well. Being aware of the sociocultural aspect of sustainability is indispensable from a tourism perspective, as studies [53,54] have shown that environmental and sociocultural dimensions constitute the aggregated touristic resources jointly—that is, they are the defining factors in the attractiveness of touristic destinations and the basis of touristic demand. A sustainable approach, besides its influence on the destination choices of tourists, is capable of inducing cultural change within a society according to Hediger (2000) [55], and this is what the innovative product presented in this paper could support. The sustainable approach represented by the product can be manifested in numerous ways. Traditionally, the environmental aspect was central for the designing and development of a sustainable product, and then a reflection on all three pillars of sustainability emerged [56]. The procedure contributes implicitly to the realisation of social equality and cohesion [57]. According to Santos et al. (2022) [58], in order to anchor this much more holistic interpretation of sustainability, a new paradigm needs to be applied in the case of touristic destinations. This theory needs to be manifested on the level of the product as well since the sustainability-related factors in product development are tightly connected to the concept of the circular economy and regenerative systems [59,60].

3. Methodology

3.1. The Aim of the Study

The relevance of the research is supported by the fact that opportunities provided by technological solutions supporting tourism are significantly increasing in today's digitalisation. This goes hand in hand with the challenges indicated by tourism megatrends and the need to provide solutions that support the safe and sustainable satisfaction of growing travel needs. The theoretical gap this paper aims to cover is that, although the interdisciplinary field of tourism is related to market-oriented fields such as marketing, more practical, design-oriented approaches to research and development are underrepresented.

The starting point of our interdisciplinary research was the following question: how can research and product development that support sustainable value creation be realised in the intersection of tourism, design, and marketing? Our goal is for theory and practice to manifest in an integrated form, and beyond research findings, provide a useful starting point for the creation of a concrete product in the future. This paper is the extension of our previous research [10,18], with our investigation extended to the subject of sustainability. While previously, [18] our objective was the reduction of pandemic-related health risks, this study and its findings aim to contribute to new tourism in a way that is sustainable in the long term, in the form of a concrete product concept that can be utilised in a practical way.

3.2. Research Question

The formulation of the final research question was supported by the process of creative dialogue based on Designcommunication [12]. DIS:CO *"is a unique perspective where 'communication' emerges simultaneously to problem seeking and solving and is coded into the product, service or procedure created"* [11] (pp. 233–234) and thus it is a design framework that can be utilised in various fields of society. The next figure (Figure 1) shows that, firstly, the common pool of researchers' and designers' subject-related questions was established in relation to the fields involved and their intersections. This method of formulating questions is in accordance with the approach of DIS:CO, as it can be understood as a form of creative connection building and creative behaviour that manifests on the level of dialogue between individuals [17].

The following dilemmas, connected to the various approaches and their intersections, assisted the process of formulating the research question through a designers' and creators' dialogue:

1. What will be the new norm of tourism?
2. How can design support achieving the new norm of tourism?
3. How can we design future-proof products in today's analog and digital worlds?
4. How can we create a sustainable product?
5. What types of products and services are there consumer demand for?
6. How can the new norm of tourism serve consumer needs?

The final research question was distilled as a synthesis of the above general questions:

How can sustainable product design and development support the prosperity of different tourism products in the new normal?

Sub-question used for qualitative research:

Q1: What are the important aspects of digital and analog lifestyles?

Sub-question used in DIS:CO product development process:

Q2: How can the aspired value (derived from Q1) manifest through a product concept?

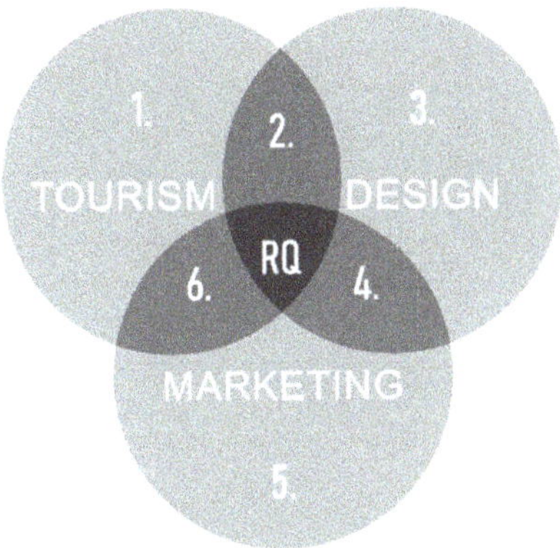

Figure 1. Mapping the overlapping interdisciplinary fields of tourism, design, and marketing. Source: author's own editing.

3.3. Research Strategy

For the exploration of the research question we formulated, we established a research plan that utilised several qualitative procedures and was multimethodological according to the definition of Morse (2003) [61] (Figure 2). It can be said that such an investigation was the combination of different methodologies, which in this case were qualitative. Thus, according to Harrison and Reilly (2011) [62], our work can be considered a multi-method integrative research project. It can be defined as a triangulated research model due to the involvement of several researchers and the various methods utilised [63].

The utilisation of a qualitative strategy is suited to understanding the interpretations of participants and can be used to explore the creation of meaning related to the phenomena. During qualitative research, the researcher observes the phenomena of the world in their natural environment and attempts to understand them based on the meanings that people assign to them [64].

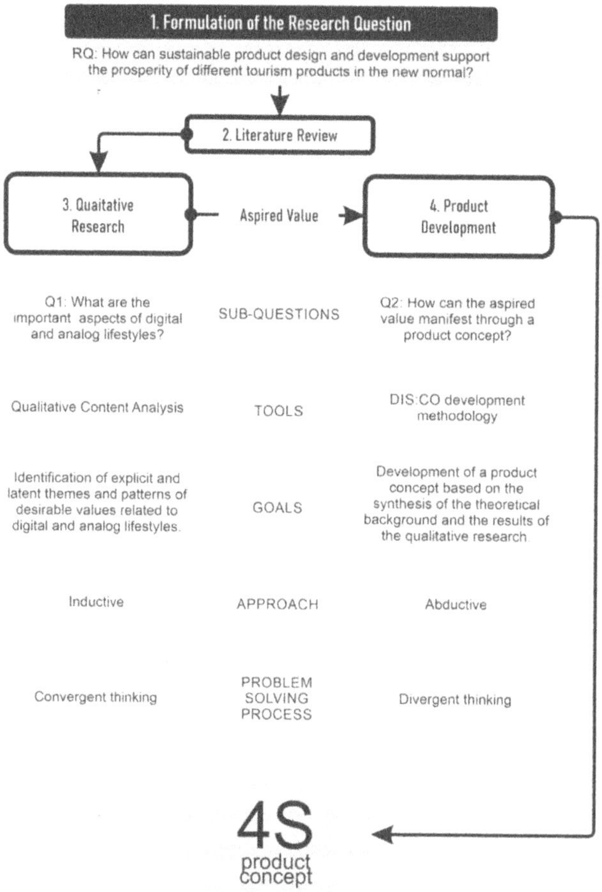

Figure 2. Research design. Source: author's own editing.

3.4. Data Collection and the Characteristics of the Sample

Our qualitative research was conducted on the basis of essays titled "Digital past, analog future" written by our respondents. The collection of the data that formed the basis for the study began before the pandemic, in 2018, and 150 essays were processed by September 2021. The data were derived from the essays of 1–2 pages of higher education students aged 20–25, written either in Hungarian or in English, either typed or handwritten. Due to the transformation of the socioeconomic environment induced by the pandemic, new opportunities arose for utilising our already existing sample. We analysed 100 essays from the pre-pandemic era and 50 from the period between 2020 and the end of data collection. In recent years, the subject of digitalisation has gained heightened relevance worldwide at virtually all levels of the operation of societies. Thus, the sample units of one of our research projects related to analog and digital product creation that was launched before COVID-19 has also been included in this research. As we had data both from before and during the pandemic, we had the opportunity to observe the changes in data patterns between these two periods. After the disclosure of the findings, we highlighted the phenomena that emphasised this change.

3.5. Analysis Phase–QCA

We analysed the data collected during our primary research through qualitative content analysis. The nature of the investigation was inductive, that is, the researcher started from the data and progressed upwards, and any theories were only guidelines. Researchers looked for, organised, and elevated to a more abstract level any patterns, categories, or themes observable in the empirical data. Qualitative content analysis (QCA) is a systematic analytical procedure. Its aim is to let us draw consequences regarding the patterns that are present in the text. This analytical method can be utilised for uncovering subjective interpretations through the identification and coding of themes and patterns found in the data [65]. According to Mayring (2000) [66], it can be said that during the utilisation of the QCA method, texts are analysed within the context of communication, and following pre-determined steps. These steps can be summarised as follows (Figure 3):

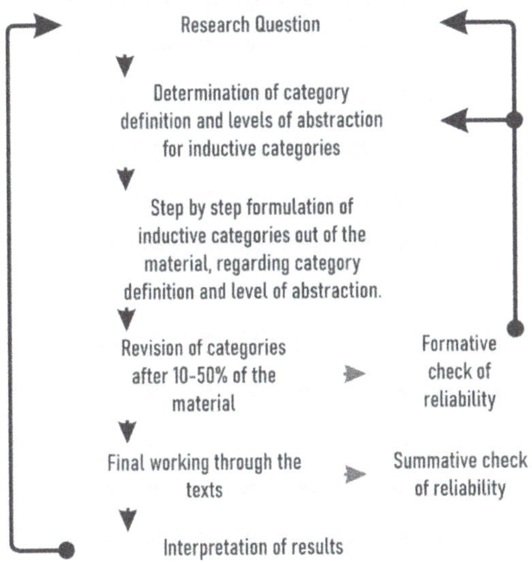

Figure 3. Steps of systematic qualitative content analysis. Source: author's own editing based on [66].

The QCA process is fixed. Raw data are categorised by themes that can be discovered based on patterns observed during the interpretation of the data, and on the basis of the relations between these patterns. Thus, the essence of the inductive analytical process is the careful investigation and continuous comparison of the data by the researcher [67]. As it can be seen in Figure 3, a qualitative content analysis goes beyond the level of words. It relies on the observation of themes and patterns to bring to the surface any explicit and latent content that is hidden in the text. This way, the social reality related to the phenomenon can be extracted in a subjective yet scientific way [65].

The data are coded and analysed manually and in an analog way. During the interpretation of the results, that is, the formulation of the aspired value, interim findings that were conflicted due to the differing outcomes of the three coders were consensually reduced to factors that were relevant for answering *Q1*. As an interpretation of the most relevant themes and patterns derived this way, the aspired value that formed the basis of the product concept was defined, and this was transformed into a product concept through the DIS:CO development framework.

3.6. Development Phase–DIS:CO Development Framework

Designcommunication (DIS:CO) is an approach and practice for designers and creators. Cosovan (2009) [12] refers to it as a philosophy that manifests through theory and practice.

The term is a compound word, in which design is understood as artistic design, creative design, and creative behaviour, while communication is understood as the designer's and creator's creative connection to building intent. The term is a patented expression and procedure. The method, originally distilled from the experiences of design art practice, has contributed to the creation of numerous internationally recognised products, such as the Red Dot Design Award-winning DSI Salt Inhaler, Teqball, Coco Dice, and Nosiboo products [68–71]. DIS:CO, beyond being a contemporary alternative to the widely known concept of Design Thinking [11], is also excellent for scientific research and for developing social and economic innovation [72]. Galla (2021) [73] and Horváth et al. (2018) [74] successfully utilised the method in secondary and university education and research. The utilisation of the method is justified by the fact that DIS:CO is a holistic framework in which the designer's toolkit is built upon the criteria of human needs, technological possibilities, and business success during the development of a creative product or procedure [11] (p. 233). Its most relevant element for the purposes of this study is the three-fold system of relations that can be described the following way (Figure 4):

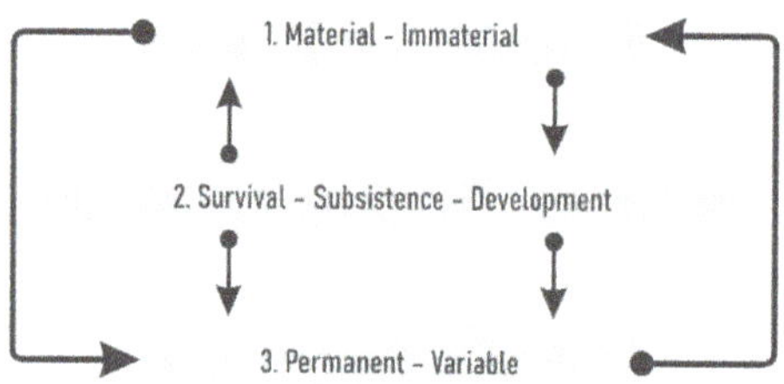

Figure 4. Cornerstones of DIS:CO research and development method. Source: author's own editing based on [12,16,17,70].

During the ritual of designing, the sequential development process of the listed dimensions is carried out in an unfixed order, in an iterative way: investigation of (1) Material–Immaterial qualities, the assessment of how (2) Survival–Subsistence–Development aspects emerge in the phenomenon, and the examination of (3) Permanent–Variable elements. This way, during the design phase, our research findings manifested in the form of a theoretical product concept, providing a basis for practical value creation as well.

4. Discussion

4.1. Findings of the Qualitative Research

We deemed the findings of our research (starting in 2018 and still ongoing) relevant for the formation of our concept. This longitudinal research provides an opportunity for the parallel investigation of pre-COVID-19 and post-COVID-19 results. However, a comparative analysis was not among the goals of the study. We considered the number of mentions and the frequency-of-mention throughout the whole sample, the direct interpretations of the statements of respondents, and the connections of themes observed in the text for the determination of the weight of particular themes, which increased with time. The following are the relevant themes and patterns identified this way, originally from the pre-COVID-19 era but also appearing in the post-COVID-19 era: (1) a rise in analog experience-seeking, (2) the recognition of the importance of in-person contact, (3) a lack of analog knowledge transfer and value creation, (4) the complementary role of digitalisation, (5) the connection between virtuality and separation, (6) topics related to a sustainable community lifestyle.

The definition of the meaning ranges of themes and a few illustrative examples (see Figure 5):

1. Analog experience-seeking: collecting experiences that require mobility, are real-time and are gained in a physical environment. Examples: 14, 7;

2. In-person contact: the reinforcement of personal relationships and the establishment of new ones in real-time and in a physical environment. Examples: 2, 81;
3. Analog knowledge transfer and value creation: the sharing of one's knowledge and experience with no transfer medium between individuals. Examples: 23, 112;
4. The complementary role of digitalisation: digital solutions support the acquisition of and the immersion in in-person experiences, but they cannot replace them. Examples: 56, 14, 99;
5. The connection between virtuality and separation: digital communication platforms can only alleviate the experience of physical distance partially, the quality of relationships changes. Examples: 56, 7;
6. A sustainable community lifestyle: the consideration of the wellbeing and sociocultural values of interrelated groups. Examples: 24, 114

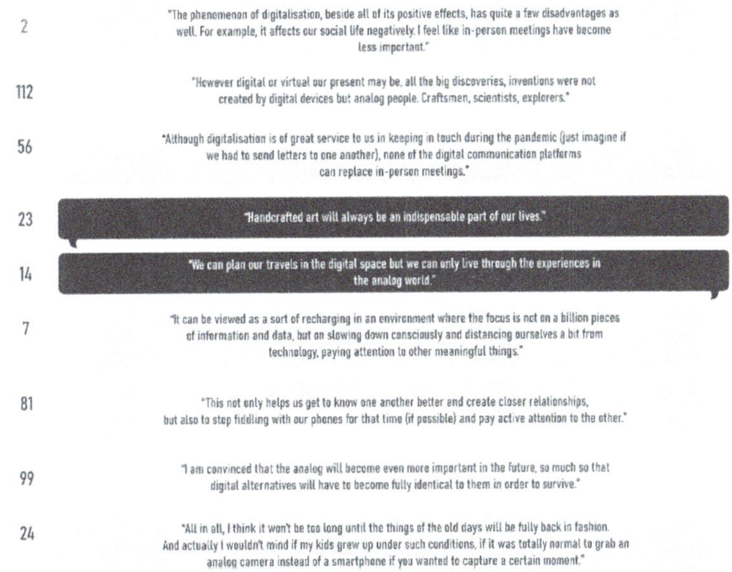

Figure 5. Text-based examples to illustrate the identified themes. Source: author's own editing.

Based on the findings of the qualitative analysis and as a summary of these, the aspired value (the input of the next phase of our research) was determined. During the fourth phase of the research, the DIS:CO product development process, we aimed to create this aspired value through an open problem-solving strategy. As Dorst (2011) [75] (p. 524) describes the development process, *"we have to create a 'working principle' and a 'thing' (product, service, system) in parallel"* that can help create the aspired value we are looking for. We define the aspired value the following way: *Instead of virtual experiences, analog, in-person experiences are desired. Digitalisation should play a complementary role besides an analog (offline) lifestyle, supporting analog experiences.* In our experience-seeking society, besides the safety-related aspect of digital solutions, their role in the acquisition of desired in-person experiences is also particularly valued. Based on the results of this research phase, we were able to shape our concept in the product development phase described below.

4.2. Results of Product Development

The central element of our product development idea was adopted from extended analysis of our previous product concept [10,18]. Keeping with the bridge allegory used in B-bridge, the designer's connection building was built upon the identification of the pillars. Our goal was to integrate the words in the Safe–Secure–Smart tagline as product attributes, surrounded by the holistic approach of sustainability, and this is how the idea of an innovative product that supports tourism emerged. The result of our sequentially progressing research and theoretical creative process was the 4S concept. While it includes the dimensions of environmental and economical sustainability, our product is predominantly built on the idea of community wellbeing and value creation facilitated by sociocultural sustainability.

4.2.1. Safe–Secure–Smart–Sustainable: The Concept of 4S Travelling

The fight against the COVID-19 pandemic is proving to be successful, with vaccinations being utilised worldwide. As a result, tourism, one of the most important strategic sectors, has been able to resume. However, new waves of the pandemic caused by new variants make returning to the pre-COVID-19 state of affairs difficult. It is uncertain if we will ever be able to fully overcome this pandemic. There is also the risk of a global outbreak of another similar virus, which could repeatedly impede the operation of the economy. The current situation requires the introduction of new solutions and new tools and demands that we review and innovatively rethink our existing toolkit. The goal is to create a method, environment, and community lifestyle that is safe for everyone involved and that allows for a sustainable and continuous operation while curbing the spread of the virus. One of the keywords for this endeavour is contactless. As a response, based on our previously created B-bridge concept, we wish to develop a new tool concept that would facilitate fully contactless getting around for tourists of all types with any motivation behind their visit. The device stores data in the central application. These can be shared with a third party, and are part of an authentication protocol that may enable smooth entry to airports, transport vehicles, conferences, and other types of events that have potentially high numbers of visitors and therefore where the risk of infection is consequently higher. Additional data, such as previous illnesses or vaccination status can be added to the data package through an authentication process. The emergence of wireless technologies and the evolution of sensors worn on the human body have paved the way for customisable, wearable health devices. The control systems worn on the wrist continuously provide physiological data and information about the general state of health of the individual. Systems that monitor vital signs reduce health-related costs and improve the quality of life by preventing disease [76].

Tourists can actively use the device in a given destination; it offers various services and a wide range of information through an application. The device supports travel planning and management in its active mode and evokes travel craving in its passive status. The data generated during its use provide essential information for marketing and tourism researchers, and also contribute greatly to the mapping of unconventional tourism. The primary areas in which the product can be particularly helpful are the following: transport, public transport, cultural events, shopping, etc.

4.2.2. Description of the Device

The services and basic toolkit of the device we designed are identical to those of the product we created for the B-bridge concept, with the difference that it includes an additional sensor. This is an analog thermometer that is able to measure body temperature with precision. In the original concept, the device was fixed onto a bracelet, but it could also be fastened to the strap of an existing smartwatch. Utilising available technologies, heart rate values that may foreshadow certain illnesses can be detected on the wrist as well. An example of this is heart rate changes that occur during a state of rest, namely, heart rate variability (HRV). Through its measurement and the averaging of the results, certain illnesses can be reliably predicted before the onset of symptoms. COVID-19 is one of these illnesses. In this stage, patients, although unaware of their condition, are already able to transmit the disease. Consequently, our device may play a role in combatting and preventing COVID-19 and other viral infections. Through storing the data processed by the device in an anonymous system and where it undergoes further processing, information can be gained about the infection level of a particular location or event, which is useful for users and organisers alike.

A large quantity of personal data is generated through the use of the device, which would be treated as sensitive data, in accordance with moral and security requirements. Stakeholders would be provided the necessary information about this, and accept the terms before they started using the device. The more versatile the available data the more efficient the operation of the system. We established two categories for the incoming and processed data: the first for the so-called anonymous data that ensure the operation of community services, the other category for detailed personal data. These are required, among other things, for organising medical care if needed and potentially for other forms of assistance or position detection. These data would not be public and would only be disclosed to the relevant authorities, with the consent of the user, and may only be utilised in an emergency or if immediate medical care was needed [77].

A worldwide pandemic that induces lockdowns affects human relationships, both on a personal and a community level. Becoming better equipped for managing or preventing a global pandemic requires the cooperation of multiple scientific fields, which can be assisted by digitalisation. We need to collaborate to reach our goal: we need to create a digital environment in which all parties feel comfortable and safe. Ethical and reliable data management is a part of this. We need to make use of the advances of digitalisation in order to be able to retain our analog values.

4.2.3. The Use of the Device in Tourism

The device we designed has the potential to make all touristic segments safer. The possibilities of its practical use will be presented through two theoretical examples. The meetings industry is one of the sectors that suffered severe damages due to the COVID-19 pandemic, and its fast revitalisation is key for the national economy. Each of the themes identified through our qualitative research is present in the field of business events; thus, this area is optimal for illustrating the practical application of the device. We have chosen youth tourism for the other tourism product because the pandemic has also affected the education plans of students in international higher education. Student mobility induces beneficial economic and sociocultural processes in host countries, and therefore supporting this mobility is important.

Theoretical example 1 (tourism product 1): Events contribute to the evolution of tourism and to the economic growth of destinations considerably [78]. The meetings industry, the fastest growing branch of tourism, has been affected by the dramatic changes that occurred in the wake of the sudden outbreak of the COVID-19 pandemic. To comply with the regulations set out by governments, huge numbers of events were cancelled or postponed [23,79], as in the case of the Tokyo Olympics. The strict travel restrictions also affected cultural, business, sports, and mega-events [23], while individual business travel has remained permitted throughout the pandemic. In-person business events could not be held for a long period of time. Subsequently, they could only be held with tight restrictions and some business events were transferred to virtual platforms. With regard to virtual events, both speakers and participants expressed that they missed in-person interactions. The device designed by us is excellent for supporting the participation of business travellers in business events, such as conferences and conventions, because it informs its wearer about their state of health even before the start of the trip, and in the case of illness, the user can connect to the event online, reducing their own and their employer's costs as well as the risk of transmitting any diseases. Beyond the monitoring of the state of health, the device also supports business travellers through additional information and contact options: the application provides travel-related information (such as the entry regulations of different countries, transport, and accommodation-related information), and it can also be used to contact the organisers of an event. Users can activate the device in their own homes (1st activation point) or at the airport (2nd activation point), and it functions as an active smart device throughout their journey (Figure 6). The device is also fitted with a QR code scanner that facilitates contactless check-in into any event, accommodation, or other activity and, if approved, all data are registered in the system. The application helps with navigation, keeping to schedules, and organising transport if needed, while keeping health safety the highest priority. If the device detects signs of illness during the trip, it notifies the user discreetly and instructs them about the next steps. The user can make the decision whether to share that information and with whom. The application helps organise medical care if needed or navigates the user to the nearest testing centre in the given destination.

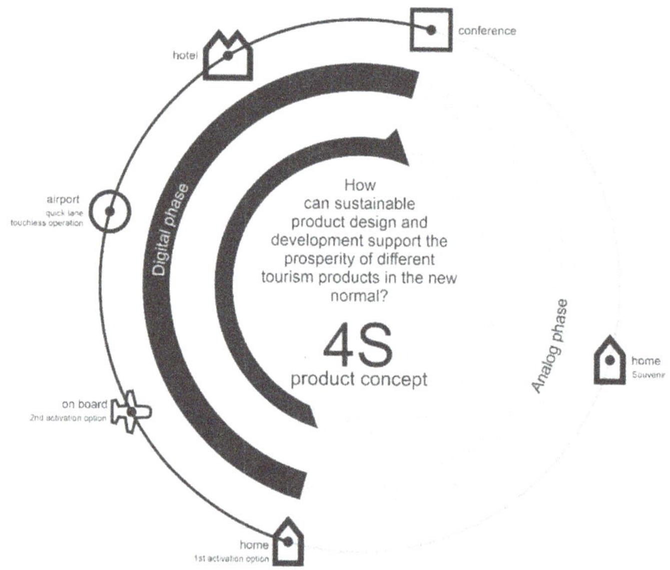

Figure 6. Theoretical example: product use by conference participants. Source: authors' own editing.

Theoretical example 2 (tourism product 2): Youth tourism (with special attention to students in education) plays a prominent role in all big cities [80]. The international mobility of students has grown continuously in recent years; however, this segment has not been left unaffected by the COVID-19 pandemic either [81]. In-person education has been replaced by online learning, especially in the case of universities [82]. Although certain studies [83,84] have found that students, following the initial shock, have adapted to the situation and have even discovered the advantages of online education, a combination of in-person and online education is a more favourable scenario for them [82]. As a result of the changes, many students have adjusted their international education plans [81]. The device designed for students, beyond strengthening their trust in the host country, provides them with practical help in transport, studying, administration, and navigation. Students receive and activate the device in their own homes (activation point), this way it can support them even before and during their travel. The device continuously monitors the student's state of health, and in the case of illness (high body temperature is the most frequent sign of infectious diseases), it notifies them, informs them about the next steps to be taken, and also supports them in connecting to the host country's online education systems. This device is not only useful for international students but can be made available to all students in order to ensure successful operation. This device can replace multiple devices, as it is capable of storing data and managing bidirectional data traffic. The device facilitates contactless identification and communication efficiently; thus, it can replace student ID cards and library passes (among other things) and frees the parties involved from unnecessary physical contact.

5. Conclusions and Implications

This paper has been written based on a non-conventional research and design approach since qualitative marketing research tools have been utilised and combined through the product development process completed within a DIS:CO framework. In our previous research projects, using explorative research tools, we aimed to identify analog values; the lack of these analog experiences is difficult to manage even for individuals entirely socialised in a digitalised world. Through the coding and analysis of essays written by individuals aged 20–25, in addition to numerous retrograde objects and concepts, we saw the emergence of a concept that appeared with high frequency both in the case of digitalisation-oriented individuals and analog-oriented individuals. This concept was the importance of personal relationships. Human beings are predominantly analog social beings that need in-person contact [85]; the natural penetration of digitalisation had already overshadowed real, in-person personal relationships, and then an even more intensive use of digital devices following the outbreak of the COVID-19 pandemic further widened the distance between members of various groups. The relevance of our project is supported by these findings. It is in the interest of communities to regain the same collective standard of living that we had previously. Our project can also serve this cause, because if the goal is reached, it can facilitate safe community coexistence through minimal compromise, contributing to the general wellbeing of the community and to the stability of tourism—the investigated sector.

Beyond the basic concept, with its extended range of services, this device can also be utilised in tourism segments other than the two theoretical examples discussed, such as health tourism, cultural tourism, sports tourism, and shopping tourism, because cross-border shopping plays an important role in the Hungarian economy [86]. The device facilitates and simplifies entry into countries, transport planning, checking in into accommodation facilities, entering and participating in cultural and business events, and in the case of an emergency, the provision of healthcare.

During the development process, we prioritised staying rooted in the three dimensions of sustainability [48]. During the conception of the device, we defined characteristics that contributed to its sustainability. Its modular layout, the utilisation of durable materials, the high quality of manufacturing, and its reusability guarantee an increased lifespan. The purpose of this is a reduction in its ecological footprint. If new generations of the product will be created during future developments, those will be compatible with previous versions wherever possible.

COVID-19 has awakened humanity to the fragility of the social life and lifestyle of our society. At the same time, the fight against it will hopefully be successful, and we can soon restore our lifestyle and quality of life to its previous level. However, no one can guarantee that another pandemic will not break out. Our product can assist not only problem solving, but prevention as well, hence contributing to sustainability.

6. Limitations and Future Research

Due to the interdisciplinary nature of the presented process that combines research and development, it carries more risks than studies that utilise conventional procedures. We can identify certain limitations, and if they are addressed, other research projects combining similar procedures, or even a possible continuation to this study, can yield better reliability and utility. In conclusion, we wish to highlight three factors that we have identified as limitations.

The first one is connected to qualitative research. As for the production of the sample, the age of respondents remained within a narrow range relative to the whole of the population; thus, the results gained in this phase provide an insight into the investigated phenomenon only from the perspective of a specific generation. Respondents became acquainted with digital technologies that entered their lives at approximately the same age and through similar socialisation processes. Thus, the significance of these and the attitudes towards them, while being somewhat varied, were presumably homogenous in comparison to the potential consumer segment of the product concept formulated during the design phase. The inclusion of other age groups and subjects, with more varied attitudes towards digitalisation, could contribute to the extension of the 4S concept to a broader audience.

The second potential research direction would be the adjustment of the details of the concept to specific market demands. Throughout the study, we successfully combined knowledge of tourism, a marketing-oriented research approach, and a designer's toolkit, since our primary goal was interdisciplinary value creation. However, a marketing-oriented review of the findings may reveal the fact that although the product concept we have formulated marginally explores the physical evidence and process dimensions of the marketing mix, the currently most developed concept of the product may need to be further supplemented, so that it can constitute a truly marketable product that is created according to scientific standards and supported by research. During the development of the whole of the marketing mix, specific target-market-related market research findings should also be utilised. Such information was not yet available to us in this phase, as the 4S product concept currently only exists in theory.

One of the important novelties the product carries is the interdisciplinary nature of the process through which it was created. Moreover, a noteworthy element is that the topics of tourism services and travel safety occur concurrently in the development of a wearable product. As an experimental approach, it is an interesting finding and useful starting point for value-centred product design and product development. However, to reach the level of social innovation, and to provide a basis for future research, it needs to manifest in a material form. A third future research direction would be the manufacturing of the prototype, the minimum viable product level production and post-production testing of complementary services and conducting research with the involvement of users.

Author Contributions: G.M., K.B. and B.F. contributed substantially to conducting the research and drafting this manuscript. The data for the qualitative research is sourced from the contribution of the first author. All authors have read and agreed to the published version of the manuscript.

Funding: The dissemination of research outcomes was in part supported by The Hungarian Tourism Association Foundation (MTSZ). We are grateful to the Sándor Imre Foundation of Marketing Communications to support our article's English translation.

Institutional Review Board Statement: Ethical review and approval were waived for this study due to the inquiry was conducted in accordance with the educational framework of Marketing-, Media, and Designcommunication Department of Corvinus University Budapest.

Informed Consent Statement: Informed consent was obtained from all subjects involved in the study.

Conflicts of Interest: The authors declare no conflict of interest.

References

1. Miettinen, S. *Designing the Creative Tourism Experience*; University of Art and Design Helsinki: Helsinki, Finland, 2007.
2. Stickdorn, M. Service Design in Tourism. In *Designing Services with Innovative Methods*; Miettinen, S., Koivisto, M., Eds.; Taik Publications: Helsinki, Finland, 2009.
3. Tussyadiah, I.P. Toward a Theoretical Foundation for Experience Design in Tourism. *J. Travel Res.* **2013**, *53*, 543–564. [CrossRef]
4. Johnson, S.; Boone, P. From Lockdown to Locked in, Here's What Post-Pandemic Travel Could Look Like. 2020. Available online: https://www.weforum.org/agenda/2020/05/coronavirus-lockdown-travel-tourism/ (accessed on 22 January 2022).
5. United Nations World Tourism Organization (UNWTO). Impact Assessment of the COVID-19 Outbreak on International Tourism. 2020. Available online: https://www.unwto.org/impact-assessment-of-the-covid-19-outbreak-on-international-tourism (accessed on 22 January 2022).
6. De Vos, J. Transportation Research Interdisciplinary Perspectives The effect of COVID-19 and subsequent social distancing on travel behavior. *Transp. Res. Interdiscip. Perspect.* **2020**, *5*, 100121. [CrossRef] [PubMed]

7. Irimiás, A.; Michalkó, G. Hosting while being hosted: A perspective of Hungarian migrant hospitality workers in London, UK. *Tour. Hosp. Res.* **2016**, *16*, 172–183. [CrossRef]
8. Mitev, A.; Irimiás, A. Travel Craving. *Ann. Tour. Res.* **2020**, *90*, 103111. [CrossRef] [PubMed]
9. Freire-Gibb, L.C.; Lorentzen, A. A platform for local entrepreneurship: The case of the lighting festival of Frederikshavn. *Local Econ.* **2011**, *26*, 157–169. [CrossRef]
10. Megyeri, G.; Boros, K.; Fekete, B. Hidakból teremtett emlékek, avagy kísérlet egy innovatív tárgy megalkotására a B-bridge koncepción keresztül. *Tur. Bull.* **2020**, *20*, 4–12. [CrossRef]
11. Cosovan, A.; Horváth, D.; Mitev Ariel, Z. A designkommunikáció antropológiai megközelítése. *Replika* **2018**, *106*, 233–245. [CrossRef]
12. Cosovan, A.; (DISCO). Co&Co Communication, 2009.
13. Zieliński, G.; Studzińska, M. Application of design-thinking models to improve the quality of tourism services. *Zarzdzanie I Finans. J. Manag. Financ.* **2015**, *13*, 133–145.
14. Robbins, P.; Devitt, F. Collaboration, creativity and entrepreneurship in tourism: A case study of how design thinking created a cultural cluster in Dublin. *Int. J. Entrep. Innov. Manag.* **2017**, *21*, 185–211.
15. Sándorová, Z.; Repáňová, T.; Palenčíková, Z.; Beták, N. Design thinking-A revolutionary new approach in tourism education? *J. Hosp. Leis. Sport Tour. Educ.* **2020**, *26*, 100238. [CrossRef]
16. Cosovan, A. VY. *Vez.-Bp. Manag. Rev.* **2017**, *48*, 4–5. [CrossRef]
17. Cosovan, A.; Horváth, D. Emóció–Ráció: Tervezés–Vezetés: Designkommunikáció. *Vez.-Bp. Manag. Rev.* **2016**, *47*, 36–45. [CrossRef]
18. Megyeri, G.; Boros, K.; Fekete, B. 3S Traveling–Turizmus a poszt-COVID19 érában. *Tur. Bull.* **2021**, *21*, 52–62. [CrossRef]
19. Hall, C.M.; Scott, D.; Gössling, S. Pandemics, transformations and tourism: Be careful what you wish for. *Tour. Geogr.* **2020**, *22*, 577–598. [CrossRef]
20. Li, Z.; Zhang, S.; Liu, X.; Kozak, M.; Wen, J. Seeing the invisible hand: Underlying effects of COVID-19 on tourists' behavioral patterns. *J. Destin. Mark. Manag.* **2020**, *18*, 100502. [CrossRef]
21. Bieger, T.; Laesser, C. *The Future of Tourism with and Potentially After SARSCoV-2: Continuous Small Steps and Drawbacks towards a Temporary New "Normal"*; AIEST: St Gallen, Switzerland, 2020.
22. Dupeyras, A.; Haxton, P.; Stacey, J. *The Covid-19 Crisis and Tourism: Response and Recovery Measures to Support the Tourism Sector in OECD Countries*; OECD, G20 Policy Insights: Paris, France, 2020.
23. Gössling, S.; Scott, D.; Hall, C.M. Pandemics, tourism and global change: A rapid assessment of COVID-19. *J. Sustain. Tour.* **2021**, *29*, 1–20. [CrossRef]
24. Rogerson, C.M.; Baum, T. COVID-19 and African tourism research agendas. *Dev. South. Afr.* **2020**, *37*, 727–741. [CrossRef]
25. Zátori, A.; Michalkó, G.; Nagy, J.T.; Kulcsár, N.; Balizs, D. The tourist experience of domestic VFR travellers: The case of Hungary. *Curr. Issues Tour.* **2019**, *22*, 1437–1459. [CrossRef]
26. Steen Jacobsen, J.K.; Farstad, E.; Higham, J.; Hopkins, D.; Landa-Mata, I. Travel discontinuities, enforced holidaying-at-home and alternative leisure travel futures after COVID-19. *Tour. Geogr.* **2021**. [CrossRef]
27. Weman-Josefsson, K. Perspectives of Life in Sweden During the COVID-19 Pandemic. *J. Clin. Sport Psychol.* **2021**, *15*, 80–86. [CrossRef]
28. Alvarez-Sousa, A.; Paniza Prados, J.L. Visitor Management in World Heritage Destinations before and after Covid-19, Angkor. *Sustainability* **2020**, *12*, 9929. [CrossRef]
29. Clair, A. Homes, Health, and COVID-19: How Poor Housing Adds to the Hardship of the Coronavirus Crisis. Social Market Foundation. 2020. Available online: https://www.smf.co.uk/commentary_podcasts/homes-health-and-covid-19-how-poor-housing-adds-to-the-hardship-of-the-coronavirus-crisis/ (accessed on 26 September 2021).
30. Crompton, J.L. Motivations for pleasure vacation. *Ann. Tour. Res.* **1979**, *6*, 408–424. [CrossRef]
31. Guo, X.; Sun, Z. A Novel Evaluation Approach for Tourist Choice of Destination Based on Grey Relation Analysis. *Sci. Program.* **2016**, 1812094. [CrossRef]
32. Hsu, T.-K.; Tsai, Y.-F.; Wu, H.-H. The preference analysis for tourist choice of destination: A case study of Taiwan. *Tour. Manag.* **2009**, *30*, 288–297. [CrossRef]
33. Li, Y.; Song, H.; Guo, R. A Study on the Causal Process of Virtual Reality Tourism and Its Attributes in Terms of Their Effects on Subjective Well-Being during COVID-19. *Int. J. Environ. Res. Public Health* **2021**, *18*, 1019. [CrossRef]
34. Choi, Y.; Hickerson, B.; Lee, J. Investigation of the technology effects of online travel media on virtual travel experience and behavioral intention. *J. Travel Tour. Mark.* **2018**, *35*, 320–335. [CrossRef]
35. Shokouhyar, S.; Shokoohyar, S.; Sobhani, A.; Gorizi, A.J. Shared mobility in post-COVID era: New challenges and opportunities. *Sustain. Cities Soc.* **2021**, *67*, 102714. [CrossRef]
36. Sultan, M.-T.; Sharmin, F.; Badulescu, A.; Stiubea, E.; Xue, K. Travelers' Responsible Environmental Behavior towards Sustainable Coastal Tourism: An Empirical Investigation on Social Media User-Generated Content. *Sustainability* **2021**, *13*, 56. [CrossRef]
37. Roman, M.; Kosiński, R.; Bhatta, K.; Niedziółka, A.; Krasnodębski, A. Virtual and Space Tourism as New Trends in Travelling at the Time of the COVID-19 Pandemic. *Sustainability* **2022**, *14*, 628. [CrossRef]
38. Milgram, P.; Kishino, F. A taxonomy of mixed reality visual displays. *IEICE Trans. Inf. Syst.* **1994**, *77*, 1321–1329.

39. Kounavis, C.D.; Kasimati, A.E.; Zamani, E.D. Enhancing the tourism experience through mobile augmented reality: Challenges and prospects. *Int. J. Eng. Bus. Manag.* **2012**, *4*, 10. [CrossRef]
40. Hutchins, B.; Andrejevic, M. Olympian Surveillance: Sports Stadiums and the Normalization of Biometric Monitoring. *Int. J. Commun.* **2021**, *15*, 363–382.
41. FIPRA. Despite Devastating Blow, COVID-19 Gives Tourism Industry A Chance to Redeem Itself. 2020. Available online: https://fipra.com/update/despite-devastating-blow-covid19-gives-tourism-industry-a-chance-to-redeem-itself/ (accessed on 28 March 2021).
42. Rahimizhian, S.; Irani, F. Contactless hospitality in a post-COVID-19 world. *Int. Hosp. Rev.* **2021**, *35*, 293–304. [CrossRef]
43. Kuklina, M.; Trufanov, A.; Krasnoshtanova, N.; Urazova, N.; Kobylkin, D.; Bogatyreva, M. Prospects for the Development of Sustainable Tourism in the Okinsky District of the Republic of Buryatia. *Sustainability* **2021**, *13*, 8042. [CrossRef]
44. World Commission on Environment and Development (WCED). *Our Common Future*; Oxford University Press: New York, NY, USA, 1987.
45. Buckley, R. Sustainable tourism: Research and reality. *Ann. Tour. Res.* **2012**, *39*, 528–546. [CrossRef]
46. Iniesta-Bonillo, M.A.; Sánchez-Fernández, R.; Jiménez-Castillho, D. Sustainability, value, and satisfaction: Model testing and cross-validation in tourist destinations. *J. Bus. Res.* **2016**, *69*, 5002–5007. [CrossRef]
47. Ko, T.G. Development of a tourism sustainability assessment procedure: A conceptual approach. *Tour. Manag.* **2005**, *26*, 431–445. [CrossRef]
48. Solís-Radilla, M.M.; Hernández-Lobato, L.; Pastor-Durán, H. The importance of sustainability in the loyalty to a tourist destination through the management of expectations and experiences. *Sustainability* **2019**, *11*, 4132. [CrossRef]
49. Rahmoun, M.; Baeshen, Y. Marketing tourism in the digital era and determinants of success factors influencing tourist destinations preferences. *Asia-Pac. Manag. Account. J.* **2020**, *16*, 163–181. [CrossRef]
50. Rodríguez-Díaz, M.; Espino-Rodríguez, T. Determining the sustainability factors and performance of a tourism destination from the stakeholders' perspective. *Sustainability* **2016**, *8*, 951. [CrossRef]
51. Gössling, S.; Peeters, P. Assessing tourism's global environmental impact 1900–2050. *J. Sustain. Tour.* **2015**, *23*, 639–659. [CrossRef]
52. Veiga, C.; Santos, M.C.; Águas, P.; Santos, J.A.C. Sustainability as a key driver to address challenges. *Worldw. Hosp. Tour. Themes* **2018**, *10*, 662–673. [CrossRef]
53. Hanafiah, M.H.; Zulkifly, M.I. Tourism destination competitiveness and tourism performance: A secondary data approach. *Compet. Rev.* **2019**, *29*, 592–621. [CrossRef]
54. Kamel, N. Examining the mediating role of celebrity endorsement in green advertisements to improve the intention of Egyptian Millennials towards environmental behaviours in tourist destinations. *Tour. Manag. Stud.* **2020**, *16*, 7–21. [CrossRef]
55. Hediger, W. Sustainable development and social welfare. *Ecol. Econ.* **2000**, *32*, 481–492. [CrossRef]
56. Clark, G.; Kosoris, J.; Hong, L.N.; Crul, M. Design for sustainability: Current trends in sustainable product design and development. *Sustainability* **2009**, *1*, 409–424. [CrossRef]
57. Karlsson, R.; Luttropp, C. EcoDesign: What's happening? An overview of the subject area of EcoDesign and of the papers in this special issue. *J. Clean. Prod.* **2006**, *14*, 1291–1298. [CrossRef]
58. Santos, M.C.; Veiga, C.; Santos, J.A.C.; Águas, P. Sustainability as a success factor for tourism destinations: A systematic literature review. *Worldw. Hosp. Tour. Themes* **2022**. [CrossRef]
59. Geissdoerfer, M.; Savaget, P.; Bocken, N.M.; Hultink, E.J. The Circular Economy–A new sustainability paradigm? *J. Clean. Prod.* **2017**, *143*, 757–768. [CrossRef]
60. Kirchherr, J.; Reike, D.; Hekkert, M. Conceptualizing the circular economy: An analysis of 114 definitions. *Resour. Conserv. Recycl.* **2017**, *127*, 221–232. [CrossRef]
61. Morse, J.M. Principles of mixed methods and multimethod research design. In *Handbook of Mixed Methods in Social and Behavioral Research*; Tashakkori, A., Teddlie, C., Eds.; Sage Publication: Thousand Oaks, CA, USA, 2003; pp. 189–208.
62. Harrison, R.L.; Reilly, T.M. Mixed methods designs in marketing research. *Qual. Mark.Res. Int. J.* **2011**, *14*, 7–26. [CrossRef]
63. Denzin, N.K. *The Research Act*, 3rd ed.; Prentice Hall: Englewood Cliffs, NJ, USA, 1989.
64. Denzin, N.K.; Lincoln, Y.S. *Handbook of Qualitative Research*; Sage Publication: Thousand Oaks, CA, USA, 2000.
65. Zhang, Y.; Wildemuth, B.M. Qualitative analysis of content. In *Applications of Social Research Methods to Questions in Information and Library Science*; Wildemuth, B.M., Ed.; Libraries Unlimited: Westport, CT, USA, 2009; pp. 308–319.
66. Mayring, P. *Qualitative Inhaltsanalyse: Grundlagen und Techniken*, 7th ed.; Deutscher Studien Verlag: Weinheim, Germany, 2000.
67. Patton, M.Q. *Qualitative Research and Evaluation Methods*; Sage Publication: Thousand Oaks, CA, USA, 2002.
68. CO&CO; Cosovan, A. DSI-Red Dot. 2012. Available online: https://www.red-dot.org/project/dsi-29724 (accessed on 22 January 2022).
69. CO&CO; Cosovan, A. Tequball—Red Dot. 2015. Available online: https://www.red-dot.org/project/teqball-1-8088 (accessed on 22 January 2022).
70. CO&CO; Cosovan, A. Coco Dice—Red Dot. 2011. Available online: https://www.red-dot.org/project/cocodice-corner-contact-game-system-28164 (accessed on 22 January 2022).
71. CO&CO; Cosovan, A. Nosiboo—Red Dot. 2014. Available online: https://www.red-dot.org/project/nosiboo-32116 (accessed on 22 January 2022).

72. Cosovan, A.; Horváth, D. Tervező művész(ek) a közgazdászképzésben. In *EMOK XXII. OrszágosKonf.—Hitel. Értékorientáció A Mark*; Fehér, A., Kiss, V.Á., Soós, M., Szakály, Z., Eds.; Debreceni Egyetem Gazdaságtudományi Kar: Debrecen, Hungary, 2009.
73. Galla, D.D. Gyerekek. Otthon. Együttműködés. Kreativitás. A Designkommunikáció, Mint a Soft Készségek Fejlesztését Támogató Oktatásmódszertani Eszköz. Ph.D. Thesis, Corvinus University of Budapest, Budapest, Hungary, 2021. [CrossRef]
74. Horváth, D.; Cosovan, A.; Horváth, D.; Lachin, N. Tanulás-munka interface. A valós idejű találkozások jelentősége a digitális oktatási környezetben. *Vez.-Bp. Manag. Rev.* **2018**, *49*, 67–77. [CrossRef]
75. Dorst, K. The core of 'design thinking' and its application. *Des. Stud.* **2011**, *32*, 521–532. [CrossRef]
76. Srikrishnan, M.R.; Archana, N.; Niresh, J. The Evolution of Smart Wrist Band by Using Sensors. Journal of Siberian Federal University. *Eng. Technol.* **2020**, *13*, 525–535.
77. Prince, P.B.; Lovesum, S.P.J. Privacy Enforced Access Control Model for Secured Data Handling in Cloud-Based Pervasive Health Care System. *SN Comput. Sci.* **2020**, *1*, 239. [CrossRef]
78. Seraphin, H. COVID-19: An opportunity to review existing grounded theories in event studies. *J. Conv. Event Tour.* **2020**, *22*, 3–35. [CrossRef]
79. Ozili, P.K.; Arun, T. Spillover of COVID-19: Impact on the Global Economy. *SSRN Electron. J.* **2020**, 3562570. [CrossRef]
80. United Nations World Tourism Organization (UNWTO). *Youth Travel Matters Understanding the Global Phenomenon of Youth Travel*; World Youth Student & Educational Travel Confederation (WYSE): Madrid, Spain, 2008; ISBN 978-92-844-1239-6.
81. Mok, K.H.; Xiong, W.; Ke, G.; Cheung, J.O.W. Impact of COVID-19 pandemic on international higher education and student mobility: Student perspectives from mainland China and Hong Kong. *Int. J. Educ. Res.* **2021**, *105*, 101718. [CrossRef]
82. Jaworek, M.A. The Effects of Temper Traits and Study Method (Full-Time vs. Extramural) on Polish Students' Adaptability to Online Learning as a Result of COVID-19. A Pilot Study. *Sustainability* **2021**, *13*, 14017. [CrossRef]
83. Bączek, M.; Zagańczyk-Bączek, M.; Szpringer, M.; Jaroszyński, A.; Wożakowska-Kapłon, B. Students' perception of online learning during the COVID-19 pandemic. A survey study of Polish medical students. *Medicine* **2021**, *100*, e24821. [CrossRef]
84. Chakraborty, P.; Mittal, P.; Gupta, M.S.; Yadav, S. Opinion of students on online education during the COVID-19 pandemic. *Hum. Behav. Emerg. Technol.* **2020**, *3*, 357–365. [CrossRef]
85. Megyeri, G.; Horváth, D.; Cosovan, A. Analóg Lények Vagyunk Egy Digitális Világban–Avagy a Digitalizáció Következményei és Lehetőségei Kutató és Tervezői Szemszögből. In *Az Egyesület a Marketing Oktatásért és Kutatásért XXV.; Ismerjük a Vevők? A Vásárlás Pszichológiája; Országos Konferenciájának előadása*; Veres, Z., Sasné Grósz, A., Liska, F., Eds.; Pannon Egyetem: Veszprém, Hungary, 2019; pp. 711–720. ISBN 978-615-00-58.
86. Michalkó, G.; Timothy, D.J. Cross-border shopping in Hungary: Causes and effects. *Vis. Leis. Bus.* **2001**, *20*, 4–22.

sustainability

MDPI

Article

Subjective Bodily Experiences of Island Cyclists in Different Contexts: The Case of Hainan Island, China

Minhui Lin [1,*] and Haichao Xu [2,*]

[1] College of Forestry and Landscape Architecture, South China Agricultural University, Guangzhou 510642, China
[2] Tourism College, Hunan Normal University, Changsha 410081, China
* Correspondence: linminhui@scau.edu.cn (M.L.); xuhaic@hunnu.edu.cn (H.X.)

Abstract: A single subjective bodily experience is at the core of sports tourism activities, but the current literature on sports tourism largely ignores both the continuous and ephemeral experiential processes of individuals in mobile sports activities from a bodily perspective. In this study, we developed a "context–body–perception" framework and selected a sample of tourists from Hainan Island, China, in order to explore the embodied experience of cycling tourists, using a qualitative approach. We found that the contexts encountered by island cyclists could be divided into a human context and nonhuman context. The human context included the companion context, pan-companion context, and host context, and the nonhuman context comprised the natural context, mediated facility context, and digital technology context. The cyclists' physical experiences and perceptions in multiple different contexts were inseparable from each other, and both were embedded in a specific context through the five senses, through the state of body and activity, through emotions and memories, and through interaction with a specific context, all of which formed a dynamic feedback system. Through bodily practices in different contexts, cyclists acquired meaningful representations of their bodies, social relationships, and self-worth. The findings of this study can enrich the study of embodiment in sports and recreation areas, as well as provide an initial foray into bodily research in island-based cycling.

Keywords: bicycle tourism; island cyclists; embodied experiences; contexts; Hainan

Citation: Lin, M.; Xu, H. Subjective Bodily Experiences of Island Cyclists in Different Contexts: The Case of Hainan Island, China. *Sustainability* **2022**, *14*, 10176. https://doi.org/10.3390/su141610176

Academic Editors: Dallen J. Timothy, Gábor Michalkó and Anna Rita Irimiás

Received: 27 July 2022
Accepted: 15 August 2022
Published: 16 August 2022

1. Introduction

The development of sports tourism is recognized in many countries as a way to better avoid the seasonality of tourism [1,2]. Bicycle tourism has been one of the most important niche markets for sports due to the greater flexibility, accessibility, and environmental friendliness of the activity itself [3]. It first emerged in Europe and the United States during the late Victorian and Edwardian boom in Irish tourism and has become increasingly popular around the world ever since [4,5]. In the context of tourism, the bicycle is not simply a means of transportation or equipment for sport, but a medium of interaction between cyclists and the environment as well [6]. Furthermore, because most bicycle rides take place through transitions between places, cycling shows great potential for integration with other tourism activities [7]. Although cycling has received much attention within sports research, there are few research perspectives that combine the sport and tourism attributes of cycling.

Another key aspect of cycling and tourism that has received little attention is the cycling mobility experience. Combined with the definition of cycling tourism put forward by Ritchie, the cycling mobility experience refers to the use of bicycles as the main means of transportation during the mobility process, or where bicycles play a very important role during a tour [8]. Current studies have deconstructed various cycling phenomena around the themes of cycling infrastructure [9], the socio-economic impacts of cycling [10,11], cyclists' motivations and preferences [12,13], and cyclists' embodied practices [14]. Although some scholars have explored the embodied practice aspects of cycling, most of these research findings have focused on the subjective experience more so than on the body of travelers [15]. However, the bodily experience is most memorable and important, which can be triggered by the variability of the environment and encounters with human and nonhuman elements during a ride [16]. These encounters in turn frame diverse contexts, giving rise to different meanings that constitute the whole cycling experience.

Unlike tourism by conventional transportation, bicycle tourism, which is more autonomous and embodied, is an unconventional mobility experience that allows cyclists to relate to a wide range of places and space [17,18]. However, such a research theme has not received sufficient attention [19]. Most previous studies of tourism and transportation have separated travel and transportation; however, the new mobility paradigm challenges this view [20]. It argues that no point or space is completely isolated and advocates a relational turn, emphasizing the connections among elements [19,21]. However, studies on the bicycle tourism experience are still described in a more traditional, static way [3,12]. Bicycle tourism not only has the typical human–land connection characteristics, but also allows cyclists to actively experience the natural and human environment of cycling in depth [17–19]. It also allows them to gain both explicit and tacit knowledge in the process [18]. Therefore, we adopt an unconventional mobility perspective and a relational geography approach to explore the relationship among context, body and mind, and perception in bicycle tourism. This study can further compensate for the neglect of previous studies on the tourism transportation experience. It also complements the study results on the relationship between tourism activities and subjects under the transformation of the new mobility paradigm.

Against this research background, this article attempts to explore further, through qualitative research, which contexts island-based cyclists encounter in their cycling using independent bicycle tourists on Hainan Island, China. What are the processes of the cyclists' embodied practices in these contexts? What kind of meaningful representations does the embodied practice of cycling bring to these cyclists? The specific research process is as follows. First, we found the gap in previous studies based on a literature review and constructed the "context–body and mind–perception" three-dimensional theoretical framework that guides this paper. Second, this study collected data through a mobile ethnographic approach, and the authors also accumulated a large amount of primary data from several Hainan rides. The data cover the types of contexts generated, the outcomes of the physical and mental perceptions, and the learning and understanding of explicit and tacit knowledge in multiple contexts. Finally, we analyzed the themes that emerged from the data in conjunction with the theoretical framework to arrive at our perspective.

2. Literature Review

2.1. The Embodied Experiences: Context, Body, and Perception

The field of phenomenology and the emergence of embodied cognition influenced by phenomenology have pushed people to rethink the role of the body [22,23]. Concern for the concept of embodiment originated from Heidegger's concept of "being" and Merleau-Ponty's concept of "embodied subjectivity", which affirms that the body and mind are one [24]. Phenomenology emphasizes the body as the subject that shapes our perception and cognition in various environmental forms. As Merleau-Ponty argues, the body is not only a part of space, but is also integrated with the mind as a prerequisite for people to perceive space [25].

In the process of understanding space in the context of the body, human beings are capable of producing numerous contexts based on their cognition, and these contexts can be used to help perceive the world through certain rules [26]. More precisely, the term context as used here refers to a tendency to structure different factual elements by combining them in multiple different ways by forming different contexts to reflect different information through different relationships [27]. In the course of their activities, subjects form different contexts by establishing connections with human and nonhuman elements, and the individual bodily sensations and perceptions formed in these contexts reflect their characteristics [28].

These contexts are not completely neutral but are influenced by the process of embodied practice, subject identity, and cultural contexts as well [29]. This notion of context is in line with what Polanyi calls cognitive conditioning, where he argues that there is a very complex field emergence mechanism behind the cognitive activities that occur in us [30]. The knowledge acquired in such contexts is distinguished between explicit and tacit knowledge [31,32]. In Polanyi's work, tacit knowledge and explicit knowledge are two opposing concepts. The former refers to "silent," "unspoken" knowledge, while the latter refers to "explicit," "articulate," and "clear" knowledge [32]. In contrast to explicit knowledge, which can be expressed in words, diagrams or formulas, the acquisition of tacit knowledge is based on individual behavior [31,33–35]. It is always associated with a specific situation and has unconscious characteristics, and it is difficult to be expressed clearly in words, language, images, and other explicit forms [20,31,36]. Among human practical activities, there is focal and subsidiary consciousness. In his book, *Personal Knowledge*, Polanyi also mentions that the realization of individual skills in cycling is a state of "not knowing how it happens", and in cycling tourism, too, individuals have a great deal of unspoken and unexpressed tacit knowledge [31,36].

While the exploration of embodied contexts is still ongoing, researchers generally accept that the way a subject's body structure, senses, and motor systems operate affects his perception of the world, the way he thinks, and the content of the various contexts he forms about space [37]. For tourists, the "context–body–perception" interaction has also been reported in some tourism studies. Liu argues that honeymooners form their initial impressions of the honeymoon destination through the operation of five senses (the senses of vision, hearing, smell, taste, and touch) [38]. The involvement of the body and the functioning of the bodily senses form the important experiences in honeymoon tourism within the confines of a tourism context, which in turn contributes to the generation of tourism cognition, and the results of bodily functioning and perception, in turn, have a counter effect on this tourism context through the mediation of individual behavior. Thus, the body and mind, perception, and context in the embodied experience of honeymoon tourism are organically integrated. Mertena found that tourists create human and nonhuman contexts during the flow of a train ride as well [39]. Here, the human context includes emotional experiences formed by talking with companions and playing games, the rhythm of the train machinery, and the bodily movement experience brought about by the train track. In this nonhuman context, the train as a transportation intermediary brings the tourist unique experiences when passing through different geographical environments.

In summary, the dynamic relational mechanism established between perception, the body, the environment, and various contexts provides a way of thinking about the process by which humans make sense of the world. Constraints serve as connections between elements of a context, and they may be natural laws, conventions, logical rules, empirical analysis, or law-like correspondences [26]. Thus, the body and a subject's perceptual outcomes are both referents and constituents.

2.2. *Bicycle Tourism*

Sports tourism experiences are generated by the interaction of the two more fundamental experiences of sport and tourism, and embodied experience is also a particularly important part of the tourism experience [40]. The tourist experience is at the core of what draws people to engage in tourism, and it covers four dimensions: individual emotional experience, experiential learning, physical ability enhancement, and spiritual transformation [41]. The mobile experience of cycling is different from the traditional touring mobile experience [42]. For example, cycling requires more use of the legs, and the visual and auditory sensory experience is a complex, dynamic process [43]. In addition, bicycle tourism gives people a certain sense of control, which is expressed in the individual's active regulation of body and senses during the cycling process. Take the travel rhythm as an example: the speed of travel depends entirely on an individual's physical capabilities and will. As long as the cyclist wishes, he can interrupt the trip at any time, reduce the cycling speed to zero, and enjoy the scenery.

Differences in cycling type also bring about differences in mobility experiences. For long-distance cyclists, the perception of "space" and "place" plays an important role in the cyclist's embodied tourism experience [44]. Riding not only gives people a mobile perspective on people, things, and objects but also constructs a panoramic mobility experience together with other senses such as touch and smell [45]. Riders can also become emotionally connected through mobility [46]. Furthermore, for cycling mobility, the speed formed by the interaction between the bicycle and the outside world is an important stimulus for the cyclist's emotional arousal. Meanwhile, a change in speed and velocity also can trigger different physical experience output results [47].

In summary, a cyclist's travel experience is complex and dynamic and comes from the continuous process of collision and contact with various elements of that particular mode of travel [48,49]. Although some scholars have focused on analyzing cyclists' embodied experience, their research has been more empirical, which can lead to insufficient theoretical development. Therefore, it is necessary to examine the various contexts that cyclists generate during their activities, from a physical perspective but also from their encounters and perceived outcomes when riding in these specific types of environments.

3. Methodology

3.1. Study Area

For this article, we chose Hainan Island in southeastern China as a case study. Hainan Island is known as the "Hawaii of the East" and is one of the world's most important tropical destinations, with rich tourism resources, well-organized cycling, and excellent infrastructure [50]. As of 2020, including the third top international cycling event in Asia, the "Tour de Island", China's Hainan Island has held fourteen international road cycling races, and its high profile as a result of the "Tour" race has continued to attract cycling enthusiasts from both China and abroad, creating a strong cycling tourism atmosphere. With its newly created, island-wide "Sports Tourism Demonstration Zone", the Hainan Provincial Government has also now incorporated the island-wide cycling tourism industry into its new round of overall tourism planning [50]. Hainan cycling routes are primarily divided into the East, West, and Central routes (see Figure 1), each with its own characteristics and charm. The East route has the most densely populated towns and has a deep cultural heritage; the West route is relatively challenging, slightly less convenient, and has relatively small towns; and the Central route is more difficult to ride and requires climbing up hilly and even mountainous terrain.

Figure 1. Diagram of the Hainan cycling routes.

3.2. Research Framework and Data Collection

From the literature review, it is clear that the physical sensations and inner perception of tourists are inextricably linked to the environment they experience. The embodied experience of a cyclist is influenced by a variety of situations arising from the interaction of subject and object [51]. More precisely, tourists learn about a place through practice and establish relationships with it [52]. This active participation creates a dual relationship between individual and place by using tourists' physical sensations and psychological perceptions as tools. In other words, tourists will encounter different situations in their tour that will affect their emotions, memories, senses, etc. In turn, this multifaceted embodied experience simultaneously keeps the context constantly changing.

This article combines the objective space, body space, and the third space (perceptual space) proposed by Merleau-Ponty, and uses the conceptual framework of an embodied experience, including body and mind, perception, and context (Figure 2), to explore the process of diverse bicycle tourism experiences [25]. Our conceptual framework emphasizes that the interaction between body, perception, and situation influences tourists' meaning construction and cognitive outcomes in tourism spaces and explores the specific effects of diverse situations on tourists' physical experience and perceptual outcomes.

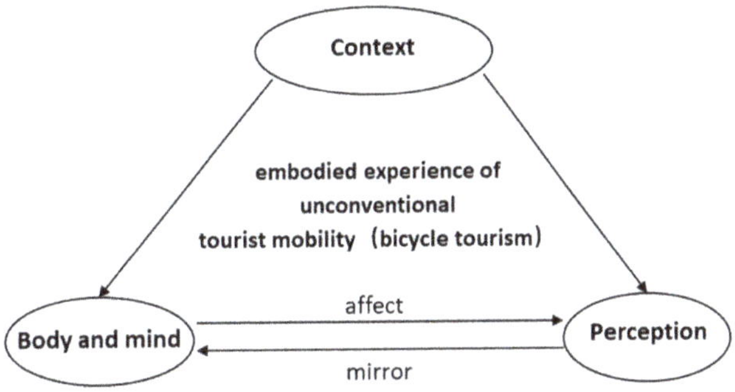

Figure 2. The "context–body–perception" embodied experience framework.

In the conceptual framework we use, the context is consistent with the Gestalt context mentioned by Polanyi. Both refer to the field generated by the subject during the participatory activity. Explicit knowledge and tacit knowledge are mixed in the rider's physical and mental feelings and perception. At the consciousness level, according to Polanyi's view, this paper argues that explicit knowledge in bicycle tourism includes information about the destination, riding skills, and knowledge of other equipment. The tacit knowledge includes the individual's perception of the destination based on his or her own experience, and the individual's unspeakable cycling experience formed through the concrete practice of cycling. The individual's body and the bicycle cooperate during the cycling process. Finally, explicit knowledge and tacit knowledge are interconnected and integrated.

This study focuses on mobile ethnography to collect data. Nonrepresentational theory suggests that there is a lot of information to be found in people's behavior and nonverbal mannerisms [53]. Therefore, to learn more about nonverbal content, a researcher should be physically present in the research setting to pick up on the original intent of the subjects [54]. Furthermore, the mobile ethnography method allows researchers to shift between different contexts and participatory experiences randomly with the cycling "companions" and "pan-companions" with whom they are on the move. This can allow a researcher to conduct semi-structured interviews to obtain tangible representations of physical and psychological feelings, companion relationships, and electronic media use during the ride. It can also help him to observe and collect nonrepresentational content, such as the physical characteristics, emotional changes, and companion emotions of cyclists [55].

The field researcher for this article interviewed and observed a total of 23 cyclists in depth, subsequently excluding 3 invalid samples, to obtain a total of 20 high-quality, in-depth interview samples for follow-up analysis. The interviewees' information is shown in Table 1. In addition, the authors also self-reflected during several rides and recorded their reflection results, and they were also able to obtain some high-quality web texts from other cyclists' blogs (35 people in total) that involved the sharing of photos, event experiences, and moods and feelings during their rides. The collected original texts were integrated and theorized with text analysis methods, and the key categories extracted from the texts were then linked several times with the analytical framework through qualitative methods to help form generalized conclusions.

Table 1. Profiles of the interviewed cyclists.

Code	Residence	Gender	Age	Traveling Method	Interview Time (Times)
F1-LWL	Guangzhou	male	26	companionship	3 h (2)
F2-XC	Guangzhou	male	24	companionship	2 h (3)
F3-NCC	Beijing	male	26	companionship	3 h (2)
F4-TJH	Shenzhen	male	31	companionship	1 h (1)
F5-NJY	Guangzhou	female	46	personal	1 h (2)
F6-SC	Hangzhou	male	38	companionship	2 h (2)
F7-DDM	Hangzhou	female	34	companionship	1 h (2)
F8-TXL	Guangzhou	male	23	personal	0.5 h (1)
F9-YJ	Shantou	female	39	companionship	1 h (2)
F10-JM	Shenzhen	male	21	companionship	3 h (2)
F11-CYC	Chengdu	male	41	personal	0.5 h (1)
F12-XF	Guangzhou	female	20	personal	2 h (2)
F13-NJD	Shenzhen	male	33	companionship	2 h (2)
F14-DJ	Guiyang	female	21	companionship	3 h (2)
F15-FYT	Guangzhou	male	48	personal	1.5 h (2)
F16-LY	Zhuhai	female	34	companionship	0.5 h (1)
F17-ZMM	Shanghai	female	22	personal	1 h (2)
F18-XB	Liuzhou	male	26	companionship	1.5 h (2)
F19-ZXH	Shanghai	male	35	companionship	1 h (2)
F20-LYH	Changsha	male	24	companionship	0.5 h (1)

4. Findings

4.1. Cyclists' Embodied Experiences in a Human Context

The "companion context" is one of the most important human contexts for cycling. Riding companions are informal groups formed to engage in cycling activities, and the behavior and personality of members have an important impact on the physical and mental experience of the cyclists involved [56]. Importantly, however, not all individuals in such cycling teams have the same purpose and power. For example, the lead rider (F2-XC) controls the team's movement speed and human–ground interaction activities, which may take more energy than simply enjoying the scenery along the ride. The riders in the middle tend to be emotionally stable and their body movement speed is derived from the riders in front of them. Riders at the end of the fleet tend to be tense due to the possibility of disengaging from the team at any time. Because of the unstable nature of cycling activities, the position and power of the cycling team may change with emergencies, and the change in the role of cycling companions may be influenced by the cyclists' physical qualities, age, gender, and family structure [57].

The term "pan-companion" refers to cyclists who have a spatial and temporal intersection with the individual cyclists in Hainan in this study, and encounters between "pan-companions" may occur at nodes or during a linear journey. During a journey, accompanied by pan-companions day after day, pan-companions may become cycling companions. However, in most cases, the physical interaction between them is simply due to their relational distance; for example, a simple social interaction with a thumbs up and a wave while on the road, as F19-ZXH said:

> *Every time we meet cyclists on the road, we will give a thumbs up, shouting to each other to cheer.*

What follows is a typical example of tacit knowledge being transformed into explicit knowledge and changing the individual's behavior. In the process of cycling, cyclists often experience incidental events that are like "people in the mirror", reflecting on themselves in the flowing landscape composed of others and adjusting their physical ridging behavior, as one of the lead cyclists (F11-CYC) mentioned:

> *In that section of the seaside highway section, we saw the trash left by other cyclists and felt that it would be harmful to marine life, so I reminded my companions to behave [in a] civilized [manner] throughout the ride.*

Additionally, the human context also includes the "host context". Unlike traditional tourism, the large number of stopover nodes in cycling activities promotes the generation of special human–ground interaction experiences [19]. To ensure the completion of cyclists' activities, residents provide materials for cyclists' physical and mental experiences in different venues and bring cyclists rich stimulation for their vision, hearing, smell, taste, and touch. When cyclists enter a series of town nodes from the sparsely populated cycling trails, the first sensation they encounter is the visual change, from the natural landscape to the more human landscape of the town, such as the streets, local food stores, ethnic minority residents, and special transportation, and all become objects of the cyclists' gazes. Sensory experience is accompanied by the accumulation of explicit and tacit knowledge; for example, unlike the food culture of the hometown of cyclists, Hainan's tropical climate helps the locals create unique and delicious dishes that bring wonderful treats to the rider's mouth and help restore and heal the rider's body and mind, as cyclist F12-XF pointed out:

> *After the meal, the tour leader took us to a dessert store to eat Hainan Qing Tonic Liang and mango intestinal noodles, which not only neutralized the greasiness of dinner, but also took away the fatigue of the day.*

Accommodation is also an important part of the cyclists' experience throughout the ride. For cyclists who sleep in the wind, only simple facilities and clean, comfortable accommodation are needed, but for others the services and facilities of the local hostel also bring multiple sensory experiences to cyclists.

4.2. Cyclists' Perceptions in a Human Context

Embodiment theory asserts that a subject's perception is generated by the experience of the body interacting with the environment and that the body plays a key role in this process [58]. In both the "companion context" and the "pan-companion context", travel companions can have a positive arousal effect on unknown and challenging activities. In other words, riding companions and "pan-companions" can help individuals with negative emotions during long rides. In addition, in team riding, the competition and cooperation among riding companions can enrich individual physical practice, strengthen relationships, and form a sense of group identity and shared memories. Both explicit and tacit knowledge plays an equally motivating role here; for example, the explicit knowledge represented by cheers and slogans of encouragement, and the tacit knowledge conveyed by the silent leader of the pioneers. Cyclist F2-XC recalled:

> *There was a long uphill climb that I was ready to give up on, it was already 7 pm, almost dark and I was extremely exhausted. But with the encouragement of my fellow leaders and the encouragement of the group ahead of me, I persevered little by little over the rainforested mountains to reach the destination.*

Of course, in any companion context, conflicts may occur among cyclists sparked by events such as route finding, destination selection, and unexpected accidents during the ride, and negative emotions such as anger, rage, hard feelings, and sadness may occur, affecting the way cyclists behave and feel psychologically among their companions.

In the "host context", tourists can empathize with certain elements of the destination context through diverse embodied practices, triggering their past emotions or memories. Along the long-distance ride in Hainan, the information gained from the various senses of the cyclists in the "host context" always evoked microscopic feelings from the cyclists, contrasting and reflecting on their original life. This is because of the implicit knowledge embedded in the scenes of island life, which is silently conveyed to cyclists, as cyclist F11-CYC said:

> *On the ride, you can always see the Hainan countryside surrounded by coconut groves. Every house has a hammock, and everyone will go to the teahouse everywhere to drink tea and chat. It seems that people do not have any worries, which reminds me of the leisurely days before I started work.*

Some of the scenes that cyclists see in the "host scenario" make them feel relaxed, comfortable, or full of admiration, and also evoke their memories of a "stress-free" and "leisurely" life. In addition, the individual behavior of the hosts also plays an important role in the emotion generation and memory recall of the cyclists. Sometimes cyclists may even inadvertently participate in local festivals, religious ceremonies, and other folk activities. These host individuals and diverse folk activities can remind cyclists of the similarities and differences in their cultural backgrounds in the context of local images and cultural perceptions.

4.3. Cyclists' Embodied Experiences in a Nonhuman Context

The nonhuman context consists of a natural context, mediated facility context, and digital technology context. First, the "natural context" is one of the most important nonhuman contexts that riders experience. The contemporary natural aesthetics theory believes that people need to perceive the world through the five senses to make the true aesthetic characteristics of natural objects appear [59,60]. Hainan, with its unique natural geographic location and multi-ethnic intersection, has formed a diverse natural landscape that is distributed in different forms in the spatial and temporal range of the Hainan cyclist's path.

However, there are times when natural environmental conditions exceed the limits of the body; for example, in some extreme weather (such as typhoons) cyclists have to leave the field temporarily and give up the physical and mental experience of island riding. In addition, the elements of the natural environment are constantly changing during the flow of the ride, triggering the body to pay attention to the natural environment and driving the body to change the rhythm of its movement. For example, the direction of the mountains, the distribution of the sea, and the status of plants and animals in the activities not only bring cyclists a diverse sensory experience but also limit the position, direction of movement, and forward speed of the cyclists' body, as cyclist F11-CYC said:

> *The section to Changjiang Li Autonomous County was hell for the body: 130 km, headwinds, many slopes, and an overall rising altitude. [I was] so physically and mentally exhausted that I doubted my life, and all the body's senses were numbed in the final sprint.*

Second, a " mediated facility context" in a "nonhuman context" refers to the context created by the cyclist during the bicycle experience. Transportation plays a mediating role in the mobility of a traveler and influences the extent to which he/she is involved in the road space. In contrast to automobiles, the mobility of bicycles can shape the experience of particular human–terrestrial interactions during the cyclist's activity [47]. A cyclist's experience generation is accomplished in part through the bicycle, and this mediated experience is more susceptible to factors such as the weather, road conditions, and the cyclist's abilities. However, a cyclist stays at more nodes during a riding tour compared to mass transit, so the cyclist's body interacts with the external space more frequently, presenting him with more travel options as a result. For example, cyclists are free to choose their riding routes, from national highways, provincial roads, county roads, and even country roads, and they get to form visual, auditory, and taste sensations different from those of mass tourism. In addition, cyclists also commonly experience discomfort in specific parts of the body due to bike overuse, such as wrist, knee, and hip pain, as cyclist F2-XC recalled:

> *On the first day of the ride, after less than 50 km, I started to feel vague pain in my leg muscles after almost a year of no exercise, and my buttocks started to burn because of the hard saddle of the bike. My knees were not ideal even though I had brought knee pads.*

Third, the "digital technology context" also belongs to the "nonhuman context". The rapid development of digital technology, the Internet, and mobile devices has changed travelers' behavior and influenced both the content and methods of travel experiences [61]. As a result, digital contexts require riders to use more explicit knowledge to engage. In particular, a mobile phone, as a vessel for "total media", with various functions such as a camera, Internet, music, and navigation, plays an important role in cyclists' activities. Research has shown that rhythmic music can make exercisers more relaxed and persistent [62], and the "cooperation" between cell phones and headphones can provide cyclists with just such an external rhythmic power. Furthermore, the music, through the auditory system, can have a regulatory effect on the cyclists' functions, spurring the auditory and motor states to interact. Music not only relaxes the muscles and nerves but also gives cyclists a broader spatial and temporal connection. A cyclist's body and mind, perception, and context all fit together in the rhythm of embodied movement to achieve a harmonious state of cycling. In addition to communication devices, some cyclists also carry GoPro sports cameras, bicycle recorders, drones, and other devices. These can provide technical support for recording the whole process of cycling, extending and enriching the limited sensory sensations of the cyclist's body.

4.4. Cyclists' Perception in a Nonhuman Context

Merleau-Ponty argues that we in the world must experience and perceive things with a priori knowledge and a biological body as a mediator [25]. For the cyclist, the "natural context" provides the raw material for perception, while the mediating aspects and digital technologies equipped on a cyclist's body enrich the results of his perception. Therefore, cyclists have the opportunity to feel happy as long as the subject perceives the beauty of nature [63]. The natural context is not only the explicit knowledge of appearance, but also has the potential to bring about unexpected behavioral adjustments in the acceptance of tacit knowledge by the riders. For example, in the process of natural landscape transformation, a rider's body and mind in the repetitive and ordinary monotonous natural landscape are more likely to be affected by the negative feelings of exhaustion and dullness brought about by riding long distances. Cyclists are also more likely to have negative emotions, such as frustration, regret, sadness, feelings of difficulty, and helplessness.

A diverse combination of natural landscapes or landscapes with greater impact than what a cyclist is used to may allow a cyclist to show greater willingness to ride continuously during a ride, and to relax physically and mentally in a wonderful riding environment, as cyclist F6-SC said:

> *On the road to Wanning, the scenery is really great, first the endless blue sea, and then the lush [greenery]. The tropical rainforest behind, in such a shocking scene, feet pedals are also more powerful, and the fighting spirit is also higher even if tired but also uplifting.*

In addition, in contrast to mere static images, the flowing natural landscape during the ride itself can also have a significant impact on an individual cyclist's vision, which is conducive to triggering the psychological tendency of those engaged in cycling activities to want to continue. Because of the special mode of transportation and the formation of the "intermediary context", shuttling in Hainan's various remote corners can also form a different perceptual experience for cyclists who travel there. For example, the end of the journey of the cyclist will not lose the flow of space and may still bestow profound observation and reflection. Cyclist F11-CYC said:

> *After ten days of cycling around the island, passing through villages, towns, and then cities along the way, you can feel the development of the times and the change of people's needs, as if you are witnessing the sculpting process of history for space.*

During long road trips, the bicycle creates a mobile experience of both "connection and separation". The "connection" refers to the cyclist's ability to travel around Hainan by bicycle as a means of transportation, to see the natural and human landscape at a more appreciable speed of movement, to perceive space and the landscape in depth, and develop a love and preference for space and place. As bicyclist F5-NJY believes:

I have been in the city for too long, and I feel that riding in such a good ecological environment can make me and nature become one.

"Separation" is a way of movement in which the cyclist, although deep in the cycling space, always uses the bike as an intermediary to travel and gaze, and there is often a sense of separation from the outside world, as cyclist F19-ZXH said:

I am sitting and observing the environment around the bike.

However, throughout a long ride, a cyclist may also become integrated with the bicycle, gaining a deeper sense of road space and physical experience. The practice of cycling requires coordination between nearly all parts of the body, and during the cycling process, a cyclist's body and mind can gradually become more in tune with the bicycle as intermediary equipment.

Cyclists use their bikes to perceive changes in the road environment as well. Changes in road material, slope, route direction, and type may also trigger emotional changes and memories for cyclists. Additionally, the convenient and low-cost communication method of the Internet, supported by digital technology context, has facilitated cyclists' travel-sharing behavior. The cyclists recorded their images and "self-gaze" during the ride through cameras, cell phones, and other media, while other users of social media liked to comment on what they shared, giving cyclists a sense of accomplishment, satisfaction, and other positive emotions. In addition, the cyclists used electronic products to record the sensations of the body while also strengthening the cyclist's sense of conquest and control over the natural space of the ride. Cyclist F2-XC mentioned:

What I read in a magazine said that the motivation for humans to take pictures during travel is similar to hunting activities in the primitive society period. The long lens of the camera is like a pair of wild eyes in the jungle, aiming at the prey, only now the prey is replaced by a photo.

The space and place in the cycling journey are recorded in multiple media with digital technology support and form a unique cycling space between individual recollection and group memory sharing in the future. With the selection of new landscapes to explore, more riding spaces may be filled with riding places that are full of emotions and subjects' perceptions, thus drawing a treasure map of riding places.

4.5. Value Generation of an Embodied Experience in Different Contexts

In the tropical landscape of Hainan Island, cyclists arm themselves with cycling tools such as bandanas and sunglasses. In this way cyclists can travel through different types of geographic areas and road systems, temporarily distancing themselves from their original social identities and immersing themselves in cycling activities. In a differentiated environment such as this, cyclists can reawaken their anesthetized, mechanized bodies and feel the purest natural projections and reactions of their bodies, such as breathing, muscle aches, and bodily pains. In addition, the body goes through a "rhythm cycle" consisting of a painful period, an adjustment period, a climax period, and a final sprint period during a long ride.

In addition, deep social interaction is important to cyclists in ephemeral spatial experiences [50]. In long-distance cycling, cyclists and their companions gradually construct inner connections and their mutual relationships shift from the weak relationship of strangers to the stronger relationship of acquaintances. Moreover, cyclists open themselves to each other and construct both a self-identity and group identity. Therefore, cycling activities go beyond physical movement and become a ritual of personal relationship production and emotional connection in the cycling space.

In the process of bodily practice, cyclists experience multiple levels of suffering and pain and generate a new self with a tough character and multiple types of cognition. During a ride, cyclists unconsciously socialize with other cyclists or the host, which involves the exchange of explicit knowledge such as local tourist information. Moreover, a cyclist's body can serve as a social mirror image, and the better body shape and appearance achieved by cycling activities may inspire praise and commendation by other individuals inside the cyclist's social network; that is, the objective experience of cycling can trigger a rider experiencing truth [17] and complete self-growth while awakening body consciousness.

5. Discussion

Because they employ a mode of mobility different from the usual means of travel of the general public, cyclists are freer to choose the content and intensity of human–place connections and to face more complex contexts, resulting in the special embodied experiences. Therefore, this paper explored the types of contexts encountered by cyclists during island cycling, as well as the physical experiences and perceived outcomes encountered by cyclists in different contexts (Figure 3). Our specific findings are reflected in the following areas.

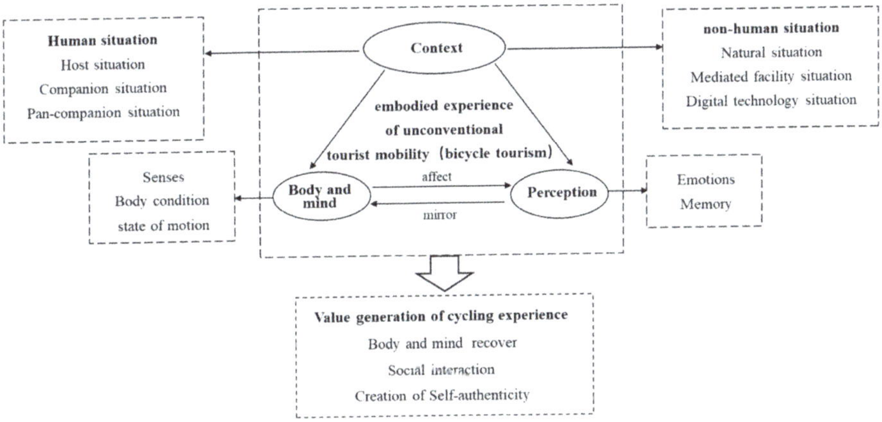

Figure 3. Diagram of the embodied experience mechanism of cycling in Hainan.

First, the contexts encountered by cyclists in the Hainan traffic circle include human and nonhuman contexts, and the combination of these contexts constitute their cycling experience system. The environmental variables that limit cyclists' physical experience and perceived outcomes include the road network characteristics, the natural environment, the social environment, and the building environment. In addition, cyclists gain embodied experiences in mobility ranging from visual experiences to multisensory experiences [19], and this paper also analyzed the micro-practices of embodied senses such as taste, smell, and hearing in mobility experiences one by one. Han found that the main components with the greatest perceived value for cyclists were tourism attractions, accessibility, amenities, and complimentary services [9], and this study identifies the key elements that influence embodied experience and uses this as a basis for classifying contexts into human and nonhuman contexts. Among these contexts, the human context included the companion context, pan-companion context, and host context, and the nonhuman context included the natural context, mediated facility context, and digital technology context.

It is worth mentioning that during the Hainan ride, cyclists gained explicit and tacit knowledge in the embodied experience of different contexts. These two kinds of knowledges are interchanged in certain behaviors. Polanyi argues that an individual's consciousness is composed of focal awareness and subsidiary awareness in the context to which the behavior is associated [31,32,34–36]. Associated with the bicycle tour context, the "focal awareness" is the target point of the act of cycling, while the body's senses are used as "subsidiary awareness" to serve the "focus awareness" of cycling [31,34]. However, in order to accomplish the action of "cycling", both "focus awareness" and "subsidiary awareness" are indispensable, and it is not necessary to say which is more important. The basis of an embodied experience is based on bodily sensations, and from the above analysis, it is clear that bodily sensations are used as a kind of "subsidiary awareness", which Polanyi calls "tacit knowledge". This "tacit knowledge" is integrated into the action of "cycling", thus forming the embodied experience of the rider's final ride.

Second, as mentioned by Shipway, there is a unique interaction between the cyclist, the cycling environment, and the activity, but the mechanism or process of this interaction has rarely been explored [64]. This paper finds that different contexts generated by the cyclists together with their bodies and perceptions constitute a dynamic system. As for long-distance cyclists, the local community plays an important role in the cyclist's travel experience, as a node for cyclists to resupply and rest [18]. In this study, we also found that hosts are important actors in providing embodied experiences for cyclists, providing them with the basic physical needs such as food, clothing, and shelter. Embodied experiences in a nonhuman context mostly cover the direct and indirect contents of the natural environment, intermediary facilities, and digital technology products that help cyclists gain physical experiences and perception. The natural landscape not only serves to heighten a rider's multi-sensory experience but also limits the direction, rhythm, and energy efficiency of his body, while the changes in the grade and combination of the natural landscape itself can trigger emotional changes. This finding also agrees with Han, Meng, and Kim's research that natural elements, such as plentiful natural resources and weather, are important experiential elements in cyclists' mobile visuals [9,65]. As the main mediating facility during cycling, the bicycle is the main tool for cyclists' mobility and observation, and cyclists gain unique physical and mental experiences by using it to interact with the outside world. In addition, while previous studies have rarely explored cyclists' use of digital technology to enrich a specific context and their embodied experiences, this study expanded cyclists' context perspectives and developed encounters within a digital technology context that not only broaden the cyclists' sensory experiences and physical abilities but also facilitate further storage of the nonrepresentational contents, such as memory and emotions, for the cyclists' subsequent reconstitution of their cycling trips as historical experiences.

Third, there is still a lack of attention paid to the interpretation of the cycling space in shaping the experience of cyclists, but in this paper we made a series of explanations of the cyclists' experience in their relevant space [64]. For example, during the cyclists' entire ride, the places along the route were not only physical spaces, but also generative spaces filled with individual emotions and memories. The cyclists were far away from the space of daily life, free from the atomized and disciplined body of daily life, and returned to the life-state body, a form of embodied practice that Levers called "anti-rhythmic" [66]. Human beings are social animals and need to enter into social relationships. Riding with a partner or multiple partners not only enhanced the emotions of the fellow riders during the ride but also formed beneficial social interactions with many hosts, satisfying cyclists' innate human interpersonal needs. This finding is similar to that of Xu's study where cycling was found to help consolidate old social relationships and possibly also enable new ones to be established during brief social interactions [50]. Unlike the mobile stage deliberately shaped by cycling events, independent cyclists in Hainan, China, were more random and freer in their choice of context. For many individual cyclists, cycling is an embodied practice driven by their random combination of routes, landscape choices, and accompanying intrinsic motivations, and the stage it creates lacks the political intentions of large-scale sports promotion and local showcasing of sports events [45]. Instead, for these riders, cycling is more of a self-actualization process for individuals pursuing their dreams and discovering their true selves.

6. Conclusions

Understanding the different embodied experiences of sports tourism requires exploration of specific sports tourism contexts. Therefore, to explore the embodied experiences of independent bicycle tourists in tropical islands, this study used qualitative methods to analyze the human and nonhuman contexts and their bodily sensations and perceptual outcomes in different contexts. We believe that the cyclist's body and mind are inseparable, and that the two become one in different contexts. Moreover, these unique physical sensations and perceptual outcomes both help cyclists restore their physical and mental rhythms, strengthen social relationships, and discover a new aspect of themselves. This paper further enriches the application of existing embodiment theory in tourism experiences and expands the study of the integration of sports and tourism from a micro analytic perspective. In addition, we also incorporated Polanyi's theory in our study, giving examples of explicit and tacit knowledge that cyclists may acquire during their rides.

Tourism riding experiences, from the embodied perspective, are the result of the interaction of a series of related elements, such as perception, body, and context, in the tourism process. Therefore, tourist destinations that feature cycling may attract more tourists if they optimize the various contexts encountered by cyclists, create a more personalized cycling environment, develop cycling routes, and regularly maintain cycling infrastructure. Moreover, bicycle tourism product providers may want to focus on the design of cycling routes so that they switch between diverse landscapes, highlighting the regional characteristics of the landscape to bicycle tourists. In addition, governments or other entities who wish to promote the bicycle tourism market may want to form a complete spectrum of bicycle tourism products, as high-intensity bicycle tourism activities may not be suitable for all people, and there is a need to develop a diverse range of cycling tourism products to meet the requirements of the different groups.

This paper makes a useful exploration of island-based cycling experiences through a three-dimensional embodied theoretical framework, but the breadth and depth of the original materials obtained in this article are limited. In addition, the article lacks an analysis of the differences in context types and embodied experiences of the three main existing cycling routes on Hainan Island. There are several directions for future research. First, suitable distance and psychological testing equipment can be used to obtain various aspects of the cyclists' physical data during cycling to help reveal the dynamic relationship between cycling distance and cyclists' physical and mental rhythms more scientifically. Second, the similarities and differences in the physical experience and perceived results of individuals under different mobility modes, such as self-driving tours, motorcycle tours, and hiking tours, can also be explored. Finally, further research is needed on the physical sensations of cyclists in various nodal stops to explore the physical and perceptual differences between these stops and the cycling process itself.

Author Contributions: Conceptualization, M.L. and H.X.; methodology, M.L. and H.X.; validation, H.X.; formal analysis, M.L. and H.X.; investigation, M.L. and H.X.; resources, M.L.; writing—original draft preparation, M.L. and H.X.; writing—review and editing, M.L. and H.X.; supervision, M.L.; funding acquisition, M.L. All authors have read and agreed to the published version of the manuscript.

Funding: This research was funded by the National Natural Science Foundation of China, grant number 41701143, and Guangdong Basic and Applied Basic Research Foundation, grant number 2021A1515011186.

Institutional Review Board Statement: Ethical review and approval were waived for this study. due to the national law "Regulations for Ethical Review of Biomedical Research Involving Humans" do not require the full review process for the data collection from adults who have adequate decision-making capacity to agree to participate.

Informed Consent Statement: Informed consent was obtained from all subjects involved in the study.

Data Availability Statement: Not applicable.

Conflicts of Interest: The authors declare no conflict of interest.

References

1. Hinch, T.D.; Higham, J.E. Sport tourism: A framework for research. *Int. J. Tour. Res.* **2001**, *3*, 45–58. [CrossRef]
2. Higham, J. Sport tourism as an attraction for managing seasonality. *Sport Soc.* **2005**, *8*, 238–262. [CrossRef]
3. Bakogiannis, E.; Vlastos, T.; Athanasopoulos, K.; Vassi, A.; Christodoulopoulou, G.; Karolemeas, C.; Tsigdinos, S.; Kyriakidis, C.; Noutsou, M.-S.; Siti, M. Exploring motivators and deterrents of cycling tourism using qualitative social research methods and participative analytical hierarchy process (AHP). *Sustainability* **2020**, *12*, 2418. [CrossRef]
4. Oosterhuis, H. Cycling, modernity and national culture. *Soc. Hist.* **2016**, *41*, 233–248. [CrossRef]
5. Griffin, B. *Bad Roads Will Absolutely Nip in the Bud the New Development?: Cycling Tourism in Ireland in the Late Nineteenth and Early Twentieth Centuries*; Liverpool University Press: Liverpool, UK, 2016; pp. 187–206.
6. Lamont, M. Reinventing the Wheel: A Definitional Discussion of Bicycle Tourism. *J. Sport To.* **2009**, *14*, 5–23. [CrossRef]
7. Bull, C.J. Racing cyclists as sports tourists: The experiences and behaviours of a case study group of cyclists in East Kent, England. *J. Sport To.* **2006**, *11*, 259–274. [CrossRef]
8. Ritchie, B.W. Bicycle tourism in the South Island of New Zealand: Planning and management issues. *Tour. Manage.* **1998**, *19*, 567–582. [CrossRef]
9. Han, H.; Meng, B.; Kim, W. Bike-traveling as a growing phenomenon: Role of attributes, value, satisfaction, desire, and gender in developing loyalty. *Tour. Manage.* **2017**, *59*, 91–103. [CrossRef]
10. Petino, G.; Reina, G.; Privitera, D. Cycling Tourism and Revitalization in the Sicilian Hinterland: A Case Study in the Taormina–Etna District. *Sustainability* **2021**, *13*, 10022. [CrossRef]
11. Vujko, A.; Dragosavac, M.; Kisin, N.; Maksimović, B. The influence of cyclo-tourism on health and economic destination development. Facta Universitatis. *Ser. Phys. Educ. Sport* **2017**, *15*, 227–237.
12. Ritchie, B.W.; Tkaczynski, A.; Faulks, P. Understanding the motivation and travel behavior of cycle tourists using involvement profiles. *J. Travel Tour. Mark.* **2010**, *27*, 409–425. [CrossRef]
13. Malchrowicz-Mośko, E.; Młodzik, M.; León-Guereño, P.; Adamczewska, K. Male and female motivations for participating in a mass cycling race for amateurs. the skoda bike challenge case study. *Sustainability* **2019**, *11*, 6635. [CrossRef]
14. Zhang, C.; Zhang, X. Constructing a mobile travel experience model:Empirical study of cyclists travelling to Tibet. *Geogr. Res.* **2017**, *36*, 2332–2342.
15. Xie, Y.J.; Fan, Y.M.; Tourism, S.O.; University, H. Tourist experience on a body perspective: Grounded theory analysis on trekking writings and interviews. *Human Geogr.* **2017**, *32*, 129–137.
16. Wilson, H.F. On geography and encounter: Bodies, borders, and difference. *Prog. Hum. Geog.* **2017**, *41*, 451–471. [CrossRef]
17. Wang, R.H.; Ma, Z.X. The Interaction between Objective Authenticity and Existential Authenticity: A Study on Cycling Tourism on the Sichuan-Tibet Line. *J. Sichuan Norm. Univ.-Soc. Sci.* **2020**, *47*, 64–73.
18. Wang, X.J. Host-Guest Interaction and Human-place Relationship Production in Road Travel. *J. N Minzu Univ.* **2020**, *6*, 71–76.
19. Wang, X.; Sun, J.; Huang, X. Mediation, Body and Emotion: Mobility Experiences in Road Travel on Sichuan-Tibet Highway. *Sci. Geogr. Sin.* **2019**, *39*, 1780–1786.
20. Shaw, J.; Hesse, M. Transport, geography and the 'new' mobilities. Transactions of the Institute of British. *Geographers* **2010**, *35*, 305–312.
21. Doel, M. Post-structuralist geography: A guide to relational space by Jonathan Murdoch. *Ann. Assoc. Am. Geogr.* **2007**, *97*, 809–810. [CrossRef]
22. Hockey, J.; Collinson, J.A. Grasping the phenomenology of sporting bodies. *Int. Rev. Sociol. Sport* **2007**, *42*, 115–131. [CrossRef]
23. Larkin, M.; Eatough, V.; Osborn, M. Interpretative phenomenological analysis and embodied, active, situated cognition. *Theory Psychol.* **2011**, *21*, 318–337. [CrossRef]
24. Heidegger, M. *Being and Time*; SCM Press: London, UK, 1962.
25. Merleau-Ponty, M. *Phenomenology of Perception*; Routledge & Keagan: New York, NY, USA, 1962.
26. Devlin, K. Situation theory and situation semantics. In *Handbook of the History of Logic*; Elsevier: North Holland, The Netherlands, 2006; Volume 7, pp. 601–664.
27. Seligman, J.; Moss, L.S. Situation Theory. In *Handbook of Logic and Language*; Elsevier: North Holland, The Netherlands, 1997.
28. Braun, B. Environmental issues: Writing a more-than-human urban geography. *Prog. Hum. Geog.* **2005**, *29*, 635–650. [CrossRef]
29. Stebbins, R.A. Studying the definition of the situation: Theory and field research strategies. *Can. Rev. Sociol.* **1969**, *6*, 193–211. [CrossRef]
30. Zhang, Y.-B. The Significance of Being in the Overall Context of Cognitive Gestalt: An Interpretation of Polanyi's Personal Knowledge. *Study Explor.* **2019**, *293*, 1–10.
31. Polanyi, M. *Personal Knowledge*; Routledge: London, UK, 2012.
32. Zhao, S.Y.; Hong, X.N. The Dialectical Relationship between Explicit Knowledge and Tacit Knowledge. *Stud. Dia. Nat.* **2001**, *10*, 21–23.
33. Héder, M. Explicit Knowledge in the Philosophies of Harry Collins and Michael Polanyi. In *The Periodical of the Michael Polanyi Liberal Philosophical Association*; The University of Chicago Press: Chicago, IL, USA, 2012; Volume 45. Available online: https://polanyiana.org/articles/polanyiana-2012-1-2-21_mihly-hder-explicit-knowledge-in-the-philosophies-of-harry-collins-and-michael-polanyi.pdf (accessed on 11 August 2022).
34. Polanyi, M. *The Tacit Dimension*; Routledge: London, UK, 1962; pp. 135–146.
35. Polanyi, M. *Study of Man*; The University of Chicago Press: Chicago, IL, USA, 1958; Volume 11.

36. Jie, Q.I.; Zhong-Lin, L. What is Tacit Knowledge Exactly?—Intensive Study on Michael Polanyi's Deep-rooted Thinking. *Stud. Dia. Nat.* **2009**, *25*, 46–51.
37. Gallese, V.; Lakoff, G. The Brain's concepts: The role of the Sensory-motor system in conceptual knowledge. *Cogn. Neuropsychol.* **2005**, *22*, 455–479. [CrossRef]
38. Liu, Y.; Wang, S.; Jie, X. In Experiential honeymoon tourism study from perspective of embodied cognition. In Proceedings of the 2019 5th International Conference on Information Management (ICIM 2019), IEEE, Cambridge, UK, 24–27 March 2019; pp. 327–331.
39. Ilze, M. Tourists' Embodied Transport Experiences of Travelling by Train. 2015. Available online: https://e-space.mmu.ac.uk/617320/1/Ilze%20Mertena%20PhD%202015.pdf (accessed on 8 February 2022).
40. Bouchet, P.; Lebrun, A.M.; Auvergne, S. Sport tourism consumer experiences: A comprehensive model. *J. Sport To.* **2004**, *9*, 127–140. [CrossRef]
41. Aho, S.K. Towards a general theory of touristic experiences: Modelling experience process in tourism. *Adv. Psychol. Sci.* **2001**, *56*, 33–37. [CrossRef]
42. Pan, J.; Hai, L.I. Place attachment and Sports Tourism: Experience Generation of "Dynamic Attachment". *Sports Sci.* **2018**, *39*, 63–69.
43. Themen, K.; Popan, C. Auditory and visual sensory modalities in the velodrome and the practice of becoming a track cyclist. *Int. Rev. Sociol. Sport* **2022**, *57*, 618–633. [CrossRef]
44. Spinney, J. A place of sense: A kinaesthetic ethnography of cyclists on Mont Ventoux. *Environ. Plann. D* **2006**, *24*, 709–732. [CrossRef]
45. Mutter, S. Kin-aesthetics, ideology, and the cycling tour: The performance of territory in the Israeli Giro d'Italia. *Mobilities* **2022**, *17*, 501–516. [CrossRef]
46. D'Andrea, A. Neo-nomadism: A theory of post-identitarian mobility in the global age. *Mobilities* **2006**, *1*, 95–119. [CrossRef]
47. Scuttari, A. Tourism experiences in motion. Mobile, visual and psychophysiological methods to capture tourists "on the move". *Tour. Manag. Perspect.* **2021**, *38*, 100825. [CrossRef]
48. Chin, J.W.; Masucci, M.; Johnson, J. Keep it rolling: Examining the San José Bike Party with methods on the move. *Qual. Res. Sport Exerc.* **2020**, *12*, 159–174. [CrossRef]
49. Chiu, W.; Leng, H.K. Let's Go Cycling: An Analysis of Tourists' Experience on Online User-Generated Content. 2017. Available online: https://repository.nie.edu.sg/bitstream/10497/19757/1/IJTC-3-1-30.pdf (accessed on 15 February 2022).
50. Xu, H.; Yuan, M.; Li, J. Well-being of cycling tourists in Hainan, China. *Sport Soc.* **2019**, *22*, 1449–1465. [CrossRef]
51. Lee, C.-F.; Huang, H.-I. The attractiveness of Taiwan as a bicycle tourism destination: A supply-side approach. *Asia Pac. J. Tour. Res.* **2014**, *19*, 273–299. [CrossRef]
52. Crouch, D. Places around us: Embodied lay geographies in leisure and tourism. *Leis. Stud.* **2000**, *19*, 63–76. [CrossRef]
53. Andrews, G.J.; Chen, S.; Myers, S. The 'taking place'of health and wellbeing: Towards non-representational theory. *Soc. Sci. Med.* **2014**, *108*, 210–222. [CrossRef] [PubMed]
54. Finlay, L. Negotiating the swamp: The opportunity and challenge of reflexivity in research practice. *Qual. Res.* **2002**, *2*, 209–230. [CrossRef]
55. Büscher, M.; Urry, J. Mobile methods and the empirical. *Eur. J. Soc. Theory* **2009**, *12*, 99–116. [CrossRef]
56. Zhang, C.; Wong, I.K.A.; Zhang, X.; Fyall, A. From Online Community to Offline Travel Companions: Technology-Mediated Trust Building and Ad Hoc Travel Group Decision Making. *J. Hosp. Tour. Res.* **2020**, *44*, 1101–1125. [CrossRef]
57. Sersli, S.; Gislason, M.; Scott, N.; Winters, M. Easy as riding a bike? Bicycling competence as (re)learning to negotiate space. *Qual. Res. Sport Exerc.* **2021**, *14*, 268–288. [CrossRef]
58. Ignatow, G. Theories of Embodied Knowledge: New Directions for Cultural and Cognitive Sociology? *J. Theor. Soc. Behav.* **2010**, *37*, 115–135. [CrossRef]
59. Chatterjee, A.; Vartanian, O. Neuroscience of aesthetics. *Ann. N. Y. Acad Sci.* **2016**, *1369*, 172–194. [CrossRef]
60. Barwich, A.-S. Up the nose of the beholder? Aesthetic perception in olfaction as a decision-making process. *New Ideas Psychol.* **2017**, *47*, 157–165. [CrossRef]
61. Buhalis, D.; Law, R. Progress in information technology and tourism management: 20 years on and 10 years after the Internet—The state of eTourism research. *Tour. Manag.* **2008**, *29*, 609–623. [CrossRef]
62. Huang, M.; Zhang, M. Cognition of place in embodied practice: Non-representational theory and Nanjing Marathon. *Geogr. Res.* **2019**, *38*, 1355–1366.
63. Eaton, M.M. *Aesthetics and the Good Life*; Associated University Presse: Plainsboro Township, NJ, USA, 1989.
64. Shipway, R.; King, K.; Lee, I.S.; Brown, G. Understanding cycle tourism experiences at the Tour Down Under. *J. Sport To.* **2016**, *20*, 21–39. [CrossRef]
65. Han, H.; Meng, B.; Kim, W. Emerging bicycle tourism and the theory of planned behavior. *J. Sustain. Tour.* **2016**, *25*, 292–309. [CrossRef]
66. Lefebvre, H. *Rhythmanalysis: Space, Time and Everyday Life*; Bloomsbury Publishing: London, UK, 2013.

sustainability

Article

Suitability Evaluation of Popular Science Tourism Sites in University Towns: Case Study of Guangzhou University Town

Wei Guo, Da-Fang Wu *, Yue Li, Feng-Xi Wang, Yong-Qi Ye, Hua-Wei Lin and Chi-Fang Zhang

School of Geography and Remote Sensing, Guangzhou University, Guangzhou 510006, China; guowei990905@gmail.com (W.G.); 15625126772@163.com (Y.L.); 18929655472@163.com (F.-X.W.); yyq2392339622@163.com (Y.-Q.Y.); vivi_23_warwick@163.com (H.-W.L.); zhangcf1116@163.com (C.-F.Z.)
* Correspondence: wudafang@gzhu.edu.cn; Tel.: +86-020-39366890

Abstract: With the advent of the knowledge economy, universities have increasingly important roles in cities. However, some universities and cities are still fragmented, and popular science tourism is one way to solve this. The purposes and destinations of popular science tourism differ from those of traditional tourism. Consequently, their resources have a high development value for ecological and human resources. However, research on popular science tourism in China is still in its infancy. Here, we studied popular science tourism from a geographical perspective, selecting scale capacity, environmental level, resource level, location, and service conditions as factors to construct a judgment matrix to calculate the weights of indicators at various levels. Analytic hierarchy was used to build a popular science tourism evaluation system for Guangzhou University Town as a case study and the suitability of each tourist destination in the university town was evaluated. The results show that west Guangzhou University Town is more suitable for popular science tourism than the east, which possesses more value for popular science tourism development. Finally, we give recommendations for developing popular science tourism in Guangzhou University Town.

Keywords: unconventional tourism; popular science tourism resources; development suitability; evaluation system; geographic information system; Guangzhou University Town

Citation: Guo, W.; Wu, D.-F.; Li, Y.; Wang, F.-X.; Ye, Y.-Q.; Lin, H.-W.; Zhang, C.-F. Suitability Evaluation of Popular Science Tourism Sites in University Towns: Case Study of Guangzhou University Town. *Sustainability* **2022**, *14*, 2296. https://doi.org/10.3390/su14042296

Academic Editors: Anna Rita Irimiás, Dallen J. Timothy and Gábor Michalkó

Received: 14 January 2022
Accepted: 9 February 2022
Published: 17 February 2022

1. Introduction

1.1. The Development of Science Tourism and the Role of Universities

With the improvement of education and the cultural level of tourists, traditional sightseeing tourism can no longer meet the personalized and diversified needs of tourists; thus, science tourism, which is a knowledgeable, educational, engaging, and entertaining form of tourism, has emerged to supplement traditional tourism. Traditionally, schools and universities are places where people learn and conduct research [1,2]; however, few consider them to be potential tourist destinations. In spite of this, popular science tourism has emerged with the purpose to gain knowledge and learn more about emerging technologies, rather than entertaining and relaxing; this is also known as "spiritual fulfilling tourism" [3]. Universities are at the forefront of scientific development and can fulfill people's needs; however, universities often do not provide accommodation outside the university grounds. If universities can work with residents of the surrounding area, then university towns can become a new kind of fully serviced tourist destination and become unique growth points for the tourism industry.

In 2015, China set out development goals for the next five years of building a moderately prosperous society, becoming an innovative country, and becoming a world power in science and technology. For these goals, the popularization of science is a key aspect and has great social significance. The "13th Five-Year Plan for Building a National Science Popularization and Innovation Culture" (from here on referred to as "the Plan"), formulated by the Ministry of Science and Technology and the Central Propaganda Department as part of the "Outline of the National Strategy for Innovation-driven Development" and the "13th Five-Year Plan for National Science and Technology Innovation" emphasizes the importance of the popularization of science. The Plan states that science popularization can enhance the level of social civilization, improve the quality of the ecological environment, promote the spirit of science and popularizing scientific knowledge in society as a whole, raise awareness of science and technology and scientific quality, and improve the ability of the population to solve practical problems and participate in public affairs.

Some scholars have defined science tourism resources to be the key to science tourism [4–6], believing that science tourism resources are "the sum of various natural and social factors that occupy a certain geographical space in nature and human society, and have special scientific and cultural values, and can be attractive to tourists, and can be exploited by the tourism industry, and produce scientific and cultural dissemination effects." Scholars have provided suggestions for the development of science tourism in specific cities, considering the science tourism resources in the area [7–9]. Science tourism industry projects started earlier and have achieved more than universities. Industrial science tourism is the most successful form of science tourism, mainly involving automobile manufacturing enterprises, such as Hyundai in South Korea, Toyota in Japan, Renault in France, and Volkswagen in Germany. However, with the rapid changes in science and technology, science tourism is gradually moving toward cutting high technology. The high-tech café project is a current research hotspot and universities are the best carriers for it [10]. In recent years, with the advent of the knowledge economy, universities have taken on additional functions and responsibilities, such as being sustainable parts of the city's futures, cultural icons and symbols, hearts of urban innovation, and suitable places for employment [11]. Therefore, campuses now emphasize the links among creativity, higher education, and urban life. These physical and functional links with the wider community and urban fabric can help improve the competitiveness, productivity, and cultural strength of a city [12,13] and should be considered before a university or university town is built. Some cities and their universities provide us with good examples, as shown in Table 1.

In general, as well as playing roles in teaching students, universities are important for the development of their respective cities. From the above analysis of the relationship between renowned universities and cities, it can be concluded that the beneficial relationship between cities and universities is mutual [21]. The knowledge economy has had the effect of breaking down traditional boundaries between campus and city. Modern universities are the most fertile intellectual environments and are no longer specialized cells; they are becoming accessible, giving ordinary people access to the most cutting-edge knowledge and technology [22]. In China, universities are still in their early stages of development, with campuses cut off from their cities and are gradually facing the challenge of transformation. One way to help integrate the campus and the city is through science tourism.

For the development of science tourism in China, most scholars in recent years have focused on the existing natural resources in their study areas, such as wetlands and parks, and designed science tourism routes based on these [23–26], while ignoring the role that universities could play in science tourism.

As pioneers of academic science and technology research, university towns are the birthplace of many recent scientific and technological achievements. For example, a university in a university town in Guangdong Province demonstrated the latest technology in pollution control to the public based on its own environmental technology innovation platform [27]. As a result, compared to conventional tourism, science tourism is more focused on the dissemination of knowledge and technology and often offers a more in-depth experience. Moreover, some research show that due to COVID-19, people are now more likely to choose short trips, while focusing more on destinations that are more than just leisure [28,29]. As a result, popular science tourism, especially in a university town that is close to the center of the city, becomes a good choice for most people. Travel companies and universities can design travel products based on various technologies already available, such as tracking through personal registration information, government, or analysis based on reviews of travel software [30–32].

Table 1. Some famous universities and their role and functions in cities.

University	City	Roles and Functions
Chicago University, Northwest University, and University of Illinois at Chicago	Chicago, United States of America	A leisure center for the city's inhabitants, a central place for the development of the city's scientific and technological strength, an important cultural center and tourist destination for the city. Creates many jobs and contributes to the economic development of Chicago [14,15].
Cambridge University	Cambridge, United Kingdom	A model of city–university integration, the University of Cambridge has broken down the traditional boundaries between campus and city in its construction. The city provides the environment for the university to develop, and the university provides the city with a humanistic foundation and heritage, enhancing its potential for development. In the current and future plans, the University of Cambridge serves as part of the heart of the city, providing services and leisure to the surrounding community, while also using advanced technology to replicate the traditional Cambridge scene, allowing more people to choose Cambridge or Cambridge University as a travel destination [16].
University of Technology Sydney	Sydney, Australia	Located in the heart of Sydney, it has a complete infrastructure and is designed without clear boundaries with the surrounding neighborhoods (i.e., no fences). Therefore, the school provides a recreational space for the surrounding neighborhoods and allows for the initial practice and popularization of advanced technologies, which can be combined with practical technological improvements while promoting them. In addition, the school improves the sophistication of urban governance through alliances with planning and municipal authorities, while also repositioning the city to external investors and knowledge workers [17].
Arizona State University	The city of Phoenix, United States of America	The university is an important part of the development of the City of Phoenix, providing jobs, the latest technology, and training for the city's workforce, as well as providing the nearby community with a recreational area of shopping, dining, and other functions. In turn, the city provides funding, space, and policies to support the university's development [18].
Hönggerberg site of ETH Zürich	Zürich, Switzerland	The transformation from a traditional campus to a community center is a typical example of campus-led regional development. The campus is surrounded by a high density of housing and shops, providing a recreational space for the surrounding population (e.g., Leerpark Dordrecht, West) and promoting high-tech experimentation [19].
Washington University	Seattle, United States of America	The University of Washington has strong ties to the surrounding community and downtown. It is also a center for weekend recreation for city residents. As part of the urban landscape, the University of Washington provides residents with a place to relax and promotes all aspects of the campus and gives citizens a taste of the latest technology [20].

University towns have considerable potential as foundations for science tourism. Guangzhou, an innovative city in the southern coastal region of China, is equipped with many excellent educational resources at all levels, from primary schools to universities, owing to its advanced level of economic and social development. However, these resources are in urgent need of development and are expected to be planned based on local realities [33]. In addition, some scholars have pointed out that tourism can have edifying and practical functions [34] and science tourism, as a form of tourism that integrates knowledge and tourism, is better suited to these functions. If university towns are included in the design of popular science tourism routes, the distance between the public and universities can be reduced. Subsequently, the advantages of university towns in promoting science to the public can be introduced.

In Guangzhou, representative science resources, such as the South China Botanical Garden and the Guangdong Science Centre, have been developed in recent years, gradually becoming essential attractions for science tourism in Guangzhou; however, universities do not currently play a role. Guangzhou University Town, which contains many universities and resources for science tourism, is a pioneer in the development of university towns in China. Using Guangzhou University Town as the starting point for science tourism using university resources in China, its influence on science tourism can be used to inform other university towns in China on how to promote science tourism and expand the scope of science tourism in China. In addition to university resources, Guangzhou University Town contains a wealth of other types of tourism resources, such as Guangdong Science Centre, Lingnan Impression Park, and the Wetland Park, making it easy to explore the linkage between university resources and other natural or humanities and social science resources in the development of science tourism.

In general, science tourism outside of China is more developed than that in China [35,36], which is still in its infancy. Although science tourism projects started earlier in certain countries, these are mainly industrial projects, and they are gradually moving towards high technology. During our survey, we found that though some universities held these activities for a few years, there is a lack of data about science tourism in universities, especially visitor data. This is due to the fact that most universities do not pay attention to registering and keeping data on the number, age, and gender of visitors when organizing science tourism, but only as a basis for verifying the presence of visitors. It can be found that the current university science tourism, as an emerging unconventional tourism project, is in great need of formal guidance and training in terms of supervision, record keeping, and other areas.

There is an urgent need to strengthen theoretical research and technical improvements to fully explore the potential of the science tourism industry and to enhance the value of science popularization.

1.2. Land Suitability Evaluation

In terms of land suitability evaluation, different scholars have provided unique insights. In his theory on land suitability evaluation, McHargue advocated that a suitability analysis should be carried out according to the natural properties of the land to ensure the value of the land to the natural environment and introduce the economic value of the land. This viewpoint is also applicable to the suitability evaluation of science and technology tourism sites [37].

At present, suitability assessments of tourism sites generally begin from the perspectives of mountain topography, climate comfort, and land use [38], mainly from a natural standpoint to examine its impact on various aspects of tourism. Suitability assessments have been used for assessing pension tourism [39], urban greenways [40], and rural development [41]; however, there has been few studies on popular science tourism. In terms of methodology, geographic information system (GIS) has been used as the primary tool for evaluating the suitability of tourism development from the perspectives of tourism comfort [42], tourism competitiveness [43], and ecological sensitivity [44]. Previous studies generally evaluated tourism from a single perspective; however, some studies combined various factors to form a set of evaluation criteria and applied these criteria to assess suitability for development in a specific area of science tourism. In contrast to previous studies where weighting was based on expert assignments, the assignments in these studies were made using the hierarchical analysis method, with experts adjusting the weights objectively. The difference in importance of each element in the evaluation system can be derived and can be combined with the spatial analysis method and buffer zones to provide references for the construction of supporting facilities for science tourism sites, as well as provide a more objective and comprehensive evaluation of existing science tourism sites. Therefore, this paper will take universities as the carriers of science tourism, explore their unique science resources, and objectively evaluate their science value and development potential.

Taking Guangzhou University Town as a case study, this study combines science tourism with tourism resources to evaluate the suitability of science tourism sites, focusing on university resources for science tourism. The weights of each factor are determined by constructing matrices, which are then combined with hierarchical analysis to evaluate their potential for development and use. Popular tourism sites are then graded, and several tourism routes are designed to provide a theoretical basis for the rational development of popular science tourism in Guangzhou University Town and to determine the types of tourism activities and scale that university towns can provide in popular science tourism.

2. Overview of the Study Area, Research Methodology, and Data Sources

2.1. Overview of the Study Area

At present, although Guangzhou's tourism industry is highly developed, the number of attractions developed in recent years is relatively small and the focus is on cultural tourism (e.g., the construction of the "Super Wenheyou"). Guangzhou University Town is located in the eastern part of Guangzhou, on the island of Xiaoguwei, Xinzao Town, Panyu District, with a planned area of 43.3 km^2. The university town houses ten universities, historical and cultural heritage, and museums. It is currently the most significant investment project among university towns in China [45]. The rich tourism land resources in the town are the basis for tourism development, as well as the basic agglomeration unit for spatial tourism organizations. Lingnan history, ancient folk culture, Guangdong Science Center, and university construction in Guangzhou University Town complement each other, forming rich and diverse tourism resources in the city [46]. The landscape resources of university towns are classified as cultural tourism and historical sites [47]. The functions of the tourist sites in the university town include sightseeing tourism, which contains Lingnan vernacular and Lingnan folk culture and special tourism, such as enriching knowledge and increasing the experience of tourism [48,49]. Therefore, Guangzhou University Town has high scenic quality and aesthetic value, extending and assimilating the city's historical and cultural resources on three levels: material cultural layer, behavioral cultural layer, and conceptual cultural layer [50]. Figure 1 is a schematic diagram of the distribution of various blocks in Guangzhou University Town.

2.2. Evaluation Index Selection

The quality of science tourism sites mainly depends on the indicators of the extended attributes of tourism resources, tourism areas, and location conditions within the tourism sites [51]. In this study, based on existing evaluation methods [52–56], combined with relevant graphic information collected by the natural resources and planning departments and relevant textual information, the evaluation factors of different scales were selected from the macro and micro perspectives and combined with the actual situation of the university town area to establish an evaluation system for the development of science tourism sites. The following five items were selected as indicators for the evaluation of the quality of popular science tourism sites in Guangzhou University Town: (1) scale capacity, including the ratio of the area of land available for tourism to the area of all tourism land (expressed as S1); (2) environmental level, mainly the vegetation coverage of tourism sites; (3) resource level, including natural science popular tourism resources and humanistic science popular tourism resources; (4) service conditions, including infrastructure (communication, power supply, water supply), service conditions (accommodation and other services), and equipment conditions; and (5) location conditions, including accessibility and traffic conditions (road conditions, convenience, safety).

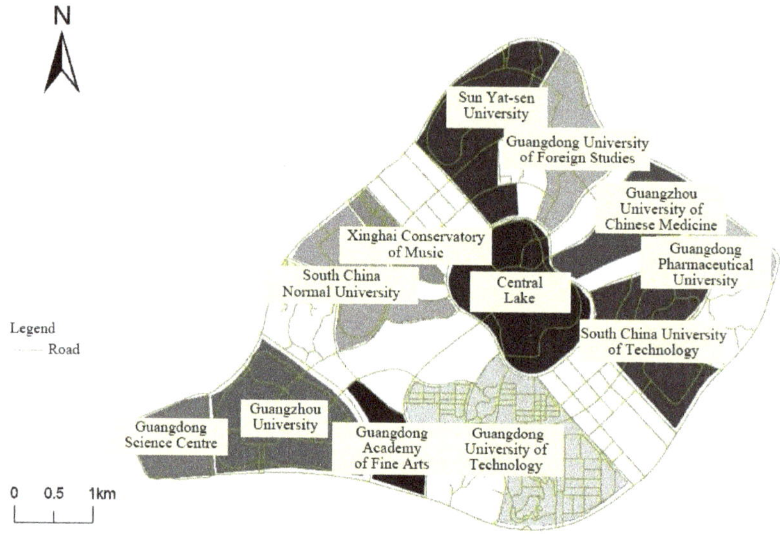

Figure 1. Distribution of areas of Guangzhou University Town.

2.2.1. Size and Capacity

The tourism environmental capacity was determined according to the area capacity method considering theories in ecology, environmental protection, tourism, sociology, economics, and education, and national regulations and relevant norms, which determine tourism environmental capacity (e.g., Environmental Protection Law of the People's Republic of China) [57–62]. The three methods of estimating tourism environmental capacity are the area capacity method, the route capacity method, and the chokepoint capacity method. Each of these methods are more applicable to different tourism scenic areas and tourism environments. Therefore, when conducting tourism environment capacity estimation, it is necessary to first determine the type of tourism functional area or tourism environment and then determine the basic parameters according to the characteristics of different tourism environments. Science tourism in Guangzhou University Town includes ten famous universities, cultural monuments from the Han and Tang dynasties to the present day, and other high-quality resources. The attractions are mainly distributed in blocks and the touring method is primarily based on land, so it is suitable to use the area capacity method for estimation.

$$C = (A/a) \times D$$

where C is the daily environmental capacity in person-times, A is the visitable area in m^2, a is the reasonable area or minimum area that each visitor should occupy in m^2, and D is the turnover rate (time the site is open, or the time required to complete the site).

The area capacity method calculates the daily environmental capacity of the touring area according to the formula based on indicators, such as the tourist area of the science tourism area, space reasonably occupied by visitors, opening hours, and time required for touring.

2.2.2. Environmental Level

Combined with relevant studies [63–66], since the vegetation in the university town is arranged by each university according to the campus pattern and the needs of science popularization, the normalized vegetation index (NDVI), a method of calculating the vegetation coverage degree by combining the red and near-infrared bands according to the spectral characteristics of vegetation, was used to calculate the vegetation distribution of Guangzhou University Town and to assess the greening degree of each science popularization tourist site.

$$NDVI = (NIR - R)/(NIR + R)$$

where NIR is the reflection value in the near-infrared band and R is the reflection value in the red band. The calculated NDVI index is less than 0 for water bodies, 0 for buildings, and greater than 0 for vegetation and these values can be used to classify the land use of Guangzhou University Town into these three types.

2.2.3. Resource Level

The resource level refers to the current science tourism resources in the tourism area. Combining the results of recent research on science popularization tourism resources in China [67] and the characteristics of the object and the scale of the study, the science popularization tourism resources in this study were defined as either facilities with knowledge science popularization value (e.g., science popularization bases, exhibition halls, ecological gardens, and historical and cultural sites) or activities with knowledge science popularization value in fixed places (e.g., student bijou exhibitions, stalls displaying the achievements of associations). Popular science tourism resources are an essential indicator for assessing whether a tourist destination has the value of popular science tourism. The importance of popular science tourism development in tourist destinations with popular science tourism resources is often higher than that in those without popular science tourism resources. If a tourist destination has fewer science tourism resources, it is necessary to invest in tourism development to enhance the value of tourism in the destination.

2.2.4. Service Conditions

Combined with the characteristics of science tourism in Guangzhou University Town and based on the principles of scientificity and operability, this study used catering as an evaluation factor to evaluate the level of service conditions in each block of Guangzhou University Town. Considering that universities differ from traditional tourism industry subjects, some elements from previous studies on science tourism [25,68] were not suitable as reference factors for evaluation. However, since popular science tourism has tourism at its essence, catering was included in the evaluation of service conditions as the evaluation standard. The primary method of the assessment was the use of Arcadia. The preliminary evaluation method used ArcGIS tools to create three levels of buffers on each tourist area and overlay them with the dining area to obtain the distance between each tourist location and the nearest dining area.

2.2.5. Location Conditions

In evaluating the location conditions of science tourism, the convenience of transportation plays a crucial role as it affects tourists' choice of science tourism destination [69–71]. Guangzhou University Town is included in a four-in-one urban transportation system, which includes "Guangzhou highway transportation (Beijing-Zhuhai Expressway and Central City Express Line), rail transportation (Metro Line 4 and Metro Line 7 'cross intersection'), general road transportation, and broadband network information transportation". The diverse transport needs of university towns can be met to the greatest extent possible within these network systems. The university town is ideally located in a central position between some of Guangzhou's major scenic spots, such as the Whampoa Military

Academy, Yu Yin Shan Fang, Bao Mo Yuan, Xiang Jiang Safari Park, Chang Long Happy World, Pazhou International Convention, and Exhibition Center, and Guangzhou Bio Island, making it a unique tourist location. Therefore, this study used GIS to calculate the nearest distance by road from each plot of Guangzhou University Town to University Town North Station, University Town South Station, and Nansha Express University Town Exit to analyze the locational relationship.

2.3. Determination of Evaluation Index Weights and Consistency Test

2.3.1. Construction of Judgment Matrix

We used the "1 to 3 scale method" [71–76] to construct the judgment matrix, with 1, 2, and 3 indicating that the indicator is equally important, more important, and obviously more important than the other when comparing two indicators; the inverse shows the opposite comparison of the two indicators. According to the evaluation hierarchy model, judgment matrices of the suitability of science tourism sites and the location conditions were constructed. Professors from several schools and organizations, including Guangzhou University, South China Human Geography, Urban Development Research Centre, and Nanning Normal University, were then invited to discuss and determine the final assigned values (Table 2).

Table 2. Hierarchical model and index weights for the suitability evaluation of popular science tourism in Guangzhou University Town.

General Objective A Level	Evaluation Indicator B Level	Evaluation Factor C Level
Suitability of the science tourism site	Service conditions (0.143)	Catering (0.143)
	Location conditions (0.143)	Higher Education Mega Centre North Station (0.047)
		Higher Education Mega Centre South Station (0.047)
		Highway entrance and exit (0.0047)
	Scale capacity (0.143)	Environmental degree (0.143)
	Resource level (0.428)	Number of attractions (0.428)
	Environmental level (0.143)	Vegetation cover (0.143)

2.3.2. Calculating the Weights of Each Evaluation Index

The nth root of the product of the elements of each row of the judgment matrix were calculated to obtain the square root vector T_i:

$$T_i = \sqrt[n]{\prod_{k=1}^{n} X_{ik}}, (i = 1, 2, \ldots, n) \quad (1)$$

where n is the number of evaluation factors and X_{ik} (i = 1, 2, . . . , n; k = 1, 2, . . . , n) is the judgment value obtained by comparing the relative importance of the ith indicator with that of the kth indicator.

The square root vector T_i was normalized to obtain the relative weight value of each indicator in the layer close to a particular indicator in the upper layer.

$$W_i = \frac{T_i}{\sum_{i=1}^{n} T_i}, (i = 1, 2, \ldots, n) \quad (2)$$

2.3.3. Consistency Test

The results obtained using the hierarchical analysis method were tested for consistency across the judgment matrices to ensure the reasonableness and reliability of the results.

The maximum characteristic root λ_{max} was calculated using:

$$\lambda_{max} = \sum_{i=1}^{n} \frac{1}{nW_i} \sum_{i=1}^{n} X_{ij} W_j \quad (3)$$

Consistency indicators were calculated using:

$$I_C = {(\lambda_{max} - 1)} \Big/ {(n-1)} \quad (4)$$

Stochastic consistency ratio R_C was calculated using:

$$R_C = {I_C} \Big/ {I_R} \quad (5)$$

I_R is an average random consistency indicator. If $R_C < 0.100$, the judgment matrix passed the consistency test. The R_C of the evaluation hierarchy model constructed in this study was 0.000, which passed the consistency test.

2.3.4. Calculation of Suitability Scores

Based on the weights of the obtained evaluation indicators, the scores of the different evaluation factors for each tourism site in the university town were determined. The sum of the indices for each tourism site was derived according to the following formula:

$$A = \sum_{i=1}^{n} P_i A_i \; (i = 1, 2, \ldots, n) \quad (6)$$

where A is the index sum of the evaluation factors of a tourism plot, P_i is the weight of the ith evaluation factor, A_i is the score of the I evaluation factor, and n is the number of evaluation factors. Once the evaluation factor index and the evaluation factor of each plot were obtained, the potential level criteria could be assigned separately to determine the level of each plot.

2.4. Data Sources

This study used 2018 index data of Guangzhou University Town obtained from geographic data platforms, including 2018 Guangzhou Landsat 8 remote sensing images obtained from the geospatial data cloud free download, with an accuracy of 30 m × 30 m, fieldwork, and questionnaires. Data on science resources were collected from the official websites of universities during university town and field visits. Data on tourism resources were collected by visiting tourist attractions in the town and recording the opening situation and number of attractions. The dataset contained nearly all the data on science and tourism resources. After researching the distribution of catering, it was found that due to the strict management of the university town, almost all food and beverage outlets were concentrated in several areas, as discussed in Section 3.4.

Interviews with pedestrians were obtained through field research conducted on multiple days from November to December 2020. We counted the number of people at the main entrances and exits of each university during weekends during the same period to ensure the scientific validity of the data. In addition, we randomly distributed questionnaires and collected 120 valid questionnaires through various means. The results of the questionnaire are presented in Section 3.6.

3. Results

3.1. Comparative Analysis of Daily Environmental Capacity

The environmental capacity of science tourism in Guangzhou University Town was calculated according considering the minimum area of 601 m^2 per visitor as stipulated in the Planning Code for Scenic Areas; the number of visitors that can be accommodated in each plot is shown in Figure 2, with a total capacity of 129,108 visitors per day. The areas with high daily environmental degrees were mainly located in the Central Lake, Guangdong University of Technology, Guangzhou University, Guangzhou Chinese Medicine and Pharmaceutical, and Sun Yat-sen University plots. The Guangzhou University, Sun Yat-sen University, and South China University of Technology plots had advantages in terms of visitor capacity, whereas the Guangzhou Academy of Fine Arts and the western part of Guangzhou International Innovation City plots had the lowest daily environmental capacity and could not accommodate many visitors, which is a disadvantage for these plots as prospective science tourism bases.

Figure 2. The daily environmental capacity of Guangzhou University Town (601 m^2 per person).

3.2. Comparison of Environmental Level

Table 3 of the land-use of Guangzhou University Town shows that the proportions of the areas of vegetation, buildings, and water bodies in Guangzhou University Town are 40%, 52%, and 8%, respectively. The total proportion of green areas and water bodies was 48%. Overall, Guangzhou University Town has high vegetation cover and a high degree of greenery, which is suitable for tourism and leisure.

Table 3. Land use of Guangzhou University Town.

Tourism Block	Vegetation (%)	Buildings (%)	Water Bodies (%)
Overall	40	52	8
Guangdong Science Centre	18	66	15
Lingnan Impression Park	67	28	4
The second Mausoleum Museum of the Southern Han Dynasty	70	29	1
Guangzhou University	22	68	10
Guangdong University of Technology	36	48	16
Guangzhou Academy of Fine Arts	22	56	22
South China University of Technology	24	65	11
Guangzhou University of Chinese Medicine—Guangdong Pharmaceutical University	15	76	9
Guangdong University of Foreign Studies	15	78	7
Sun Yat-sen University	23	72	5
Xinghai Conservatory of Music—South China Normal University	30	61	9
Guangzhou International Innovation City West	38	53	8
Central Lake Park	50	35	15
Wan Jutou Wetland Park	76	23	1
Chikan Wetland Park	77	22	1
Bei Gang Wetland Park	76	17	7

The land use of each plot shows that the environmental levels of natural science resources were significantly higher than those of human science resources. The highest percentage of vegetation was found in park-type plots, such as Wan Jutou Wetland, Chikan Wetland, and Bei Gang Wetland, all located in the outer ring. The Guangdong Science Centre, International Innovation City West, and various other universities had lower percentages of vegetation. The plots with the lowest greenery rate were Guangzhou University of Chinese Medicine, Guangdong Pharmaceutical University, and Guangdong University of Foreign Studies.

3.3. Distribution of Science Tourism Resources

The science tourism resources table distribution shows that human science tourism resources are the most critical science resource in Guangzhou University Town. The current natural science tourism resources in the university town are scarce and the degree of development is low. The plots of Guangdong Science Centre and Lingnan Impression Park are rich in science tourism resources. They have the lowest development difficulty and investment cost for developing as science tourism bases. The plots of Guangzhou University, Guangzhou University of Chinese Medicine—Guangdong Pharmaceutical University, Central Lake Park, and Xinghai Conservatory of Music—South China Normal University have some science tourism resources. In contrast, the Wan Jutou Wetland, Chi Kan Wetland, and Bei Gang Wetland plots have the lowest science tourism resources and are the least developed and, therefore, have the most severe difficulty in developing science tourism (Table 4).

Table 4. Distribution of Popular Science Resources in Guangzhou University Town.

Tourism Plots	Science Resources	Total Number
Guangdong Science Centre	Low Carbon and New Energy Vehicles Science Experience Hall, People and Health Hall, Green Home Hall, Innovation Space Hall, Digital Paradise Hall, Materials Park Hall, Transport World Hall, Children's World Hall, Experiments and Discoveries Hall, Lingnan Science and Technology Exhibition, Flying Dreams Pavilion, Sense and Think Pavilion, Science Square, Robotics Lab, AR Lab, 3D Printing Creative Design Lab, Solar Cell Production Lab	17
Lingnan Impression Park	Old Wine Shop, Old Barber Shop, Lion Dance Club, Lingnan Art and Culture Exhibition, Canton Embroidery Museum, Nun's House, Guangdong Broadcasting Museum, Old Newspaper House, Big House, Old Cinema, Old Photo Gallery, Guangdong TV's "Chen Mengji" Filming Base	12
The second Mausoleum Museum of the Southern Han Dynasty	Southern Han History Exhibition Hall, Archaeological Research Specimen Showroom, Public Archaeological Simulation Centre	3
Guangzhou University	Planetarium Science Base, Optoelectronic Science Base, Science Building North Building Student Works Exhibition, School History Exhibition Hall, Clubs Achievement Exhibition, Performing Arts Centre, He Shijie Gymnasium	7
Guangdong University of Technology	Industry 4.0 Experimental Base, Society Achievement Exhibition, Cultural Exchange Centre, Creative Center of Guangdong University of Technology	4
Guangzhou Academy of Fine Arts	Art Museum Sculpture Branch, Art Museum of Guangzhou Academy of Fine Arts, Association Achievement Exhibition, Liang Mingcheng Sculpture Garden	4
South China University of Technology	Exhibition of students' innovative design works, exhibition of club achievements, South China University of Technology Library	3

Table 4. *Cont.*

Tourism Plots	Science Resources	Total Number
Guangzhou University of Chinese Medicine -Guangdong Pharmaceutical University	Guangdong Pharmaceutical University Human Science Education Centre, Guangdong Pharmaceutical University Chinese Medicine Herbarium and Medicinal Botanical Garden, Guangdong Digital Home Interactive Application Engineering Laboratory, Guangdong Museum of Chinese Medicine, Guangdong Pharmaceutical Society Achievement Exhibition, Guangdong Chinese Medicine Society Exhibition	6
Guangdong University of Foreign Studies	Octagon House, Club Achievement Display	2
Sun Yat-sen University	Statue of Dr. Sun Yat-sen, Forest of Three Quotations, Society Achievement Exhibition	3
Xinghai Conservatory of Music-South China Normal University	Xinghai Conservatory of Music Concert Hall, Conservatory of Music hall of South China Normal University, Beigang Villagers' Traditional Residence, Sports Park of College town Campus of South China Normal University, Xinghai Club Achievement Exhibition, Club Achievement Exhibition of South China Normal University	6
Guangzhou International Innovation City West	Guangzhou National Archives, Guangzhou International Innovation City Exhibition Hall, Guangzhou Patriotic Education Base	3
Central Lake Park	Guo Long Guo Clan Ancestral Hall, Bauhinia Garden of Guangzhou University Town, Central Lake Cultural Square of Guangzhou University Town, Guangzhou Literature and Art Creation Centre, Sports Centre of Guangzhou University Town	5
Wan Jutou Wetland Park	Waterfowl Habitat Wetland	1
Chikan Wetland Park	Waterfowl Habitat Wetland	1
Bei Gang Wetland Park	Waterfowl Habitat wetland	1

3.4. Location of Tourist Sites and Catering Areas

The catering distribution map (Figure 3) shows that catering areas in Guangzhou University Town are primarily distributed between the central ring and the inner ring, and that the tourist plots located in this zone have good catering service conditions. All university plots have catering areas and, thus, had the best catering service conditions. Among non-university plots, the plot of Central Lake Park, a relatively high-traffic park in the middle of the island adjacent to several universities, had a catering area within 100 m. The second Mausoleum Museum of the Southern Han Dynasty is close to the living quarters of the South China Normal University and the Guangdong University of Industry; as a result, catering areas can be found within 200 m. The Guangzhou International Innovation City West is close to the Guangzhou Academy of Fine Arts and Guangdong University of Industry and, thus, has catering areas within 300 m. Most of the remaining plots, such as Lingnan Impression Park, Wan Jutou Wetland Park, Chikan Wetland Park, and Bei Gang Wetland Park, are located on the edge of the outer ring, but the living quarters of the major universities are generally located between the inner and middle rounds. In addition, several villages with catering areas are located between the center and outer rings, making it difficult to reach catering areas within 300 m from these plots.

Figure 3. Distribution of catering areas of Guangzhou University Town.

3.5. Accessibility of Tourist Areas

The location table of the Guangzhou University Town tourist area (Table 5) shows that Central Lake Park has the best location conditions and is particularly close to the metro station of Higher Education Mega Center south and north. In contrast, the plots of Guangdong Science Centre, Guangzhou University of Chinese Medicine—Guangdong Pharmaceutical University, Wan Jutou Wetland Park, Chi Kan Wetland Park, and Bei Gang Wetland Park were far from the metro station and highway exit, giving them the worst location conditions and making them relatively inaccessible for tourists.

Table 5. Guangzhou University Town Tourism Area Location.

Tourism Plots	Higher Education Mega Centre North Station	Nansha Harbour Expressway	Higher Education Mega Centre South Station
Guangdong Science Centre	4.4	2.8	4.5
Lingnan Impression Park	3.3	4.6	1.5
The second Mausoleum Museum of the Southern Han Dynasty	1.9	2.1	2.3
Guangzhou University	2.9	0.8	2.5
Guangdong University of Technology	2.9	2.9	0.8
Guangzhou Academy of Fine Arts	3.8	1.1	2.3
South China University of Technology	3.3	3.3	0.8
Guangzhou University of Chinese Medicine—Guangdong Pharmaceutical University	2.7	4.2	2.6
Guangdong University of Foreign Studies	1.7	4.3	2.5
Sun Yat-sen University	0.9	3.7	3.1
Xinghai Conservatory of Music—South China Normal University	1.1	2.2	3
Guangzhou International Innovation City West	2.9	1.7	2.4
Central Lake Park	1.1	2.6	1.7
Wan Jutou Wetland Park	4.3	3.5	5.5
Chikan Wetland Park	2.7	5.4	3.9
Bei Gang Wetland Park	3.5	7.2	3.6

3.6. Synthesis of Evaluation Results and Validation

3.6.1. Comprehensive Analysis of Evaluation Results

In this study, through field research, data analysis, expert empowerment, and theory on science tourism resources and supporting facilities, such as transportation and catering of universities and tourist attractions in the university town, a three-level system of potential level (level 1), landscape resource category (level 2), and functional type (level 3) was used to evaluate the science tourism sites in Guangzhou University Town. According to the actual situation of the university town, three tourism place potential levels were determined, that is, tourism places with high (Class I), medium (Class II), and low (Class III) tourism potential (Figure 4). Each university town site was scored according to the evaluation mentioned above. Then, the sum of the evaluation factor scores of the university town sites was calculated. The average sum of the sites' scores was used as the criterion for the corresponding potential level. The sum of the scores for each factor was the weighted index sum. Tourist sites with sums of evaluation factor scores higher than 3.1 were classified as Class I, those with scores from 2.4–3.1 were Class II, and less than 2.4 were Class III.

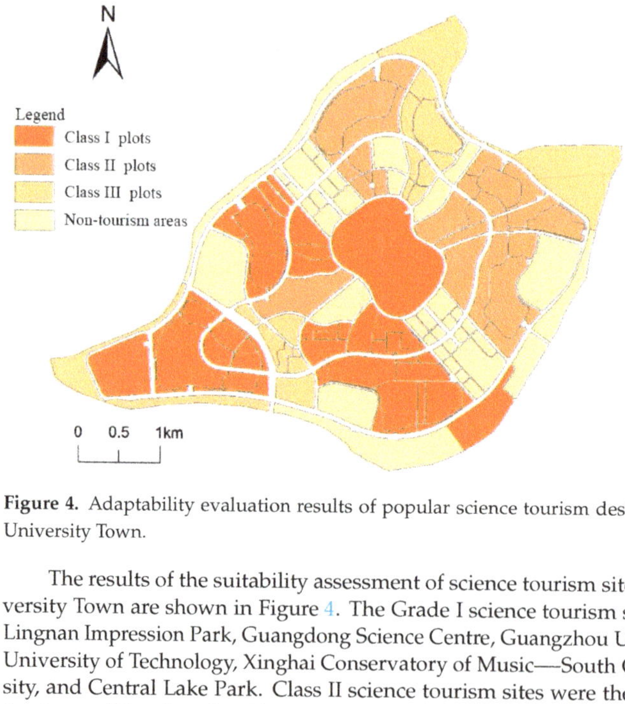

Figure 4. Adaptability evaluation results of popular science tourism destinations in Guangzhou University Town.

The results of the suitability assessment of science tourism sites in Guangzhou University Town are shown in Figure 4. The Grade I science tourism sites were the plots of Lingnan Impression Park, Guangdong Science Centre, Guangzhou University, Guangdong University of Technology, Xinghai Conservatory of Music—South China Normal University, and Central Lake Park. Class II science tourism sites were the plots of Guangzhou Academy of Fine Arts, Sun Yat-sen University, Guangzhou University of Chinese Medicine, Guangdong Pharmaceutical University, and South China University of Technology. The remaining plots were Class III science tourism sites. The overall spatial distribution shows a generally higher science tourism suitability score to the west of the central lake than to the east and a rough northeast–southwest symmetry.

3.6.2. Comprehensive Verification of Evaluation Results

Grade I tourist sites had more obvious advantages in terms of science tourism resources and catering service conditions than Grade II and III tourist sites, and these were reflected in several factors with higher weights in the evaluation list, especially the number of science tourism resources.

In addition, the convenience of transportation was an essential factor affecting the rating of science tourism sites. For example, although the plot of Guangzhou University of Chinese Medicine–Guangdong Pharmaceutical University was comparable to Class I tourism sites in terms of science tourism resources, it is far away from important transportation hubs, such as metro stations and highways, and so did qualify as a Class I tourism site. The overall traffic layout of Guangzhou University Town has three ring roads: an inner ring, middle ring, and outer ring, and the two subway stations in the south and north of the university town are symmetrically distributed. However, the Nansha Harbor Expressway is in the southwest of the island and has a very close traffic connection with downtown Guangzhou; therefore, science tourism sites in the southwest of the island, such as the plots of Guangzhou University and Guangdong Science Centre, had relatively more convenient road transport and their scores in terms of location conditions were relatively higher.

This study found that some Class I tourist sites were not reasonably developed, and that the development of universities is still in the early stages. Field research to different science tourism locations and supporting facilities was conducted on the weekends and 12:00–13:00 noon peak times to study the flow of people and average values from several visits were used. The average peak flows of Central Lake Park, Guangdong University of Technology, Guangdong Science Center, Gogo Xintiandi Shopping Center, and Chikan Wetland Park were 482, 8, 102, 2712, and 2 people, respectively. These results show that the number of people going to universities for tourism activities is much lower than that of other Class I tourist destinations, such as the Guangdong Science Center. This shows that science tourism in universities still needs to be developed and implemented.

In addition, this study conducted a research questionnaire on the current transportation situation, tourism resources, tourism experience, food service, and publicity of the university town from the perspective of having traveled to Guangzhou University Town/living in Guangzhou University Town/not having traveled to Guangzhou University Town, to understand the 120 respondents' awareness and participation. Fifty-five of respondents had visited various science tourism sites, while 17, who had not visited, said they had not heard of any science tourism sites on the island. Sixty respondents thought that the popularity of science tourism in the university town was average and needed to be improved. Regarding travel options to various attractions, 55 tourists who had visited the city said they used buses, subways, and cycling more often to get to the university town. The results shows that the most common duration for popular science tourism is roughly one day, which shows the same trend as other unconventional tourism, such as visiting friends and relatives tourism (VFR) [77]. We suggest that this is because the purpose of traveling is different from conventional purposes. While VFR may not have clear goals for traveling, for popular science tourism, people are determined to learn rather than relaxing. This has been dubbed "travel craving" by some scholars [78].

The overall satisfaction rate for the convenience of transportation on the island was 3.84, indicating that there is still a need to strengthen the construction of public transit and public roads to provide more convenient and fast travel conditions for tourists. In addition, respondents usually dined in commercial areas, such as Gogo Xintiandi (the commercial center) and various dining halls in the university. However, the overall satisfaction with the general dining conditions in the university town was 3.71, indicating that there is much room for improvement in the dining conditions. The questionnaire shows that more people preferred the center of Lake Park, Guangdong Science Center, and Lingnan Impression Park due to their rich tourism resources, various characteristics, and convenient transportation. At the same time, wetland parks, such as Wan Jutou, Chi Kan, and Bei Gang were almost unattended because of their remote location and relative inaccessibility in terms of transportation conditions. Most visitors spend half a day in these science tourism places, with only a few cases of two days or more. Based on this, Guangzhou University Town's science tourism planning should focus more on enriching playable resources to attract more off-island and foreign tourists.

4. Discussion

4.1. Suggestions for Science Tourism Development

This paper suggests the following for the development of science tourism in Guangzhou University Town:

1. More roads and public transportation systems should be built to strengthen the connections with the city and improve the layout of transportation in the university town to be more symmetrical. Although the town has a complete public transportation system, the current connection between Guangzhou University Town and downtown Guangzhou is limited to two methods: by car through Nansha Port Expressway and Guanzhou Tunnel or by subway through Metro Line 7 and Line 4 of Higher Education Mega Center North and South Station. In general, the road layout is asymmetrical, and the subway station is in the center of the University Town traffic circle, meaning that areas at the edge of the university town do not benefit from the convenience and have greater commuting times. Therefore, the future infrastructure development of the university town should focus on the development of roads, the construction of highways similar to the Nansha Port Expressway in the northern part of the University Town, and the opening of subway stations, such as "University Town East and West," which can balance the uneven traffic and distribution of Guangzhou Metro stations. These developments would also better connect the university with the Guangzhou Beltway, resulting in the university town being less isolated.
2. Efforts should be made to improve supporting service facilities on the island, plan the layout of service facilities and break the boundary to form a piece of the area with nearby villages and islands. There is some land near the universities in Guangzhou University Town that still need to be developed. The places providing catering, services, accommodation, and leisure in Guangzhou University Town are limited and are mainly concentrated in the Gogo Xintiandi shopping mall, the general commercial center, and the natural villages. Universities can only provide certain catering services and the provision of other services is more limited. Compared with the shopping malls, the natural villages provide different quality of services. The quality of services offered by universities and colleges varies, making it difficult to attract tourists to stay overnight and extend their visits. Therefore, the spatial distribution and types of land to be developed should be fully considered. Comprehensive shopping malls, parking lots, hotels, and other service facilities should be constructed for several sites that are evaluated as Class I tourist sites but have relatively low traffic (e.g., Guangdong Science Center) and those with fewer resources in terms of spatial distribution (e.g., wetland parks, northern areas of university towns). At the same time, the university town should break its boundaries between it and nearby villages and islands (Xiaozhou Village, Changzhou Island), strengthening the connections between them. This would result in the economy of the area growing rapidly and becoming an important part of the city.
3. The range of guest source radiation should be expanded. The primary source of tourists for Guangzhou University Town is Guangzhou citizens, nearby villagers, and students in the university town. While nearby villagers and students are the main groups, the university town should aim to attract more tourists from outside the island. First, the university town should increase publicity and establish a tourism company led by the government to integrate all the science tourism resources and hold activities to promote the university town science tourism, such as a college in Guangzhou University Town hosting a geology summer camp. This would attract more people from Guangzhou and the surrounding areas to visit for weekends, leisure and vacation, recreation, and fitness. Second, the university town should build more complete scientific, educational, cultural, health, and sports facilities and use the Internet and other means to attract more tourists.

4.2. Outlook

This paper attempted to build a suitability evaluation index system for science tourism in Guangzhou University Town from the perspective of tourism mode. However, there were some limitations. First, many indicators affect science tourism in university towns and the results differ when different indicators are chosen. Second, the evaluation method had some uncertainties. The subjectivity of scoring weights is difficult to avoid and can affect the accuracy of the final evaluation results to varying degrees. Therefore, this requires improvement in subsequent research. Third, the evaluation of the suitability of science tourism is extremely critical in the tourism planning of Guangzhou University Town; however, because of the enormous workload of surveys and calculations, there is much information not readily available, which will lead to errors in the estimation of the number of visitors in the planning period. If this error is too large, construction planning for water, electricity, tourist beds, commercial service facilities, and other facilities becomes meaningless, and the general layout loses its basis. Therefore, it is a demanding and challenging task, but is one of great significance.

As previously discussed, university towns have the potential to be important parts of popular science tourism because they have tourism resources that are suitable for fulfilling the curiosity and demand of many people. Although university towns are built in many cities and some leaders are aware of the value of these university towns, most remain to be developed. Some environmental and social questions also need to be considered because there are not only universities and students in the university town but also villages and residents. The development of science tourism can bring social problems, such as conflicts between students, teachers, residents, and outsiders and insufficient coordination of city administration can lead to environmental damage and the accumulation of rubbish. These issues need to be considered for the development of science tourism at university and non-university tourism sites. Thus, managing this type of unconventional tourism needs great consideration. Moreover, it cannot be ignored that human traffic flow is massively different between holidays and regular times. Therefore, how the government aids merchants during the summer and winter holidays to survive in the off-season should be discussed.

5. Conclusions

This study used remote sensing data and field investigation to analyze the resource value and environmental level of tourism sites considering five factors: scale capacity, ecological level, resource level, location condition, and service condition. We constructed a set of indicators to evaluate the suitability of the science tourism sites in Guangzhou University Town by combining the hierarchical analysis method and the opinions of experts to assign weights to the five factors. The suitability of the development of science tourism sites in Guangzhou University Town was evaluated through measurement and field surveys, and the main findings were as follows.

1. In the classification of the suitability of the development of popular science tourism sites in Guangzhou University Town, the five evaluation indices had higher weights in terms of resource level and location conditions. This means that the resource level was most critical factor affecting the evaluation of popular science tourism sites, followed by service conditions. In the future, the development of science tourism sites should prioritize the resource level of the region; however, other elements should still be considered.
2. Overall, the grades of popular science tourism sites of the western part of the University Town were higher than those of the eastern part, demonstrating the uneven development of the western and central regions of the University Town and the eastern regions. This uneven development is particularly evident in the number of tourism resources and the grades of popular science tourism sites, which is due to the more developed transportation in the western part.
3. The degree of development and use of tourist sites does not only depend on the abundance and characteristics of resources but also on a range of socioeconomic,

locational, and environmental conditions associated with tourism development. The development of science tourism in university towns should optimize the layout of transportation, catering, and other elements while also improving the infrastructure and service level in areas rich in attractions. They are combined with the actual situation in the town of Guangzhou University.

4. Popular science tourism is an essential part of the developing industry of unconventional tourism and Guangzhou University town is rich in resources for this industry according to our survey and dataset. Our questionnaire revealed that people prefer to only stay for one day when visiting university towns. Therefore, we suggest that the most suitable duration for popular science tourism is one day, which shows the same trend as VFR, another kind of unconventional tourism.

Author Contributions: Conceptualization, W.G., D.-F.W., F.-X.W.; Methodology, W.G., D.-F.W., F.-X.W.; software, D.-F.W., F.-X.W., Y.L.; validation, W.G.,H.-W.L. and Y.-Q.Y.; formal analysis, W.G., D.-F.W., F.-X.W.; investigation, H.-W.L., C.-F.Z.; resources, W.G., D.-F.W., F.-X.W.; data curation, Y.L., Y.-Q.Y., H.-W.L., C.-F.Z.; writing—original draft preparation, W.G., Y.L., H.-W.L.; writing—review and editing, D.-F.W., F.-X.W., Y.-Q.Y.; visualization, W.G., F.-X.W.; supervision, D.-F.W.; project administration, D.-F.W. All authors have read and agreed to the published version of the manuscript.

Funding: This research was funded by Guangdong Planning Office of Philosophy and Social Science under the project called "Disciplinary Co-projects for 2020 under the Thirteenth Five—Year Plan" (No. GD20XYJ32); by The Department of Education of Guangdong Province, under the "2020 Research Project under the Thirteenth Five—Year Plan, Special Research area on the Construction of Guangdong—Hong Kong—Macao Greater Bay Area" and "The Silk Road", entitled Evaluation of Popular Science Tourism Sites in University Towns – Taking Guangzhou University Town as an Example" (No. 2020GXJK199), and under the Guangdong Higher Education Teaching Reform Project, entitled "The lesson "Real Estate Management"—Online and offline blended teaching practice and innovation." (No. 462), by Guangzhou University under the "On—campus research projects (research category)" project (No. YJ2021007) and the "Innovation Training Program for University Students" project, titled "Evaluation of Popular Science Tourism Sites in University Towns—Taking Guangzhou University Town as an Example" (No. 311).

Institutional Review Board Statement: Not applicable.

Informed Consent Statement: Not applicable.

Data Availability Statement: Not applicable.

Acknowledgments: All authors would like to thank Zhu Hong and his team at Guangzhou University for advice on testing methods and discussions, and staff at parks and shopping centers in the university town, which provided us with much important information.

Conflicts of Interest: The authors declare no conflict of interest.

References

1. Eccles, J.S.; Roeser, R.W. Schools as Developmental Contexts During Adolescence: Schools As Developmental Contexts. *J. Res. Adolesc.* **2011**, *21*, 225–241. [CrossRef]
2. Thomas, K.J.; Napolitano, P.H. Educational privilege: The role of school context in developing just-world beliefs among Brazilian adolescents: Role of School Context In Bjw. *Int. J. Psychol.* **2017**, *52*, 106–113. [CrossRef] [PubMed]
3. Jarratt, D.; Sharpley, R. Tourists at the seaside: Exploring the spiritual dimension. *Tour. Stud.* **2017**, *17*, 349–368. [CrossRef]
4. Weng, G.; Li, L. The Coupling Coordination Degree and Spatial Correlation Analysis on Integrational Development of Tourism Industry and Cultural Industry in China. *Econ. Geogr.* **2016**, *36*, 178–185.
5. Edelheim, J.R. With The Simpsons as Tour Guides: How Popular Culture Sources can Enhance the Student Experience in a University Tourism Unit. *J. Hosp. Tour. Manag.* **2009**, *16*, 113–119. [CrossRef]
6. Wang, N.; Zhong, Y.; Li, S. The evaluation system of forest park tourism resources for pro-propaganda sciences based on the Analytic Hierarchy. *J. Cent. South Univ. For. Technol.* **2015**, *35*, 139–143.
7. He, Z. Search on the Evaluation of Popular Science Tourism Resources for Geoparks: Based on Danxiashan Global Geopark. Ph.D. Thesis, China University of Geosciences, Beijing, China, 2015.
8. Benson, A. *Chapter 11—Research Tourism: Professional Travel for Valuable Discoveries*; Novelli, M., Ed.; Niche Tourism Butterworth-Heinemann: Oxford, UK, 2005; pp. 133–142. [CrossRef]

9. Yin, S.; Liu, H. Research on governance evolution path of scenic mountain spots under the background of international tourism organisation. *J. Nat. Resour.* **2021**, *36*, 1316–1333.
10. Vargas-Sánchez, A. Industrial Tourism: Opportunities for City and Enterprise. *Ann. Tour. Res.* **2011**, *38*, 1203–1204. [CrossRef]
11. Dalton, L.C.; Hajrasouliha, A.H.; Riggs, W.W. State of the Art in Planning for College and University Campuses: Site Planning and Beyond. *J. Am. Plan. Asoc.* **2018**, *84*, 145–161. [CrossRef]
12. den Heijer, A.C.; Curvelo Magdaniel, F.T.J. *Campus–City Relations: Past, Present, and Future*; Meusburger, P., Heffernan, M., Suarsana, L., Eds.; Geographies of the University; Springer International Publishing: Berlin/Heidelberg, Germany, 2018; pp. 439–459. [CrossRef]
13. Tomasi, S.; Paviotti, G.; Cavicchi, A. Educational Tourism and Local Development: The Role of Universities. *Sustainability* **2020**, *12*, 6766. [CrossRef]
14. Mennel, T. The city as campus: Urbanism and higher education in Chicago. *Plan. Perspect.* **2012**, *27*, 156–158.
15. Haar, S. *Urbanism and Higher Education in Chicago*; University of Minnesota Press: Minneapolis, MN, USA, 2010. Available online: http://ebookcentral.proquest.com/lib/cityuhk/detail.action?docID=678658 (accessed on 22 December 2010).
16. Coulson, J.; Roberts, P.; Taylor, I. *University Trends: Contemporary Campus Design*, 2nd ed.; Routledge: Abingdon, UK, 2017. [CrossRef]
17. Freestone, R.; Pullan, N.; Saniga, A. The making of a city campus. *Geogr. Res.* **2021**, *59*, 29–45. [CrossRef]
18. Friedman, D. An Extraordinary Partnership between Arizona State University and the City of Phoenix. *J. High. Educ. Outreach Engagem* **2009**, *13*, 89–100.
19. Höger, K.; Christiaanse, K.; Bindels, E. *Campus and the City: Urban Design for the Knowledge Society*; gta Verlag: Zurich, Switzerland, 2007. Available online: https://www.research-collection.ethz.ch/handle/20.500.11850/2729 (accessed on 2 February 2022).
20. Way, T. The urban university 's hybrid campus. *J. Landsc. Archit.* **2016**, *11*, 42–55. [CrossRef]
21. Hebbert, M. The campus and the city: A design revolution explained. *J. Urb. Des.* **2018**, *23*, 883–897. [CrossRef]
22. Benneworth, P.; Charles, D.; Madanipour, A. Building Localized Interactions Between Universities and Cities Through University Spatial Development. *Eur. Plan. Stud.* **2010**, *18*, 1611–1629. [CrossRef]
23. Lou, Y. Sensitivity Assessment of Ecological Landscape in Changchun Lianhuashan Mount Ecological Tourism Resort. *J. Green Sci. Technol.* **2019**, *18*, 43–47.
24. Liu, X.; Liang, L. Construction of Geopark Popular Science Tourism Evaluation System and Empirical Research– Take the Yuntai Mountain World Geopark in Henan Province as an example. *Econ. Geogr.* **2016**, *36*, 182–189.
25. Lv, Y. Study of Regional Ecological Environment Optimization Based on Urban Wetland Parks Cluster. *Plan. J. Shandong Forest Sci. Technol.* **2019**, *49*, 91–93, 114.
26. Luo, W.; Huang, L.; Yan, Z. study on the Popular Science Tourism Comprehensive Evaluation Index System of the Geological Park. *Hubei Agri. Sci.* **2016**, *55*, 4340–4343, 4348.
27. Wang, W.H.; Tang, Y.P.; Wu, F.Q. Quantitative Evaluation of Science Technology Tourism Resources Based on Geographic Information System. *Adv. Mater. Res.* **2014**, *955–959*, 3824–3827. [CrossRef]
28. Wang, K.; Liu, Y.; Mashrur, S.M.; Loa, P.; Habib, K.N. COVid-19 influenced households' Interrupted Travel Schedules (COVHITS) survey: Lessons from the fall 2020 cycle. *Transp. Policy* **2021**, *112*, 43–62. [CrossRef] [PubMed]
29. Cajiao, D.; Leung, Y.-F.; Larson, L.R.; Tejedo, P.; Benayas, J. Tourists' motivations, learning, and trip satisfaction facilitate pro-environmental outcomes of the Antarctic tourist experience. *J. Outdoor Recreat. Tour.* **2022**, *37*, 100454. [CrossRef]
30. Michalkó, G.; Balizs, D.; Kiss, É.; Huszár, S.F.; Sik, A.; Karácsonyi, D. A vidékiesség vonzásában: A nagyvárosi agglomerációk láthatatlan turizmusa Magyarországon. *Tér és Társadalom* **2017**, *31*, 44–61. [CrossRef]
31. Park, S.; Xu, Y.; Jiang, L.; Chen, Z.; Huang, S. Spatial structures of tourism destinations: A trajectory data mining approach leveraging mobile big data. *Ann. Tour. Res.* **2020**, *84*, 102973. [CrossRef]
32. Hu, M.; Li, H.; Song, H.; Li, X.; Law, R. Tourism demand forecasting using tourist-generated online review data. *Tour. Manag.* **2022**, *90*, 104490. [CrossRef]
33. Li, B.Q.; Ren, S.Z.; Fan, W. Research on the mechanism of science popularization activities based on scientific research resources—An example of the environmental protection science and technology innovation platform of Guangdong Environmental Protection Engineering Vocational College. *Mod. Vocat. Edu.* **2020**, *43*, 12–13.
34. Yuan, Z.J.; Guo, J.W.G.; Yang, Y.Y.; Zhu, H. Analysis of the spatial structure, mechanism, and social-political effects of the high-quality basic educational resources in China. *Acta Geogr. Sin.* **2020**, *75*, 318–331.
35. Cheng, L. Current Situation and Countermeasures of Popularization Tourism of Geological Relics—Taking Wansheng national geopark as an example. *J. Yili Norm. Univ. (Nat. Sci. Ed.)* **2019**, *13*, 89–92.
36. Qi, W.; Cheng, X.; Huang, Q. Study on Status, Problems, and Countermeasures of Developing Popular Science Tourism of Geoparks in Yunnan, China. *IOP Conf. Ser. Earth Environ. Sci.* **2017**, *95*, 052015. [CrossRef]
37. Ma, L.; Zhu, H. Rethinking the Role of Tourism towards Human Existence: From the Perspective of Humanism. *Tour. Trib.* **2018**, *33*, 14–23.
38. Wei, Y.; Chen, W.; Luo, S.; Pan, X.; Li, C.; Luo, Q. Karst Landscape Features of the Women Mountains and Its Tourism Potential Analysis for Poverty Alleviation. *Trop. Geogr.* **2020**, *40*, 10–26.
39. Yu, Y.; Wu, J. Suitability Evaluation of Tourism Endowment in Coastal City: A Case Study of Zhanjiang City, Guangdong Province. *Sci. Technol. Manag. Land Resour.* **2019**, *36*, 1–9.

40. Song, M. Research on Selection of City Greenways Which Based on Slow-Traveling Corridors Suitable Assessment. Master's Thesis, Sichuan Agricultural University, Chengdu, China, 2019.
41. Weng, J.; Li, X.; Zhang, H.; Wu, X. A comprehensive evaluation of rural tourism development resources in Jinjiang City. *J. Henan Inst. Sci. Technol. (Nat. Sci. Ed.)* **2020**, *48*, 42–50.
42. Wei, Y.; Hu, H. Study on the Ecotourism Comfort Evaluation System Based on GIS. *J. Shandong Agric. Univ. (Nat. Sci. Ed.)* **2021**, *52*, 149–152.
43. Shu, X.; Gao, Y.; Zhang, Y.; Yang, C. Study on the Coupling Relationship and Coordinative Development between Tourism Industry and Eco-civilization City. *China Popul. Resour. Environ.* **2015**, *25*, 82–90.
44. Chen, X.L.; Cai, X.M.; Zhu, H. Reflection on China's tourism study based on the perspective of the field of place. *Geogr. Res.* **2019**, *38*, 2578–2594.
45. Li, G.B.; Tian, Y.P. Validity Analysis of Tourism Resources Competitiveness and Tourism Development Level in Pearl River Delta. *Econ. Geogr.* **2019**, *39*, 218–224, 239.
46. Li, F.; Huang, Y.L.; Zheng, J.Q.; Li, F. An Analysis of the Difference and Integration of Perception Images in Regional Tourism Cooperation. *Tour. Trib.* **2007**, *01*, 30–35.
47. Zhang, J.Y.; Tang, G.H. An Exploration on the Construction of "Heritage Trail" in Guangzhou University Town. *South Archit.* **2017**, *02*, 117–122.
48. Luo, Y.J.; Su, Q.Q. A Study on the Characteristics of Theme Park Tourism Behavior of College Students in Higher Education–Jinan Changqing University City as an example. *Tour. Overv.* **2019**, *18*, 59–61, 64.
49. Wang, C.C.; Huang, M.S. The Spatial Models and Influential Factors of Chinese University Towns. *Econ. Geogr.* **2006**, *03*, 482–486.
50. Pan, G.X. Analysis of Further Study Tourism Resources Development of Guangzhou University Town. *J. Heilongjiang Inst. Teach. Dev.* **2012**, *31*, 199–202.
51. Liu, L. research on the development of tourism products in university cities based on tourists' satisfaction—A case study of Guangzhou University Town. Master's Thesis, Guangzhou University, Guangzhou, China, 2012.
52. Ma, T.F. Land Applicability Research in Urban Planning. *Planners* **2008**, *6*, 71–74.
53. Zhang, L.J.; Li, K. The Other Orientation of University City—"Sun" of Guangzhou Tourism. *Yunnan Geogr. Environ. Res.* **2005**, *3*, 62–64.
54. Luo, W. A Primary Discussion on the Tourist Resources Development of Fuzhou University City. *J. Heilongjiang Univ. Technol. (Compr. Ed.)* **2007**, *05*, 43–44.
55. Fang, Y.; Yin, J. A GIS-Based Evaluation on the Suitability of Slow Tourist Destination Development in Yangtze River Delta. *Tour. Sci.* **2014**, *28*, 82–92.
56. Bao, J.F.; Yang, Y. Evaluation Research on the Development Potential of Provinces Popular Science Tourism. *Tour. Forum* **2015**, *8*, 65–71.
57. Lee, C.-K.; Han, S.-Y. Estimating the use and preservation values of national parks' tourism resources using a contingent valuation method. *Tour. Mang.* **2002**, *23*, 531–540. [CrossRef]
58. Zhang, J. Research on Ecological Suitability Evaluation for Tourism Development of Yixing Bamboo Forest Park Based on GIS. Master's Thesis, Nanjing University, Nanjing, China, 2016.
59. Li, N.; Wang, J.Y.; Wang, H.Y.; Yang, X.L. Study on Ecological and Environmental Suitability of Tourism Resources Development In Lijiang River Basin of Guilin, China. *Int. Arch. Photogramm. Remote Sens. Spat. Inf. Sci.* **2020**, *XLII-3/W10*, 601–604. [CrossRef]
60. Kou, W.; Li, D.; Qin, Y.; Lu, M. Overview of the Domestic Research on Tourism Carrying Capacity. *J. Cap. Norm. Univ. (Nat. Sci. Ed.)* **2016**, *38*, 56–61.
61. Yan, H.; Xiong, H. The Access Batch and the Maximum Carrying Capacity of Scenic. *Tour. Trib.* **2013**, *28*, 83–87.
62. Zhang, X.; Zhu, Z. An Analysis of Tourism Environment Capacity in Jiuzhaigou Scenic Area. *Tour. Trib.* **2007**, *09*, 50–57.
63. Wang, Y.; Li, J.; Zhang, M. Evaluation of Tourism Environmental Carrying Capacity in Diaoshuihu National Forest Park. *Int. J. Sustain. Dev. Plan.* **2019**, *15*, 761–766. [CrossRef]
64. Sun, Q.; Zhang, N.; Liu, Z.; Liao, B. Tourism resources and carrying capacity of scenic tourism areas based on forest ecological environment. *South. For. A J. For. Sci.* **2019**, *82*, 10–14. [CrossRef]
65. Zurui, L.; Dongdong, Y.; Fengxian, T.; Jie, L. Present Situation and Prospect of Tourism Carrying Capacity Research in China. *Resour. Dev. Mark.* **2017**, *33*, 379–384.
66. Qian, J.; Shen, H.; Law, R. Research in Sustainable Tourism: A Longitudinal Study of Articles between 2008 and 2017. *Sustainability* **2018**, *10*, 590. [CrossRef]
67. Liu, H.; Zhang, Y.; Zhang, X. Monitoring vegetation coverage in Tongren from 2000 to 2016 based on Landsat7 ETM+ and Landsat8. *An. Da Acad. Bras. De Cienc.* **2018**, *90*, 2721–2730. [CrossRef]
68. Zhang, L.H. Research on Popular Science Tourism in Dunhuang UNESCO Global Geopark Based on the Tourists' Perception. Master's Thesis, China University of Geosciences, Beijing, China, 2020.
69. Bourlon, F.; Gale, T.; Adiego, A.; Alvarez-Barra, V.; Salazar, A. Grounding Sustainable Tourism in Science-A Geographic Approach. *Sustainability* **2021**, *13*, 7455. [CrossRef]
70. Liu, N.R.; Xue, W.L. Research on the Development Strategy of "Internet + All-for-One Tourism"—Taking the Dalian City of Liaoning Province as An Example. *Sci. Technol. Econ.* **2021**, *34*, 56–60.
71. Xiong, C.N. Vegetation Coverage Based on Landsat8 Images in Guangzhou. *Environ. Sci. Technol.* **2015**, *35*, 383–386.

72. Jia, K.; Yao, Y.; Wei, X.; Gao, S.; Jiang, B.; Zhao, X. A Review on Fractional Vegetation Cover Estimation using remote sensing. *Adv. Earth Sci.* **2013**, *28*, 774–782.
73. Fu, S.; Wang, W.; Wan, B. Study on Normalized Coefficient of Hebei Provincial Biology Abundances and Vegetation Abundances Based on RS and GIS. *J. Environ. Manag.* **2014**, *6*, 27–31.
74. Liu, X.; Liang, L. The Progress of Domestic Popular Science Tourism and its Enlightenment. *J. Henan Univ. (Soc. Sci.)* **2013**, *53*, 49–55.
75. Liu, C.; Zhang, R.; Wang, M.; Xu, J. Measurement and Prediction of Regional Tourism Sustainability: An Analysis of the Yangtze River Economic Zone, China. *Sustainability* **2018**, *10*, 1321. [CrossRef]
76. Xu, J.; Sun, D.; Wang, Y.; Wen, H.; Wang, D. Landslide Susceptibility Mapping of Fengjie County Based on GIS and Improved Analytic Hierarchy Process. *J. Chongqing Norm. Univ. (Nat. Sci.)* **2020**, *37*, 36–44.
77. Zátori, A.; Michalkó, G.; Nagy, J.T.; Kulcsár, N.; Balizs, D. The tourist experience of domestic VFR travellers: The case of Hungary. *Curr. Issues Tour.* **2019**, *22*, 1437–1459. [CrossRef]
78. Mitev, A.Z.; Irimiás, A. Travel craving. *Ann. Tour. Res.* **2021**, *90*, 103111. [CrossRef]

sustainability MDPI

Article

Cemeteries as a Part of Green Infrastructure and Tourism

Ágnes Sallay *, Zsuzsanna Mikházi *, Imola Gecséné Tar and Katalin Takács

Institute of Landscape Architecture, Urban Planning and Garden Art (TTDI), Hungarian University of Agriculture and Life Sciences (MATE), 1118 Budapest, Hungary; gecsene.tar.imola.csilla@uni-mate.hu (I.G.T.); takacs.katalin@uni-mate.hu (K.T.)
* Correspondence: sallay.agnes@uni-mate.hu (Á.S.); zsuzsanna.mikhazi@gmail.com (ZS.M.)

Abstract: The world's population and the proportion of it living in cities and urban areas has exploded in recent decades. In the European Union, 62% of the population lives in urban areas and 80% in suburban areas, and these proportions are projected to increase further in the coming decades. It has long been researched and proven that 'urban greenery' can play a major role in mitigating the so-called urban heat island effect, and during the COVID-19 pandemic the role of daily recreation has come to the forefront. The combined memorial, recreational, and touristic use of cemeteries can help to ensure their economic management, and thus the long-term preservation of their value. In international tourism the model of managing cemeteries as tourist attractions already exists; however, this is not yet part of conventional practice. In addition to traditional cemetery tourism (e.g., visiting the graves of celebrities or enjoying artistic treasures and values), cemeteries are used as venues for events and sports activities. In Western Europe forest and park cemeteries have been established since the 19th century, and their large green areas and open spaces are a prerequisite for their use as public parks. Thus, the use of cemeteries as public parks is a common if quite specific practice. Our aim with this article is to identify the green space values of Budapest's cemeteries, in addition to their well-known cultural and architectural significance, as well as to define the potential and means of their involvement in tourism-related activities. Another aim of our study is to raise awareness of green cemeteries within the tourism profession as potentially wider tourist attractions. We consider it important to draw the attention of decision-makers to the significance of the greenspace values when preserving or reusing closed cemeteries. Based on our work, other major cities in Hungary can identify and exploit the touristic and green space potential of their cemeteries.

Keywords: cemetery; tourist attraction; green space; urban green infrastructure; cemetery tourism

Citation: Sallay, Á.; Mikházi, ZS.; Gecséné Tar, I.; Takács, K. Cemeteries as a Part of Green Infrastructure and Tourism. *Sustainability* **2022**, *14*, 2918. https://doi.org/10.3390/su14052918

Academic Editors: Anna Rita Irimiás, Dallen J. Timothy and Gábor Michalkó

Received: 31 January 2022
Accepted: 23 February 2022
Published: 2 March 2022

1. Introduction

The role of urban green infrastructure or urban green areas is growing in cities as a result of population growth and urban intensification. In recent years, the COVID-19 pandemic has amplified the importance of open spaces for the local population, who spend more time in parks and other green spaces available in close proximity to their home. Cemeteries, similar to urban parks, represent an important part of the urban ecosystem, being the remaining semi-natural habitat of many species of plants and animals. However, there is no equivalence between cemeteries and public parks: while their functions and uses are similar in many respects, specificity could be mentioned as well. Cemeteries are part of the urban green infrastructure network, however, their use is limited in both time and in function. Although the use of cemeteries and parks has converged in recent decades, significant differences nevertheless exist between them (Figure 1).

While the use of cemeteries for recreation and tourism purposes is not yet considered conventional, the rise of 'green needs' can be expected to increase their role in the future. On average, four million people visit Budapest cemeteries every year, and on the All Saints' and All Souls' Day the number of visitors can reach 1–1.5 million individuals over 2–3 days.

However, no general statistics are available on the number of visitors coming to cemeteries with targeted touristic and recreational aims.

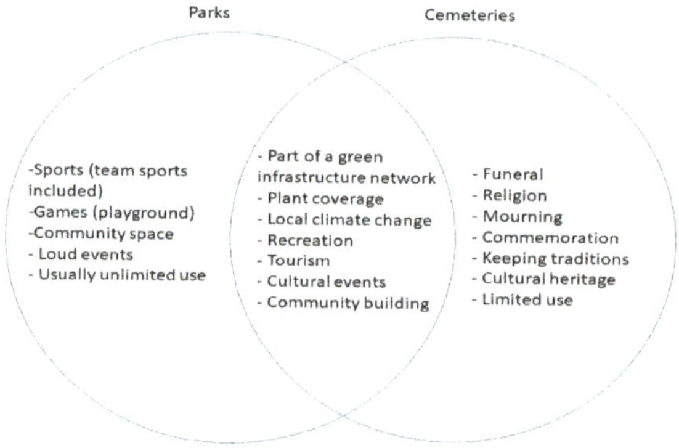

Figure 1. Differences between the functions of cemeteries and public parks (edited by the authors).

The value of cemeteries for the tourism sector lies in both the cultural aspects they offer and in the vegetation and green spaces that frame their physical realities, which can offer many services. "There are many types of cemeteries. As many as man and his follies," writes Károly Eötvös [1] (p. 175). The layout, the use of materials, and the way of commemoration can vary from simple wooden headstones to massive stone carvings and sculptures. In Hungary, there are several legendary cemeteries where families commemorate their deceased relatives according to special traditions, and these sites have become tourist attractions. In Szatmárcseke (HU), the graveyard with its nearly 600 dark headstones representing a man's head or a man lying in a boat is a unique sight [2]. In Balatonudvari, near the national road crossing the village, the local cemetery is protected due to its heart-shaped tombstones carved in white limestone [3] (Figure 2). The cemeteries in the region of Őrség (a cultural landscape area in Hungary near the Slovenian border) are a reminder of a unique burial tradition in the country. In Kercaszomor and other settlements in this region there exist some old and unique wooden headstones that were once erected for Calvinist deceased. This tradition is now extinct, and cemeteries with these sights are today only a tourist attraction [4]. In Hajdúböszörmény, in the former Calvinist graveyard known as the Nyugati Cemetery, we can see typical 19th century gravestones with their boat-shaped headstones [5].

In addition to cemeteries that are special due to their burial traditions, there are other tourist attractions related to memorial places in the country: the Historical Memorial Site of the Mohács Battle, the Budaörs Military Cemetery and Peace Park (Deutsch-Ungarischer Soldatenfriedhof) [6] (Figure 3), the British Military Cemetery in Solymár, the Turkish memorial tomb of Gül Baba in Budapest (Gül baba türbéje) [7], and the Old Christian burial chambers site in Pécs, which is on the Unesco World Heritage List [8].

1.1. *Theoretical Background*

In previous works, the authors have dealt with religious tourism [9,10] and the development, structure, and history of cemeteries [11–13]. This study now focuses on the role and potential of cemeteries for recreation and tourism. To this end, we conducted a literature review on cemeteries and recreation and cemeteries and tourism.

1.1.1. Cemeteries and Recreation

"Today, cemeteries are more than a place of reflection. They are a place of beauty and a place of history" [14]. A cemetery is a green open area [15], a "garden" with architectural and sculptural elements. It performs an ecological function and it is a permanent element of the landscape. It offers a chance for survival to many species of plants and birds, especially in cities, and natural "monuments" are often found among the many trees [16]. Because of their characteristics and location, throughout history cemeteries have often had a secondary function in addition to their primary one [17,18]. In the Middle Ages, church graveyards were often the central points in cities, and were the sites of fairs and festivities as well as being used for local parliaments, trials, preaching, miracle play performances, folk rites, executions, and demonstrations; however, the cemetery has always been a place of solemnity.

Figure 2. Heart-shaped gravestones in the cemetery of Balatonudvari. (Photo: Imola Gecséné Tar, 2016).

Figure 3. Tourists in the German–Hungarian Military Cemetery in Budaörs. (Photo: Imola Gecséné Tar, 2018).

The majority of historical cemetery complexes are park-type areas, and are endowed with recreational facilities: clean air, silence, limited urbanization, aesthetic landscape features, and favorable climatic and bioclimatic conditions [16]. Visiting cemeteries can therefore be an opportunity for recreation. "It can be a place to get one's thoughts rested and let them stretch themselves out. So, it is very good mentally. Yes, good to the eye and good for the head." (man in his 40s visiting the Old Town Cemetery) [17] (p. 1). As cities become denser, green spaces are in danger of decreasing. Evensen et al. [19] (p. 76) argue that "in densified parts of cities the cemetery may be the closest greenspace accessible for every-day use" [20] (p. 2). This may have consequences for how urban cemeteries shift from being burial spaces to becoming spaces for recreation [17,18].

1.1.2. Cemeteries and Tourism

In the twentieth century, people began to travel to sites associated with death out of curiosity rather than because of their philosophical or spiritual connotations, thus initiating the origins of dark tourism. Dark tourism is the act of travel and visitation to sites, attractions, and exhibitions that have a connection to real or recreated death, suffering, or the seemingly macabre as a main theme. Tourist visits to former battlefields, slavery-heritage attractions, prisons, cemeteries, particular museum exhibitions, Holocaust sites, and disaster locations all constitute the broad realm of 'dark tourism' [21].

Cemetery tourism (thanatourism) is a specific sub-section of dark tourism that is becoming increasingly popular [22]. Tourists wander through burial grounds with the aim of discovering the artistic, architectural, historical, and scenic heritage that often abounds in cemeteries. The changing perception of cemeteries from a place for burial towards a cultural heritage space provides several opportunities for tourism. It enables the community to explore the development of products and services that help the destination to gain new income while preserving its heritage [23].

Cemeteries are more than just resting places for the dead; they serve a practical purpose, serve as historical markers, reflect cultural values, and impress visitors with their gorgeous designs and much more. Many people find cemeteries particularly interesting for these and other various reasons. Most cemeteries welcome the public free of charge, and many offer thematic maps, brochures, smartphone apps, audio tours, or guided tours that highlight notable graves, statues, monuments, chapels and other architectural structures of the site [14]. There are many books that have been written on the topic in recent years, for example, Stories in Stone: A Field Guide to Cemetery Symbolism and Iconography, by Douglas Keister; Beautiful Death: Art of the Cemetery, by David Robinson and Dean Koontz; and Your Guide to Cemetery Research, by Sharon DeBartolo Carmack. Today, many tourist guide books include cemeteries which may attract different groups of travelers and inspire the development of newer cemeteries.

One of the 45 European Cultural Routes of the Council of Europe is the European Cemeteries Route, certified in 2010. The European Cemeteries Route refers to cemeteries as 'places of life', environments that, as urban spaces, are directly linked to the history and culture of the community to which they belong and where people can find many of their references [24].

1.1.3. Summary of the Literature Search

Based on the literature review, we have summarized the possible functions of cemeteries, which are shown in Figure 4. Many people see cemeteries as somber places with little connection to the local environment and community; however, we believe this could not be further from the truth. Cemeteries bring families together and provide insight into local history.

We have analyzed in detail the role that cemeteries can play and the needs they can meet for both local people and tourists. The basic and primary function is burial and commemoration. All other functions can only be carried out with this in mind and subordinate to it. The secondary function is linked to the recreational needs of the local population through the preservation of local natural and architectural assets. The tertiary function is to satisfy the needs of tourist visits.

Figure 4. Diverse functions of cemeteries (edited by the authors).

1.2. Presentation of the Sample Area: Green Infrastructure and Cemeteries of Budapest

In the Middle Ages, the greenspace areas of Pest and Buda developed in a similar way to those of other European cities. Within the castle walls and the city-enclosing walls, castle gardens, manorial gardens, and many small kitchen gardens were erected, while outside the walls the large meadow fields and forests enriched the landscape. The medieval cemeteries were located within the town, around the churches.

Hungary's development was severely set back by a century and a half of Turkish occupation during the 16–17th centuries. Despite this, Pest and Buda developed gradually, although in different ways. Within the walls of Buda's castle the free open spaces had been used up, and both the expansion of other urban areas and property rights were regulated according to political and military defense. In contrast, after the Turkish dominance there was more intensive development at Pest, with construction and a greater pace of transformation, although at that time civil governance was far away from creating public green spaces or parks [25] In the area of today's Budapest the practice of burial around churches had ceased after the Turkish conquest, and cemeteries similar to the present ones started to be developed in different parts of the city.

In the 18th century, due to the large number of newcomers the development of the suburbs around Pest began. At that time, few public spaces were created in the densely built-up urban areas. With the advance of civilisation, the demand for public green spaces rose, just as in other European cities. As more and more city dwellers became increasingly detached from their peasant life, they were keen for recreation and leisure. Thus, the city mayor proposed the creation of a new forestry zone, which later became the first public park in the capital (today's Városliget), as well as a promenade (Városligeti allee) connecting the urban areas to the new recreational zone. This provided a pleasant open space for the entertainment of the citizens of Pest and its surroundings. In Buda the royal gardens were open to the public, although on a limited basis, and new pleasure gardens (Városmajor, Horváth garden) were created in the late 18th century. In addition to gardens, the linear axes, alleys, promenades, and walkways played an important role in the early days of the development of public open greenspace [25,26].

The regulations on the removal of cemeteries issued in 1777 led to the opening of several new cemeteries in Buda and Pest. Two new cemeteries were opened in Buda, though both were closed at the end of the 19th century [27] (pp. 4–5) [28] (p. 235). In Óbuda, the cemetery surrounding the parish church ceased to exist in 1744. Instead, two new cemeteries were opened in 1780, along with one for the Jewish population of Óbuda [27] (p. 7). In the 18th century there were several cemeteries in Pest, and from the second half of the 18th century onwards new cemeteries were established outside the city walls. In the 1790s the first central cemetery in Pest, the Váci Road cemetery, was opened; it was eliminated around 1910. There was a Jewish cemetery next to it as well [27] (pp. 8–10).

In 1846 the Pest City Council decided to replace the old cemeteries with a large central cemetery. The Kerepesi cemetery (now National Graveyard on Fiumei Road) was opened on 15 June 1847. In 1885 the entire ensemble was declared an ornamental graveyard. In 1874, the City Council handed over a part of the Kerepesi cemetery to the Israelite community. Burials took place in the separate Jewish cemetery section in Salgótarjáni Street until 1950, and it has been preserved as a closed protected cemetery. In 1972 a tender was launched for the arrangement of the cemetery and the landscaping process, which is ongoing today. Today this cemetery has a special status as a national pantheon, although it continues to be possible to carry out public burials there [27] (pp. 10, 14, 16, 31), [28] (pp. 235–236, 238–239).

The Parliament of 1872 voted for the unification of Pest, Buda, Óbuda, and Margaret Island. The newly unified capital sacrificed for the development of its metropolitan character, following the European (Viennese and Parisien) models by removing its rural character by building new residential areas, representative axes, squares, and public institutions. In 1873, the afforestation of Gellért Hill was started. After the construction of the Chain Bridge, the Buda Hills as green open spaces have become accessible to the citizens of Pest and increasingly popular as a recreational destination. On the other hand, the hillsides were gradually urbanized by new villa areas, replacing traditional orchards and vineyards damaged in the Phylloxera epidemic [25].

Several new large cemeteries were opened in the merged capital. In Buda, the Németvölgyi cemetery was opened in 1885; it was intended to be an ornamental graveyard for all time, similar to the Kerepesi cemetery. Despite this intention, the cemetery operated for an unusually short period. The Németvölgyi Orthodox Jewish cemetery, which was opened in 1890 and used until 1961, stands on a small plot of land surrounded by a high fence. In 1894, the Farkasréti cemetery was opened in Buda, which is still in use today; it is the resting place of many famous Hungarian people. The Farkasréti Jewish cemetery is wedged between the two parts of the Christian cemetery [27] (pp. 5–6, 53, 58). A new public cemetery and a new Israelite cemetery were opened in Óbuda; however, these have now been closed down [27] (p. 7). In Pest, the expansion of the Kerepesi cemetery became impossible with the opening of the Jewish cemetery on Salgótarjáni Road, and a site for the new public cemetery had to be found elsewhere. On 1 May 1886 the New Public Cemetery of Rákoskeresztúr, Hungary's largest cemetery at 207 hectares, was opened. In 1891 the Kozma Street Jewish cemetery, Hungary's largest Jewish graveyard, was established next to the New Public cemetery, with the Rákoskeresztúr Orthodox Jewish cemetery adjoining it on the northern side [27] (pp. 32, 35, 44) [28] (p. 236).

The Fortepan photo collection contains photographs from the beginning of the 20th century until the change from the Soviet regime in Hungary in 1989, thus, the cemeteries in Budapest have been photographed beginning in the early 1900s. When we reviewed these photos, we found that the oldest photo analyzed was taken in 1900 in the National Graveyard, formerly called as Kerepesi Cemetery. There is a difference in the use of the two cemeteries from the very beginning: in case of the National Graveyard, the archive contains almost exclusively pictures of the burials and graves of famous people, while in the Farkasréti Cemetery it is the funeral motifs that appear in large number, of which the most striking is a horse-drawn hearse. For both cemeteries there are almost no photographs that do not show some form of vegetation, usually tree alleys or massive woody vegetation in the background. In the case of the Farkasréti cemetery, there was one photograph where the spatial structure and alleys were clearly recognizable (see Figure 5) in addition to the main motif of the funeral process.

Figure 5. Alleys such as this one define the structure of the Farkasréti cemetery (Photo: Fortepan/Horváth Miklós dr, ID N°129338, Date: 1955).

At the turn of the 19–20th century the function and use of public parks evolved, with crowds from various social levels, coming to walk, play skate, and seek active or passive outdoor recreation. At this time, the general condition, maintenance, appropriate use, and thus popularity of the green spaces in the capital were all outstanding. Many public parks as well as the largest urban parks were enriched and transformed with newly integrated functions such as early childrens' playgrounds [25]. In 1910 the Óbuda cemetery, which is still in use today, was opened, and a Jewish cemetery was opened next to it in 1922 [27] (p. 68). In 1919 the Budapest Municipal Cemetery Institute (Budapest Székesfővárosi Községi Temetkezési Intézet), the predecessor of the existing Budapest Funeral Institute (Budapesti Temetkezési Intézet, BTI) began its operations. From 1949 onward, the cemeteries previously owned by the Catholic church were taken over by the capital [28] (pp. 236, 239).

Budapest gained its current extension in 1950 when several surrounding small villages were annexed to the capital and the present-day outlying districts were formed. The previously independent municipalities had their own cemeteries; therefore, in 1950 Budapest possessed 87 cemeteries [28] (p. 236). In the 1950s, several small cemeteries were closed. Most were liquidated, although some cemeteries were reopened in the second half of the 20th century; in some cases, urn cemeteries were built on their sites. Those former village cemeteries that were suitable for expansion remain in operation today [12].

After the Second World War and from the 1950s onwards, development and settlement of housing estates began in earnest, along with the creation of significant public greenspace. At that time, these residential areas had a wide range of functions as public green spaces and a very high green surface ratio. On the other hand, the number of touristic developments in the urban forests increased dramatically [29]. After the change of regime in 1989 the structure of the city became more dense and the role of green spaces, including cemeteries, became more important, mainly due to the loss of former industrial areas and their conversion into office buildings and housing estates (Table 1).

Today, there are fifteen public cemeteries and four Jewish cemeteries in use in Budapest (Figure 6, Table A1). The public cemeteries are managed by the Budapest Funeral Institute, with the exception of the National Graveyard on Fiumei Road (Kerepesi cemetery). Due to its special status, this latter has been managed by the National Heritage Institute since 2016. The operating Israelite cemeteries are managed under ecclesiastical care. The closed cemeteries are partly municipality-owned and partly church-owned, and their management is mostly unresolved.

Table 1. The changing role of cemeteries and other urban green spaces (by authors, 2021).

Period	Urban Green Areas	Cemeteries
until the end of the Turkish occupation	small vegetable gardens, private residential gardens, extended agricultural land	cemeteries next to churches, burial places
1700s	increasing density of urban development leading to an emerging need for public green spaces and parks	cemeteries appear on the outskirts of the settlement, independently of churches
1800s	first public parks, large boulevards with alleys, afforestation, public squares and pedestrian areas with numerous trees	large new cemeteries established on the outskirts of the city. with the burial function remaining primary
1872–1945	more leisure activities in public parks; the first public playgrounds	large new cemeteries opened on the outskirts of the city due to rapid urban development, in which the graves of famous people become urban pilgrimage sites.
1945–1990	in parallel with the construction of housing estates, public green spaces around them are created; the first protected green areas are designated in the city, and park woodlands become significant	former village cemeteries around Budapest annexed to the capital by the creation of peripheral areas (many of which have been dismantled, although a few remain in use today), either as extensions or as urn cemeteries; cemeteries popular for recreational purposes thanks to attractive mature vegetation
1990s	new residential and office developments created on brownfield sites, increased building density, growing demand for open greenspace	tourist use of cemeteries, mainly guided walks around the graves of famous people and tertiary usage such as classical music concerts and exhibitions
2020	pandemic closures have pushed people even more towards urban green spaces, for which demand has increased, and many have become overused	during lockdowns, people spend part of their recreational time in cemeteries, and mourning also 'attracts' more and more people to cemeteries

Figure 6. Overview map of functioning cemeteries in Budapest. Cemeteries managed by BTI (Budapest Funeral Institute) [30]. Functioning cemeteries (green): 1. Angeli Road Urn Cemetery, 2. Cemetery in Budafok; 3. Cemetery in Cinkota; 4. Cemetery in Csepel; 5. Cemetery in Pesterzsébet;

6. Farkasréti Cemetery; 7. Old Cemetery in Kispest; 8. Cemetery in Kispest; 9. Cemetery in Pestszentlőrinc; 10. Megyeri Cemetery; 11. National Graveyard on Fiumei Road (from 2016 managed by NÖRI); 12. Óbudai Cemetery; 13. Cemetery in Rákospalota; 14. Tamás Street Urn Cemetery; 15. New Public Cemetery in Rákoskeresztúr; 16. Cemetery in Csömör (outside the administrative boundaries of Budapest). Closed cemeteries (red): 1. Cemetery in Rákosszentmihály; 2. Véka Street Cemetery; 3. Cemetery in Albertfalva; 4. Ganz Street Cemetery; 5. Nagykőrösi Road Cemetery; 6. Göcsej Street Cemetery; 7. Bocskai Street Cemetery.

1.3. Green Infrastructure Network of Budapest and the Role of Cemeteries in the Capital Today

In Budapest, as in other major European cities, the effects of climate change are increasingly noticeable and are expected to intensify in the coming decades. Forecasts indicate an increase in the number of heatwave days and tropical nights, which will make everyday life much more difficult for city residents [31].

The so-called urban heat island effect is a microclimatic phenomenon in large cities, where the temperature in built-up urban areas is significantly higher than in the suburban and rural areas surrounding the city. The causes of this phenomenon are complex; the most important are:

- lack of natural evaporative surfaces
- concrete and asphalt surfaces in cities absorb more solar radiation than they reflect
- vertical surfaces increase the absorption of radiation
- human activity generates heat in many different ways
- pollutants greatly modify the atmosphere
- changes to meteorological conditions such as wind direction, humidity, visible sky, precipitation, and radiation conditions [32–34].

Around 65% of Budapest's territory (34 thousand ha) is covered by green space and vegetation, and almost 2% of the city's territory is parkland; 40% of these green areas are managed by the Municipality of Budapest (420 ha), while the rest are owned or maintained by the district municipalities. Nearly 6000 ha, or 11% of the city's territory, is forested. The distribution of green spaces in Budapest is not even; in some inner-city districts (VI, VII) there is less than 1 m^2 of public park per inhabitant, while in the suburban districts there is good coverage of green spaces thanks to park forests. In Budapest, there is an average of 25 m^2 of forested parkland and only 6 m^2 of public park or garden per inhabitant. There is uneven use of green spaces, with well-positioned parks with cultural and historical value being constantly overcrowded [35].

Cemeteries play a dominant role in the urban structure due to their large surface area and, as a result of their function, their typically high proportion of green space. In addition to their memorial function, cemeteries represent a significant green space value, their green spaces being a key element of the capital's green space system [36] (Figure 7).

Both functioning and closed cemeteries are an important part of the capital's green infrastructure network. Cemeteries play an important role in the green infrastructure network of Budapest both because of their size (467 ha) and because of their high proportion of green space. The proportion of green space in cemeteries is well above the 40% defined in the OTÉK (Országos Településrendezési és Építési Követelmények, the National Town Planning and Building Requirements), even if the vegetation on the graves is not taken into account. The National Cemetery on Fiumei Road is the second-largest cemetery in Budapest, with the highest green cover (67%) [37]. For the cemeteries in operation, the Green Space Intensity Value ranges from 58% (Budafok cemetery) to 91% (New Public Cemetery), which is very high for the city. This value is largely influenced by the large mature trees in the cemeteries. Both the proportion of green areas and the ZFI value are higher for closed cemeteries; however, in many cases they are affected by the presence of weed species due to lack of maintenance, which negatively affects their use and value.

Figure 7. Chestnut tree alley in the National Graveyard on Fiumei Road (Photo: Imola Gecséné Tar, 2021).

Green spaces in cemeteries provide a range of ecosystem services to the population, such as improving air quality, enhancing the local climate, and providing aesthetic and recreational value [38]. In Hungary, the recreational use of cemeteries is not considered conventional; however, many people take advantage of the recreational opportunities offered by cemeteries. In addition to commemorative structures, almost all cemeteries have valuable mature tree species, tree lines, and rich biodiversity, which makes cemeteries an important part of biodiversity conservation [39,40]. In many cemeteries, both cemetery managers and visitors are already making serious efforts to preserve biodiversity; for example, in Farkasréti cemetery, bird feeders and bird boxes have been installed in several places (Figure 8).

The use of cemeteries as public green spaces contributes to the green coverage of the capital. The Action Area X of the Radó Dezső Plan, approved in 2021, concerns the cemeteries of Budapest, with the objective of "Protection, quality renewal and continuous professional maintenance of green infrastructure; development of an ecological approach to burial; efficient development and co-planning of public cemeteries; recreational use of functioning cemeteries; public use of closed cemeteries; temporary use of disused cemeteries as green spaces." The main tasks related to the achievement of these objectives, which are relevant to the present research, are "Better use of the recreational potential of existing cemeteries; public use of closed cemeteries; temporary green infrastructure use of disused cemeteries" and "Promotion of alternative burial methods; designation of urban areas suitable for forest burial" [35] (p. 62). The Capital intends to complete the planned improvements to the cemeteries by 2027.

Figure 8. "Bird feeder" in Farkasréti cemetery (Photo: Ágnes Sallay, 2021).

1.4. Touristic Activities in Cemeteries

For thousands of years, cemeteries have had the fundamental role of providing a burial place for the deceased and a place of remembrance for the bereaved. This primary function has been complemented in recent decades by secondary and tertiary functions such as providing adequate green spaces for local climate, recreation, and tourism [41]. This expansion of functions has implied a number of conflicts; how can all these traditional and new functions be managed in parallel? How can the primary needs be fully satisfied alongside the secondary and tertiary functions? Can the duty of remembering the deceased persons in cemeteries be reconciled with tourism? In many cases, the human need to say a dignified farewell to the dead has been eroded during the COVID-19 pandemic, traumatizing society and increasing the demand for cemetery use. The lockdowns due to the pandemic have affected the way cemeteries are used, with many people finding cemeteries a suitably quiet place to relax and remember their loved ones; even if they were not buried in the cemetery, they could visit in proximity. For those cemeteries where the green assets and vegetation were well established, the cemetery has become a place to meet nature (Figure 6).

In many places throughout Western Europe, the daily recreational use of cemeteries by the local population has become a general need and constant demand; in addition to the traditional walking activities in cemeteries, other forms of exercise such as running (in many places with separate tracks available on site), silent sports such as yoga, and dog walking are allowed under regulated conditions.

Among the tertiary touristic functions, guided walks are the most widely accepted, with visitors introduced to the values and stories of the cemeteries, mainly built heritage and grave marker interpretation. The natural or green values of cemeteries are rarely highlighted; rather, the guides include ideas about the flora and fauna of the cemeteries in other guided walks, for example in Highgate cemetery. Other tourism-related uses have been raised and implemented in different cemeteries; theater performances and concerts are more and more commonly organized, although these activities divide society and often provoke opposition from the local population, for whom the cemetery is a place of remembrance. There are various guided tours in the largest cemeteries of Budapest, where visitors can learn about the local history, the gravestones, and the famous people who rest in these cemeteries (Table 2). However, there are only a minority of guided tours that interpret the cemeteries as green spaces, presenting their living flora, and there are very few guided walks especially offered for children.

Table 2. The topics of guided walks in the largest cemeteries of Budapest (author's table with use of [42–44] sources).

Organizing Institute or Company	The Name or the Theme of Walks and Guided Tours
Sétaműhely Ltd.	Trees and headstones—guided walks in the Farkasréti Cemetery On the trails of Kohanites
Imagine Budapest (Sétapálca Ltd.)	A past encased in stones—a tour of the Salgótarjáni Street Jewish Cemetery
Nemzeti Örökség Intézete (NÖRI) National Heritage Institute	Parcels of artists "Faster, Higher, Stronger"—Olympic medalists in the National Graveyard "On hidden pathways"—cycling tour in the National Graveyard on Fiumei Road Celebrated gipsy violinist First World War—Military tombs 19th century's prime ministers Musician souls The artists of the "Nyugat" generation Under the spell of "Thália" The masters of Hungarian painting The death cult of the socialism The secrets of the plot 301—What do the graves tell us? The Prime Ministers of the 20th century Inventors and engineers In the footsteps of great travelers The gravestone artworks of Béla Lajta in the Jewish cemetery on Salgótarjáni Street Budapest, Budapest, so wonderful The life of an artist in sculptures Sacred depictions in funerary art of tombs In the footsteps of the Piarist Fathers and their disciples Mausoleums and the National Graveyard on Fiumei Road Those doomed to oblivion—Specific interpretation walk Graves in the garden—plants and symbols in the cemetery "That soul in me..."—Literary walk in the cemetery Immortal art—Sculptors and sculptures Remembering 1848–49 Ferenc Deák and his generation Memento '56 History of the Jewish cemetery on Salgótarjáni street Women, Muses, Destinies Women on the trail of success Gastro moods Family walk with the little ones Go, Hungarians! Athletes and sports leaders Irregular school lessons in the cemetery Secrets of an oasis in the heart of the city—The National Graveyard on Fiumei Road (guided walk in English language)
Budapesti Temetkezési Intézet (BTI) Budapest Funeral Institute	Guided walks in the Farkasréti, Óbudai and Rákoskeresztúri Cemeteries

The touristic use of cemeteries in Western Europe was already evident in the early 2000s. The Südwest-Kirchhof in Stahnsdorf, in the immediate vicinity of Berlin, is one of the first representatives of a landscape-style forestry cemetery, and the third largest in Germany. The cemetery, which was almost abandoned at the end of the Second World War, was the subject of a strategic plan, a programme drawn up in the early 2000s by the cemetery's advocacy association. After recognising the cemetery's cultural, historical, and artistic importance along with its natural value, the primary objective was to bring it back into public consciousness while introducing new ways of usage. Attracting community programmes, frequent invitations of the press, and a widespread sensitisation and awareness to unique burial opportunities have brought the result that the cemetery has now been reborn. It has become a popular destination for Berliners, as have, the cemetery gardens in the city center, which are operating as active elements in the green space infrastructure as a place for recreation. The function of the funeral chapel has been extended to include cultural functions such as conferences, exhibitions, and concerts. The cemetery employs volunteers and trainees who contribute to the success of the actions through their maintenance, research, and guide work. The cemetery's management has discovered and uses the opportunities offered by the media and the internet; periodic press conferences on the cemetery's cultural and historical significance, its artistic values, its current events, and even a promotional film have all been regularly created [11,45].

In terms of touristic use, the Central Cemetery of Vienna (Zentralfiedhof Wien) is a particularly successful example. In addition to the traditional use of the cemetery for burial and commemoration purposes, a number of improvements with diverse goals have been made in recent years:

- Classical music and rock concerts have been organized
- Sports facilities were developed, e.g., a running track was arranged in the cemetery,
- and electric bicycles for hire and carriage rides are offered as new services
- Regular night-time opening during the month of October to give visitors a night-time view of the cemetery
- Music walks offered, with live music accompanying the walkers
- Painting course offered, where visitors can create their own artwork souvenir of the cemetery
- A cemetery museum has been opened where visitors can learn about the history of the cemetery and the famous people buried there
- Nature-focused guided walks and camps for children
- A café and gift shop have been opened at the cemetery gate

The cemetery has responded to emerging digital needs as well; anyone can take part in a self-guided walk with a QR code (Hearonymus programme), and if required, visitors can find out information about the available burial places from a digital database. Visitors can even open a digital memorial link for a deceased loved one, with the possibility of storing commemoration materials, photos, and videos about them.

All these improvements to the Central Cemetery of Vienna have been carried out in close and regular consultation with the public, taking into account their opinions and requests, to the great satisfaction of the people of Vienna as shown by the fact that the cemetery was awarded the "best place in Vienna prize" in 2020 [46].

Following foreign trends, alternative ways of using cemeteries have already appeared in Hungary. The best example of this is the National Graveyard (Fiumei Road Cemetery), where visitors can take part in numerous thematic walks which introduce them to the stories of emblematic persons buried in the cemetery and of other people with interesting background histories. A more 'sporting' form of visit is exists where tourists can sign up for a bike tour of the cemetery. In this largest cemetery of Budapest, it is common to see mums walking with baby strollers, young people picnicking or studying, or sportsmen exercising, running, or cycling. Because the cemetery was designed and executed as a large green park area from the beginning of its creation and is planted with many valuable trees and alleys, the recreational popularity is not surprising. The urban wildlife established in this cemetery is quite rich, and the site is comparable to a large-scale arboretum with its 56 hectares. It is important to mention the Memorial Museum operating in the National Graveyard, where visitors can learn about different aspects of Hungarian cemetery culture and burial traditions through permanent and temporary exhibitions [13].

2. Materials and Methods

This research focuses on the cemeteries of Budapest, analyzing them based on data collected during site visits and a research process employing relevant photograph and map sources. The theoretical foundations were established through literature sources on the following topics: (1) cemeteries–recreation and cemeteries–tourism relationship; (2) green spaces–green space infrastructure and the history of cemeteries in Budapest; (3) tourism activities in cemeteries–opportunities and foreign models. The main questions of our study were then formulated as follows: (1) what are the motivations of visitors for visiting cemeteries, and what are the values of each cemetery related to visitor motivations? and (2) what are the cemetery visiting habits and needs of the Hungarian population?

To answer the first question, we used literature sources to prepare an aggregation of cemetery visitor motivations followed by a survey of the attractions associated with these motivations in each cemetery. Visiting a cemetery can be either the main purpose of a trip or a secondary purpose, as a complementary part of it. As cemeteries have a wide range of attractive aspects, there are many motivations for visiting them; using the work of Tomašević (2018) [47], Moreno (2018) [48] and Pécsek (2015) [49] as a starting point, we defined the following motivation categories based on literature research and our own experience:

(1) Basic motivation: visiting family graves. This includes visiting, searching one's own or others' family roots and genealogy. A place of confrontation with mortality.
(2) Cultural motivation: the cemetery as an open-air museum, a repository of material and spiritual treasures.
    (i) Graves of famous people (kings, presidents, politicians, artists, writers, poets, actors, scientists, etc.)
    (ii) Architectural curiosities (emblematic constructions, crypts, mausoleums, etc.)
    (iii) Sculptural curiosities
    (iv) Places of interest, or a place of personal interest (e.g., the cemetery is featured in a film or a book), ideal for practicing a hobby (e.g., photography, painting)
    (v) A "must see" place, i.e., it is on the World Heritage List or on Tripadvisor
(3) National sentiment: visiting memorial parks, military cemeteries; participating on different on-site memorial events
(4) Admiration/love of nature: visiting cemeteries as other parks and arboretums. Special botanical and garden architecture interests, or more generally, a reason for escaping from urban pressures and providing relaxation and recreation
(5) Education and research: visits driven by an interest in history and culture. Visitors especially come to learn more about the history of the cemetery, the place where it was established, and the people who are buried there
(6) Religious motivation (pilgrimage): visiting graves of significant religious/ecclesiastical personalities (e.g., popes, church leaders) or tombs of religious and/or historical significance (e.g., the tomb of Jesus in Jerusalem). A place of confrontation with mortality
(7) The cemetery is an open-air leisure site with organized programs; visiting a concert or a performance in a cemetery
(8) No motivation: unplanned visits. No specific interest in the special features and offerings of the cemeteries. Curiosity, "let's see what we can find here" attitude

To answer the second question, we conducted a questionnaire survey in September–November 2021 among the population of Hungary with the intention of uncovering their cemetery visiting habits. In Hungary, the number of visits to cemeteries associated with All Saints' Day and Day of the Dead is high [50]; therefore, the questionnaire was distributed during this period. We hoped this recent experience would increase the response rate and lead to more accurate answers. The questionnaire was completed in electronic form and was available online at https://forms.gle/PaZcbcRFidn4G1wKA (accessed on 1 November 2021). We wanted to know how often people visit cemeteries and for what reasons. We asked what activities not related to the basic memorial functions of cemeteries respondents considered to be acceptable.

The structure of the questionnaire can be found in the Appendix B. The questionnaire consisted of certain main questions and diverse supplementary questions. The main questions directly addressed the research topic, while the supplementary questions were asked in order to increase the reliability of the information obtained. For this reason, these latter were general demographic questions used to determine gender, age, education, etc. The questionnaire included open, semi-closed, and closed questions. In the case of the closed questions, respondents were asked to tick one of the pre-selected options. Semi-closed questions included an 'other option' category next to the offered ones, which allowed respondents to complete the list themselves if they did not find a suitable pre-written answer(s). Another example of a semi-closed question is where, in addition to yes/no answers, respondents were asked for the reasons behind their choice. For open questions, respondents were given the opportunity to formulate their answers in their own words.

The sampling was completely random. No attempt was made to limit or influence completion or to narrow down the pool of respondents. The aim was to obtain as wide a range of ages and interests as possible filling out the questionnaire and to determine the overall aspects of the typical visit. Due to the COVID-19 pandemic, we distributed the questionnaire online through professional and personal platforms, websites, and social media. The survey was successful, as 213 people completed it. Our objective was achieved in that the respondents were diverse in terms of gender, age, place of residence, education, and employment background.

In order to increase the reliability of the survey, another questionnaire was included for those who organized or led visits to cemeteries (Appendix C). This was used as a control, as some of the questions overlapped with the questionnaire for the general public. From these responses we gained better insight into the opportunities and problems from a tourism-specific point of view.

3. Results

With appropriate promotion, urban cemeteries could become an integral part of urban green tourism, as has already been the case for cemeteries abroad, e.g., in Paris and Prague. Visitors to urban cemeteries spend relatively long periods of time, even several hours, at these sites. The offered leisure activities, such as walking, photography, etc., contribute to a positive visitor experience. The traditional authenticity of cemeteries in terms of their extension, spatial structure, architecture, landscaping, artistic value, and local historical background, is sufficient to maintain their tourist appeal and visitors do not expect major improvements to infrastructure or attractions. Thus, from the visitor point of view, cemeteries do not require any major investment to enhance their attractiveness other than promotion of their highlights and conservation of their existing value (Figure 9).

3.1. Results of Research on Visitor Motivations

Visiting a cemetery might be the main purpose of traveling or a part of regular touring, or a primary or secondary motive for traveling, and might be driven by emotions of variable intensity. During our research we identified eight motivating elements in visitors to Budapest cemeteries. We identified the potential tourist attractions within each motivation category; then, through written sources, maps, and field visits, we identified which cemeteries in Budapest have which tourist attractions (Table 3).

Figure 9. Guided walk in the National Graveyard on Fiumei Road (Photo: Imola Gecséné Tar, 2021).

Table 3. Visitor motivations, with related attractions and values in the cemeteries of Budapest (edited by authors, 2021).

Visitor Motivations	Attractions	Representative Cemeteries
1. Basic motivation	tombs and crypts of families	All functioning cemeteries
2. Cultural motivation		
2.1. graves of famous people	graves of kings, presidents, politicians, artists, writers, poets, actors, scientists, etc.	National Graveyard on Fiumei Road Farkasréti Cemetery New Public Cemetery in Rákoskeresztúr Farkasrét Jewish Cemetery Óbudai Jewish Cemetery Kozma Street Jewish Cemetery Salgótarjáni Street Jewish Cemetery
2.2. architectural curiosities	emblematic constructions, crypts, mausoleums, etc.	National Graveyard on Fiumei Road Farkasréti Cemetery Kozma Street Jewish Cemetery Salgótarjáni Street Jewish Cemetery Budafoki temető
2.3. sculptural curiosities	cross, statue, carving	National Graveyard on Fiumei Road Farkasréti Cemetery Óbudai Cemetery Salgótarjáni Street Jewish Cemetery Óbuda Jewish Cemetery Budafoki Cemetery Kisszentmihályi Cemetery
2.4. places of interest, or a place of personal interest, ideal for practicing a hobby (e.g., photography, painting)	the cemetery is featured in a film or a book	
2.5. a "must see" place	it is on the World Heritage List or on Tripadvisor	
3. National sentiment	memorial parks, military cemeteries graves being attached to wars, monuments of wars	National Graveyard on Fiumei Road New Public Cemetery in Rákoskeresztúr
4. Admiration/love of nature	Special botanical and botany or zoology value	National Graveyard on Fiumei Road Farkasréti Cemetery New Public Cemetery in Rákoskeresztúr Óbudai Cemetery
5. Education and research	curiosity of settlement history	National Graveyard on Fiumei Road Salgótarjáni Street Jewish Cemetery Closed cemeteries outskirt, earlier communal cemeteries
6. Religious motivation (pilgrimage)	tombs of religious and/or historical significance, calvary	All Jewish Cemetery National Graveyard on Fiumei Road
7. Local	organised programs, concert, or a performance	National Graveyard on Fiumei Road
8. No motivation		All functioning cemeteries

It is important for cemetery managers, travel agency staff, and tourism organizations to use standardized classification categories in order to target and reach visitors with the right interests. An important role in promotion should be TripAdvisor ranking; the high position of cemeteries as places to visit in several destinations proves the significance of cemeteries as tourist attractions. Cemeteries should make a proper and thorough analysis of all possibilities in order to increase their attractiveness and accessibility as well as awareness about their cultural, historical and natural importance. Further research could be aimed towards practical implementation of the cemetery concept as a tourist attraction with a focus on travel agents and local authorities and on enhancing their understanding of cemeteries as tourist products which can enable them to create new programs and find new markets.

3.2. Results of Questionnaire Survey

The results of the two questionnaire surveys are presented separately in the following subsections.

3.2.1. Results of Visitor Questionnaire Survey

The majority of respondents were aged between 25 and 60 years (25–40 years 21.6%, 40–50 years 31.5%, 50–60 years 26.8%), female (83.6%), with a university/college degree (84%) and living in Budapest (54.9%) or in the Budapest agglomeration (14.6%). According to our survey, 37.1% of respondents visited a cemetery only once a year and 15.5% even less frequently; 29.1% visited quarterly, and 13.6% monthly.

First, we wanted to know whether the frequency of respondents' visits had changed due to the pandemic closures. For the majority (73.2%), there was no change; 23.9% visited cemeteries less frequently, and only 2.8% visited more often. Those who went to the cemetery more often than before were asked why. They were given the option of indicating more than one answer and were given the option of giving an individual answer; 32% expected to find peace of mind from their visits, 28% increased their knowledge by visiting the cemetery, 24% were able to walk and move around the cemetery at a reasonable distance during the closure period, and 18% had a cultural programme.

The purpose of the visit was explored from several perspectives: (1) Why do people usually visit cemeteries? and (2) Do they visit cemeteries during their travels? In both cases, several answers were possible, and individual answers were possible as well. The answers were quite varied. The primary purpose of the visit was "to visit the graves of my loved ones/relatives", at 81.7%; 21% and 27% of respondents, respectively, chose "to remember the deceased" (not necessarily in the cemetery where they are buried), "to visit the graves of famous people" (Figure 10), "to see the statues and buildings in the cemetery", and "to walk around, to relax". Individual responses included an interest in cemetery culture, walking the dog, chestnut/peanut picking, learning about the history of the village, peace and quiet, and greenery.

Of the respondents, 35.7% did not visit cemeteries during domestic trips. The rest (based on multiple choice) mainly visited the graves of relatives and friends (36.6%). Visiting historical sites (28.2%) and the graves of famous people (24.4%) were popular as well. The interest in visiting works of art (17.4%) and vegetation (17.8%) in the cemetery were almost the same, while 10.8% said they visited buildings/construction. Individual responses included "just looking around", "for the atmosphere" (tranquility + green), "to visit cemeteries that are unique/special in some way", "for cultural purposes", "to learn about local history", "as a tourist attraction", and "for the gravestones/inscriptions".

Fewer people visited cemeteries when traveling abroad, with 44.6% of respondents not visiting cemeteries. The majority of people who visited cemeteries abroad went to historical sites (35.2%) or to the graves of people they know (26.3%). Individual responses were dominated by the desire to learn about the cemetery culture of other countries and to visit unique/special/noteworthy cemeteries, as well as by the desire for green space and a peaceful atmosphere.

Figure 10. The tomb of János Arany in the National Graveyard on Fiumei Road (Photo: Imola Gecséné Tar, 2021).

Of the respondents, 40.8% had visited a cemetery for recreation. In addition to walking (74%) and sightseeing (74%), reading (5.2%) was the most common activity (multiple choice). Individual responses varied widely in their choice of activities, including photography (statues or Halloween), sports (running, cycling), guided walks, birdwatching, and concerts.

Of the survey respondents, 31.9% had attended a cemetery-related event. Based on individual responses, these includes any kind of commemoration (national holiday, World War II, Day of the Dead), wreath-laying, and guided walks. An equal number of respondents attended theater events and concerts. Only 10.8% participated in an event unrelated to the basic function of the cemetery. These included interesting events such as a photo lessons, drawing, nature walks, "the night of nightingales", book readings, exhibitions, and concerts, of which the latter two were the most frequently mentioned.

Well-kept cemeteries are important to visitors (92%), and this influences the time they spend there (65.7%). In terms of vegetation, the presence of trees/woods is considered the most important (93%, based on multiple choice), while shrubs (61.5%), grassy areas (60.6%) and flower beds (53.5%) play a significant role.

To conclude the questionnaire, we asked respondents what conditions they thought should be met in order for a cemetery to be suitable for recreation. Several answers could be ticked, and individual comments were possible. Of the answers we offered, the highest-ranked was green space (79.3%), followed by good accessibility (57.3%), open space between graves (57.3%), and a closed cemetery (11.7%). Among the individual responses, several respondents said that they did not consider cemeteries suitable for recreation. However, there were many constructive suggestions. Respondents felt it was important that recreational activities should not in any way interfere with the primary function of the cemetery and those who go there to commemorate the deceased. Several respondents mentioned basic infrastructure elements such as water supply, toilets, lighting, benches, waste bins, and security. They stressed the importance of being close to nature and the view, tidiness, well-kept graves, and the existence of quality green spaces. The creation of a park-like greenspace and memorial garden as a recreational area could be an important aspect of cemetery planning in the future. However, the first and most important thing to do is to change attitudes.

3.2.2. Results of Guide Questionnaire Survey

The questionnaire for tour guides was filled in by seven respondents, most of them contracted guides for the managing companies of Budapest's cemeteries (NÖRI and BTI), which provide regularly-conducted guided walks in the cemeteries they manage (namely,

National Graveyard on Fiumei Road, Jewish Cemetery on Salgótarjáni Road, Farkasréti Cemetery, Óbuda Cemetery, and Budafok Cemetery). The responses clearly showed that guided tours of cemeteries are in increasing demand among Budapest residents and tourists alike. Visitors go on guided walks mainly due to the cultural aspects of graves and memorials of famous people buried there and the stories associated with them, and additional attractions such as artistic sculptures and buildings make for a very charming experience. The wildlife, both flora and fauna, in the cemeteries are an additional attraction that adds color to the walks. During touristic use of cemeteries, the operators pay particular attention to ensuring that the main functions of the cemeteries are not compromised; thus, guided walks are organized during periods when burials are not in operation. During the guided walks, several companies use equipment such as earphones to provide adequate audibility while ensuring that loud speech does not disturb people visiting and caring for the graves of their loved ones.

Cemetery guides declare that the tourist potential of these facilities depends largely on the maintenance level of the green spaces which, especially during the growing season, provides an attractive background for the graves; in the summer, the shade of the trees makes the walks more pleasant for visitors. The development of appropriate infrastructure is essential for touristic exploitation; accessibility, parking, and restrooms are needed in sufficient quantities and quality. In the local metropolitan context of Budapest, a need for gift shops and catering facilities has not yet been formulated by visitors.

In addition to the primary functions of cemeteries, professional guides consider thematic walks and cultural activities (e.g., exhibitions, concerts) as possible secondary uses for memorial parks and graveyards.

4. Discussion

The responses to the two questionnaires show that people in Hungary are not yet open enough to attending activities in cemeteries other than attending and caring for the graves of their deceased relatives or commemorating honorable persons. This is somewhat contradicted by the fact that many people and a wide variety of programmes have taken part in cemeteries, which shows that people are not closed to the possibility. A change in attitude, as mentioned by one of the respondents, may be a reason for that, as we may currently be in an early phase of slow transformation. However, the variety of organized programmes listed in the questionnaire responses shows that cemetery managers are open to new potentialities. The Budapest Funeral Institute is explicitly open to challenges, although they are aware that changes should be made slowly and steadily, as sudden reforms can have the opposite effect [51]. A free small train tour has been organized in the Fiumei cemetery to complement the already-existing irregular school lessons. While the Day of the Bells took time to become established, people gradually accepted the idea of alternating childrens' programmes with light music and instrumental concerts in a cemetery. The Fiumei út Cemetery is a popular place for people to walk, push a pram, play ball with their children, or just lie on the grass with a blanket. It is home to the Museum of Funerary Art, Hungary's only specialized collection of funerary and memorial art. The crematorium in Csömöri Graveyard has a café where relatives can 'wait' for their loved ones to be buried in peace.

Our research clearly shows that the advanced types of cemetery tourism forms in Hungary belong today to the unconventional category. The touristic use of cemeteries is a non-conventional tourism industry sector because cemetery visitors mostly visit graves without using registered tourism services, but they still strengthen the tourism service industry by travelling to the site and using other services near the cemetery. Meanwhile, the responses indicate that this issue should be discussed and convenient options and solutions acceptable to all parties should be explored.

Nevertheless, we can state that there exists real cemetery tourism in Hungary. Not only do cemeteries in the capital increasingly appearing on the list of tourist attractions; those in the countryside with an interesting history and local graves of famous Hungarian people do as well. According to Gábor Móczár, Director General of the National Heritage Institute (NÖRI), the main activity of facilities such as the National Graveyard is no longer to provide burial services; it is rather to present the diverse cultural values and promote dignified remembrance. For foreigners, the biggest attraction in cemetery tourism might be exploring and enjoying the local works of art. In June 2021, the NÖRI joined the Association of European Significant Cemeteries, a European network of historic and thematic cemeteries that includes the two cemeteries they manage (the National Graveyard on Fiumei Road and the Jewish Cemetery on Salgótarjáni Road). At the same time, the managing institution has joined the European Cemeteries Route, a more operational European cooperation community in terms of mutual travel programs, funding, and professional contacts, which has put Hungary on the map of European cemetery tourism in this respect [52]. A free app called FiumeiGuide has been developed which, in addition to presenting 150 priority highlights in the graveyard, currently allows users to plan their own personalized thematic tour.

The answers to the main questions of our study were provided by the capabilities and constraints identified during our research on the touristic use of Budapest cemeteries.

(1) In terms of tourist attractions and sights, the surveyed cemeteries are well-endowed. However, there is an urgent need to collect and identify these assets and values as thematic tourist attractions and to make their relevant information available to visitors. It is important to bear in mind that the cemeteries were not created for tourism purposes. They have gradually become tourist attractions throughout history. Thus, their primary function should not be negatively affected by their use for tourism and recreation. Although cemeteries are not yet conventional touristic destinations in Hungary, they have the potential to become more attractive in the future; their natural, built, and spiritual value and the tangible and intangible heritage they involve reinforce each other. Thus, cemeteries can be identified as open-air recreational areas with several attractions, where the past and present of a settlement as well as its development can be explored in a relatively small area during an individual or guided walking tour. Another beneficial factor is that most municipal cemeteries are open for a wide range of hours and are usually free of charge for visits. Attractiveness can be further enhanced by improving organized cultural programs and events on site. Depending on the cemetery, this could be the development of an existing attraction or the creation of a new one by unique or complex product improvement. This latter may include singular interventions as enhancements of the immediate surroundings around a main attraction, as well as any green space development.

(2) We have found a number of shortcomings in the tourist infrastructure of the cemeteries surveyed on the basis of both our personal site visits and the questionnaire responses. Tourism infrastructure in this case refers to the whole of the welcoming facilities, i.e., those infrastructural elements that can be used by the general public and that enable touristic activities. The improvement of basic tourism services and their background infrastructure is essential and covers, in particular, the development of utilities and pathway systems in a graveyard. This is important for the needs of both tourists and everyday visitors. For recreational and touristic purposes, it is important to provide the essential conditions for a longer stay. For example, there is a fundamental need to provide more benches and basic services such as restrooms. Improving accessibility in general is usually linked to complex touristic and infrastructural development, and is certainly a critical issue. The development of other services that serve mainly tourists, such as souvenir shops, should be preferably developed only after basic service needs have been satisfied.

(3) There are major gaps in the use of visitor management tools. In the case of historic cemeteries, reconciling their original function with touristic uses requires due diligence. Problems related to touristic use can be avoided or solved by using appropriate management methods. For example, problems are often caused by lack of infrastructure (toilets, cleaning facilities, catering facilities), resulting in pollution

and litter issues, or by tourists/visitors disturbing mourners and commemorators with their constant presence, talking, or general loudness. Solutions could include the preparation of visitor management tools, planning the movement of visitors in advance, better managing the potential negative impact of visits, and avoidance or mitigatation of negative consequences. Passive exhibition techniques appropriate to the character of the cemetery may be good practice for visitor education, although in most cemeteries only an overview map of the cemetery layout is displayed. There are rarely any signs or printed materials (brochures, guides, and more detailed special publications) providing information on the specific values and attractions of the cemeteries and visualizing their locations. For larger cemeteries, it is necessary to develop thematic itineraries for guided tours or self-exploration walks. During self-guided tours, visitors can follow a pre-edited brochure along the way.

5. Conclusions

Based on our research, we can conclude that cemeteries, as green open spaces in major cities, play a highly-important role in urban greenspace systems and infrastructure such as local climate control, daily recreation, and tourism services. Cemeteries provide profoundly essential ecosystem services for both the residents of surrounding urban areas and for visitors. Among the ecosystem services provided by cemeteries, their regulating services have a main role in the adjustment of climate issues, while their cultural services contribute to the preservation of traditional and cultural heritage aspects as well as to the recreational and touristic function of these sites.

The green surfaces of cemeteries are traditionally defined by tree alleys which delimit the pathways separating the burial plots, as well as by grassland areas and flower beds between the graves; these are, as a whole, important for visitors. As the vegetation becomes older and larger in well-established and maintained sites, the shading effect of trees becomes increasingly important, as they provide tolerable urban climate conditions for visitors in the summer. Large cemeteries with mature trees can have temperatures during the summer several degrees lower than in the surrounding areas, providing a pleasant resting place for visitors. Our questionnaire research clearly shows that the level of maintenance and care for the vegetation in cemeteries is essential for both visitors and tourist guides, two target groups whose opinions must not be neglected.

Despite the recognised green value of urban cemeteries, the presentation of their semi-natural vegetation and habitat is unfortunately little-developed in Hungary, although this could help to diversify guided walks and even enlarge environmental education functions in these cultural complexes. However, any tourist development should prioritize the fundamental needs of the local population, namely, unhindered commemoration and care of graves. Hopefully, in the short term and with the right development accompanied by public education, remarkable cemeteries with significant green spaces can play a satisfactory role in future green tourism in urban areas in Hungary.

The use of cemeteries for tourist purposes must remain in balance with their traditional memorial function. We have examined the possible tourist exploitation of different types of cemeteries (e.g., closed/inactive or active cemeteries, gridded, and landscape-style cemeteries) within an ethical framework (Figure 7). We have found that in addition to their architectural and artistic value, the green surface intensity of a cemetery largely contributes to their wider potential for tourism uses.

The greenspace potential of a cemetery consists of several factors. If the vegetation cover and plant use are adequate, the ecological importance of the area is evident. However, only cemeteries with a sufficient size and maintenance level are suitable for other green space functions such as recreation purposes. Cemeteries with a smaller area may be suitable for daytime recreation, while cemeteries with a larger surface area can be visited from a greater distance, similar to urban public parks. It is important to note here that the primary memorial function of cemeteries should not be compromised, which is why closed cemeteries are more suitable for recreation than functioning cemeteries. It is important to mention that, even in a closed cemetery, recreational activities should be allowed only after due consideration. Walking, reading, and quiet contemplation do not disturb a functioning cemetery; however, certain sports activities and noisy programmes that attract large crowds are not even conceivable in a closed cemetery. Some non-intensive sports (e.g., running, yoga) or cultural events that attract fewer people and do not result in damage either physically or spiritually may be allowed in a closed cemetery.

The recreational and touristic use of cemeteries are partly interconnected, as the green potential of a cemetery determines the interest in the cemetery (e.g., visits for botanical values). In terms of tourist use, priority should be given to those cemeteries in Budapest that can already offer other attractions in addition to their primary function. A cemetery is most attractive for tourism if it has suitable potential beyond its original function, i.e., visitors seek out its special architectural, natural, or artistic values, the burial places of famous people, or sites of cultural and historical interest.

Author Contributions: Conceptualization, Á.S. and ZS.M.; methodology, Á.S. and ZS.M.; writing—original draft preparation, Á.S., ZS.M., I.G.T. and K.T.; validation, Á.S., ZS.M. and I.G.T.; formal analysis, K.T.; investigation, Á.S., ZS.M. and I.G.T.; resources, Á.S., ZS.M. and I.G.T.; data curation, ZS.M.; writing—review and editing, Á.S., ZS.M., I.G.T. and K.T.; visualization, Á.S., ZS.M. and I.G.T.; supervision, Á.S.; project administration, Á.S. All authors have read and agreed to the published version of the manuscript.

Funding: The dissemination of research outcomes was partly supported by The Hungarian Tourism Association Foundation (MTSZ).

Institutional Review Board Statement: Not applicable.

Informed Consent Statement: Not applicable.

Data Availability Statement: Not applicable.

Conflicts of Interest: The authors declare no conflict of interest.

Abbreviations

NÖRI	Nemzeti Örökség Intézete—National Heritage Institute
BTI	Budapesti Temetkezési Intézet—Budapest Funeral Institute
MAZSIHISZ	Magyarországi Zsidó Hitközségek Szövetsége—Federation of Hungarian Jewish Communities

Appendix A

Table A1. Operating cemeteries in Budapest: an overview table (by authors, 2021).

Operating Public Cemeteries			
Name of the Cemetery	**Location in Budapest**	**Cemetery Management**	**Features**
Cemetery in Óbuda (Óbudai temető)	3rd district	BTI	guided tours
Tamás Street Urn Cemetery (Tamás utcai urnatemető)	3rd district	BTI	urn cemetery on the site of the former Békásmegyeri cemetery
Megyeri Cemetery (Megyeri temető)	4th district	BTI	extension of the former village cemetery
National Graveyard on Fiumei Road = Kerepesi cemetery (Fiumei úti sírkert = Kerepesi temető)	8th district	NÖRI	national pantheon, guided tours
New Public Cemetery in Rákoskeresztúr (Rákoskeresztúri Új Köztemető)	10th district	BTI	the largest cemetery in Hungary, guided tours
Farkasréti Cemetery (Farkasréti temető)	11th district	BTI	the largest cemetery in Buda, guided tours
Cemetery in Rákospalota (Rákospalotai temető)	15th district	BTI	extension of a former village cemetery
Cemetery in Cinkota (Cinkotai temető)	16th district	BTI	extension of a former village cemetery

Table A1. *Cont.*

Operating Public Cemeteries			
Name of the Cemetery	**Location in Budapest**	**Cemetery Management**	**Features**
Cemetery in Pestszentlőrinc (Pestszentlőrinci temető)	18th district	BTI	extension of a former village cemetery
Old Cemetery in Kispest (Kispesti Öreg temető)	19th district	BTI	reopened former village cemetery
Cemetery in Kispest (Kispesti temető)	19th district	BTI	extension of a former village cemetery
Cemetery in Pesterzsébet (Pesterzsébeti temető)	20th district	BTI	extension of a former village cemetery
Cemetery in Csepel (Csepeli temető)	21st district	BTI	extension of a former village cemetery
Cemetery in Budafok (Budafoki temető)	22nd district	BTI	former village cemetery
Angeli Road Urn Cemetery (Angeli úti urnatemető)	22nd district	BTI	urn cemetery on the site of the former Nagytétényi cemetery
Operating Jewish cemeteries			
Name of the cemetery	Location in Budapest	Cemetery management	Features
Jewish Cemetery in Óbuda (Óbudai izraelita temető)	3rd district	MAZSIHISZ	
Jewish Cemetery in Kozma Street (Kozma utcai izraelita temető)	10th district	MAZSIHISZ	the largest Jewish cemetery in Hungary
Orthodox Jewish Cemetery in Gránátos Street (Gránátos utcai ortodox izraelita temető)	10th district	MAZSIHISZ	
Farkasréti Jewish Cemetery (Farkasréti izraelita temető)	11th district	MAZSIHISZ	

Appendix B

Questionnaire on the use of cemeteries (visitor questionnaire)

I. Questions about the respondent:

- age: 18–25 y, 25–40 y, 40–50 y, 50–60 y, 50–70 y, 70 over
- sex: female/male
- education: 8 primary school, secondary school, university/college
- place of residence: Budapest, Budapest agglomeration, large city, medium-sized city, small town, village, other options to define:

II. Questions on cemetery visiting habits:

- How often do you visit a cemetery?
- weekly
- several times a month
- on a quarterly basis
- yearly
- less frequently

III. Has the frequency of your cemetery visits changed due to the pandemic closures?

- yes, I've been there less often
- yes, I have gone there more often
- No, it has not changed

IV. If you went to the cemetery more often during the closures, what was the reason? (multiple answers are possible)

- I expected spiritual comfort from the visits

- It was possible to walk and move around the cemetery at a reasonable distance during the closures
- The visit to the cemetery provided a cultural programme
- I expanded my knowledge with what I saw in the cemetery
- other options to define:

V. What is the purpose of your visit to a cemetery? (multiple answers are possible)

- visit the graves of my loved ones/relatives
- remember the deceased (not necessarily in the cemetery where they are buried)
- visit the graves of famous people
- get to know statues, artworks and buildings in cemeteries
- for walking and relaxing

VI. Do you visit cemeteries when you travel in your home country? (you can tick more than one answer)

- yes, I mainly visit the graves of relatives and friends
- yes, I visit cemeteries as historical sites
- yes, I visit the graves of famous people
- yes, I visit artworks and statues in cemeteries
- yes, I visit buildings and other architectural elements
- yes, I am interested in the green surface and vegetation of cemeteries (protected and special species)
- yes, other options to define:
- no

VII. Do you visit cemeteries when you travel abroad? (you can tick more than one answer)

- yes, I mainly visit the graves of relatives and friends
- yes, I visit cemeteries as historical sites
- yes, I visit the graves of famous people
- yes, I visit artworks and statues in cemeteries
- yes, I visit buildings and other architectural elements
- yes, I am interested in the green surface and vegetation of cemeteries (protected and special species)
- yes, other options to define:
- no

VIII. Have you ever visited a cemetery for recreational purposes?

- yes
- no

IX. If yes, what were you doing there? (multiple answers are possible)

- walking
- reading
- contemplating
- doing sport activity (e.g., running)
- other options to define:

X. Have you ever participated in a cemetery-related event (e.g., commemoration, other ceremony)?

- yes
- no

If yes, what was it?

XI. Have you ever participated in an event not related to the cemetery's core function (e.g., concert, exhibition, bird ringing activity, etc.)?

- yes
- no

If yes, what was it?
XII. Is the maintenance of the cemetery important to you?

- yes
- no

XIII. Is the time spent in the cemetery influenced by the current physical condition of the cemetery or by the abundance of green spaces there?

- yes
- no

XIV. What kind of vegetation do you consider important in a cemetery, apart from the plants on the graves?

- trees, alleys
- flower beds
- shrubs
- grasslands

XV. What conditions do you think must be provided by a cemetery to be attractive for recreation? (multiple answers are possible)

- good accessibility
- it should be a closed cemetery (with no more active burials)
- free space between graves
- good coverage of green areas
- other options to define:

Appendix C

Questionnaire for tourist guides on the use of cemeteries
I. In which cemetery do you usually organize guided tours?

- National Graveyard on Fiumei Road (Fiumei úti temető/Kerepesi úti temető)
- Farkasréti Cemetery (Farkasréti temető)
- Jewish Cemetery in Salgótarjáni Road (Salgótarjáni úti zsidó temető)
- other options to define:

II. How often do you organize guided walks?

- weekly
- once a month
- several times in a month
- a few times a year

III. Do you organize guided walks specifically dedicated to the cemetery/ies or as part of a more complex tourist programme?

- I organize themed cemetery walks on different sites
- I also bring tourists to cemeteries on my sightseeing walks.
- I organise walks only in the National Graveyard on Fiumei Road (Fiumei úti temető/Kerepesi úti temető)
- I organise walks only in the Farkasréti Cemetery (Farkasréti temető)
- I organise walks only in the Jewish Cemetery on Salgótarjáni Road (Salgótarjáni úti zsidó temető)
- other options to define:

IV. How many people visit cemeteries with you each year?

- Under 100 people
- Between 100–200 people

- Between 200–500 people
- More than 500 people

V. What age group of people usually takes part in your organized walks?

- mostly aged between 18–30
- mostly aged between 30–50
- mostly aged between 50–70
- mostly over 70

VI. What are the main elements you are showing during your walks? (multiple answers are possible)

- the graves of famous people
- the statues in the cemetery,
- notable buildings
- old/notable trees
- other options to define:

VII. During guided trips abroad, are you taking groups to a cemetery?

- yes, we visit historical sites
- yes, we visit the graves of famous people
- yes, we visit works of art and statues in cemeteries
- yes, we visit them for their architectural values (e.g., buildings)
- yes, I highlight the vegetation of the cemeteries (protected and special species)
- no, I don't have these kind of activity

VIII. Is the maintenance of a cemetery important for your interpretation as a guide?

- yes
- no

IX. What kind of vegetation do you consider important for guided walks in a cemetery, apart from the plants on the graves? (multiple answers are possible)

- trees, alleys
- flower beds
- shrubs
- grassland areas

X. Is the time spent in the cemetery being influenced by the current condition of the cemetery or by the amount of green space?

- yes
- no

XI. What services would be needed in cemeteries for guided groups? (multiple answers are possible)

- parking places
- restroom, toilet
- benches
- souvenir shop
- catering facilities
- other options to define:

XII. Would you organize any events in the cemeteries that are not strictly related to the primary function of the cemetery?

- yes
- no

XIII. If yes, what would it be?

- concert

- exhibition
- theatre performance
- other options to define:

XIV. What conditions do you think must be provided by a cemetery to become a suitable place for recreation? (multiple answers are possible)

- having good accessibility
- being a closed cemetery
- having enough free spaces between graves
- having good coverage of green spaces
- other options to define:

References

1. Eötvös, K. *A Balatoni Utazás Vége [The End of the Balaton Trip]*; Vitis Aureus Bt.: Budapest, Hungary, 2007; 175p.
2. Szatmárcsekei Csónakos Fejfás Református Temető [Calvinist Cemetery in Szatmárcseke, Hungary]. Available online: https://reformacio.mnl.gov.hu/orokseg/szatmarcsekei_csonakos_fejfas_reformatus_temeto (accessed on 12 January 2022).
3. Balatonudvari—Szívalakú Sírkövek [Balatonudvari—Heart-Shaped Gravestones]. Available online: https://www.balatonudvari.hu/balatonudvari-telepules/telepulesunk-kepekben/szivalaku-sirkovek/ (accessed on 12 January 2022).
4. Sökfás Temetők az Őrségben [Cemeteries in the Őrség, Hungary]. Available online: http://guide2visit.eu/vonzerok-reszletek/sokfas-temetok-az-orsegben (accessed on 12 January 2022).
5. Hajdúböszörmény, Kopjafás Temető - Csónakos Fejfás Temető [Cemeteries in Hajdúböszörmény, Hungary]. Available online: https://www.turautak.com/cikkek/latnivalok/szakralis--vallasi-helyek/hajduboszormeny--kopjafas-temeto-csonakos-fejfas-temeto.html (accessed on 12 January 2022).
6. Katonai Temető, Budaörs [Military Cemetery, Budaörs, Hungary]. Available online: https://airbase.blog.hu/2020/10/29/katonai_temeto_budaors (accessed on 12 January 2022).
7. Gül Baba Türbéje. [Tomb of Gül Baba's, Budapest, Hungary]. Available online: https://gulbabaalapitvany.hu/ (accessed on 12 January 2022).
8. Világörökség Pécs. [World Heritage Site at Pécs, Hungary]. Available online: https://www.pecsorokseg.hu/ (accessed on 12 January 2022).
9. Sallay, Á.; Máté, K.; Mikházi, ZS.; Csemez, A. *A Kálváriák Szerepe a Balaton-Felvidék Turizmusában. [The role of calvarias in the tourism of the Lake Balaton region.] In Turizmus 3.0 (Turizmus Akadémia 10)*; Kátay, Á., Michalkó, G., Rátz, T., Eds.; Kodolányi János Egyetem: Orosháza: Budapest, Hungary, 2019; pp. 72–85, ISBN 978-615-5075-47-6. ISSN 1786-2310.
10. Mikházi, ZS.; Csemez, A.; Máté, K.; Sallay, Á. The role of calvaries in Hungarian religious tourism. *J. Tour. Chall. Trends Tour. Relig.* **2016**, *9*, 65–92, ISSN 1844-9742.
11. Gecséné Tar, I.; Hajdu Nagy, G. Südwest-Kirchhof, Stahnsdorf—Egy Temető Újjászületése. *[Südwest-Kirchhof, Stahnsdorf—A Cemetery Rebirth.] Tájépítészet* **2005**, *6*, 8–11.
12. Gecséné Tar, I. *Történeti Temetők Magyarországon*; Doktori értekezés, Budapesti Corvinus Egyetem: Budapest, Hungary, 2012.
13. Gecse-Tar, I.; Takács, K.; Bechtold, Á. The Hungarian National Graveyard (Budapest) as a public park. *Teka Kom. Urban. Archit.* **2016**, *44*, 187–193. Available online: http://teka.pk.edu.pl/index.php/tom-xliv-2016-2/ (accessed on 12 January 2022).
14. Prater, T. Cemetery Tourism—A New Trend? Available online: https://travelinspiredliving.com/cemetery-tourism-new-trend/ (accessed on 5 January 2022).
15. Quinton, J.M.; Duinker, P.N. Beyond burial: Researching and managing cemeteries as urban green spaces, with examples from Canada. *Environ. Rev.* **2019**, *27*, 252–262. [CrossRef]
16. Tanaś, S. The cemetery as a part of the geography of tourism. *Turyzm* **2004**, *14*, 71–87. [CrossRef]
17. Skår, M.; Nordh, H.; Swensen, G. Green urban cemeteries: More than just parks. *J. Urban. Int. Res. Placemaking Urban Sustain.* **2018**, *11*, 362–382. [CrossRef]
18. Deering, B. In the Dead of Night. In *The Power of Death: Contemporary Reflections on Death in Western Society*; Blanco, M.J., Vidal, R., Eds.; Berghan Books: New York, NY, USA, 2014; pp. 183–197.
19. Evensen, K.H.; Nordh, H.; Skaar, M. Everyday use of urban cemeteries: A Norwegian case study. *Landsc. Urban Plan.* **2017**, *159*, 76–84. [CrossRef]
20. McClymont, K.; Sinnett, D. Planning Cemeteries: Their Potential Contribution to Green Infrastructure and Ecosystem Services. *Front. Sustain. Cities* **2021**, *3*, 789925. [CrossRef]
21. Marinkovic, L. Issues of Croatian Touristic Identity in Modern Touristic Trends. *J. Tour. Res. Hosp.* **2016**, *5*. Available online: http://www.scitechnol.com/peer-review/issues-of-croatian-touristic-identity-in-modern-touristic-trends-VJGq.php?article_id=4907 (accessed on 12 January 2022). [CrossRef]
22. Millán, M.G.D.; Naranjo, L.M.P.; Rojas, R.D.H.; De La Torre, M.G.M.V. Cemetery tourism in southern Spain: An analysis of demand. *Tour. Hosp. Manag.* **2019**, *25*, 37–52. [CrossRef]

23. Pliberšek, L.; Vrban, D. Cemetery as Village Tourism Development Site. In *4th International Rural Tourism Congress, Congress Proceedings*; 2018; pp. 194–209. Available online: https://www.fthm.uniri.hr/images/kongres/ruralni_turizam/4/znanstveni/Plibersek_Vrban.pdf (accessed on 12 January 2022).
24. European Cemeteries Route. Available online: https://www.coe.int/en/web/cultural-routes/the-european-cemeteries-route (accessed on 12 January 2022).
25. Nagy, K. *Közparkok és Közkertek Használatának Vizsgálata és Értékelése [Survey and Evaluation of the Use of Public Parks and Gardens]*; Kandidátusi értekezés, Kertészeti és Élelmiszeripari Egyetem: Budapest, Hungary, 1997; pp. 3–10.
26. Balogh, P.I. A Szabadterek Szerepváltozása a Nagy Európai Városmegújításokban. [The Changing Role of Open Spaces in the Great European Urban Renewal.]. Ph.D. Thesis, BCE, Budapest, Hungary, 2006; pp. 9–35.
27. Csernus-Lukács, L.; Triff, V.; Zsigmond, J. *Budapesti Temetők [Cemeteries in Budapest]*; A mi Budapestünk Sorozat: Budapest, Hungary, 1999.
28. Seléndy, S.Z. *Temetőkert [Cemetery gardens]*; Mezőgazdasági Kiadó: Budapest, Hungary, 1972.
29. Bakay, E. Lakótelepek Szabadtérépítészete 1945–1990 Között Budapest Példáján. [Open Space Architecture of Housing Estates in Budapest 1945–1990.]. Ph.D. Thesis, BCE, Budapest, Hungary, 2012.
30. Budapest Környezeti Állapotértékelése [Budapest's Environmental Assessment]. 2021. Available online: https://budapest.hu/Lapok/2020/budapest-kornyezeti-allapotertekelese.aspx (accessed on 5 January 2022).
31. European Environment Agency Report: Urban Adaptation to Climate Change; European Environment Agency: Copenhagen, Denmark. 2012. Available online: https://www.eea.europa.eu/publications/urban-adaptation-to-climate-change (accessed on 12 January 2022).
32. Rizwan, A.M.; Dennis, L.Y.C.; Liu, C. A review on the generation, determination and mitigation of Urban Heat Island. *J. Environ. Sci.* **2008**, *20*, 120–128. [CrossRef]
33. Kleerekoper, L.; Van Esch, M.; Salcedo, T.B. How to make a city climate-proof, addressing the urban heat island effect. *Resour. Conserv. Recycl.* **2012**, *64*, 30–38. [CrossRef]
34. Martin, P.; Baudouin, Y.; Gachon, P. An alternative method to characterize the surface urban heat island. *Int. J. Biometeorol.* **2014**, *59*, 849–861. [CrossRef] [PubMed]
35. *Radó Dezső terv: Budapest Zöldinfrastruktúra Fejlesztési és Fenntartási Akcióterve [Radó Dezső Plan: Budapest's Green Infrastructure Development and Maintenance Action Plan]*; Főváros Önkormányzata: Budapest, Hungary, 2021.
36. Anna, D.; Ewa, K.-B. How to Enhance the Environmental Values of Contemporary Cemeteries in an Urban Context. *Sustainability* **2020**, *12*, 2374. [CrossRef]
37. Budapest Környezeti Állapotértékelése 2019–2020 [Budapest's Environmental Assessment 2019-2020], II.7. Zöldfelület-Gazdálkodás. Available online: https://budapest.hu/Documents/BKAE/2019-2020/27_BK%C3%81%C3%89-2020_II-7-Z%C3%B6ldfel%C3%BClet-gazd%C3%A1lkod%C3%A1s.pdf (accessed on 10 January 2022).
38. Costanza, R.; d'Arge, R.; de Groot, R.; Farber, S.; Grasso, M.; Hannon, B.; Limburg, K.; Naeem, S.; O'Neill, R.V.; Paruelo, J.; et al. The value of the world's ecosystem services and natural capital. *Nature* **1997**, *387*, 253–260. [CrossRef]
39. Smith, A.D.; Minor, E. Chicago's Urban Cemeteries as Habitat for Cavity-Nesting Birds. *Sustainability* **2019**, *11*, 3258. [CrossRef]
40. Konic, J.; Essl, F.; Lenzner, B. To Care or Not to Care? Which Factors Influence the Distribution of Early-Flowering Geophytes at the Vienna Central Cemetery (Austria). *Sustainability* **2021**, *13*, 4657. [CrossRef]
41. Nordh, H.; Evensen, H.K. Qualities and functions ascribed to urban cemeteries across the capital cities of Scandinavia. *Urban For. Urban Green.* **2018**, *33*, 80–91. [CrossRef]
42. Sírkerti Séták [Graveyard Walks]. Available online: https://fiumeiutisirkert.nori.gov.hu/sirkerti-setak (accessed on 5 January 2022).
43. Sétaműhely Séták [Sétaműhely Walks]. Available online: https://setamuhely.hu/seta (accessed on 5 January 2022).
44. Budapesti Temetkezési Intézet. Available online: https://www.btirt.hu (accessed on 5 January 2022).
45. Südwestkirchhof Stahnsdorf. Available online: www.suedwestkirchhof.de (accessed on 10 January 2022).
46. Friedhöfe Wien. Available online: www.friedhoefewien.at (accessed on 5 January 2022).
47. Tomašević, A. Cemeteries as tourist attraction. *Tur. Posl.* **2018**, 13–24. [CrossRef]
48. Moreno, A.S. Cemetery Tourism: Visitors' Motivations for Visiting the Rakowicki Cemetery in Kraków. Master's Thesis, Universitat de Girona, Girona, Spain, 2018; p. 74.
49. Pécsek, B. City Cemeteries as Cultural Attractions: Towards an Understanding of Foreign Visitors'Attitude at the National Graveyard in Budapest. *Deturope* **2015**, *7*, 44–61. Available online: http://www.deturope.eu/img/upload/content_42123976.pdf (accessed on 5 January 2022). [CrossRef]
50. Rátz, T. Mindenszentek és Halottak Napja Szerepe a Magyar Lakosság Utazási Magatartásának Befolyásolásában. [The role of All Saints' Day and Day of the Dead in influencing the travel behaviour of the Hungarian people.]. *Tur. Bull.* **2006**, *X/3*, 27–35.
51. Balajti, P. Élettel Töltenék Meg a Temetőket [Fill the Cemeteries with Life.]. Available online: https://www.borsonline.hu/aktualis/2013/08/elettel-toltenek-meg-a-temetoket (accessed on 12 January 2022).
52. Nemzeti Örökség Intézete. Sajtószemle: Ősz a Fiumei úti sírkertben. [National Heritage Institute. Press Review: Autumn in the Fiumei út Cemetery.] (2021. Október 30). Available online: https://intezet.nori.gov.hu/hirek/osz-a-fiumei-uti-sirkertben/ (accessed on 5 January 2022).

sustainability

MDPI

Article

The Changes in the Demographic Characteristics and Spatial Structure of Tourism Demand in the West Balaton Region's Spa Cities

László Könnyid [1], Zsuzsanna Váradi [1,*], Zsombor Nagy [2], Noémi Ilyés [3] and Orsolya H. Horváth [4]

1 Hungarian Tourism Agency, 1027 Budapest, Hungary
2 Faculty of Economic Sciences, Corvinus University of Budapest, 1093 Budapest, Hungary
3 Hungarian Tourism Association Foundation, 1027 Budapest, Hungary
4 Promo Zone Trend Ltd., 8380 Hévíz, Hungary
* Correspondence: zsuzsanna.varadi@mtu.gov.hu

Abstract: The last two years, the period of the pandemic, have brought a significant change in the tourism of Hungary, which has been developing unbroken until then. The year 2019 broke all the peaks that were interrupted by the pandemic. This particularly affected our spa towns of international significance, including the examined settlements, Hévíz and Zalakaros. The aim of the study is to show what changes have taken place in the development of the number of visitors in the cities that have been based mainly on foreign traffic until then, what territorial reorganization has taken place in terms of sending areas, and what new target groups with modified attitude have emerged. In this study, we analyzed in detail the databases of the National Tourist Data Center, which has been operating since July 2019, and the monthly database of the Hungarian Central Statistical Office. The special, so-called unconventional tourism is carried out on the one hand by the methodology of statistical data collection and on the other hand by the explored tourism behavior. According to our results, it is clear that due to the domestic traffic, a completely new target group (age group and status) appeared in the two spa towns, their sending areas affected the metropolitan suburban zones, and the target group was high-status, younger guests. In our opinion, this offers a new opportunity for spa towns to generate more sustainable, future-oriented guests with a focus on local values, creating a new supply structure and image, as well as messages.

Keywords: medical tourism; Hungary; postcovid; change of tourism supply; sustainable way; change of visitor flow

Citation: Könnyid, L.; Váradi, Z.; Nagy, Z.; Ilyés, N.; Horváth, O.H. The Changes in the Demographic Characteristics and Spatial Structure of Tourism Demand in the West Balaton Region's Spa Cities. *Sustainability* **2022**, *14*, 10531. https://doi.org/10.3390/su141710531

Academic Editors: Anna Rita Irimiás, Dallen J. Timothy and Gábor Michalkó

Received: 4 April 2022
Accepted: 11 August 2022
Published: 24 August 2022

1. Background

The need for sustainable tourism arose long before the COVID-19 global epidemic. As early as 2005, United Nations published a guide on sustainable tourism, which still provides the basics of the sustainable goal developments of the World Tourism Organization. The guideline—written within the framework of the United Nations Environment Program—addresses policy-makers and calls attention to creating sustainable tourism development projects and policies in order to protect the environment and local areas from overconsumption and exploitation [1]. Thus, the focal points of the World Tourism Organization's work regarding sustainable tourism development are the following: Optimal use of the environment, maintaining healthy ecological processes, respecting the host communities, and ensuring viable economic operations [2]. The United Nations also designated the year 2017 as the International Year for Sustainable Tourism Development in order to encourage the policy-makers to contribute to the United Nations Sustainable Development Goals (SDGs).

The threat of over-tourism seems real and more pressing than ever, even in the light of the coronavirus epidemic. Moreover, the epidemic—as a magnifier—actually exacerbates the problems that already existed regarding the tourist's overconsumption. While the lack of tourists meant economic difficulties and hard times for international tourism, it also gave the opportunity for locals to regain back their living space. There was already a tension between the ambitious tourism development projects and the long-term sustainability, and a lot of the development projects left out social relations. When these kinds of projects met with failed regulations of natural resources, that made the situation even worse. The role of globalization in tourism growth is unquestionable. The unlimited access to information, the modernization of transport, free markets—that led to the expansion of international hotel chains—and the significant demand for touristic services had a huge impact on tourism growth [3]. UNWTO's Tourism Dashboard shows incredible growth between 2009 and 2019. While in 2009, the UNWTO registered 897 million international tourist arrivals, in 2019, this number was 1466 million [4]. 2009 is an important year, as according to UNWTO, this was the year when international tourism receipts exceeded the number of international tourist arrivals permanently [5]. Not to mention the non-conventional aspects of tourism mobility like one-day visitors, VFR tourism, which can hardly be detected in tourism statistics and is difficult to measure.

The rapid urbanization, affordable travel and visa services, technological advances, and new business models are all behind this rapid and monumental growth [5]. The cheap flight tickets and constantly growing air transportation made the exotic destinations more reachable, and new consumer trends have appeared too. The considerable demand for touristic services accelerated development and made long-haul and foreign travels extremely popular.

The constantly increasing demand exceeded the level of sustainable tourism development. It is a legitimate question whether the socio-economically more vulnerable settlements, and especially the people living there, can actually benefit from the increasing demand. Different strategy-making levels have detached tourism as an economic sector from socio-economic and environmental impacts. Before the COVID-19 pandemic, there was already a call for more sustainable, smart, and local cities. Sustainability has gained greater importance as it contributes to today's cities by making the lives of residents easier and offering sustainable and responsible management of natural resources [6]. This backs up an argument about the carrying capacity of a tourism destination: What is the maximum number of visitors that is acceptable without destroying the environment and bothering local communities [7]? Carrying capacity is linked to the tourist activities done within the region. These impacts have to be within the ecological carrying capacity of the given territory [8].

The economic contribution of tourism has become so significant that the collapse of the industry has pointed out some serious problems that the COVID-19 pandemic could change in terms of sustainable tourism development. On the other hand, these changes should have been done long before the crisis, as the excessive growth means more vulnerability in the case of those countries and settlements where the main source of revenue is mass tourism.

The pandemic brought lockdowns, social distancing, travel bans, flight cancellations, and numerous health and safety risks, which obviously decreased the number of travels. The pandemic shows signs that it could bring long-term changes in the industry and reimagine sustainable and responsible tourism [9]. Achieving sustainable tourism and, at the same time, maximizing growth is always a challenge [10] as usually, the cheap and convenient options are not the sustainable ones. It is also important to note that the tourism sector also significantly contributes to global greenhouse gas emissions [11].

Due to the lockdowns and difficult travel regulations during the pandemic, the demand for domestic products has emerged. Tourists who usually choose long-haul and foreign travel routes were forced to travel within the country borders. This phenomenon led to more rational consumption. Tourism consumption starts when the traveler leaves his permanent home until he arrives home [12]. It would be naive to expect rational behavior regarding sustainability from tourists who spend their free time, which often ends up in unreasonable decisions regarding the type of transportation, accommodation, shopping, and sightseeing. Despite the COVID-19 pandemic, the younger generation still considers that the waste situation, pollution, use of non-renewable resources are the main challenges that sustainable tourism faces [13]. Domestic travels—forced by the crisis—resulted at least in less air pollution. Nevertheless, distance is a critical factor in tourism. Variables such as gender, time factor, education, number of children, seasonality, and weather, largely determine the distance of the travels [14].

In the case of a small country like Hungary, where domestic flights do not exist, it seems to be more environmentally friendly. Achieving high-value tourism is an old desire in Hungary's tourism, which means attracting high-spending tourists and increasing the revenue from tourism expenditure instead of increasing the number of tourists. This could result in much more sustainable growth.

The Hungarian Tourism Agency is a governmental organization responsible for the development and public administration of tourism in Hungary. The organization's main mission is to gain competitiveness of Hungary as a tourism destination and to promote the country in order to be more attractive for international, domestic, and business travelers. The Hungarian Tourism Agency defines the tourism development strategy, supervises the utilization of EU funds and domestic budgetary sources dedicated for tourism development and manages the tourism brand of Hungary [15]. As part of the governmental tasks, the organization issued the National Tourism Development Strategy in 2017, which includes all the guidelines regarding the tourism development in Hungary until 2030. The strategy was updated in 2021 due to the COVID-19 circumstances, and the National Tourism Development Strategy 2.0 version was born. The strategy places a strong emphasis on sustainability and responds to the ever-changing trends in tourism. Due to the destination approach, territorial aspects are also prominent, providing a framework for defining the profile of each touristic region. This also helps specific regional branding. According to the strategy's vision, by 2030, the tourism sector in Hungary will be the driving force for sustainable economic development by achieving quality experience, innovative solutions, and a strong national tourism brand. The strategy dedicates a whole chapter to sustainability as a strategic direction, in line with international standards. One of the long-term goals is the introduction of a trademark developed along sustainability criteria that is widely applicable in the tourism sector. Sales efficiency is essential for sustainable growth. In addition to this, professional marketing activities are indispensable to exploit the potential of domestic tourism. Data-driven and targeted marketing is needed as it is extremely important to restore demand as soon as possible after the significant decline in tourism as a result of the epidemic crisis [16].

It can be clearly stated that tourism has many social and economic effects and advantages. This is especially true for the field of spa and health tourism, that generates longer stays and, therefore, high expenditures. Spa and health tourism is extremely important in countries like Hungary, which are on the one hand, weak in natural resources but, on the other hand, rich in natural thermal water. Under the territory of Hungary, there is a presence of 70% of thermal water, which is a remarkable gift. There are many settlements spread across Hungary that are blessed with thermal water that differ in terms of size, accommodation facilities, and faculties.

This study reveals two different dimensions of tourism typology. One of them are the characteristics of the receiving areas (Hévíz and Zalakaros), and the other are the characteristics of the visitors on the demand side. According to Candela and Figini [17], it is debatable that the destinations are at the core of the tourism system, and they also argue that choosing a destination is at the center of decision-making. Health tourism, that is examined in this study, is one of the most popular tourism products in Hungary, which can be spotted in those settlements where thermal water is available. Analyzing its generated effects is quite important in the context of the changes caused by the COVID-19 pandemic. Previous studies have already examined the socio-economic changes in the case of bath regions, e.g., the case of Kehidakustány, another well-known spa settlement in Hungary. It turned out that those cities which are successful in health tourism have the possibility to gain a higher level of income, which results in a major economic role [18]. Obviously, higher income also contributes to the social development and satisfaction of the local society.

2. Data Collection and Research Methods

Our study uses quantitative methods based on secondary analysis of tourism statistics. There are several possible methods for measuring tourism demand. In the case of Hévíz and Zalakaros, which are typically multi-day destinations with a relatively long stay of the guests, accommodation statistics provide reliable information on the volume of tourism demand. The primary source of the examined data is the National Tourism Data Center (NTAK) database. The digital platform operating under the supervision of the Hungarian Tourism Agency allows the anonymous traffic statistics of all accommodation establishments in the country to be displayed in real time. From the data received directly through the accommodation management software, analyses are made to make the data-driven decisions needed to support the tourism profession. NTAK also has a significant role in the whitening of the tourism sector, as the territorially competent local governments and the National Tax and Customs Office also have an overview of the data areas that are relevant to them. The NTAK receives statistical data from the individual accommodation management, catering and ticketing software. For our study, we obtained data gathered between 1 January 2020 and 31 December 2021. To present the accommodation offer of Zalakaros and Hévíz, the capacity data of the accommodations registered in the NTAK digital platform (number of accommodations and offered rooms), which operates with complete coverage in the two settlements. To examine the characteristics of tourism demand, guest traffic data were aggregated according to the age groups and place of residence of guests, in the case of both Zalakaros and Hévíz. Regarding the applied methodology, it was a challenge that the NTAK database would be completely operational only from the beginning of 2020, therefore, the analysis examining the multi-year perspective is based on the freely available database of the Hungarian Central Statistical Office (CSO), which is limited to changes in the volume of tourism traffic and does not cover the demographic and geographical characteristics of domestic demand. As, unfortunately, there are no detailed data on the spatial pattern of domestic demand for the period before the Covid, other alternative methodological approaches are warranted to investigate this in the future.

3. Study Area

The research focuses on two Hungarian spa towns—Hévíz and Zalakaros—due to their international importance. They are located in the western part of Lake Balaton Touristic Region, the most important tourist destination in Hungary (Figure 1). The tourist area of Lake Balaton has the largest accommodation capacity after Budapest (60,600 rooms) and, accordingly, the most visited rural tourist area, with a total of 5874 thousand guest nights, of which 67.38 percent are domestic, 32.62 percent foreign (31% German, 11% Czech, 11% Russian and 10% Austrian). Its main tourist products are waterfront holidays and thermal baths (source: CSO 2019). Lake Balaton has appreciated in recent years, and by the summer of 2021, we could see the operation of the previous successful years almost entirely.

The area has a quite good location and geographical position: The accessibility of the area is mainly determined by the E71 (M7) motorway and Hévíz-Balaton Airport, which is of exceptional importance for the West-Balaton destination, but the north-south transport possibilities need to be improved.

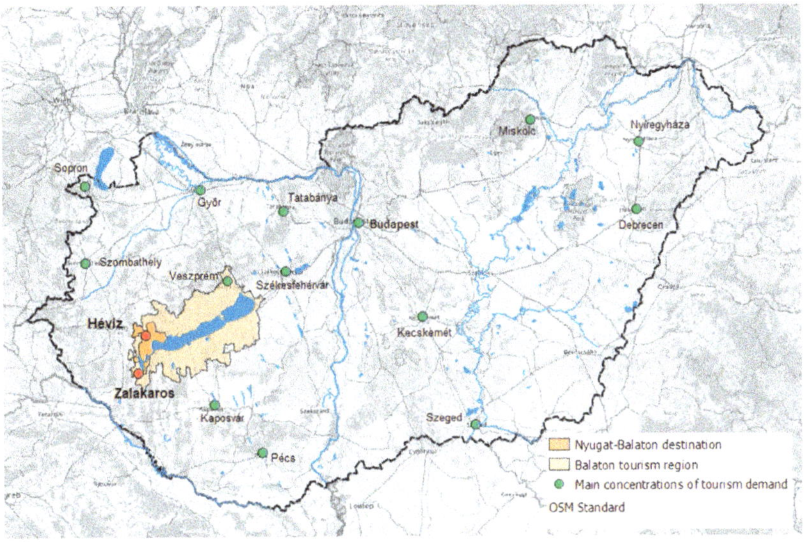

Figure 1. The sample area of the research.

Hévíz, as a small town, has approx. 4800 inhabitants, with international second-home owners—mostly German and Russian—as well. The basis of tourism in Hévíz is the world-famous and unique Lake Hévíz and the medical tourism services based on it. The lake has a water surface of 44,400 square meters and is the world's largest naturally warm, biologically active spa lake. The medicinal water erupts from the spring crater at a depth of 38 m, its average summer temperature is 33–35 °C, but it does not fall below 23 °C in winter, so it is suitable for outdoor swimming all year round. The water contains sulfur, alkali bicarbonate, slightly radon ingredients, which can primarily cure various rheumatic, musculoskeletal, and gynecological diseases [19]. The central spa was built in 1795 on the initiative of Count György Festetics, the owner of the area. In recognition of the development, Hévíz was declared a spa settlement in 1911. On 1 January 1952, the Hévíz State Spa Hospital was established in the buildings of the spa, which is now the St. Andrew's State Rheumatology and Rehabilitation Hospital. Due to this, Hévíz has become the largest spa in the country for the treatment of musculoskeletal rheumatic diseases. In 2015, Hévíz created its own trademark for the accreditation of therapies based on the healing effects of Lake Hévíz. Institutions that have the necessary professional, medical background, and medical infrastructure for the treatment (e.g., spa pool, weight bath, mud packer) are entitled to wear the trademark. There are currently nine certified spa hotels under the trademark that provide treatments under sanitary conditions under medical supervision. The last 30 years saw a greater development than ever before: Many new restaurants and hotels have been built. Recent spa and accommodation developments have aimed to increase the quality of care for patients recovering here.

Zalakaros—with approx. 2100 inhabitants—has also become one of the most important Hungarian international spa destinations. The flourishing of the settlement can be clearly linked to the construction of the spa. In 1962, during the hydrocarbon exploration in Zala, hot thermal water was found in the settlement. The spa opened in 1965, and the settlement underwent a significant development, the plans for the development and arrangement of the village and the recreation area were completed in a decade. The year of 1975 was decisive, when the building of the indoor bath was handed over, so the seasonal nature of the bath in Zalakaros ceased to exist and the basic conditions for the medical utilization of the thermal water were established. In the following years, the spa continued to expand, and new pools were built, significantly increasing the water surface. In 1978, the Granite

Spa was declared a recognized spa. The spa developments of the last 10 years have focused on the needs of the guests with children, families and young adults looking for wellness and adventure bathing, pushing the development of medical products into the background. In parallel with this, the capacity of tourist accommodation has been constantly increasing, new hotels and private accommodation have been established in the settlement.

4. The National Significance of the Two Cities, Based on Visitor Traffic and Capacity Data

The region of Western Balaton is one of the most visited destinations in Hungary. Hévíz and Zalakaros are the two most important spa resorts in the country in terms of available accommodation services. There were 653 accommodation establishments in Hévíz in July 2021, offering a total of 3920 rooms, while in Zalakaros, there were 2332 rooms available. The offer of both settlements is dominated by hotels, with more than half of the capacity being supplemented by private accommodation and other accommodation. Despite the fact that the attractiveness of both settlements is focused on medicinal water, the two settlements address fundamentally different target groups. Hévíz is an internationally known spa town with a significant foreign tourist turnover (675.3 thousand), while Zalakaros was the number one medical tourist destination in the country in terms of domestic tourism in 2019, with 508.1 thousand domestic guest nights. In the years before the epidemic, more than half of Hévíz's guest traffic was generated by foreign tourism, while in Zalakaros, less than a quarter of the guests were foreign (Figure 2).

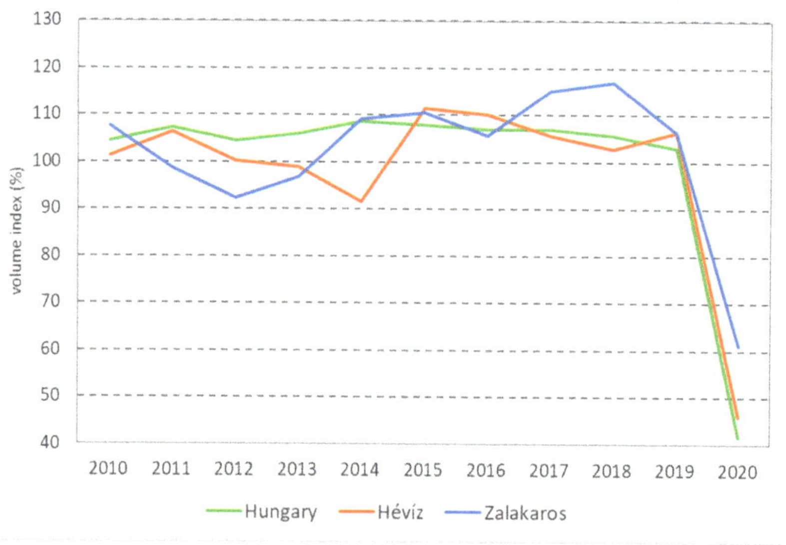

Figure 2. Volume index of the number of guest nights registered in commercial accommodation establishments between 2010 and 2020. Own editing based on the CSO Information Database.

5. Impact of the COVID-19 Epidemic on Tourism in the Region

Before presenting the changes caused by Covid, it is worth looking at the state of the Hungarian tourism sector and the two settlements before the pandemic started. The 2010s were about the prosperity of the Hungarian tourism industry, with a steadily increasing performance year by year. Concentrating on the two settlements examined, the guest turnover of Hévíz increased by more than a fifth, the growth of the tourism performance of Zalakaros exceeded 60% in the period between 2010 and 2019. In 2020, the Covid epidemic hit the Hungarian tourism sector hard, causing serious economic and social changes, especially in areas with a strong tourist profile, such as the West Balaton region. The negative effects of the epidemic on tourism performance were severe in both settlements. According to the data from the Hungarian Central Statistical Office, the number of foreigners visiting Hévíz fell by almost a fifth in 2020 compared to 2019, but there was also a decrease of about 40% in the case of domestic demand. In Zalakaros, the foreign guest turnover decreased by almost 70%, while the number of domestic guest nights decreased by about 30%. Due to the drop in demand and the uncertainty surrounding the conditions of international tourism, the spa settlements had to reach new target groups after the lifting of the epidemiological restrictions. In our analysis, the national, demographic, spending, and territorial characteristics of the volume of guest traffic in Hévíz and Zalakaros for the period following the outbreak of COVID-19 are presented.

6. Results

In 2020, compared to previous years, a drastic decline in foreign demand could be observed, and based on turnover data, the number of foreign guests returned only partially to the two settlements in 2021. In the case of Hévíz, with the loss of the previously dominant foreign target groups, domestic demand gave approximately two-thirds of the guest nights in 2020 and 2021, while the share of domestic demand continued to grow in Zalakaros, which has traditionally been strong on domestic guests (Table 1).

Table 1. Share of domestic and foreign market in guest turnover (%).

	Hévíz			Zalakaros		
	2019	2020	2021	2019	2020	2021
Domestic	41	69	65	76	89	86
Foreign	59	31	35	24	11	14

Source: Own editing based on the CSO and NTAK databases.

The common feature of the guests of the two settlements is the high proportion of guests using wellness and health tourism services (MTÜ, 2016). In addition to the similarities, however, the two spa towns address significantly different target groups. Nearly two-thirds of Hévíz's guests are people over the age of 45, whose main motivations are rest and health. In contrast, Zalakaros is characterized by a younger group of guests, with more than 60% of guests in Zalakaros under the age of 45. A significant proportion are couples between the ages of 35 and 45, who often arrive on long weekends, and those who go on family vacations to Zalakaros. The twenty-year-olds are underrepresented in both settlements. It can also be said that the demographic characteristics of foreign and domestic demand are significantly different for both settlements. In the case of foreign markets, well-defined target groups can be observed (Figure 3).

The base of Hévíz's foreign target market is provided by the senior age group, which can be characterized by a long stay and is looking for high-quality health tourism services. Two-thirds of the income from foreign tourism of accommodation establishments in Hévíz is related to people over 55 years of age, mostly people over 65 years of age. In Zalakaros, however, foreign guests are mainly couples between the ages of 35 and 44, who are primarily attracted by wellness services. More than half of Zalakaros' income from foreign tourism can be attributed to this age group. The group of foreign visitors to the two settlements is not homogeneous, different foreign target groups can be identified on the basis of nationality, length of stay, demographic characteristics, and spending habits.

The German market provides a significant part of the senior population visiting the area. The German seniors visit both settlements in large numbers, spending six to seven days at the destination. As the only place in Hungary, one of the dominant target groups of Hévíz is the Russian guest group, which is mainly a group of solvent, middle-aged, and elderly couples with a long stay. Russian demand is quite concentrated in Hévíz, there is no interest in the wider region. The number of Russian guests has been declining in recent years. By far, the German and Russian guests spend the most on accommodation. Due to the geographical proximity and the good value for money of the destination, many people come to the area from the border areas of Austria. In the case of the Czech and Slovak guests, it can be clearly observed that the older age groups visit Hévíz, while a higher proportion of people between the ages of 35 and 44 visit Zalakaros. Guests from the Czech Republic, Poland, and Slovakia are not only looking for premium accommodation, they have a significant need for mid-range services as well. Czech guests are characterized by average, Slovak and Polish guests by low unit spending, although the number of active families with medium and higher status is increasing from the Czech and Slovak sending markets. The stagnation or moderate decline of the sending markets (Germany and Russia), which provide the largest volume of guest traffic in the region, may be offset by the growth of these markets, which have additional potential and are less concentrated in Hévíz and visit the wider destination in large numbers (Table 2). Other countries sending a significant number of guests include Ukraine, Switzerland, and Romania, which also have growth potential. In the pre-Covid period, groups of visitors from outside Europe, mainly from China and Israel, which generated a significant amount of traffic in Hévíz, almost completely disappeared from the region in 2020 and did not return in 2021.

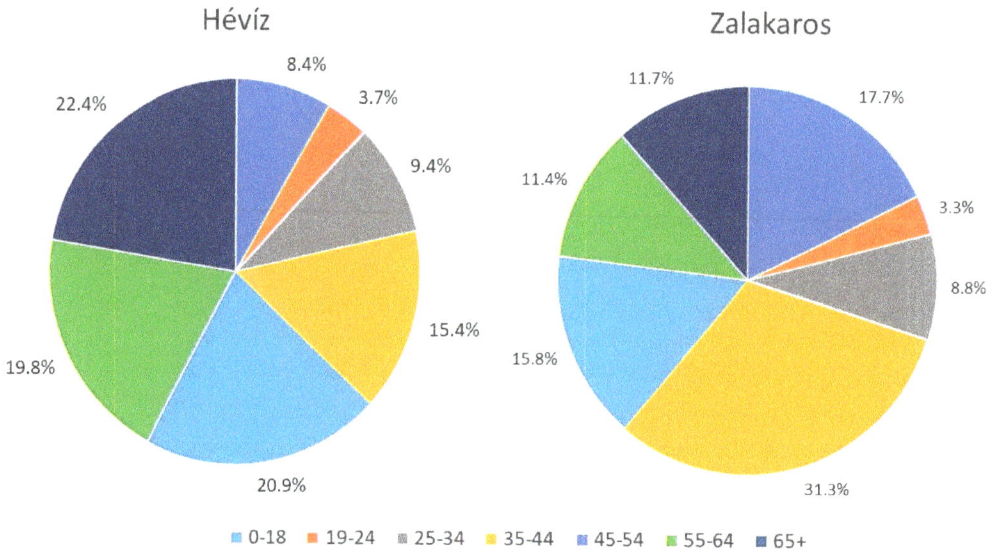

Figure 3. Guest turnover by age groups in Hévíz and Zalakaros, 2021. Source: Own editing based on the NTAK database.

It is important to emphasize that the specific expenditure of foreign guests is significantly higher than that of domestic guests. In the case of Hévíz, the expenditure per guest was 87.3% higher in the case of Zalakaros and 41.2% higher among foreign guests in 2021. The greatest contrast is in the comparison of foreign and domestic seniors. Foreign guests over the age of 65 spend 2.5 times as much on average in Hévíz and 2.2 times in Zalakaros as Hungarian guests. The difference is much smaller among the younger age groups. Data from recent years, on the other hand, show that this long-standing finding that foreign visitors are highly profitable for caterers appears to be declining on the basis of data from recent years. Partly as a result of the epidemic situation and state measures to encourage domestic tourism, a large number of solvent Hungarian guests have appeared, and their returning guests may be of strategic importance to the spa settlements.

Table 2. Top 5 sending market in Hévíz and in Zalakaros, 2021.

Municipality	Country of Origin	Number of Guests	Number of Overnight Stays	Average Length of Stay	Spending Per Guest (Thousand HUF)
Hévíz	Germany	12,687	77,423	6.1	118.4
	Russia	5589	36,677	6.6	169.8
	Czech Republic	8711	30,501	3.5	72.2
	Austria	7959	25,791	3.2	80.1
	Slovakia	4576	13,332	2.9	65.8
	Abroad total	51,822	232,559	4.5	
Zalakaros	Czech Republic	8535	31,476	3.6	76.7
	Germany	3745	26,007	7.3	101.3
	Austria	1888	6558	2.9	74.1
	Slovakia	1581	4596	2.7	55.1
	Poland	733	2339	2.8	32.8
	Abroad total	18,584	77,146	4.2	

Source: Own editing.

Domestic demand is becoming more and more important in the tourist traffic of the West-Balaton destination, which is significantly more diverse in terms of demographics than foreign visitors. Among the domestic tourists of Hévíz, in addition to the traditionally large number of seniors, the 35–44, 45–54, and 55–64 age groups also come to the spa settlement in similar numbers. The over-55 age group stays almost exclusively in hotels, while the share of private accommodation and other types of accommodation is high (30–40%) for the younger age groups. Regarding the spending per guest, a significant difference can be identified between the groups of guests under 35 (HUF 43,329) and those over 35 (HUF 54,758). There is no significant difference in the comparison of the 10-year age groups of those over 35 years of age. In the case of Zalakaros, the largest proportion of domestic guests are between the ages of 35 and 44 (30%), while the second largest age group is under the age of 18. It is clear from the data that Zalakaros is a typical destination for domestic family holidays, mainly during the summer months of the season. Families do not only stay in hotels, almost 20% of them choose private accommodation. In the age group between 35 and 44, another solvent target group arriving without children can be identified, which is characterized by high specific spending (HUF 73,289) compared to the other age groups, which is catching up to the spending intensity of foreign guests. The number of visitors to this target group increased almost one and a half times in Zalakaros in 2021 and 2020, while the number of visitors with families did not change significantly based on the number of registered guests under the age of 18. In the summer of 2021, the capacity utilization showed values above 80%, domestic demand in 4- and 5-star hotels was able to make up for the lost foreign guest traffic in the summer. This is the young and middle-aged group, who presumably travel abroad, are characterized by high spending, are less price sensitive, but chose domestic destination in the summer of 2021 due to the uncertainty caused by Covid. On the other hand, domestic demand showed moderate interest in lower-category hotels. Pre- and post-season traffic lagged behind, thus strengthening the seasonality that used to be generally less common in spa towns.

The spatial pattern of tourism demand in Hungary is less a Covid-related phenomenon than a phenomenon determined by the urbanization processes and changes in the spatial pattern of development of the past decades, which in Hungary has two main dimensions: The metropolitan–rural (center–periphery) contrast and the west–east regional inequalities, it means the division of the country is quite sharp [20]. Due to the strongly unipolar spatial structure of Hungary [21], the metropolitan area of Budapest, like in the case of many other socio-economic phenomena, plays a prominent role in terms of the territorial distribution of tourism demand. The distribution of domestic demand in Zalakaros and Hévíz by destination area is represented by dot density maps (see Figures 4 and 5), where one point represents 10 guests. As shown by spatial data, the majority of domestic guests—24% in the case of Hévíz and 17% in the case of Zalakaros—come from Budapest. The capital city and its growing agglomeration are increasingly dominant in terms of economic, technological, cultural, and human resources [22], and as a result, the purchasing power and potential target groups for spa towns are also predominantly concentrated in the metropolitan region. Satellite towns around Budapest with a high standard of living are among the most important sending areas of the West Balaton region: Not only because of the better geographical location, but the higher status population could offer to reach these settlements for holidays.

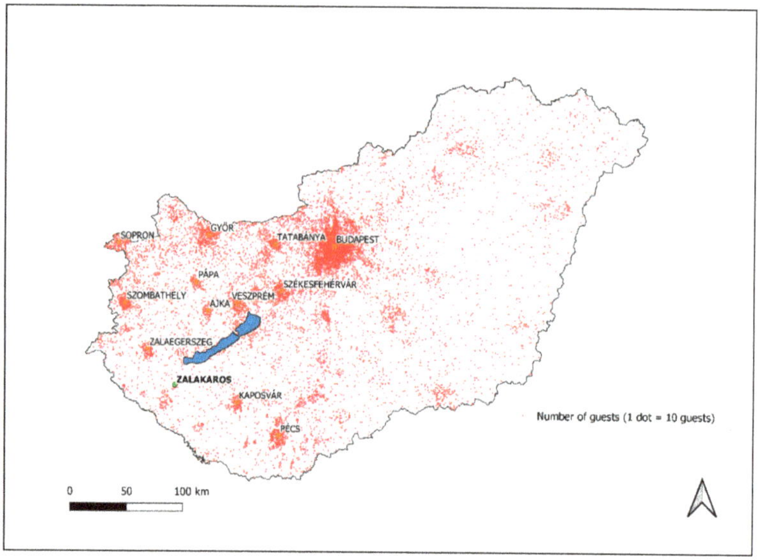

Figure 4. Spatial distribution of domestic guests registered in Zalakaros, 2021 (1 dot = 10 guests).

Comparing the domestic traffic of the two cities (Figures 4 and 5), it can be stated that Zalakaros is much more of a destination with regional significance, with an area extending to the west of the Danube (Transdanubia), while Hévíz is a nationally known and visited destination. Apart from the Budapest area and the city of Pécs, the main Hungarian sending cities of Zalakaros are exclusively Transdanubian cities (Székesfehérvár, Győr, Szombathely, Veszprém, Sopron and Tatabánya). Important sending towns—such as Veszprém, Zalaegerszeg, Ajka, Pápa, and Kaposvár—are situated within 100 km of the destination. In the case of Hévíz, Pécs, Győr, Székesfehérvár, Sopron, and Szombathely are dominant markets as well, but Hévíz is also visited in large numbers from the cities of Northern Hungary (Miskolc), the Northern Great Plain (Debrecen, Nyíregyháza), and the Southern Great Plain (Kecskemét). We can conclude that the attractiveness of Hévíz extends to a wider area, geographic proximity seems to be less important for the visitor flows of Hévíz. Looking at the sending areas of the region in the light of the spatial structure of Hungary, it can be seen that the two settlements tend to target more developed urban agglomerations with a higher standard of living, which is also reflected in the increasing trend of the expenditure structure.

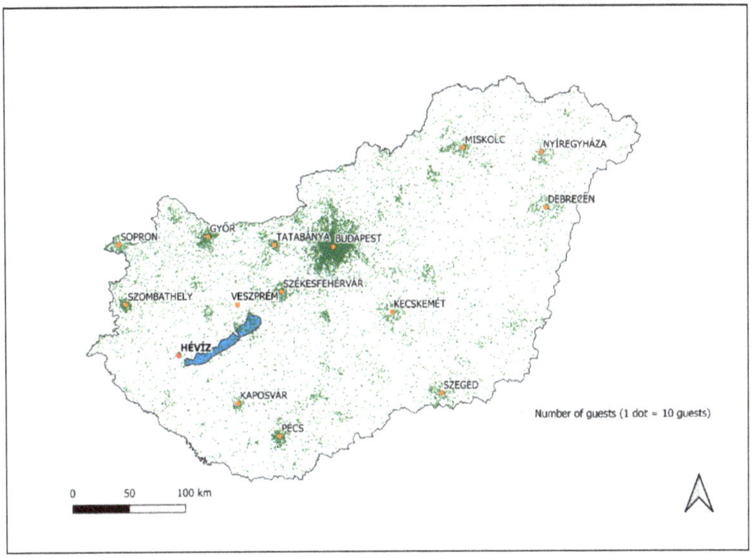

Figure 5. Spatial distribution of domestic guests registered in Hévíz, 2021 (1 dot = 10 guests).

In the case of Hévíz, the analysis of the new type of database made the reorganization of the target groups even more visible, which resulted primarily in 2020 from the fact that the guests of the classic spa treatments of the pre-pandemic period (German seniors and Russian spa guests) became unavailable due to the restrictions, and a proportionally larger share of the thus reduced traffic reached the traffic of domestic and neighboring countries. The data analysis clearly highlights the need for a drastic transformation of the target groups, as well as the creation of a new image of Hévíz. The city's new marketing strategy responded to this change, on the basis of which, sports and slow tourism, as well as the Post Covid Care program, are currently in focus, still addressing target groups with a higher ability to pay [23]. However, it is important to mention that the transformational effect of the processes of the past years on the long-term target group cannot yet be established based on the data from 2021.

In the case of Zalakaros, in the years prior to the pandemic, the spa town already placed family-friendly experience bathing and wellness bathing in the focus of its product development and marketing strategy instead of being an exclusive health resort mainly aimed at young, family customers from domestic and neighboring countries [24]. The travel restrictions affected these target groups less, so they were able to further strengthen the already started communication and act effectively in the changed tourism market. In the case of Zalakaros, the change in the target group can already be clearly seen from the statistical data from 2021, which is also helped by the fact that the investments affecting the young age group were realized in connection with the urban development (walking route around the artificial thermal Lake and Ecopark, thus increasing the settlement as a family-friendly tourist destination appearance.

7. Conclusions

The last two years have caused a significant change in the life of the two settlements, which clearly made it necessary to rethink the target groups. From the data of returning guests, it is clear that people's desire to travel is unbroken. Despite the restrictive measures, they are looking for travel opportunities. As a result, the role of stable domestic demand is becoming more important for individual destinations.

Hévíz is the number one tourist settlement in Hungary after the capital, however, it was most severely affected by the pandemic, with a 62% shortage of visitors. One of the main reasons for the large decline is the slower return of senior (over 65) foreign visitors to the destination. This is intensifying this year, as the Ukrainian–Russian conflict is clearly making its mark on the lack of the current main sending market—in the case of Hévíz—which must be filled with a new strategy by renewing the supply market and brand of the settlement. Here we can clearly relate to the issue of sustainability: Is the settlement able to renew itself, address the domestic market sufficiently, adjust its offer to the Hungarian—primarily high-status clientele, establish a network connection in the region—one of the elements of which is sustainability in connection with domestic travel? In the city of Hévíz, the service providers have started their activities in this direction, i.e., a completely new target group: In addition to the seniors, the family has become the most important potential target group, targeting those interested in hobby and professional sports tourism. In addition, it has launched a new health tourism product package called "Post Covid Care" that offers standardized treatment packages using existing medical experience and medical infrastructure as part of private health care in a quality hotel environment. For Hévíz, due to its distinctive medical tourism image, adult-friendly, age-restricted tourism products can also be a competitive advantage, which can also be unique, attractive products for the target groups of the domestic and surrounding countries.

In the city of Zalakaros, the service providers were in a less vulnerable position to the markedly present foreign sending market, so the transformation, addressing the guests, offering the service providers, rethinking the brand of the city created an easier situation: Addressing the domestic target group, being active on social media, using new urban attractions and programs for the young age group clearly have the effect of changing the city's clientele. Sustainable product development opportunities should be to develop for the new domestic clientele. In Zalakaros, ecotourism (through the protected areas of the Kis-Balaton National Park) and active tourism have been further strengthened, but there are untapped opportunities to address the young adult clientele with high wellness, such as manager disease management, burn out programs, immunity boosting, and post-Covid cures.

From the development of demand and the comparison of the two settlements, it can be seen that the domestic guests are mainly couples aged between 35 and 45, looking for a wellness experience, not a treatment based on balneotherapy. This attitude could only be changed by shaping a long-term national approach.

Based on the new type of database (the National Tourism Data Center) examination, we can highlight the fact that the two settlements, Hévíz and Zalakaros, have two similar attractions, but their guest structure, destination areas, and future perspectives have changed radically in the last two years. The differences are also very clear in terms of the domestic market of the two spa towns: While in the case of Hévíz, the medicinal character generated many guests from the eastern part of Hungary, also from the big cities (which means that the people living in the big cities have higher status), in the case of Zalakaros, the sending area of the guests affects the more developed western region. This can also be attributed to the fact that in the eastern part of the country, the people living there can visit many family-friendly settlements with medicinal water, meaning that it is not worth it for them to travel to a more distant settlement, while in the case of Hévíz, the unique medicinal character generates greater and higher-status purchasing power.

The examples show that in the short term (1–2 years), destinations (without external help) cannot introduce a new product and reach a large number of new target groups, even though product development is taking place. Introduction and promotion of the market require regional or state-level resources. A suggestion for both settlements is that much more attention should be paid to the National Tourism Development Strategy 2.0. the goals of creating sustainable destinations, the development and communication in this direction at the local level, the implementation of a conscious attitude formation, as a result of which a much more conscious, high-quality clientele seeking recreation, healing, and sustainability can flow into the region.

Author Contributions: Conceptualization, L.K., Z.V., Z.N., N.I. and O.H.H.; Formal analysis, L.K., Z.V., Z.N., N.I. and O.H.H.; Methodology, L.K., Z.V., Z.N., N.I. and O.H.H. All authors have read and agreed to the published version of the manuscript.

Funding: The dissemination of research outcomes was supported by The Hungarian Tourism Association Foundation (MTSZ).

Institutional Review Board Statement: Not applicable.

Informed Consent Statement: Not applicable.

Data Availability Statement: Not applicable.

Conflicts of Interest: The authors declare no conflict of interest.

References

1. United Nations Environment Programme; World Tourism Organization. Making Tourism More Sustainable. 2005, pp. 8–9. Available online: https://digitallibrary.un.org/record/561577 (accessed on 19 January 2022).
2. Sustainable Development. Available online: https://www.unwto.org/sustainable-development (accessed on 16 January 2022).
3. Brelik, A. Globalization in Tourism. In Proceedings of the 2018 International Conference "Economic Science for Rural Development", Jelgava, Latvia, 11 May 2018; pp. 42–48. Available online: https://www.researchgate.net/profile/Agnieszka-Parlinska/publication/326450019_GRAPH_THEORY_AND_AGROBUSINESS/links/5b4e3d18aca27217ff9e97f7/GRAPH-THEORY-AND-AGROBUSINESS.pdf#page=42 (accessed on 25 January 2022).
4. UNWTO Tourism Dashboard Global and Regional Tourism Performance of Tourism International Tourist Arrivals. Available online: https://www.unwto.org/global-and-regional-tourism-performance (accessed on 16 January 2022).
5. UNWTO International Tourism Highlights: 2020 Edition. Available online: https://www.e-unwto.org/doi/pdf/10.18111/9789284422456 (accessed on 16 January 2022).
6. Casado-Aranda, L.A.; Sánchez-Fernández, J.; Bastidas-Manzano, A.B. Tourism research after the COVID-19 outbreak: Insights for more sustainable, local and smart cities. *Sustain. Cities Soc.* **2021**, *73*, 103126. [CrossRef]
7. Li, J.; Weng, G.; Pan, Y.; Li, C.; Wang, N. A scientometric review of tourism carrying capacity research: Cooperation, hotspots, and prospect. *J. Clean. Prod.* **2021**, *325*, 129278. [CrossRef]
8. Graymore, M.L.; Sipe, N.G.; Rickson, R.E. Sustaining Human Carrying Capacity: A tool for regional sustainability assessment. *Ecol. Econ.* **2010**, *69*, 459–468. [CrossRef]
9. Sigala, M. Tourism and COVID-19: Impacts and implications for advancing and resetting industry and research. *J. Bus. Res.* **2020**, *117*, 312–321. [CrossRef] [PubMed]
10. Gössling, S.; Scott, D.; Hall, C.M. Pandemics, tourism and global change: A rapid assessment of COVID-19. *J. Sustain. Tour.* **2021**, *29*, 1–20. [CrossRef]
11. Gössling, S. National emissions from tourism: An overlooked policy challenge? *Energy Policy* **2013**, *59*, 433–442. [CrossRef]
12. Min, W.; Juan, Y. Irrational Tourism Consumption of Zero Inclusive-fee and Negative Inclusive-fee: Reflection and Measurements. *Energy Procedia* **2011**, *5*, 1416–1424. [CrossRef]
13. Šaparnienė, D.; Mejerė, O.; Raišutienė, J.; Juknevičienė, V.; Rupulevičienė, R. Expression of Behavior and Attitudes toward Sustainable Tourism in the Youth Population: A Search for Statistical Types. *Sustainability* **2022**, *14*, 473. [CrossRef]
14. Wynen, J. Explaining travel distance during same-day visits. *Tour. Manag.* **2013**, *36*, 133–140. [CrossRef]
15. Hungarian Tourism Agency's Website. Available online: https://mtu.gov.hu/cikkek/kozerdek-adatok (accessed on 28 January 2022).
16. National Tourism Development Strategy 2.0, Hungarian Tourism Agency. Available online: https://mtu.gov.hu/documents/prod/NTS2030_Turizmus2.0-Strategia.pdf (accessed on 28 January 2022).
17. Candela, G.; Figini, P. The Economics of Tourism Destinations. In *The Economics of Tourism Destinations*; Springer Texts in Business and Economics; Springer: Berlin/Heidelberg, Germany, 2012. [CrossRef]
18. Zoltan, H.; Ádám, K.; Tamás, M. The Effect of Health Tourism on the Lives of the Settlements–Settlement MORPHOLOGICAL Investigations in Kehidakustány. Available online: https://www.google.com.hk/url?sa=t&rct=j&q=&esrc=s&source=web&cd=&cad=rja&uact=8&ved=2ahUKEwjEzIOdk8D5AhXTx4sBHf9JCq8QFnoECAcQAQ&url=https%3A%2F%2Fojs.elte.hu%2Findex.php%2Ftft%2Farticle%2Fdownload%2F3344%2F3068&usg=AOvVaw224JHB4PR-xORLtY_hH8PF (accessed on 20 January 2022).
19. Official Webpage the of City of Hévíz. Available online: www.heviz.hu (accessed on 13 March 2022).
20. Nemes Nagy, J. Centrumok és perifériák a piacgazdasági átmenetben. *Földrajzi Közlemények* **1996**, *44*, 31–48.
21. Kovács, Z. *Népesség- és Településföldrajz*; ELTE Eötvös Kiadó: Budapest, Hungary, 2007.
22. Kovács, Z. Városok és urbanizációs kihívások Magyarországon. *Magy. Tudomány* **2017**, *178*, 302–310.
23. Integrated Development Strategy 2017–2023—City of Hévíz. Available online: www.onkormanyzat.heviz.hu/letolt/8329 (accessed on 15 February 2022.).

24. Integrated Development Strategy 2015–2020—City of Zalakaros. Available online: http://www.onkormanyzat.zalakaros.hu/tarsadalmi-egyeztetesek-dokumentumai (accessed on 15 February 2022).

sustainability

MDPI

Opinion

Travelling the Metaverse: Potential Benefits and Main Challenges for Tourism Sectors and Research Applications

Salvatore Monaco [1,*] and Giovanna Sacchi [2]

1 Faculty of Education, Free University of Bozen-Bolzano, 39100 Bolzano, Italy
2 Faculty of Science and Technology, Free University of Bozen-Bolzano, 39100 Bolzano, Italy
* Correspondence: salvatore.monaco@unibz.it

Abstract: The paper focuses on analysing the potential benefits and challenges of the Metaverse, particularly in the field of research in the tourism and food and wine sectors. The Metaverse is part of the new generation of the internet known as web 3.0, which also includes AI, blockchain and other digital innovations. The food marketing and tourism sectors are the main fields where companies are experimenting with solutions to offer people a fully functioning immersive Metaverse experience. This paper aims to highlight the potential impact of the Metaverse on tourism sectors as well as on research activities. Open challenges concern the social acceptance, affordability, and environmental sustainability of these technologies. Research is needed on the Metaverse's ability to reduce bias and accurately simulate real experiences, as well as on tourists' perceptions, attitudes, and willingness to pay for mediated experiences. Another important issue is the management of sensitive data that will travel through the Metaverse. Looking forward, the Metaverse has the potential to become a valuable tool for advancing tourism research through virtual collaboration and interdisciplinary research projects.

Keywords: digital innovation; food; Metaverse; phygital tourism; research; tourism; web 3.0; wine

Citation: Monaco, S.; Sacchi, G. Travelling the Metaverse: Potential Benefits and Main Challenges for Tourism Sectors and Research Applications. *Sustainability* **2023**, *15*, 3348. https://doi.org/10.3390/su15043348

Academic Editors: Dallen J. Timothy, Gábor Michalkó and Anna Rita Irimiás

Received: 13 January 2023
Revised: 7 February 2023
Accepted: 9 February 2023
Published: 11 February 2023

1. Introduction

Various fields of scientific research are increasingly turning their attention toward the Metaverse, both as an object of study and as a possible new space within which to practice research.

Currently, no unambiguous or shared definition of the Metaverse can be found in the literature, but from the major published studies to date, recurring elements allow it to be described as a shared online reality accessible via the internet, in which users are represented in three dimensions and can move around, share data and information, interact and socialise through shared experiences simulated in realistic, unrealistic, and mixed environments [1–3]. Thus, the Metaverse is a new way of interacting with reality, whether real, augmented, virtual, or mixed. It is part of the so-called web 3.0, a new generation of the internet that also includes artificial intelligence, blockchain, and new frontiers of digital and phygital innovation. Users access the Metaverse via 3D viewers and have virtual experiences connected to realistic avatars, other users, objects, concerts, events, travel, and more [4–6].

Although the publicity is only recent, the first attempt to design a three-dimensional environment in which people could create their own twin identity, communicate with each other, participate in events, and use a virtual currency was Second Life, launched in 2003. Since then, increasingly advanced technological infrastructure and investors' and users' interest in more user-friendly online environments for more immersive experiences have been central elements in the further development of virtual settings. Furthermore, the many restrictions on in-person activities and physical travel due to the COVID-19 pandemic have been an impressive stimulus for the growth of the Metaverse [7,8], creating a need for novel and unconventional mobilities.

Despite the embryonic nature of the Metaverse, several organisations, both public and private, have already begun to consider how they can reorganise their existing management models. For example, several recent studies [9–11] have analysed the potential impact of the Metaverse on the health sector. They argue that the use of innovative elements, such as immersion and interactive virtual reality (VR) technologies and applications would benefit both patients and healthcare providers, especially in the public sector. They describe the Metaverse as a space in which 'virtual hospitals' can be built, where consultations, remote care, physiotherapy, disease monitoring and even surgical operations can take place. Nevertheless, some experts have expressed critical issues about the health sector that arise with the development and implementation of Metaverse technology [12,13]. For example, with the rise of telemedicine and remote consultations, it is still unverified whether the quality of care provided in the Metaverse meets the same standards as in-person medical care. Furthermore, health data are subject to strict regulation, and it is crucial to determine how these regulations apply in virtual environments.

The potential of the Metaverse has also been evaluated in relation to the education field. The transposition of education into the Metaverse could be considered an evolution of the already well-known phenomenon of distance education, which has become common practice in schools and universities since the first lockdowns. The creation of virtual classrooms along with the use of digital twins by teachers and students may contribute to the creation of a new, more engaging, technology-driven style of learning [14–16]. Additionally, in the education sector, critical issues about the Metaverse have been raised with regard to both students and teachers [17,18]. For instance, are teachers equipped with the necessary skills and knowledge to effectively deliver education in virtual environments? How user-friendly and accessible are virtual educational environments for all students and teachers? Are virtual education environments designed to effectively engage and motivate all students?

Clearly, however, the increased levels of interaction and immersion offered by the Metaverse provide unprecedented opportunities and challenges for research. Some organisations and brands seeking to better position their products and services have already taken the first steps to harness the potential the Metaverse in their research. More specifically, as highlighted in a recent review of studies dealing with the Metaverse [19], the food marketing and tourism sectors currently represent the main fields in which companies are experimenting with solutions aimed at offering people the opportunity to enter a fully functioning, immersive Metaverse somewhere between the virtual environment and the real world.

Building on these considerations, this paper aims to analyse the potential benefits and main challenges of the Metaverse, with particular reference to the field of research in the tourism and food and wine sectors.

This topic calls for sociological, economic, and environmental reflections, since the Metaverse is an emerging and evolving entity that has not yet been sufficiently investigated [20–22], despite its complexity and impacts on people and society as well as on research activities. In fact, the Metaverse is contributing to the formation of new concepts and values that do not always overlap with those found in the real world [23].

The reminder of this paper is organised as follows: Section 2 deals with issues linked to tourism activities in the Metaverse, while Section 3 focuses on travelling the Metaverse, especially through gastronomy and wine tourism experiences. Section 4 discusses implications for research applications, and in Section 5, conclusions are drawn.

2. Tourism and the Metaverse

The tourism sector comprises the goods, services, and experiences of travel, hospitality, and entertainment and the enjoyment of cultural, social, and environmental resources. As highlighted within the most recent literature on the subject [24–27], tourism experiences provide not only moments of leisure and entertainment but also situations in which subjects can form their own identity through confrontation with cultural elements other than their own.

Consequently, the Metaverse may represent an additional element capable of enriching tourism offerings beyond physical spaces. Even before the pandemic, certain digital solutions had been widely used in tourism and hospitality [28,29]. The forced immobility associated with lockdowns not only led many people to seek alternative and unconventional forms of mobility, but also incentivised many players in the tourism industry to offer mediated experiences. These give potential travellers the opportunity to experience the world and practice tourism without physically moving, in line with emerging needs related to the search for safety and sustainability [30].

In this regard, distinguishing between 'moving tourism' and 'stationary tourism' may be useful [31]. The first term denotes tourism in its conventional sense. Its main features include physical mobility, the use of transportation, and face-to-face interactions between visitors and the local population. 'Moving tourism' assumes that people travel to visit tourist destinations in person.

By contrast, 'stationary tourism' refers to tourist experiences that people can enjoy through technology, without physically moving. The word 'stationary' is commonly used in the fitness world, recalling the concept of 'stationary bicycles'. This stationary fitness equipment may be used in homes, gyms, and other indoor or outdoor locations to replicate the use of a regular bicycle for physical training, enabling people to move despite actually remaining in the same physical location. Thus, a stationary tourist can travel without changing locations in the same way that a stationary bicycle stays in place while allowing users to ride at various speeds and intensities and providing the same experience as a conventional bike.

A third way is represented by the new frontier of phygital tourism. This term refers to an innovative way to connect with the environment in which physical and virtual locations are combined to complement and reinforce one another. In phygital tourism, the intersection of digital data with physical objects and landscapes forms the foundation of the visitor experience. The process can also work in reverse: a physical action might prompt a search by interacting with sensors or machine-readable milestones that provide users with pertinent information through digital interfaces [32].

From this perspective, the Metaverse can further expand the range of possibilities for practicing 'stationary tourism', providing a good opportunity for destination marketers and scholars who intend to advance knowledge in the tourism industry by exploiting the potential of simulated environments [33]. Through virtual reality, augmented reality (AR), mixed reality (MR), telepresence, and immersive storytelling, tourism experiences in the Metaverse could indeed affect consumers' decisions in the real tourism environment. In particular, several pioneering immersive experiences have highlighted how the Metaverse is already contributing to the creation of shared value in the tourism field, especially in situations in which potential tourists are asked to express their expectations and preferences as part of research on territorial branding and the launch of new tourism services [34,35].

In the same vein, some hospitality players have already begun to create immersive and interactive hospitality spaces within the Metaverse to allow their customers to simulate the experience of their stay, with the goal of receiving feedback that can be used to improve offline offerings. In 2022 the hotel company CitizenM became the first to enter the Metaverse, creating a hotel in the Sandbox platform. In this virtual hotel, people can interact with each other and purchase goods and services, such as an exclusive collection of non-fungible tokens (NFTs), thus influencing the construction of real hotels based on their feedback. CitizenM monitors avatar behaviour to increase brand awareness and positioning and to co-design, test, and then launch real hotels in the physical world, in line with the brand's democratic approach. Similarly, the hotel chain Italian Hotel Group also entered the Metaverse to offer users the chance to experience and learn about its property before they stay there. Once they connect with the hotel in the Metaverse, users can meet staff avatars who can converse with them in real time. The staff members lead users on a tour of the rooms and hotel facilities, showing them the customisation options available in the real facility, the spa, and other services offered. Such interactions help the hotel to understand potential travellers' new and different needs and to intervene in any critical issues reported by avatar visitors.

A further field of research is also beginning to study tourists' behaviours in the Metaverse using an ethnographic perspective [36–38].

These activities represent an opportunity for tourism players to develop tourism spaces, offers, and services to be marketed in a more targeted and informed way. Similarly, these activities enable people moving through the Metaverse to become more prepared and aware consumers, encountering in advance through their simulated enjoyment the goods, products, and services that may be of interest to them during their travels [39–42].

Because of its inherent potential, the use of the Metaverse for marketing can target not only individual customers but also travel professionals, such as travel agencies, influencers, and other intermediaries.

3. Potential Metaverse Applications in Food and Wine Tourism Experiences

Gastronomy in the Metaverse may sound paradoxical, as food is strongly linked to sensory and tangible experiences. Currently, if a restaurant or a winery opts to open a store through NFTs on the Metaverse, the goal may seem purely promotional, since it would not be possible to exploit the most important senses involved in tasting, smelling, and enjoying food and/or wine. However, food and wine are much more than simply the act of eating and drinking: they involve connecting with communities, cultures, and different territories.

The recent COVID-19 pandemic represented an unprecedented crisis for the hospitality industry. Due to health and social distancing measures, many hotels, restaurants, wineries, and bars have had to close. In addition, following the financial crisis resulting from the containment measures, several hospitality companies worldwide permanently ceased their activities. Thus, innovation must be given top priority in looking for potential solutions for hotel and oeno-gastronomy businesses. In more practical terms, innovation in the food sector is not just about physical food; it is also a matter of the phygital food experience. Oeno-gastronomy in the Metaverse must extend beyond the sense of taste to enhance the enjoyment of dining through immersive experiences, even incorporating gamification technologies into the custom of eating at a table. During the pandemic, evidence of such innovation appeared in the key role played by (virtual) technologies [43,44], from instant text messaging for food orders [45–47] to the ghost, cloud, and dark kitchens initiatives, virtual wine tasting experiences [48], and virtual tours of cellars and vineyards.

Particularly in oeno-gastronomic tourism, sharing experiences with friends, family, followers, and peers became one of the most important aspects of travelling across territories. To expand such experiences, food and wine in the Metaverse may open up novel and unconventional possibilities, such as the opportunity to visit production sites, breweries or farms in different states or nations and to interact with influencers and chefs.

Since 2022, companies have begun to experiment with phygital food experiences by opening virtual outlets in which customers may place their orders by logging in using VR goggles and have food delivered to their homes. Another example is represented by the Flyfish Club, which only allows access to its physical location in New York through NFTs supported by blockchain technology.

The wine sector might also add interesting phygital features to customise wine lovers' experiences by, for instance, using VR devices to explain the wine-making process during a wine-tasting session [49]. Concerning wine tourism, some companies have proposed immersive virtual tours of the world's finest wineries, assuring that the experiences are so real that by the end of the tour, virtual tourists can smell the wine.

Currently, it is difficult to predict whether oeno-gastronomic phygital experiences will gain any interest over physical ones. Certainly, however, VR, AR, MR, and the Metaverse offer significant potential for destination promotion. Firstly, they may make it possible to erase physical barriers for those who do not have the opportunity to travel around the world, even though, at least at first, such experiences will be accessible only to a certain niche of customers. Secondly, on the one hand, tourism experiences in the Metaverse could be criticised as amplifying a stationary lifestyle, which could worsen the trend in place since the 21st century; on the other hand, immersive virtual tourism experiences may lessen the perceived risk associated with intangible services, enabling visitors to make more knowledgeable choices and have more realistic expectations for their real visit [50,51]. According to several scholars [52–54], in the post COVID-19 era, people consider their health and safety as top priorities, even if it means organizing their travel experiences differently than in past, by, for instance, leveraging virtual potentials.

Finally, adding artificial intelligence (AI) features would make it possible to customise individual travellers' experiences to new levels.

4. Discussion

The rapid spread of the Metaverse has stimulated numerous discussions about its potential to transform society. Supporters of the Metaverse describe it as an innovation that will have major impacts on people's lives in the long run, as the internet has had already [55,56]. Likewise, with the Metaverse still in its early stages, some have shared concerns about governance, ethics, privacy protection, and data reliability and accuracy.

In terms of research applications, the pioneering experiences reviewed in this article indicate that the Metaverse holds a number of possibilities for research. In the context of food marketing and tourism research, one advantage that has already emerged is the possibility of creating objects or arranging activities that are unrealistic or that are not yet available in the marketplace, thus allowing users to experience goods and services that cannot be experienced in reality. This feature expands the possibility of allowing potential customers to evaluate multiple prototypes of the same product or service at a lower cost than creating them in reality for use and evaluation [57].

However, testing vineries, hotels, and restaurants on the Metaverse might be different from the real experience, generating a sense of dissatisfaction induced by the mediated simulation. Moreover, the anticipation could affect the originality and freshness of tourists' experiences and relationships with territories [53]. In this regard, the attractiveness of real destinations could be undermined when users develop an unhealthy level of attachment to the Metaverse, due to the possible feeling of addiction, as virtual reproduction could become increasingly addictive and captivating.

The new possible forms of interaction between players and users foreshadow a scenario in which, for data and information collection, traditional channels are giving way to interactive, anthropomorphic AI agents that engage in personalised interactions with customers, such as virtual 3D AI agents in VR or holograms in AR [58]. Moreover, as a space that can be used freely, the Metaverse makes participation in research more accessible and inclusive in several respects. First, all subjects can ideally take part in such research, overcoming the physical limitations that exclude certain people, such as those with mobility impairments. In the Metaverse everyone can create their own avatar and explore simulated environments freely. Similarly, distant or hard-to-reach places can be reproduced in the Metaverse. As a result, research can benefit from the participation of those who otherwise would have opted out due to distance or travel costs. This circumstance also makes research more sustainable. The Metaverse can help reduce carbon emissions by making travel by plane, train, and personal transportation unnecessary for activities that were previously only feasible in person [59]. Additionally, access to the Metaverse provides a strategy to decongest certain areas, freeing them from seasonal over-tourism [60]. Finally, the trend of people purchasing fewer physical items in favour of digital ones hints at a more sustainable dematerialised future [61–63].

However, some scholars [64–66] urge caution regarding the benefits for environmental sustainability, since the Metaverse, along with the growth of its user base, could result in the need for significant computing power and high bandwidth speeds, consequently increasing energy consumption, mainly through non-renewable sources. In addition, further critical considerations appear to be necessary, as certain limitations in the field of research could be identified.

The technologies still need to be perfected, the governance and rules for recruiting study participants have yet to be defined, no shared navigation standards exist, and the multiple Metaverses that current exist differ from each other in some respects.

Another concern involves regulation. Organisations and nations are not yet prepared to address privacy and security issues related to the Metaverse. More specifically, data privacy and cybersecurity regulations lag behind innovations in the Metaverse [67–69]. For example, because the Metaverse is borderless, in Europe it is still unclear how the clauses of the European Union's General Data Protection Regulation (GDPR) regarding the transfer and processing of data outside the EU can be applied.

A final issue concerns possible disparities in access to the Metaverse and its associated infrastructure and technologies. Indeed, certain populations (or parts thereof) are currently unable to access the infrastructure needed to obtain entry into the Metaverse [70]. Moreover, since access to the Metaverse is stratigraphically linked to ICT skills, using the Metaverse requires a range of technical knowledge not currently possessed by everyone. Without actions to mitigate this gap, those who cannot adapt to this change or who lack sufficient resources will be excluded from the new levels of connection and collaboration fostered by the Metaverse, and consequently, from new forms of communication, interaction, and research [71,72].

As a result, further insights into how to handle the challenges and opportunities of the Metaverse for research are needed, but these must be balanced by further research into potential new forms of social exclusion, the reliability of data, and the many ethical, behavioural, and negative impacts on vulnerable users.

5. Conclusions and Future Research Perspectives

In the transition from the pandemic to the post-pandemic era, a memorable long-term change in the way that people interact through technological applications is taking place and it will affect all domains of (virtual) life.

Starting from recent developments, reflections, applications of digital innovations, and the new frontier of the Metaverse, this paper has outlined the main tendencies in the domains of tourism and food and the consequences for research applications.

In the short term, it will likely be possible to experience unconventional and unexpected possibilities in tourism, such as walking in the ancient Roman forum as part of the initiative known as Time Machine Europe, which uses AR/VR applications to simulate hypothetical spatiotemporal 4D reconstructions with the aim of 'mapping the European social, cultural and geographical evolution across times' (Time Machine Europe website, https://www.timemachine.eu/about-us/, accessed on 12 January 2023).

From a more critical standpoint, many open questions linked to the social acceptance, economic accessibility, and environmental sustainability of such technologies remain unanswered. Tourism experiences reflect intricate socio-psychological processes, including a multisensory performance. The quality, authenticity, and sustainability of the travel destination are just as important as the social, demographic, economic, and cultural characteristics of the traveller in determining value, significance, and level of satisfaction. In this framework, digital innovation and immersive virtual practices can provide further opportunities to explore and to advance research on customers' and travellers' desires, needs, expectations, and satisfaction.

Nevertheless, while the (re)construction of immersive environments, goods, and services is becoming increasingly sophisticated, the Metaverse still proposes unrealistic objects, which may mislead users or challenge the validity of results through their lesser sense of reality [73]. For example, while sound and visual aspects tend to be relatively realistic, limitations exist in relation to atmosphere, smell, and perceived tactile sensations. Based on these considerations, research on the ability of the Metaverse to reduce or mitigate hypothetical bias in choice experiments is needed. For instance, scholars should pay attention to the reliability and validity of the evaluations of virtual experiences using psychometric scales able to capture the psychological complexity and the hidden dimensions of consumers' willingness to welcome tourism experiences in the Metaverse. Indeed, the individual perception could be biased, since it is mediated by a virtual tool and may not correspond to the possible evaluation of the real experiences. Similarly, a discussion on how to appropriately design digital environments so that they can more accurately simulate real face-to-face experiences must be advanced [74,75]. Other research directions, currently unexplored, could focus on how the Metaverse could affect the decision-making behaviour and consumption habits of tourists, and/or explore whether such technologies are applicable in other tourism niches.

In addition, further research must focus on people's perceptions, attitudes, acceptance, and willingness to pay for AR/VR/MR experiences, while also considering cross-country and multi-cultural aspects, as noted by several scholars [44,76,77].

Another critical issue is related to the management of the large quantities of sensitive data that will travel through the Metaverse until clear and shared procedures and policies are defined within the scientific community.

To conclude, the Metaverse can become an important tool for promoting and advancing tourism research through the implementation of some initiatives, such as facilitating virtual collaboration among researchers, allowing them to work together on projects and share data in real time. In encouraging cross-disciplinary collaboration, the Metaverse can bring together researchers from different fields to work together on tourism research projects.

Author Contributions: All authors (S.M., G.S.) contributed equally to this work. All authors have read and agreed to the published version of the manuscript.

Funding: The APC was funded by the Open Access Publishing Fund of the Free University of Bozen-Bolzano.

Institutional Review Board Statement: Not applicable.

Informed Consent Statement: Not applicable.

Data Availability Statement: Not applicable.

Conflicts of Interest: The authors declare no conflict of interest.

References

1. Damar, M. Metaverse Shape of Your Life for Future: A bibliometric snapshot. *J. Metaverse* **2021**, *1*, 1–8.
2. Narin, N.G.A. Content Analysis of the Metaverse Articles. *J. Metaverse* **2021**, *1*, 17–24.
3. Weinberger, M. What Is Metaverse?—A Definition Based on Qualitative Meta-Synthesis. *Futur. Internet* **2022**, *14*, 310–326.
4. Gent, E. Lessons from a Second Life before Meta, Philip Rosedale Created an Online Universe. *IEEE Spectr.* **2022**, *59*, 19. [CrossRef]
5. Lee, L.-H.; Braud, T.; Zhou, P.; Wang, L.; Xu, D.; Lin, Z.; Kumar, A.; Bermejo, C.; Hui, P. All One Needs to Know about Metaverse: A Complete Survey on Technological Singularity, Virtual Ecosystem, and Research Agenda. *J. Latex Cl. Files* **2021**, *14*, 1–66.
6. Ludlow, P.; Wallace, M. *The Second Life Herald: The Virtual Tabloid that Witnessed the Dawn of the Metaverse*; MIT Press: Cambridge, MA, USA, 2007; pp. 1–312.
7. Nabity-Grover, T.; Cheung, C.M.K.; Thatcher, J.B. Inside out and outside in: How the COVID-19 pandemic affects self-disclosure on social media. *Int. J. Inf. Manag.* **2020**, *55*, 102188. [CrossRef] [PubMed]
8. Zaman, U.; Koo, I.; Abbasi, S.; Raza, S.H.; Qureshi, M.G. Meet Your Digital Twin in Space? Profiling International Expat's 310 Readiness for Metaverse Space Travel, Tech-Savviness, COVID-19 Travel Anxiety, and Travel Fear of Missing Out. *Sustainability* **2022**, *14*, 6441. [CrossRef]
9. Damar, M. What the Literature on Medicine, Nursing, Public Health, Midwifery, and Dentistry Reveals: An Overview of the Rapidly Approaching Metaverse. *J. Metaverse* **2022**, *2*, 62–70. [CrossRef]
10. Liu, M.; Fang, S.; Dong, H.; Xu, C. Review of digital twin about concepts, technologies, and industrial applications. *J. Manuf. Syst.* **2021**, *58*, 346–361. [CrossRef]
11. Sitammagari, K.; Murphy, S.; Kowalkowski, M.; Chou, S.H.; Sullivan, M.; Taylor, S.; McWilliams, A. Insights From Rapid Deployment of a 'Virtual Hospital' as Standard Care During the COVID-19 Pandemic. *Ann. Intern. Med.* **2021**, *174*, 192–199. [CrossRef]
12. Musamih, A.; Yaqoob, I.; Salah, K.; Jayaraman, R.; Al-Hammadi, Y.; Omar, M.; Ellahham, S. Metaverse in Healthcare:Applications, Challenges, and Future Directions. In *IEEE Consumer Electronics Magazine*; IEEE: Washington, DC, USA, 2022.

13. Wang, G.; Badal, A.; Jia, X.; Maltz, J.S.; Mueller, K.; Myers, K.J.; Niu, C.; Vannier, M.; Yan, P.; Yu, Z.; et al. Development of metaverse for intelligent healthcare. *Nat. Mach. Intell.* **2022**, *411*, 922–929. [CrossRef]
14. Lee, H.J.; Hwang, Y. Technology-Enhanced Education through VR-Making and Metaverse-Linking to Foster Teacher Readiness and Sustainable Learning. *Sustainability* **2022**, *14*, 4786. [CrossRef]
15. Liu, X.; Zhang, J. Foreign Language Learning through Virtual Communities. *Energy Procedia* **2012**, *17*, 737–740. [CrossRef]
16. Tang, Y. Help first-year college students to learn their library through an augmented reality game. *J. Acad. Librariansh.* **2021**, *47*, 102294. [CrossRef]
17. Hwang, G.J.; Chien, S.Y. Definition, roles, and potential research issues of the metaverse in education: An artificial intelligence perspective. *Comput. Educ. Artif. Intell.* **2022**, *3*, 100082. [CrossRef]
18. Chen, Z. Exploring the application scenarios and issues facing Metaverse technology in education. *Interact. Learn. Environ.* **2022**. [CrossRef]
19. Dwivedi, Y.K.; Hughes, L.; Baabdullah, A.M.; Ribeiro-Navarrete, S.; Giannakis, M.; Al-Debei, M.M.; Wamba, S.F. Metaverse beyond the hype: Multidisciplinary perspectives on emerging challenges, opportunities, and agenda for research, practice and policy. *Int. J. Inf. Manag.* **2022**, *66*, 102542. [CrossRef]
20. Bibri, S.E. The Social Shaping of the Metaverse as an Alternative to the Imaginaries of Data-Driven Smart Cities: A Study inScience, Technology, and Society. *Smart Cities* **2022**, *5*, 832–874. [CrossRef]
21. Bojic, L. Metaverse through the prism of power and addiction: What will happen when the virtual world becomes more attractive than reality? *Eur. J. Futur. Res.* **2022**, *10*, 22. [CrossRef]
22. Buana, I.M.W. Metaverse: Threat or Opportunity for Our Social World? In understanding Metaverse on sociological context. *J. Metaverse* **2022**, *1*, 28–33.
23. Rehm, S.V.; Goel, L.; Crespi, M. The Metaverse as Mediator between Technology, Trends, and the Digital Transformation of Society and Business. *J. Virtual Worlds Res.* **2015**, *8*, 1–6. [CrossRef]
24. Cohen, S.A.; Cohen, E. New directions in the sociology of tourism. *Curr. Issues Tour.* **2017**, *22*, 153–172. [CrossRef]
25. Corbisiero, F.; Monaco, S.; Ruspini, E. *Millennials, Generation Z and the Future of Tourism*; Channel View Publications: Bristol, UK, 2022; pp. 1–176.
26. Sharpley, R. *Tourism, Tourists and Society*; Routledge: London, UK, 2018; pp. 1–380.
27. Urry, J.; Griego, M. *Mobilities: New Perspectives on Transport and Society*; Routledge: London, UK, 2016; pp. 1–384.
28. Happ, É.; Ivancsó-Horváth, Z. Digital Tourism is the Challenge of Future. A New Approach to Tourism. *Knowl. Horiz. Econ.* **2018**, *20*, 9–16.
29. Pencarelli, T. The digital revolution in the travel and tourism industry. *Inf. Technol. Tour.* **2020**, *22*, 455–476. [CrossRef]
30. Sigala, M.; Kumar, S.; Donthu, N.; Sureka, R.; Joshi, Y. A bibliometric overview of the Journal of Hospitality and Tourism Management: Research contributions and influence. *J. Hosp. Tour. Manag.* **2021**, *47*, 273–288. [CrossRef]
31. Monaco, S. *Tourism, Safety and COVID-19: Security, Digitization and Tourist Behaviour*; Routledge: New York, NY, USA, 2021; pp. 1–182.
32. Moreira, C.O.; Ferreira, R.; Santos, T. Smart Tourism and Local Heritage: Phygital Experiences and the Development of Geotourism Routes. In *Handbook of Research on Cultural Heritage and Its Impact on Territory Innovation and Development*; Oliveira, L., Amaro, A.C., Melro, A., Eds.; IGI Global: Hershey, PA, USA, 2021; pp. 206–232.
33. Shafiee, S.; Rajabzadeh Ghatari, A.; Hasanzadeh, A.; Jahanyan, S. Smart tourism destinations: A systematic review. *Tour. Rev.* **2021**, *76*, 505–528. [CrossRef]
34. Buhalis, D.; Harwood, T.; Bogicevic, V.; Viglia, G.; Beldona, S.; Hofacker, C. Technological disruptions in services: Lessons from tourism and hospitality. *J. Serv. Manag.* **2019**, *30*, 484–506. [CrossRef]
35. Utkarsh; Sigala, M. A bibliometric review of research on COVID-19 and tourism: Reflections for moving forward. *Tour. Manag. Perspect.* **2021**, *40*, 100912. [CrossRef]
36. Buhalis, D.; Karatay, N. Mixed Reality (MR) for Generation Z in Cultural Heritage Tourism Towards Metaverse. *Inf. Commun. Technol. Tour.* **2022**, *1*, 16–27.
37. Golf-Papez, M.; Heller, J.; Hilken, T.; Chylinski, M.; de Ruyter, K.; Keeling, D.I.; Mahr, D. Embracing falsity through the metaverse: The case of synthetic customer experiences. *Bus. Horiz.* **2022**, *65*, 739–749. [CrossRef]
38. Martins, D.; Oliveira, L.; Amaro, A.C. From co-design to the construction of a metaverse for the promotion of cultural heritage and tourism: The case of Amiais. *Procedia Comput. Sci.* **2022**, *204*, 261–266. [CrossRef]
39. Dwivedi, Y.K.; Hughes, L.; Wang, Y.; Alalwan, A.A.; Ahn, S.J.; Balakrishnan, J.; Wirtz, J. Metaverse Marketing: How the metaverse will shape the future of consumer research and practice. *Psychol. Mark.* **2022**, 1–27. [CrossRef]
40. Leung, X.Y.; Lyu, J.; Bai, B. A fad or the future? Examining the effectiveness of virtual reality advertising in the hotel industry. *Int. J. Hosp. Manag.* **2020**, *88*, 102391. [CrossRef]
41. Lin, L.P.; Huang, S.C.; Ho, Y.C. Could virtual reality effectively market slow travel in a heritage destination? *Tour. Manag.* **2020**, *78*, 104027. [CrossRef]
42. Zeng, G.; Cao, X.; Lin, Z.; Xiao, S.H. When online reviews meet virtual reality: Effects on consumer hotel booking. *Ann. Tour. Res.* **2020**, *81*, 102860. [CrossRef]
43. Yung, R.; Khoo-Lattimore, C.; Potter, L.E. Virtual reality and tourism marketing: Conceptualizing a framework on presence, emotion, and intention. *Curr. Issues Tour.* **2020**, *24*, 1505–1525. [CrossRef]

44. Buonincontri, P.; Errichiello, L.; Micera, R.; Del Chiappa, G. *Tourism, Hospitality And Culture 4.0: Shifting Towards The Metaverse*; McGraw-Hill Education: Milan, Italy, 2022; pp. 1–290.
45. Cavallo, C.; Sacchi, G.; Carfora, V. Resilience effects in food consumption behaviour at the time of COVID-19: Perspectives from Italy. *Heliyon* **2020**, *6*, e05676. [CrossRef]
46. Bisoffi, S.; Ahrné, L.; Aschemann-Witzel, J.; Báldi, A.; Cuhls, K.; DeClerck, F.; Duncan, J.; Hansen, H.O.; Hudson, R.L.; Kohl, J.; et al. COVID-19 and Sustainable Food Systems: What Should We Learn Before the Next Emergency. *Front. Sustain. Food Syst.* **2021**, *5*, 650987. [CrossRef]
47. van Dijk, M.; Morley, T.; Rau, M.L.; Saghai, Y. A meta-analysis of projected global food demand and population at risk of hunger for the period 2010–2050. *Nat. Food* **2021**, *27*, 494–501. [CrossRef]
48. Gastaldello, G.; Giampietri, E.; Zaghini, E.; Rossetto, L. Virtual Wine Experiences: Is Covid Extending the Boundaries of Wine Tourism? *Wine Econ. Policy* **2022**, 1–22. [CrossRef]
49. Flavián, C.; Ibáñez-Sánchez, S.; Orús, C. The impact of virtual, augmented and mixed reality technologies on the customer experience. *J. Bus. Res.* **2019**, *100*, 547–560. [CrossRef]
50. Griffin, T.; Giberson, J.; Lee, S.H.M.; Guttentag, D.; Kandaurova, M.; Sergueeva, K.; Dimanche, F. Virtual Reality and Implications for Destination Marketing. In Proceedings of the 48th Annual Travel and Tourism Research Association (TTRA), International Conference, Quebec City, QC, Canada, 20–22 June 2017.
51. Klein, L.R. Creating virtual product experiences: The role of telepresence. *J. Interact. Mark.* **2003**, *17*, 41–55. [CrossRef]
52. Gursoy, D.; Malodia, S.; Dhir, A. The metaverse in the hospitality and tourism industry: An overview of current trends and future research directions. *J. Hosp. Mark. Manag.* **2022**, *31*, 527–534. [CrossRef]
53. Suanpang, P.; Niamsorn, C.; Pothipassa, P.; Chunhapataragul, T.; Netwong, T.; Jermsittiparsert, K. Extensible Metaverse Implication for a Smart Tourism City. *Sustainability* **2022**, *14*, 14027. [CrossRef]
54. Corbisiero, F.; Monaco, S. Post-pandemic tourism resilience: Changes in Italians' travel behavior and the possible responses of tourist cities. *Worldw. Hosp. Tour. Themes* **2021**, *13*, 401–417. [CrossRef]
55. Koos, S. Machine Acting and Contract Law—The Disruptive Factor of Artificial Intelligence for the Freedom Concept of the Private Law. *UIR Law Rev.* **2021**, *5*, 1–18. [CrossRef]
56. Koos, S. Artifical Intelligence as Disruption Factor in the Civil Law: Impact of the use of Artifical Intelligence in Liability, Contracting, Competition Law and Consumer Protection with Particular Reference to the German and Indonesian Legal Situation. *Yuridika* **2021**, *36*, 235–262. [CrossRef]
57. Branca, G.; Resciniti, R.; Loureiro, S.M.C. Virtual is so real! Consumers' evaluation of product packaging in virtual reality. *Psychol. Mark.* **2023**, *40*, 596–609. [CrossRef]
58. Dwivedi, Y.K.; Hughes, L.; Ismagilova, E.; Aarts, G.; Coombs, C.; Crick, T.; Galanos, V. Artificial Intelligence (AI): Multidisciplinary perspectives on emerging challenges, opportunities, and agenda for research, practice and policy. *Int. J. Inf. Manag.* **2021**, *17*, 101994. [CrossRef]
59. Choi, H.Y. Working in the Metaverse: Does Telework in a Metaverse Office Have the Potential to Reduce Population Pressure in Megacities? Evidence from Young Adults in Seoul, South Korea. *Sustainability* **2022**, *14*, 3629. [CrossRef]
60. Go, H.; Kang, M. Metaverse tourism for sustainable tourism development: Tourism agenda 2030. *Tour. Rev.* **2022**. *ahead of print*. [CrossRef]
61. Davies-Filleur, C. Is the Metaverse a Tool for Sustainable Development? *Polytechnique Insights*. 20 September 2022. Available online: https://www.polytechnique-insights.com/en/braincamps/digital/metaverse-hopes-promises-and-unknowns/is-the-metaverse-a-tool-for-sustainable-development/ (accessed on 12 January 2023).
62. Townsend, S. Could The Metaverse & Web3 Save Sustainability? *Forbes*. 29 September 2022. Available online: https://www.forbes.com/sites/solitairetownsend/2022/09/29/could-the-metaverse--web3-save-sustainability/ (accessed on 12 January 2023).
63. Jauhiainen, J.S.; Krohn, C.; Junnila, J. Metaverse and Sustainability: Systematic Review of Scientific Publications until 2022 and Beyond. *Sustainability* **2023**, *15*, 346. [CrossRef]
64. Allam, Z.; Sharifi, A.; Bibri, S.E.; Jones, D.S.; Krogstie, J. The Metaverse as a Virtual Form of Smart Cities: Opportunities and Challenges for Environmental, Economic, and Social Sustainability in Urban Futures. *Smart Cities* **2022**, *5*, 771–801. [CrossRef]
65. Lee, Y. A Study on Metaverse Hype for Sustainable Growth. *Int. J. Adv. smart Converg.* **2021**, *10*, 72–80.
66. Pandey, N.; Nayal, P.; Rathore, A.S. Digital marketing for B2B organizations: Structured literature review and future research 400 directions. *J. Bus. Ind. Mark.* **2020**, *35*, 1191–1204. [CrossRef]
67. Banciu, D.; Fodorean, D.; Cirnu, C.E. Cyber Security and Human Rights Considering the Metaverse. *J. Free. Conscienc. (Jurnalul Lib. Conștiință)* **2021**, *9*, 648–654.
68. Bavana, K. Privacy in the Metaverse. *Jus. Corpus. Law J.* **2022**, *2*, 1–11.
69. Jaber, T.A. Security Risks of the Metaverse World. *Int. J. Interact. Mob. Technol.* **2022**, *16*, 4–14. [CrossRef]
70. Knox, J. The Metaverse, or the Serious Business of Tech Frontiers. *Postdigital Sci. Educ.* **2022**, *4*, 207–215. [CrossRef]
71. Zackery, A.; Shariatpanahi, P.; Zolfagharzadeh, M.M.; Pourezzat, A.A. Toward a simulated replica of futures: Classification and possible trajectories of simulation in futures studies. *Futures* **2016**, *81*, 40–53. [CrossRef]
72. Zhang, L.J. MRA: Metaverse Reference Architecture. *Lect. Notes Comput. Sci.* **2022**, *12993*, 102–120.

73. Choi, H.S.; Kim, S.H. A content service deployment plan for metaverse museum exhibitions—Centering on the combination 411 of beacons and HMDs. *Int. J. Inf. Manag.* **2017**, *37*, 1519–1527. [CrossRef]
74. Giannakos, M.N.; Sharma, K.; Pappas, I.O.; Kostakos, V.; Velloso, E. Multimodal data as a means to understand the learning experience. *Int. J. Inf. Manag.* **2019**, *48*, 108–119. [CrossRef]
75. Sharma, K.; Giannakos, M. Multimodal data capabilities for learning: What can multimodal data tell us about learning? *Br. J. Educ. Technol.* **2020**, *51*, 1450–1484. [CrossRef]
76. Lu, L.; Cai, R.; Gursoy, D. Developing and validating a service robot integration willingness scale. *Int. J. Hosp. Manag.* **2019**, *80*, 36–51. [CrossRef]
77. Tuomi, A.; Tussyadiah, I.P.; Stienmetz, J. Applications and Implications of Service Robots in Hospitality. *Cornell Hosp. Q. 419* **2021**, *62*, 232–247. [CrossRef]

MDPI
St. Alban-Anlage 66
4052 Basel
Switzerland
Tel. +41 61 683 77 34
Fax +41 61 302 89 18
www.mdpi.com

Sustainability Editorial Office
E-mail: sustainability@mdpi.com
www.mdpi.com/journal/sustainability

www.ingramcontent.com/pod-product-compliance
Lightning Source LLC
Chambersburg PA
CBHW061439210926
51850CB00008B/1555
9783036572420